Active Volcanoes of the World

About 500 active volcanoes presently exist on the Earth's surface, of which around 50 erupt each year. Volcanoes played a crucial role in the evolution of the planet and early life, and are constantly reshaping the morphology of Planet Earth. Many active volcanoes are located in dense settlement areas, with over 500 million people living in close proximity of still active or dormant volcanoes.

On one side, volcanoes provide valuable soil and rock basis for agriculture, but often the "mountains of fire" cause disastrous societal and economical disasters caused by ash clouds, lahars, lava flows, and pyroclastic flows. Eruptions are still difficult to predict, although volcanologists around the world are constantly working on new ways to understand the character and behavior of volcanoes.

Active Volcanoes of the World is an official book series of the International Association of Volcanology and Chemistry of the Earth's Interior (IAVCEI). The series aims to be a scientific library of monographs that provide authoritative and detailed reviews of state-of-the art research on individual volcanoes or a volcanic area that has been active in the last 10.000 years, e.g. the Teide Volcano or the Chiapas Region. The books in the series cover the geology, eruptive history, petrology and geochemistry, volcano monitoring, risk assessment and mitigation, volcano and society, and specific aspects related to the nature of each described volcano.

The Active Volcanoes of the World series contains single and multi-authored books as well as edited volumes. The Series Editors, Dr. Corrado Cimarelli (LMU Munich) and Dr. Sebastian Müller (Universität Mainz) are currently accepting proposals and a proposal document can be obtained from the Publisher, Dr. Johanna Schwarz (johanna.schwarz@springer.com).

More information about this series at http://www.springer.com/series/10081

Franco Tassi · Orlando Vaselli ·
Raúl Alberto Mora Amador
Editors

Poás Volcano

The Pulsing Heart of Central America Volcanic Zone

Springer

Editors
Franco Tassi
Department of Earth Sciences
University of Florence
Florence, Italy

Orlando Vaselli
Department of Earth Sciences
University of Florence
Florence, Italy

Raúl Alberto Mora Amador
Escuela Centroamericana de Geología
The University of Costa Rica
San Josè, Costa Rica

ISSN 2195-3589 ISSN 2195-7029 (electronic)
Active Volcanoes of the World
ISBN 978-3-319-02155-3 ISBN 978-3-319-02156-0 (eBook)
https://doi.org/10.1007/978-3-319-02156-0

Library of Congress Control Number: 2019930645

This Springer imprint is published by the registered company Springer Nature Switzerland AG
The registered company address is: Gewerbestrasse 11, 6330 Cham, Switzerland

Foreword

The April 2017 strong phreatomagmatic eruption of Poás dramatically changed the morphology of the active crater and temporarily stopped the usual crowd of tourists. A year later at the end of August 2018, the National Park at the Poás crater partially re-opened to the public.

Poás volcano represents an excellent site for common people worldwide to be a bit educated about "*how our Earth works*". A perfect shape of the crater, a steaming crater lake, roaring fumaroles, smell of sulphur, beautiful outcrops of volcanic sequences and intense hydrothermal alteration are just a few examples of what this volcano can offer to tourists and scientists. The observations of these geological attractions are facilitated by a comfortable way to get to the rim of the crater. It is likely that the good accessibility to the Poás volcano crater has allowed to become this site one of the most studied areas for volcanologists working on the origin and dynamics of a powerful, very active, volcano–hydrothermal system. This is a natural laboratory where in situ one can observe how Nature performs experiments on the magmatic fluid–water–rock interaction; and not only to observe but also by some way to participate by performing sampling and analyses of the products of this interaction. The "Laguna Caliente"—the Poás′ crater lake—decades ago became one of the first sites where volcanologists started to study and model the heat and mass balance of an active crater lake and made detailed and precise studies of the maximum of possible isotopic, elemental and molecular geochemical systems.

The Poás crater has been the object of many international field workshops and fieldworks for generations of specialists in Earth Sciences. This active and potentially dangerous volcano is monitored by specialists from both OVSICORI-UNA (the Volcanological and Seismological Observatory settled in Heredia) and University of Costa Rica in collaboration with many foreigner scientists. Such intensive scientific attention to the volcano has exponentially increased the knowledge about Poás, and new data are continuously produced. A very fast development of the modern analytical and field techniques requires regular "summing up", e.g. a scientific report about the latest results obtained by the scientific community related to Poás.

This book compiles a set of papers that covers a wide spectrum of topics—from a general modern view on the tectonics and geodynamics of the Costa Rican segment of the Central American Volcanic Arc to an historical essay about the place of Poás in the life of the Central American people.

A reader will find here approaches and data, which are generally interesting and important for many researchers and students in the Earth Sciences. Many thanks to the editors and authors for this excellent issue.

Yuri Taran
Institute of Geophysics, UNAM
Coyoacan, Ciudad de Mexico, Mexico
Institute of Volcanology and
Seismology FEB RAS
Petropavlovsk-Kamchatsky, Russia
Institute of Experimental
Mineralogy RAS
Chernogolovka, Moscow, Russia

Contents

Introduction

"... during the previous year's Miguel Alfaro himself had seen a vapour column, which also threw stones upwards, and around those sulphur accumulations and in the volcanic ash a blue flame burned. In the surroundings ash also found conical masses of pure sulphur, with an altitude of 3 to 4 feet. The lake then was a lot smaller; nevertheless, the water roared with more power and appeared more acidic and hotter than now. It seemed that at that time the volcano would have been a lot more active in throwing ash, and that it went gradually slowing down" (descriptions of Miguel Alfaro of 1828).

Von Frantzius (1861)

It was a privilege to edit this book on the Poás volcano (Costa Rica), which consists of 13 contributions about different aspects and approaches related to geodynamics, volcanology, seismology, fluid geochemistry, botany and zoology and legends related to one of the most active volcanoes in the world. In April 2017, when a new eruptive phase commenced, we started receiving the contributions by the authors and some of the chapters had to be modified to have as much as possible an update related to the renewal activity at Poás. This was one of the reasons why this book was a little bit delayed with respect to what was planned.

Poás is a complex strato-volcano, located in the Province of Alajuela, with an altitude of 2708 m a.s.l. and is part of the local national park. This volcano can be regarded as one of the most important touristic attractions of Costa Rica. The turquoise colour of the acidic lake (Laguna Caliente) and the intense fumarolic activity are likely the most photographed shots in the whole country.

The Poás summit is characterized by three volcanic cones: Von Frantzius, Botos and the active crater, the latter being the site where the latest eruptions occurred. The surrounding area hosted vigorous fumaroles, particularly close to a pyroclastic cone (often reported in the literature as a dome) formed in the early fifties. The fumarolic gas discharges are responsible for the absence of vegetation in several portions of the park due to the acidic gases emitted from the crater. The outlet temperatures of the fumaroles close to the pyroclastic

Fig. 1 High-temperature fumarole at the pyroclastic cone (often reported in the literature as a dome): **a** from July to September 2011 a temperature up to 900 °C was measured; **b** incandesce at the pyroclastic cone

cone (often reported in the literature as a dome), which was almost completely destroyed during an intense phreatomagmatic activity on the 22 April 2018, have shown dramatic changes from water boiling to magmatic (900 °C) temperatures, showing a spectacular incandescence (Fig. 1). Laguna Caliente is a hyper-acidic lake regarded as the most active crater lake in the world. The lake disappeared several times in the past: 1953, intermittent presence up to 1967, April 1989 and 1994, June 2017 and March 2018. Between March and August 2018, the lake was newly restored. Another peculiarity of Poás is associated with spectacular eruptions of sulphur (described since the nineteenth century), which are very rare in other volcanic areas worldwide. The typical eruptive activity at Poás is related to multiple phreatic eruptions (Fig. 2). In the eighties, the volcanologist F. D. Bennett suggested to call these events "erupciones Poasinas", to be distinguished by other ordinary phreatic eruptions.

The National Park of Poás is one of the most visited parks in Central America and possibly of the whole Latin American countries. To date, more than 400,000 people are used to visit the crater from the beautiful mirador that offers a magnificent view of active crater and Laguna Caliente. Owing to the resumed volcanic activity in April 2017, the park was interdicted to all the tourists causing a severe economic loss. Fortunately, at the end of August 2018, the park newly opened to the public.

The final versions of the 13 chapters were submitted to Springer in early September 2018 that marked the second anniversary of the premature death of Bruno Capaccioni, an eminent scientist and volcanologist but, first of all, a friend. Bruno has worked with the editors and most of the authors of this

Fig. 2 Phreatic eruption: **a** Phreatic eruption occurred on 1988 (photograph by G. Soto); **b** phreatic eruption occurred on 1 May 2010 (photograph by A. B. Castro); **c** phreatic eruption occurred on 25 May 2011 (photograph by R. A. Mora Amador)

book at Poás that was one of his favourite volcanoes. His sympathy and his love for volcanoes will always remain with all those who were lucky to know him. We do wish to dedicate this book to his memory.

He will never be forgotten!

Overview of the Tectonics and Geodynamics of Costa Rica

Paola Vannucchi and Jason P. Morgan

Abstract

Although Costa Rica is a relatively small region along the Central American Volcanic Arc (CAVA), its fascinating geology records several interesting examples of recent arc evolution. The forearc at present is in a state of subduction erosion, ranging from 'moderate' long-term rates of ~ 100 km^3/km/Ma beneath Nicoya Peninsula to 'extreme' short-duration peaks of $\sim 1{,}000$ km^3/km/Ma beneath Osa Peninsula. The margin is currently both seismogenic and tsunamogenic, with seismicity in the Osa Peninsula nucleating along one of Earth's shallowest seismogenic plate interfaces. In Costa Rica, arc volcanism has created much larger volcanic edifices than it has northward along the CAVA. Forearc deformation is ongoing and active, and associated with large-scale erosion and sediment transport towards the trench that is currently being almost entirely trapped in forearc basins prior to the trench axis. Margin evolution is also strongly linked to documented spatial and temporal variations in the incoming Cocos and Nazca Plates. These conditions have had significant consequences for the geochemical evolution of the CAVA in Costa Rica, so that Costa Rica's active geology records one of Earth's most diverse and interesting volcanic arcs.

Keywords

Costa Rica · Cocos plate · Caribbean plate · Osa Peninsula · Geodynamics

1 Introduction

Costa Rica sits at the south-east end of the ≈1,500 km—long Central America convergent margin. The Cocos plate subducts with a convergence direction N25°–N30°E with respect to the overriding Caribbean Plate at rates varying between 83 mm/yr in northern Costa Rica and 93 mm/yr in southern Costa Rica (DeMets et al. 1990; DeMets 2001) (Fig. 1). The Central America Trench ends in southern Costa Rica against the Panama Fracture Zone, which marks the boundary with the Nazca plate, and forms a triple junction between the Cocos, Nazca, and Caribbean Plates. The central Costa Rica to Panama region of the Caribbean spans four converging tectonic plates: South America,

P. Vannucchi (✉) · J. P. Morgan
Earth Sciences Department, Royal Holloway, University of London, Egham TW20 0EX, UK
e-mail: paola.vannucchi@unifi.it

P. Vannucchi
Department of Earth Sciences, University of Florence, Via G. La Pira, 4, 50121 Florence, Italy

F. Tassi et al. (eds.), *Poás Volcano*, Active Volcanoes of the World,
https://doi.org/10.1007/978-3-319-02156-0_1

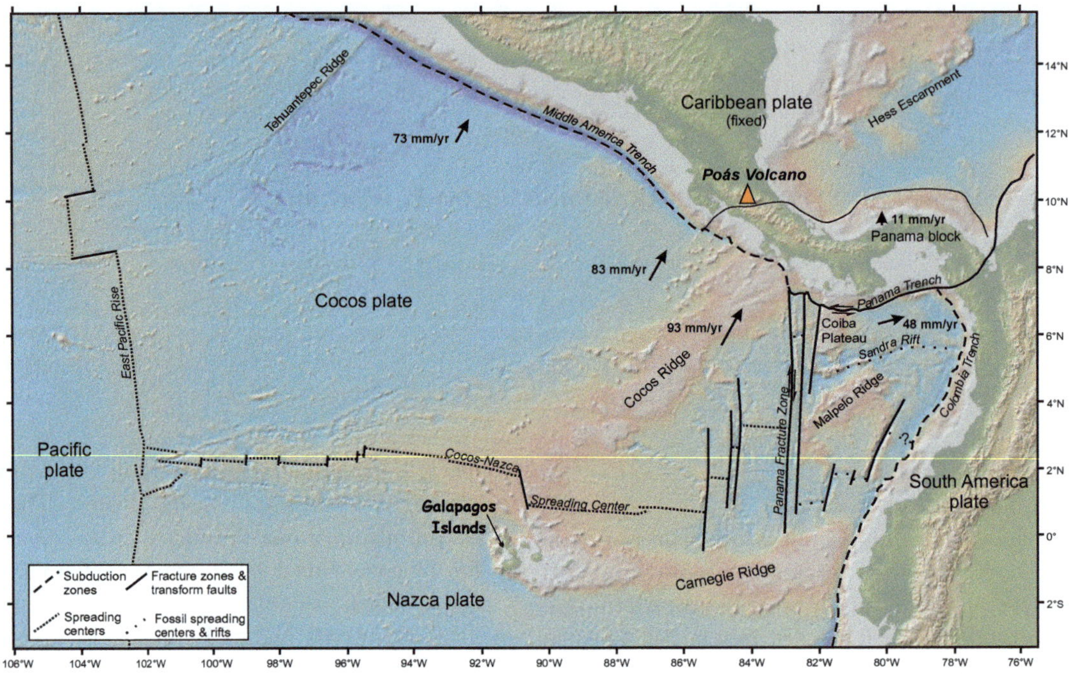

Fig. 1 Bathymetry, topography (http://www.geomapapp.org (Ryan et al. 2009)), and tectonic plate configuration in the region surrounding the Costa Rica. Arrows show plate motions with respect to the Caribbean plate (DeMets et al. 1990; DeMets 2001). Black dashed lines indicate trenches, black dotted lines indicate spreading centers, black solid lines indicate either transform faults, fracture zones or slow diffuse plate boundaries

Cocos, Nazca and the Caribbean itself (Fig. 1). The result is that this region has evolved as an independent lithospheric fragment, the Panama block, where plate motion is partitioned across a network of diffusive plate boundaries (Vergara Muñoz 1988; Marshall et al. 2000; DeMets 2001). The current motion of the Panama block is 11 mm/yr to the N with respect to the Caribbean Plate (DeMets 2001) (Fig. 1).

Poas Volcano is part of the NW-trending Central American volcanic arc (CAVA). In Costa Rica the arc extends from Orosí volcano to the NW to Turrialba volcano to the SE (Fig. 2). Between Nicaragua and Costa Rica the CAVA is offset 40 km to the SE, while a 175 km volcanic gap separates the Turrialba volcano from the Barú volcano in Panama, the last CAVA volcano to the SE (Fig. 2).

2 The Cocos Plate

During the early Neogene, the formation of the Galapagos rift system and the subsequent development of the Cocos-Nazca Spreading system (CNS) split the long-subducting Farallon plate into the Nazca and Cocos plates (Lonsdale 2005). The Cocos plate is formed at the East Pacific Rise (EPR) (Fig. 1), and at the present-day Cocos-Nazca Spreading system CNS-3. Detailed mapping of magnetic anomalies, in fact, shows two precursor spreading configurations CNS-2 and CNS-1 that imply a non-steady orientation of the CNS through time (Barckhausen et al. 1998, 2001). EPR- and CNS-crust are morphologically very different: EPR-crust is smooth and extensively cut by

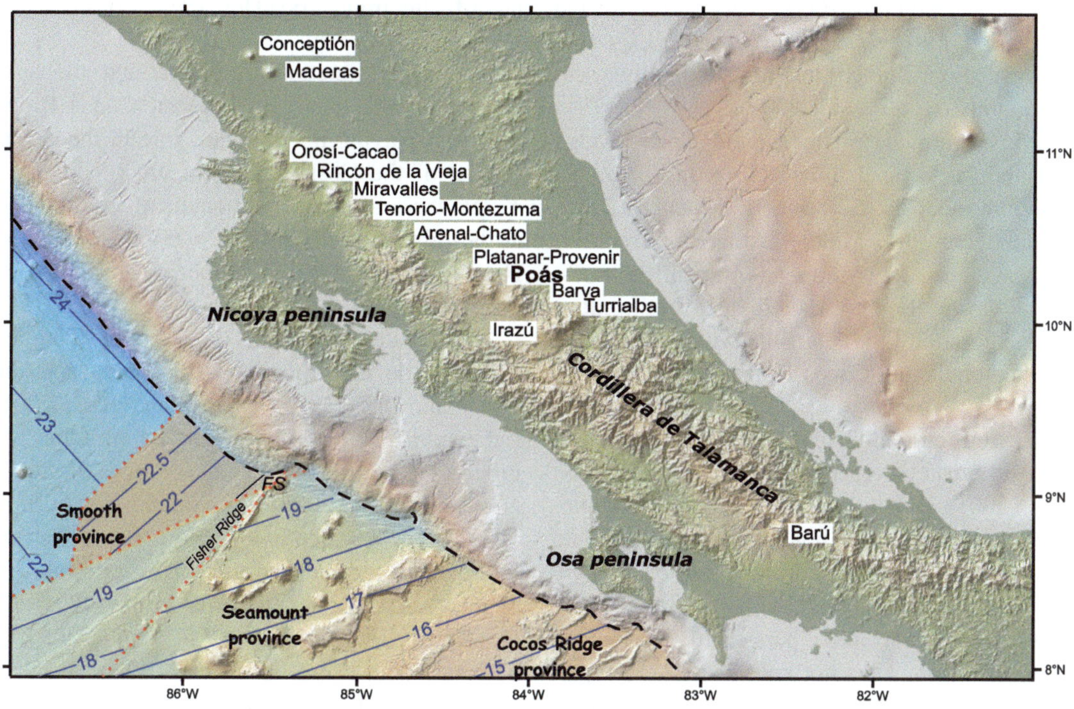

Fig. 2 Locations of volcanoes in Costa Rica and adjacent regions. Offshore Costa Rica shows a Cocos plate isochron map with ages derived from identification of seafloor spreading anomalies (Barckhausen et al. 2001). In blue is the crust generated at the East Pacific Rise, in red, orange and yellow the crust generated during the different stages of Cocos-Nazca spreading center activity. Numbers indicate crustal ages in m.y. Smooth, seamount and Cocos Ridge provinces are also indicated. FS is Fisher Seamount

faults occurring at intervals of ≈300 km generated at the ridge and reactivated at the outer-rise; CNS-crust is heavily impacted by the Galapagos hotspot activity producing a morphologically rough seafloor constellated by seamounts and the relatively buoyant Cocos Ridge (Fig. 1). Offshore Costa Rica seafloor morphology is so distinctive that it can be divided into three provinces (von Huene et al. 1995, 2000) (Fig. 2): (1) a smooth segment off north-eastern Costa Rica, (2) a seamount-dominated segment off central Costa Rica, and (3) a Cocos Ridge segment off south-eastern Costa Rica. The three provinces are also characterized by a different structure in the overlying subduction zone.

(1) The smooth province is bounded by Fisher Seamount and Fisher Ridge to the south (Fig. 2). At the trench, the crust breaks into the normal faults mentioned above and trends sub-parallel to the trench as effect of plate bending (Ranero et al. 2004). A narrow ridge offshore the central Nicoya Peninsula marks the EPR-CNS crust boundary (Barckhausen et al. 2001) with an age jump of the subducting oceanic crust from 24 Ma in northern Costa Rica to 21.5 and 22.5 Ma of the CNS-1 (Fig. 2).
(2) The seamount-dominated segment goes from south-east of the Nicoya Peninsula to the north-west of the subducting Cocos Ridge and it is 40% covered by seamounts (von Huene et al. 1995). The crust offshore central and south-eastern Costa Rica formed at the CNS-2 and has crustal ages between 19 and 15 Ma in the south (Barckhausen et al. 2001) (Fig. 2). Deep furrows and domes in the continental slope characterize the forearc slope of the seamount domain in Central Costa Rica indicating seamount subduction

(von Huene et al. 2000). This portion of the margin has generated up to M7 EQs that have been linked to the subducting seamounts (Husen et al. 2002).

(3) Moving to the southeast the Cocos Ridge subducts beneath the Osa Peninsula in the south of Costa Rica. Where the Cocos Ridge subducts, the Middle America Trench—MAT—shallows from 4,000 m in the north of Costa Rica to less than 1,000 m deep offshore the Osa Peninsula which lies above the subducting Cocos Ridge (Fig. 2).

Where it currently subducts, the Cocos Ridge is 13–14.5 Ma (Werner et al. 1999; Barckhausen et al. 2001) and it is one of the two complementary 'near-ridge' hot spot traces of the Galapagos Hotspot, the other being the Carnegie ridge on the Nazca plate, formed because the intervening CNS has persistently remained astride the Galapagos hot spot (Fig. 1) (Morgan 1978). The Cocos Ridge lies ≈2,000 m above its adjacent seafloor. Its plume-influenced oceanic crust reaches a thickness of 21 km along the crest of the ridge off Osa Peninsula (Stavenhagen et al. 1998; Walther 2003). To the east, the Panama Fracture Zone (PFZ) cuts the Cocos Ridge at the trench limiting its subducted extent to ~100 km inward from the trench (Protti et al. 1995). Here the Panama "Fracture Zone" is actually a right-lateral transform fault that is the current boundary between the Cocos and Nazca plates. East of where the PFZ intersects the trench axis, the Nazca Plate is moving to the east (N80°) towards South America where it subducts, so that this section of the Panama margin has relatively small deformation (Fig. 1).

Because the axis of the Cocos Ridge is oriented ~10° counterclockwise from the Cocos-Caribbean convergence vector (DeMets et al. 1990), it migrates slowly to the northwest along the MAT while the Panama Triple Junction migrates to the southeast (MacMillan et al. 2004; Morell et al. 2008). It has been proposed that the Panama Fracture zone truncated the Cocos Ridge around 9 Myr ago, with the Malpelo Ridge on the Nazca plate being the dislocated segment that records this truncation (Hey 1977; Lonsdale and Klitgord 1978).

The triple junction migration and the subduction of the relatively buoyant Cocos Ridge has been playing a significant role in the deformation of the Costa Rican forearc producing a ~350-km-long trench embayment culminating at the ridge axis where it is ~60-km-wide (von Huene et al. 2000; Vannucchi et al. 2013) (Fig. 1).

The plate interface is strongly coupled between the Cordillera de Talamanca and the Osa Peninsula (Fig. 2) (Norabuena et al. 2004; LaFemina et al. 2009). This contributes to infrequent, large earthquakes (Protti et al. 2001; Norabuena et al. 2004), for example the April 3, 1983 (Ms = 7.3; depth = 30 km) plate boundary thrust event located beneath the forearc inboard of the Osa Peninsula (Adamek et al. 1987), and the April 22, 1991 (Ms = 7.5; depth = 12 km) back-thrusting event, located ~100 km beneath the back-arc (Tajima and Kikuchi 1995).

Segments of the plate interface adjacent to these coupled regions experience frequent, smaller earthquakes (Protti et al. 2001; Bilek et al. 2003; Bilek and Lithgow-Bertelloni 2005). Most of the seismicity is rather shallow and strongly associated with the incoming bathymetry focusing on the incoming seamounts and faults (Bilek et al. 2003; Arroyo et al. 2014). Inboard the Cocos Ridge, the upper plate shortens considerably across the Fila Costeña thrust belt, and the regional uplift with extinction of the volcanic arc in the Cordillera de Talamanca has been used to support the currently preferred hypotheses that there has been flat subduction of the buoyant Cocos Ridge (Kolarsky et al. 1995; Protti et al. 1995; Fisher et al. 2004; Sitchler et al. 2007). Recent seismological investigations, however, show evidence of a subducted portion of the Cocos Ridge no longer than 100 km, in accordance with plate tectonic reconstructions (Protti et al. 1995), followed by a steep slab (Arroyo et al. 2003; Dinc et al. 2010; Dzierma et al. 2011). It is worth noticing that in this scenario the uplift of the Cordillera de Talamanca would not be directly caused by subduction of the Cocos Ridge as in this geometry Cocos Ridge material does not lie beneath the Talamanca.

Several dates have been suggested for the initiation of the subduction of the Cocos Ridge ranging from: 0.5 Myr based on elastic deformation studies (Gardner et al. 1992), 1 Myr based on ocean floor magnetic anomalies and plate reconstructions (Lonsdale and Klitgord 1978), 2–3 Myr from plate reconstructions (MacMillan et al. 2004; Morell et al. 2008), 3.6 Myr from on-land work on the Caribbean zone of Costa Rica (Collins et al. 1995), 5 Myr because of a sharp change in volcanic chemistry (De Boer et al. 1995), 3.5–5 Myr from thermochronological studies on the Talamanca batholits to constrain the uplift of the Cordillera de Talamanca (Grafe et al. 2002) and 8 Myr from the end of arc volcanism in the Cordillera de Talamanca (Abratis and Worner 2001). Recent drilling offshore Osa peninsula—IODP Exp. 334 and 344—revealed a rapid event of extreme subsidence of the forearc that has been interpreted as the direct damage caused by the onset of subduction of the Cocos Ridge at ≈2.2 Myr (Vannucchi et al. 2013).

2.1 Slab Imaging and Crustal Structure

The overall geometry of the Wadati-Benioff changes from Nicaragua to southern Costa Rica (Protti et al. 1995). Under the Nicaragua-Costa Rica border the seismic slab dips about 84° and reaches a maximum depth of 200 km. In central Costa Rica the seismicity is concentrated at 15 to 25–30 km depth defining a 18° dipping plate boundary (Husen et al. 2003) (Fig. 3), increasing to about 65° to a depth of 125 km. In southern Costa Rica the seismic slab seems to disappear below 50 km depth. However recent tomographic and receiver function investigations were

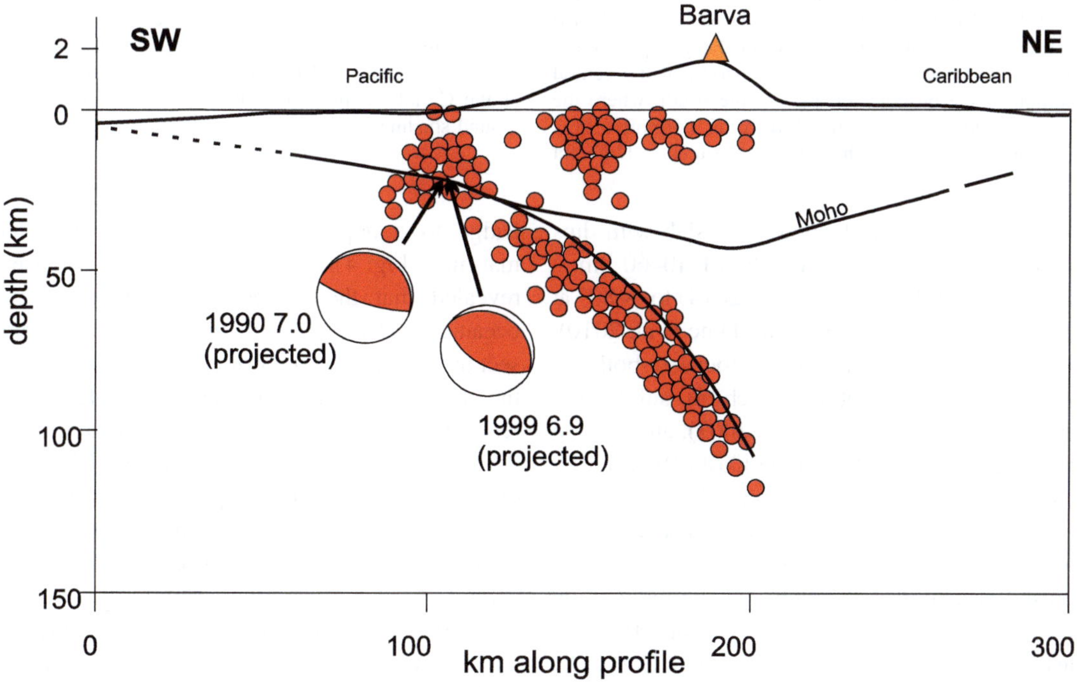

Fig. 3 Vertical section parallel to the dip of the subducting Cocos plate in the seamount-dominated segment off central Costa Rica. This section is coincident with the wide-angle seismic transect TICO-CAVA (Hayes et al. 2013—see Fig. 4 for location). The TICO-CAVA data constrain the crust and crust-mantle transition beneath the volcanic arc. Seismicity and slab depths are taken from Husen et al. (2003). Fault plain solutions of two large events within this area are also projected onto the section

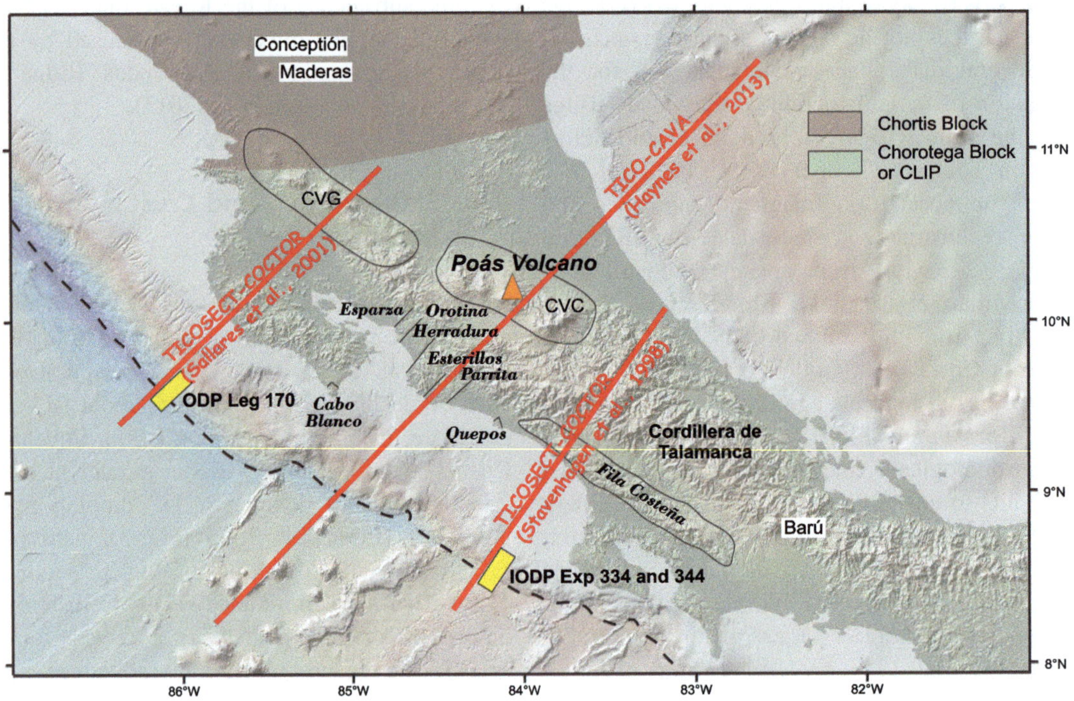

Fig. 4 Location of wide-angle seismic transects and ODP-IODP drilling expeditions discussed here. Named black line segments mark near-vertical faults perpendicular to the trench that dissect the margin in the central portion of the onshore forearc. To the south, where the Cocos Ridge subducts, the forearc is deformed in a fold-and-thrust belt—the Fila Costeña. Different *cordilleras* defining the Central American Volcanic Arc in Costa Rica: Cordillera Volcánica de Guanacaste (CVG) and the Cordillera Volcánica Central (CVC). Chorotega (green) and Chortis (grey) basement sections of the Caribbean plate are indicated by different background shadings

able to show a well-defined steep slab with dips varying from 50–55° at depths of 40–60 km—(Dinc et al. 2010; Arroyo et al. 2014) to ~80° at ~100 km below the coastline (Dinc et al. 2010).

Tomographic studies have focused both on imaging shallow features, such as subducting seamounts (Husen et al. 2002, 2003), and on slab imaging below the Nicaragua-Costa Rica border (Syracuse et al. 2008; Arroyo et al. 2009).

Seismic wide-angle data were also acquired along onshore-offshore transects to define the crustal structure of the slab and the subduction zone more in general (Fig. 4). The first transect, acquired during the TICOSECT and COTCOR projects, crossed the entire southern Costa Rica across the Talamanca Cordillera and ended in the Limon Basin on the Caribbean coast (Stavenhagen et al. 1998) imaging the subducting Cocos Ridge to a depth of about 30 km below the margin (Fig. 4). The TICOSECT transect revealed that the Cocos Ridge thickened the oceanic crust to ~15 km thick. A second transect traversed through northern Costa Rica across the Nicoya Peninsula and the active volcanic arc reaching the Nicaraguan border (Sallares et al. 2001) (Fig. 4). This transect imaged the subducting Cocos plate to a depth of 40 km beneath the Nicoya Peninsula. The most recent experiment, the TICO-CAVA transect, is located in central Costa Rica intersecting the CAVA between Barva and Irazu volcanoes (Fig. 4). TICO-CAVA is an onshore experiment designed to investigate the formation of juvenile crust (Hayes et al. 2013). The experiment suggested that the crustal thickness beneath the volcanic front in Costa Rica is ~40 km (Fig. 3). The average lower-crustal velocities, between 6.8 and 7.1 km/s, appear to be lower than modern island arc velocities, suggesting that felsic continental crust has been created at the volcanic arc in Central Costa Rica (Hayes et al. 2013).

3 The Caribbean Plate

3.1 The Forearc

Incoming plate subduction in Costa Rica has induced significant tectonic erosion, i.e. the removal of material from the base of the upper plate (Ranero and von Huene 2000; Vannucchi et al. 2001, 2003). Here, tectonic erosion occurs at different scales, and it has been detected by several investigation techniques.

Incoming seamounts left characteristic margin-perpendicular indentations and grooves on the trench and outer slope with a net subsidence record within the wake of subducted seamounts (von Huene et al. 1995, 2000) (Fig. 2). Reflection seismics showed fault-bounded mega-lenses present along the plate boundary which have been interpreted to be portions of the upper plate detached from its base (Ranero and von Huene 2000). A third independent dataset has been acquired through ocean drilling, ODP Leg 170, IODP Exp. 334 and 344 (Fig. 4). This uses the occurrence of subaerial deposits buried by deepening-upward sedimentary sequences as a recorder of the late Tertiary subsidence of the outer fore arc (Vannucchi et al. 2001, 2003, 2013). The Costa Rica forearc NW of Osa peninsula is an area where the sedimentation rates are very high reaching 1,200 m/m.y. while the adjacent trench axis remains sediment starved (Vannucchi et al. 2013).

The inner fore arc along the Costa Rican margin records a history of net uplift (Gardner et al. 1992, 2001, 2013; Fisher et al. 1998; Sak et al. 2009; Morell et al. 2012). This is in strong contrast to the net subsidence in the outer fore arc. The inner fore arc is dissected by a system of active near-vertical faults oriented at high angles to the margin (Fig. 4). These faults segment the forearc thrust belt into blocks with strongly variable uplift rates (Fisher et al. 1998; Marshall et al. 2000; Sak et al. 2009). The rivers draining this section of the Pacific slope typically follow these margin-perpendicular faults. Interestingly, the pattern of spatially variable fore arc uplift revealed by the correlation of marine and fluvial terraces along the Costa Rican Pacific coast is also spatially linked to the distribution of incoming seamounts. Uplift rates range from 1 to 8 mm/yr. The most rapid uplift occurs inboard of the Cocos Ridge within the Fila Costeña fold-and-thrust belt (Gardner et al. 1992, 2001, 2013; Fisher et al. 1998; Sak et al. 2009; Morell et al. 2012) (Fig. 4).

3.2 The Arc

In Costa Rica the CAVA is traditionally divided into two cordilleras or mountains ranges (Fig. 4): The Cordillera Volcánica de Guanacaste (CVG) and the Cordillera Volcánica Central (CVC). The Quaternary volcanoes of the CVG are from NW to SE, (see Figs. 1 and 3): Orosí-Cacao, Rincón de la Vieja, Miravalles, Tenorio-Montezuma and Chato-Arenal. Orosí-Cacao, Rincón de la Vieja, Miravalles, and Tenorio-Montezuma are shield-like stratovolcanoes build by coalescing lava flows and pyroclastic material emitted from multiple vents. While to the north, the low-relief landscape enhances the sharp morphology of the volcanoes, to the south the CVG lies behind the eroded massif of the extinct Tilarán Cordillera, masking the most active, but relatively small (15 km^3) Arenal volcano. The Volcanoes of the CVG have geochemical signatures that are transitional between the depleted mantled source and high subduction signal of the Nicaragua volcanoes and the enriched mantle source and low subduction signal of the central Costa Rica volcanoes (Herrstrom et al. 1995; Carr et al. 2003).

The CVC extends for approximately 80 km in central Costa Rica and consists of the Platanar-Provenir, Póas, Barva, Irazú and Turrialba volcanoes. Turrialba Volcano is offset 10 km northwards with respect to the other volcanoes of the Cordillera Central. These are composite shield volcanoes located northeast of the extinct Aguacate Cordillera. With peak elevations ranging from 2,000 to 3,400 m, these massive, broad-shouldered volcanoes are the largest volcanoes, in both area and volume, of the entire CAVA. Their summits exhibit wide calderas with multiple craters and transverse alignments of parasitic cones. The geochemical signal of the CVC is that of ocean-island basalt (Herrstrom et al. 1995; Carr et al. 2003).

In south-eastern Costa Rica in the Cordillera del Talamanca (Fig. 4) active arc-volcanism stopped at 5 Ma at the earliest, although 1–4 Ma pyroclastic flows (De Boer et al. 1995) and Pliocene—$\approx$1.5 Ma—adakites are widespread, but volumetrically limited (Gans et al. 2002; Goss et al. 2004; MacMillan et al. 2004). These mountains extend to almost 4000 m in elevation and they were glaciated during the Pleistocene (Protti 1996; Lachniet 2004). The Cordillera de Talamanca is composed of a suite of Neogene-Quaternary intrusive (principally granodiorites) and extrusive rocks (andesites) (De Boer et al. 1995; Alvarado 2000; MacMillan et al. 2004).

The large lateral variations in relief and geochemistry of the CAVA in Costa Rica appear to be directly related to strong lateral changes in the nature of the subducting Cocos plate (Fig. 2). The overriding Caribbean plate, in fact, does not appear to play a strong role in controlling or contaminating the volcanic products of the arc (Plank et al. 2002; Patino et al. 2000). Although the eastern margin of the Caribbean plate has a long history of subduction starting in the Late Jurassic, the on-land record of the CAVA in Costa Rica is well recorded starting from the Miocene.

3.3 The Costa Rican Basement

The Caribbean plate is formed by the combination of two main lithospheric elements (Fig. 4): the Precambrian-Mesozoic continental *Chortis block* to the north-west and the Cretaceous oceanic plateau *Caribbean Large Igneous Province—CLIP—*, also called *Chorotega block* to the south-west. The SW-NE contact between these two elements runs parallel to the Nicaragua-Costa Rica political boundary. The Chortis block is the byproduct of a long tectonic evolution with a nucleus of Paleozoic and Grenville-age metamorphic basement surrounded by Jurassic-Cretaceous ophiolitic complexes exposed along suture zones (Giunta and Orioli 2011) and late Cretaceous collided arc rocks (Rogers et al. 2007). The origin, tectonic significance and interaction of the Chortis and CLIP elements are still controversial—for example there is an ongoing argument regarding the allochthonous versus in situ origin of the Caribbean Plate (Pindell 1994; Meschede and Frisch 1998; Gahagan et al. 2007; Rogers et al. 2007; Giunta and Oliveri 2009; James 2009). According to the allochthonous hypothesis, the CLIP is the product of a Late Cretaceous period of vigorous submarine volcanism (Sinton et al. 1998) associated with the onset of the Galápagos mantle plume activity (or possibly the initiation of a new spreading center above a preexisting plume during the Late Cretaceous evolution of South American-African rifting). The emplacement of CLIP between the North and South America plates would then be the effect of its eastward migration, due to the subduction of the Farallon plate beneath the Caribbean plate, which commenced along the MAT in the Late Cretaceous (72–65 Ma) (Pindell 1994; Hoernle et al. 2004).

Subsequent dating, however, has extended the duration of CLIP volcanism to 70 m.y. (69–139 Ma) with the main volcanic events at $\sim$95–72 Ma (Hoernle et al. 2004), which makes this period far too long for the igneous province to be formed by a single plume head or plume-centered spreading center initiation event. Hoernle et al. (2004) propose that CLIP is the result of several oceanic intraplate igneous structures that were aggregated through subduction process with Galapagos hotspot being the main, but not sole factor responsible for CLIP formation. The Galapagos hot spot also formed younger ocean islands and aseismic ridge terrains now accreted to the eastern margin of the Caribbean Plate such as the 60 Ma Quepos and the

25 Ma Osa terrains (Hauff et al. 1997, 2000; Vannucchi et al. 2006).

The lithospheric structure of the Caribbean plate is directly linked to the characteristics of these two elements, but also to the great variety of the plate boundary interactions including subduction (west coast of Central America, Puerto Rico and Lesser Antilles), transform faults (Motagua-Chelungpu and Venezuela), sea-floor spreading (Cayman Trough), and continental collision (Colombian Cordilleras). Northern Costa Rica is shaped according to the classic subduction margin profile with a deep trench (≈5,000 m), an active Wadati-Benioff zone and a volcanic arc while Southern Costa Rica has a shallower (≈2,000 m) trench, a shallower Wadati-Benioff zone and an extinct volcanic arc related to the subduction of the Cocos Ridge.

4 Summary and Ongoing Questions

As is often typical in Earth Sciences, learning more about the timing of the geological and geodynamical evolution of Costa Rica has led to new questions regarding its formation and evolution. The origin of Costa Rican arc basement is still uncertain, although clearly linked in part to the origin of the Caribbean Large Igneous Province. The cessation of major arc-like volcanism along the Cordillera de Talamaca, the highest mountains along the Central American Volcanic Arc, does not appear to be directly linked to the underthrusting of the Cocos Ridge because Ridge material does not lie beneath the Cordillera de Talamanca proper, and its subduction into the Costa Rican trench occurred well after the cessation of Talamanca arc volcanism. The Panama-Nazca-Plate margin does not record simple subduction as in Costa Rica, and the ongoing westward migration of the (Cocos-Nazca) Panama Transform Fault will lead to further short-term changes in the evolution of the southern Costa Rican margin. The Costa Rican margin is already recognized as the 'type-example' for a terrestrial end-member in how subduction erosion can modify a subduction margin. We suspect that the evolution of the Cordillera de Talamanca may provide another 'type example' for how subduction processes shape the evolution of arc crust—after we can better understand what actually happened at a crustal and mantle level during this extreme event. It is even possible that this evolution is strongly linked to the recent formation of the Panama Land Bridge that has had profound effects on its surrounding marine and continental biomes.

References

Abratis M, Worner G (2001) Ridge collision, slab-window formation, and the flux of Pacific asthenosphere into the Caribbean realm. Geology 29:127–130

Adamek S, Tajima F, Wiens DA (1987) Seismic rupture associated with subduction of the Cocos Ridge. Tectonics 6:757–774

Alvarado GE (2000) Los volcanes de Costa Rica: Geología, historia, y riqueza natural. San José, Costa Rica, Editorial de la Universidad Nacional Estatal a Distancia (in Spanish)

Arroyo IG, Alvarado GE, Flueh ER (2003) Local seismicity at the Cocos Ridge-Osa Peninsula Subduction Zone, Costa Rica. In: Proceedings AGU fall meeting San Francisco, 2003, p Abstract S52F-0174

Arroyo IG, Husen S, Flueh ER, Gossler J, Kissling E, Alvarado GE (2009) Three-dimensional P-wave velocity structure on the shallow part of the Central Costa Rican Pacific margin from local earthquake tomography using off- and onshore networks. Geophys J Int 179:827–849

Arroyo IG, Grevemeyer I, Ranero CR, von Huene R (2014) Interplate seismicity at the CRISP drilling site: the 2002 Mw 6.4 Osa Earthquake at the southeastern end of the Middle America Trench. Geochem Geophys Geosyst 15:3035–3050

Barckhausen U, Roeser HA, von Huene R (1998) Magnetic signature of upper plate structures and subducting seamounts at the convergent margin off Costa Rica. J Geophys Res 103:7079–7093

Barckhausen U, Ranero CR, von Huene R, Cande SC, Roeser HA (2001) Revised tectonic boundaries in the Cocos Plate off Costa Rica: implications for the segmentation of the convergent margin and for plate tectonic models. J Geophys Res 106:19207–19220

Bilek SL, Lithgow-Bertelloni C (2005) Stress changes in the Costa Rica subduction zone due to the 1999 M_w = 6.9 Quepos earthquake. Earth Planet Sci Lett 230:97–112

Bilek SL, Schwartz SY, DeShon HR (2003) Control of seafloor roughness on earthquake rupture behavior. Geology 31:455–458

Carr MJ, Feigenson MD, Patino LC, Walker JA (2003) Volcanism and Geochemistry in Central America: progress and problems, vol 138. Inside the Subduction Factory, Geophysical Monograph-American Geophysical Union, pp 153–179

Collins LS, Coates AG, Jackson JBC, Obando JA (1995) Timing and rates of emergence of the Limón and Bocas del Toro basins: Caribbean effects of Cocos Ridge subduction? Geol Soc Am, Spec Pap 295:263–290

De Boer JZ, Drummond MS, Bordelon MJ, Defant MJ, Bellon H, Mauri RC (1995) Cenozoic magmatic phases of the Costa Rican island arc (Cordillera de Talamanca). In: Mann P (ed) Geologic and tectonic development of the Caribbean plate boundary in Southern Central America, vol 295, pp 35–55

DeMets C (2001) A new estimate for present-day Cocos-Caribbean plate motion: implications for slip along the Central American volcanic arc. Geophys Res Lett 28:4043–4046

DeMets C, Gordon RG, Argus DF, Stein S (1990) Current plate motions. Geophys J Int 101:425–478

Dinc AN, Koulakov I, Thorwart MM, Rabbel W, Flueh ER, Arroyo IG, Taylor W, Alvarado G (2010) Local earthquake tomography of central Costa Rica: transition from seamount to ridge subduction. Geophys J Int 183:286–302

Dzierma Y, Rabbel W, Thorwart MM, Flueh ER, Mora MM, Alvarado GE (2011) The steeply subducting edge of the Cocos Ridge: evidence from receiver functions beneath the northern Talamanca Range, south-central Costa Rica. Geochem Geophys Geosyst 12. https://doi.org/10.1029/2010gc003477

Fisher DM, Gardner TW, Marshall JS, Sak PB, Protti M (1998) Effect of subducting sea-floor roughness on fore-arc kinematics Pacific coast, Costa Rica. Geology 26:467–470

Fisher DM, Gardner TW, Sak PB, Sanchez JD, Murphy K, Vannucchi P (2004) Active thrusting in the inner forearc of an erosive convergent margin, Pacific coast, Costa Rica. Tectonics 23. https://doi.org/10.1029/2002tc001464

Gahagan L, Rogers DE, Mann P (2007) Chapter 8. Overview of plate tectonic history and its unresolved tectonic problems. In: Bundschuh J, Alvarado GE (eds) Central America: geology, resources and hazards. Taylor & Francis, London, pp 205–241

Gans PB, Macmillan I, Alvarado-Induni G, Perez W, Sigaran C (2002) Neogene evolution of the Costa Rican arc. Geol Soc Am 2002 Annual meeting 34:6, Denver Annual Meeting, October 27–30, 2002

Gardner TW, Verdonck D, Pinter NM, Slingerland R, Furlong KP, Bullard TF, Wells SG (1992) Quaternary uplift astride the aseismic Cocos ridge, Pacific coast, Costa-Rica. Geol Soc Am Bull 104:219–232

Gardner T, Marshall J, Merritts D, Bee B, Burgette R, Burton E, Cooke J, Kehrwald N, Protti M, Fisher MD, Sak P (2001) Holocene forearc block rotation in response to seamount subduction, southeastern Peninsula de Nicoya, Costa Rica. Geology 29:151–154

Gardner T, Fisher DM, Morell K, Cupper M (2013) Upper plate deformation in response to flat slab subduction inboard of the aseismic Cocos Ridge, Osa Peninsula, Costa Rica. Lithosphere 5:247–264

Giunta G, Oliveri E (2009) Some remarks on the Caribbean Plate kinematics: facts and remaining problems. In: James KH, Lorente MA, Pindell JL (eds) The origin and evolution of the Caribbean plate, vol 328. Geol Soc London, pp 57–75

Giunta G, Orioli S (2011) The Caribbean plate evolution: trying to resolve a very complicated tectonic puzzle. In: Sharkov EV (ed) New frontiers in tectonic research—general problems, sedimentary basins and Island Arcs. InTech, pp 221–248

Goss A, Kay S, Mpodozis C (2004) Transient "adakites" and the role of forearc subduction erosion: examples from the Central Andean and Central American margins. In: Proceedings 32nd IGC, Florence, Italy, 2004, Part 2, p 1333

Grafe K, Frisch W, Villa IM, Meschede M (2002) Geodynamic evolution of southern Costa Rica related to low-angle subduction of the Cocos Ridge: constraints from thermochronology. Tectonophysics 348:187–204

Hauff F, Hoernle K, Schmincke HU, Werner R (1997) A Mid Cretaceous origin for the Galapagos hotspot: volcanological, petrological and geochemical evidence from Costa Rican oceanic crustal segments. Geol Rundsch 86:141–155

Hauff F, Hoernle K, Tilton G, Graham DW, Kerr AC (2000) Large volume recycling of oceanic lithosphere over short time scales: geochemical constraints from the Caribbean Large Igneous Province. Earth Planet Sci Lett 174:247–263

Hayes JL, Holbrook WS, Lizarralde D, van Avendonk HJA, Bullock AD, Mora M, Ramirez C, Harder S, Alvarado GE (2013) Crustal structure across the Costa Rican Volcanic Arc. Geochem Geophys Geosyst 14:1087–1103

Herrstrom E, Reagan M, Morris J (1995) Variations in lava composition associated with flows of asthenosphere beneath southern Central America. Geology 23:617–620

Hey R (1977) Tectonic evolution of the Cocos-Nazca spreading center. Geol Soc Am Bull 88:1404–1420

Hoernle K, Hauff F, van den Bogaard P (2004) 70 m.y. history (139–69 Ma) for the Caribbean large igneous province. Geology 32:697–700

Husen S, Kissling E, Quintero R (2002) Tomographic evidence for a subducted seamount beneath the Gulf of Nicoya, Costa Rica. The cause of the 1990 Mw = 7.0 Gulf of Nicoya earthquake. Geophys Res Lett 29(8). https://doi.org/10.1029/2001gl014045

Husen S, Quintero R, Kissling E, Hacker B (2003) Subduction-zone structure and magmatic processes beneath Costa Rica constrained by local earthquake tomography and petrological modelling. Geophys J Int 155:11–32

James KH (2009) In-situ origin of the Caribbean: discussion of data. In: James KH, Lorente, MA,

Pindell JL (eds) The origin and evolution of the Caribbean plate, vol 328. Geol Soc London, pp 77–126

Kolarsky RA, Mann P, Montero W (1995) Island arc response to shallow subduction of the Cocos Ridge, Costa Rica. In: Mann P (ed) Geologic and tectonic development of the Caribbean plate boundary in Southern Central America, vol 295, pp 235–262

Lachniet MS (2004) Quaternary glaciation in Guatemala and Costa Rica. In: Ehlers J, Gibbard PL (eds) Quaternary glaciations—extent and chronology. Part III: South America, Asia, Africa, Australia, Antarctica, 2c, pp 135–138

LaFemina P, Dixon TH, Govers R, Norabuena E, Turner H, Saballos A, Mattioli G, Protti M, Strauch W (2009) Fore-arc motion and Cocos Ridge collision in Central America. Geochem Geophys Geosyst 10:Q05S14. https://doi.org/10.1029/2008gc002181

Lonsdale P (2005) Creation of the Cocos and Nazca plates by fission of the Farallon plate. Tectonophysics 404:237–264

Lonsdale P, Klitgord KD (1978) Structure and tectonic history of the eastern Panama Basin. Geol Soc Am Bull 89:981–999

MacMillan I, Gans PB, Alvarado G (2004) Middle Miocene to present plate tectonic history of the southern Central American Volcanic Arc. Tectonophysics 392:325–348

Marshall JS, Fisher DM, Gardner TW (2000) Central Costa Rica deformed belt: kinematics of diffuse faulting across the western Panama block. Tectonics 19:468–492

Meschede M, Frisch W (1998) A plate tectonic model for the Mesozoic and Early Cenozoic history of the Caribbean Plate. Tectonophysics 296:269–291

Morell KD, Fisher DM, Gardner TW (2008) Inner forearc response to subduction of the Panama Fracture Zone, southern Central America. Earth Planet Sci Lett 265:82–95

Morell KD, Kirby E, Fisher DM, van Soest M (2012) Geomorphic and exhumational response of the Central American Volcanic Arc to Cocos Ridge subduction. J Geophys Res 117:B04409. https://doi.org/10.1029/2011JB008969

Morgan JW (1978) Rodriguez, Darwin, Amsterdam, …, a second type of hotspot Island. J Geophys Res 83:5355–5360

Norabuena E, Dixon TH, Schwartz S, DeShon H, Newman A, Protti M, Gonzalez V, Dorman L, Flueh ER, Lundgren P, Pollitz F, Sampson D (2004) Geodetic and seismic constraints on some seismogenic zone processes in Costa Rica. J Geophys Res 109(B11): B11403. https://doi.org/11410.11029/12003JB002931

Pindell JL (1994) Evolution of the Gulf of Mexico and the Caribbean. In: Donovan SK, Jackson TA (eds) Caribbean geology: an introduction. University of the West Indies Publishers Association/University of the West Indies Press, Kingston, Jamaica, pp 13–39

Protti R (1996) Evidencias de glaciación en el Valle del General (Costa Rica) durante el Pleistoceno tardio. Rev Geol Am Central 19(20):75–85

Protti M, Guendel F, McNelly K (1995) Correlation between the age of the subducting Cocos plate and the geometry of the Wadati-Benioff zone under Nicaragua and Costa Rica. In: Mann P (ed) Geologic and tectonic development of the Caribbean plate boundary in Southern Central America, vol 295, pp 309–343

Protti M, Guendel F, Malavassi E (2001) Evaluación del potencial sísmico de la Península de Nicoya, Heredia, Costa Rica. Editorial Fundacíon Universidad Nacional

Ranero CR, von Huene R (2000) Subduction erosion along the Middle America convergent margin. Nature 404:748–752

Ranero CR, Weinrebe W, Grevemeyer I, von Huene R, Reichert C (2004) The relation between tectonics, fluid flow and seismogenesis at convergent erosional margins. EOS (Trans) American Geophysical Union, vol 85, no 47, Fall Meeting Suppl, p S43D-01

Rogers RD, Mann P, Emmet PA (2007) Tectonic terranes of the Chortis block based on integration of regional aeromagnetic and geologic data. Geol Soc Am, Spec Pap 428:65–88

Ryan WBF, Carbotte SM, Coplan JO, O'Hara S, Melkonian A, Arko R, Weissel RA, Ferrini V, Goodwillie A, Nitsche F, Bonczkowski J, Zemsky R (2009) Global multiresolution topography synthesis. Geochem Geophys Geosyst 10:Q03014. https://doi.org/10.1029/2008GC002332

Sak PB, Fisher DM, Gardner TW, Marshall JS, LaFemina PC (2009) Rough crust subduction, forearc kinematics, and Quaternary uplift rates, Costa Rican segment of the Middle American Trench. Geol Soc Am Bull 121:992–1012

Sinton CW, Duncan RA, Storey M, Lewis J, Estrada JJ (1998) An oceanic flood basalt province within the Caribbean plate. Earth Planet Sci Lett 155:221–235

Sitchler JC, Fisher DM, Gardner TW, Protti M (2007) Constraints on inner forearc deformation from balanced cross sections, Fila Costena thrust belt, Costa Rica. Tectonics 26:TC6012. https://doi.org/10.1029/2006tc001949

Stavenhagen AU, Flueh ER, Ranero CR, McIntosh KD, Shipley T, Leandro G, Schulze A, Dañobeitia JJ (1998) Seismic wide-angle investigations in Costa Rica—a crustal velocity model from the Pacific to the Caribbean. Zbl Geol Paläontol 1:393–408

Syracuse EM, Abers GA, Fischer K, MacKenzie L, Rychert C, Protti M, Gonzalez V, Strauch W (2008) Seismic tomography and earthquake locations in the Nicaraguan and Costa Rican upper mantle. Geochem Geophys Geosyst 9:Q07S08. https://doi.org/10.1029/2008gc001963

Tajima F, Kikuchi M (1995) Tectonic implications of the seismic ruptures associated with the 1983 and 1991 Costa Rica earthquakes. In: Mann P (ed) Geologica and tectonic development of the Caribbean plate boundary in Southern Central America. Geol Soc Am, Spec Pap 295:327–340

Vannucchi P, Scholl DW, Meschede M, McDougall-Reid K (2001) Tectonic erosion and consequent collapse of the Pacific margin of Costa Rica: combined implications from ODP Leg 170, seismic offshore data, and regional geology of the Nicoya Peninsula. Tectonics 20:649–668

Vannucchi P, Ranero CR, Galeotti S, Straub SM, Scholl DW, McDougall-Ried K (2003) Fast rates of subduction erosion along the Costa Rica Pacific margin: Implications for nonsteady rates of crustal recycling at subduction zones. J Geophys Res 108 (B11):2511. https://doi.org/2510.1029/2002JB002207

Vannucchi P, Fisher DM, Gardner TW, Bier S (2006) From seamount accretion to tectonic erosion: formation of Osa Mélange and the effects of Cocos Ridge subduction in southern Costa Rica. Tectonics 25: TC2004. https://doi.org/10.1029/2005tc001855

Vannucchi P, Sak PB, Morgan JP, Ohkushi K, Ujiie K, Scientists IES (2013) Rapid pulses of uplift, subsidence, and subduction erosion offshore Central America: implications for building the rock record of convergent margins. Geology 41:995–998

Vergara Muñoz A (1988) Tectonic patterns of the Panama Block deduced from seismicity, gravitational data and earthquake mechanisms: implications to the seismic hazard. Tectonophysics 154:253–267

von Huene R, Bialas J, Flueh E, Cropp B, Csernok T, Fabel E, Hoffmann J, Emeis K, Holler P, Jeschke G, Leandro MC, Perez Fernandez I, Chavarria SJ, Florez HA, Escobedo ZD, Leon R, Barrios LO (1995) Morphotectonics of the Pacific convergent margin of Costa Rica. In: Mann P (ed) Geologic and tectonic development of the Caribbean Plate boundary in southern Central America. Geol Soc Am, Spec Pap 295:291–307

von Huene R, Ranero CR, Weinrebe W, Hinz K (2000) Quaternary convergent margin tectonics of Costa Rica, segmentation of the Cocos Plate, and Central American volcanism. Tectonics 19:314–334

Walther CHE (2003) The crustal structure of the Cocos ridge off Costa Rica. J Geophys Res 108(B3), p. art. no. 2136. https://doi.org/10.1029/2001jb000888

Werner R, Hoernle K, van den Bogaard P, Ranero CR, von Huene R (1999) Drowned 14-m.y.-old Galapagos archipelago off the coast of Costa Rica: implications for tectonic and evolutionary models. Geology 27:499–502

Geochemical and Geochronological Characterisation of the Poas Stratovolcano Stratigraphy

Paulo Ruiz, Sara Mana, Esteban Gazel, Gerardo J. Soto, Michael J. Carr and Guillermo E. Alvarado

Abstract

In this chapter the stratigraphy of Poás volcano by using geological, petrographical, geochronological and geochemical analyses on the volcanic products erupted during the last 600 ka is defined. The northern flank of Poás consists of the following Units and Members: Río Sarapiquí, La Paz, Puente de Mulas Member (possibly from another source, though interdigitated with Poás products), Río Cuarto Lavas, Von Frantzius, Congo, Bosque Alegre and Laguna Kopper. The products on the southern flank are the Colima Formation, La Paz, Tiribí, Achiote, Poasito, Sabana Redonda and Poás Lapilli Tuff. The central part of the volcano is made up of the Poás Summit Unit, which includes the Main and Botos craters. Rock composition varies from basalts to dacites. During the last 600 ka the content of K_2O and other oxides (e.g. TiO_2 and P_2O_5) and trace elements (e.g. Zr, Ba) has varied significantly through time, implying the presence of two geochemical end-members since the beginning of the magmatic activity at Poás: (1) the Sabana Redonda Geochemical Component ($TiO_2 > 1\%$) enriched in HSFE and other trace elements, present in La Paz Andesites, Lavas Río Cuarto, Poasito, Sabana Redonda, Poás Lapilli Tuff and Botos crater lavas and (2) The Von Frantzius Geochemical Component ($TiO_2 < 0.8\%$) that was recognized in the lavas from the Main Crater, Von Frantzius, Achiote, Bosque Alegre, Congo and Botos. Lavas related to both magmatic components have coexisted over time as indicated by those found at Botos and the Main Crater. Units possibly related to a common vent, i.e. La Paz, Achiote and Main

Electronic supplementary material The online version of this chapter (https://doi.org/10.1007/978-3-319-02156-0_2) contains supplementary material, which is available to authorized users.

P. Ruiz (✉)
Laboratorio de Materiales y Modelos Estructurales, Universidad de Costa Rica, Ciudad de la Investigación, Finca 2, 15501-2060 San José, Costa Rica
e-mail: paulo.ruizcubillo@ucr.ac.cr

S. Mana
Department of Geological Sciences, Salem State University, 352 Lafayatte Street, Salem, MA 01970, USA

E. Gazel
Earth and Atmospheric Science, Cornell University, 2122 Snee Hall, Ithaca, NY 14853-1504, USA

G. J. Soto
Terra Cognita Consultores, San José, Costa Rica

M. J. Carr
Department of Earth and Planetary Sciences, Rutgers University, 610 Taylor Rd., Piscataway, NJ 08854, USA

G. E. Alvarado
Instituto Costarricense de Electricidad (ICE), San José, Costa Rica

F. Tassi et al. (eds.), *Poás Volcano*, Active Volcanoes of the World,
https://doi.org/10.1007/978-3-319-02156-0_2

Crater, the percentages of K_2O and TiO_2 decreased through time.

Keywords
Poás volcano · Volcano stratigraphy · Volcano growth · Subduction zone

1 Introduction

Subduction processes recycle crustal material and sediments back into the mantle, directly influencing both the segregation of continental crust and the geochemical variability in the mantle (e.g. Plank and Langmuir 1993; Rudnick and Gao 2003; Hofmann 2003; Ryan and Chauvel 2014; Gazel et al. 2015). The Central American Volcanic front is characterized by drastic chemical variations, making it an ideal location to study the role of the subduction processes in the generation of juvenile continental crust and element recycling processes (Carr et al. 1990; Vogel et al. 2004; Saginor et al. 2011, 2013). In this chapter, we present the results related to one of the most active volcanoes of Costa Rica in the last decade: Poás, which can be regarded as the pulsing heart of Central America Volcanic Zone.

Poás volcano is a complex stratovolcano located in the southern end of the Central American Volcanic front (Vannucchi and Mason Chapter "Overview of the Tectonics and Geodynamics of Costa Rica"). It is one of the five active volcanoes of Costa Rica along with Rincon de la Vieja, Arenal, Turrialba, Irazu and belongs to the Central Volcanic Range (CVR). Located at Lat 10° 11′N and Lon 84° 13′W, it is just 30 km NW of the capital, San José, which, together with the other major cities of the Central Valley, hosts ~2.1 million inhabitants (Fig. 1). Due to its proximity to major cities, Poás volcano has a relevant scientific and sociological importance and in the past three decades several studies focused on the geology, geophysics and geochemistry of this apparatus. This has led to make considerable progress in understanding its historical eruptive cycles and present activity (e.g. Thorpe et al. 1981; Casertano et al. 1983; Prosser and Carr 1987; Cigolini et al. 1991; Rymer et al. 2000, 2009; de Moor et al. 2016; Mora Amador et al. Chapter "Volcanic Hazard Assessment of Poás (Costa Rica) Based on the 1834, 1910, 1953–1955 and 2017 Historical Eruptions"). In addition, other studies regarded the flanks of the volcano (e.g. Tournon 1984; Borgia et al. 1990; Soto 1999; Alvarado and Salani 2004; Gazel and Ruiz 2005; Carr et al. 2007; Montero et al. 2010; Alvarado et al. 2011). The work by Ruiz et al. (2010) was the first attempt to complete a geological map that studied the volcano as a whole edifice, from its flanks at ~400 m a.s.l. up to its summit at 2708 m a.s.l.

Here, new results of a comprehensive field mapping, geochemical analyses, Ar–Ar dating, high-resolution orthophotos and LiDAR data study of Poás volcano (Fig. 1) are presented. The main goal of this chapter was to understand the changes that this complex stratovolcano has experienced in its last two stages of activity (during the last 600 ka) by characterizing petrologically and geochemically the various volcanic units that compose this massif and tying them to their relative time and space. The level of detail of the mapping involved in this chapter, together with a relatively complete geochemical and geochronological database, provided a unique opportunity to study the evolution of a complex composite volcano in Central America providing a better understanding on how volcanoes grow in a volcanic front.

2 The Study Area

Poás volcano units reported in this chapter enclose an area of ~477 km^2, 337 km^2 of which are covered by new LiDAR data and high-resolution orthophotos. The study area is limited by the Río Cuarto maar to the north, and the Alajuela thrust fault escarpment to the south (Fig. 1c). The rivers Tambor and Sarchí and those of Toro and Sarapiquí (Fig. 1c) define the eastern and western limits, respectively. The geologic formations and members can be subdivided into three sectors: north, south and central

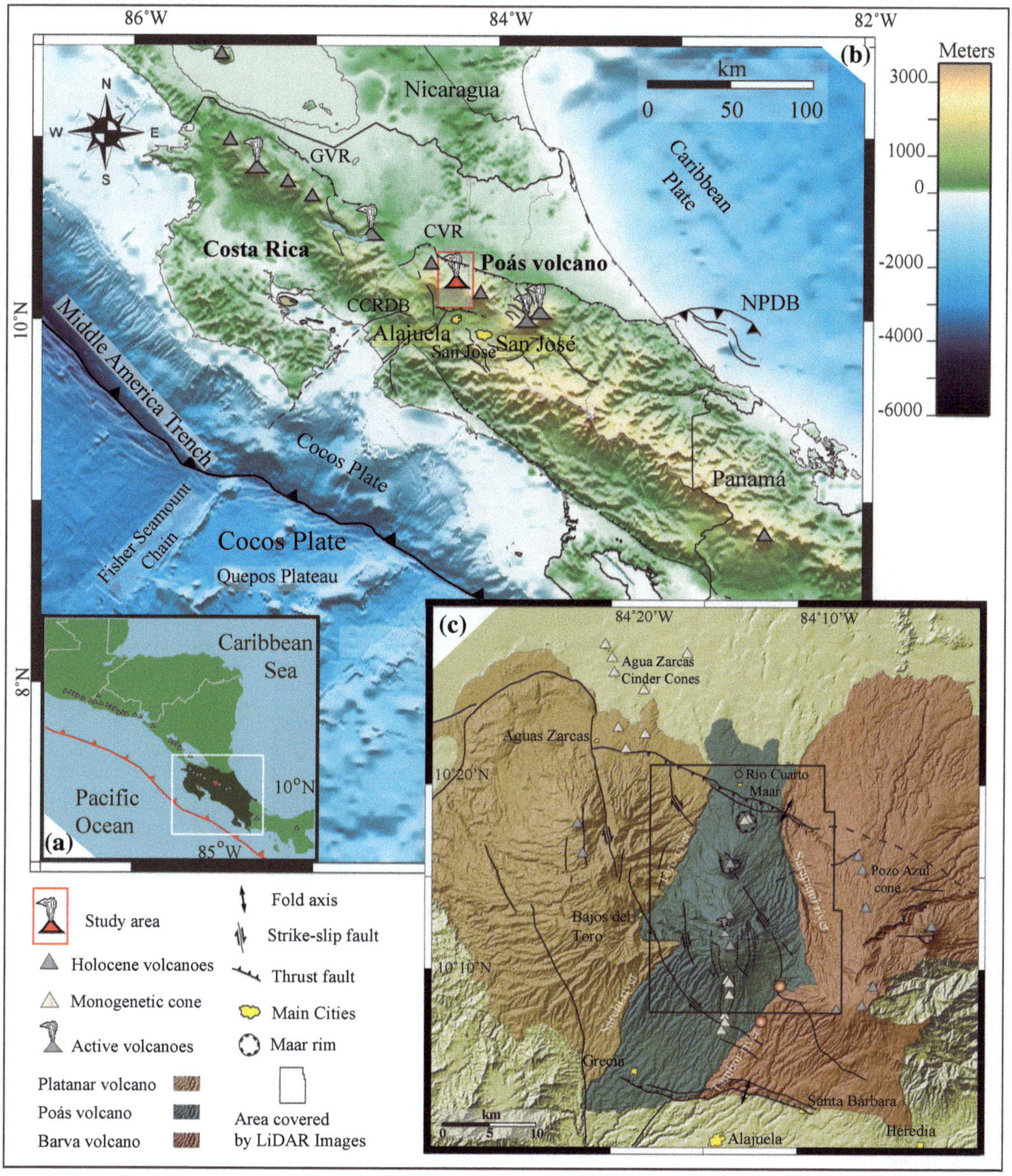

Fig. 1 **a** Map of the Central American Volcanic front and its tectonic setting; **b** digital elevation map of Costa Rica, showing the main volcanoes as dark grey triangles, with active volcanoes labeled with a plume. Bathymetry is from Ranero et al. (2005); **c** Poás volcano and its main tectonic structures from Montero et al. (2010). The black rectangle shows the area covered by LiDAR data used to build the presented digital elevation models

(present vent), based on the location of the vents from which they were expelled. Each one of these is composed by a different sequence of lavas, pyroclastic flows and fall deposits, and lahars among other volcanic materials. This chapter mostly focuses on the characterization of the various lava units mostly based on field and geomorphological observations (Fig. 2).

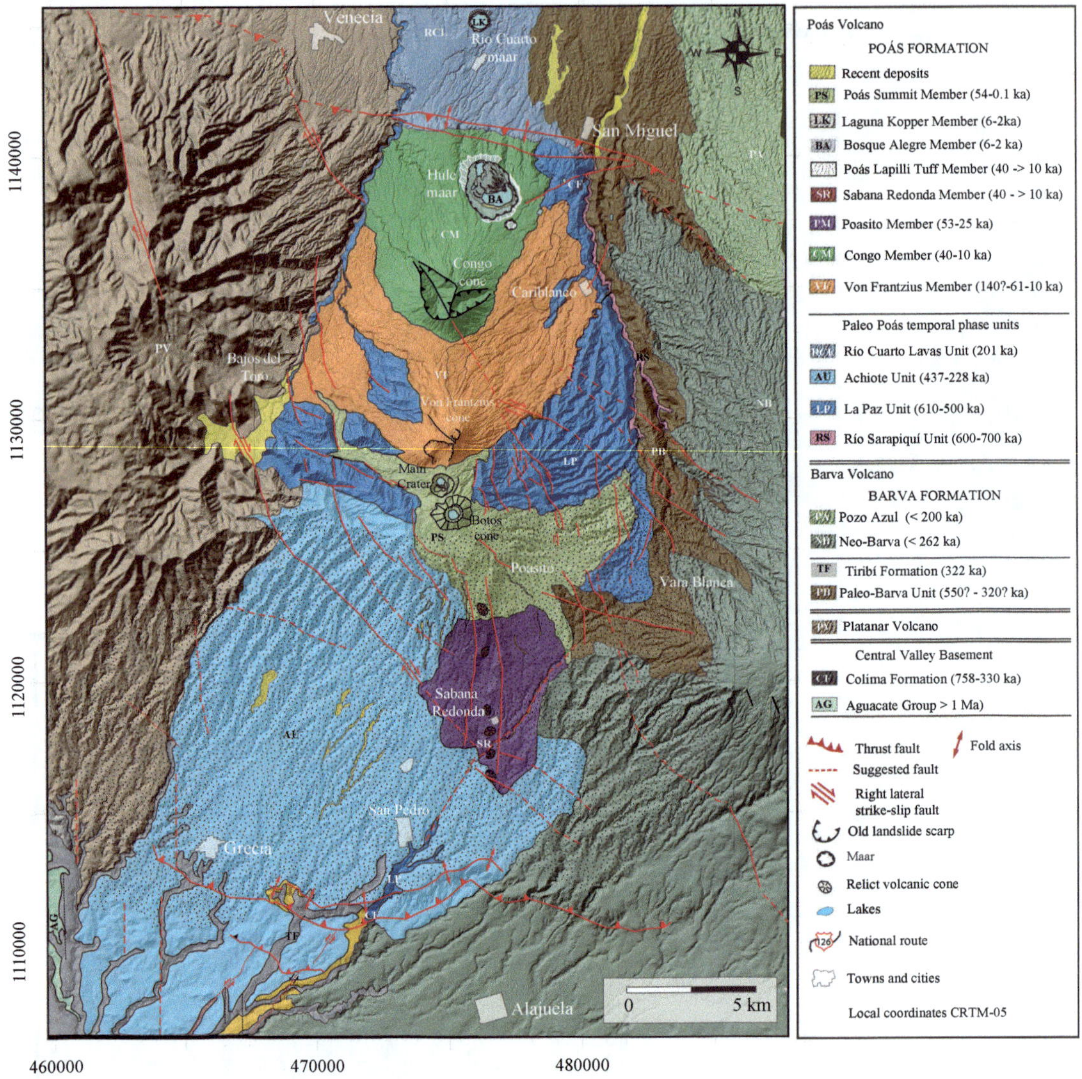

Fig. 2 Geological sketch map of Poás Volcano based on Prosser (1983), Alvarado and Climent (1985), Borgia et al. (1990), Rojas (1993), Alvarado and Carr (1993), Soto (1999), Campos et al. (2004), Gazel and Ruiz (2005), Montes (2007), Montero et al. (2010), Ruiz et al. (2010), Soto and Arredondo (2007), Huapaya and Rojas (2012), and new fieldwork. It is to note the location of Hule and Río Cuarto maars in the northern slopes of the volcano

2.1 Regional Setting

The CVR of Costa Rica is part of the Central American Volcanic front that extends parallel to the Middle American Trench from Guatemala to Costa Rica (Fig. 1). Its volcanic activity is due to the subduction of the Cocos Plate under the Caribbean Plate (Vannucchi and Mason Chapter "Overview of the Tectonics and Geodynamics of Costa Rica"). The convergence rate increases from ~ 83 mm yr^{-1} off southern Nicaragua to ~ 89 mm yr^{-1} off southern Costa Rica (DeMets et al. 2010). The lavas from Poás volcano as well as those from other volcanoes from central Costa Rica have an anomalous OIB signature when compared to the rest of the Central America

Volcanic arc and different models explained this peculiarity (e.g. Herrstrom et al. 1995; Russo and Silver 1994; Feigenson et al. 2004; Goss and Kay 2006; Hoernle et al. 2008). The latest model presented by Gazel et al. (2009) presented clear evidence that the observed OIB signature is derived from the Galapagos hot spot tracks subducting beneath Costa Rica and Panamá.

The summit of the Poás stratovolcano edifice is at 2708 m a.s.l. The currently active vent is located in the Main Crater, also known as Laguna Caliente. This crater holds one of the world's most acidic natural lakes, with a pH near zero and lies along a large ($\sim$35 km long) N-S running volcano-tectonic cortical fracture (Prosser 1983; Soto and Alvarado 1989). There are other volcanic structures located in this north-south trending fissure that are geomorphologically distinctive and developed in different time periods and volcanic episodes from separated vents, e.g. the pyroclastic cones of Sabana Redonda, the Von Frantzius, Congo and Botos cones, and the explosive craters (maars) of Hule, Pata de Gallo and Río Cuarto. The scarps of the Alajuela and San Miguel reverse faults bound the northern and southern flanks of this complex stratovolcano. There are other tectonic structures (in particular strike-slip faults) to the east and west flanks of the volcano, characterized by historical destructive seismic activity. The major historical earthquakes occurred in 1772 (Mw 6.0), 1851 (Mw 6.0), 1888 (Mw 6.0), 1911 (Mw 6.0), 1912 (Mw 5.5), 1955 (Mw 6.1), and the Cinchona earthquake 2009 (Mw 6.2). All of them were located within the cities of Bajos del Toro, Fraijanes, Vara Blanca, and Poás (Peraldo and Montero 1999; Montero et al. 2010; RSN 2009).

The Main Crater of Poás has frequently been active during the last 200 years, with eruptions characterized by periodic phreatic explosions (Alvarado 2009; Mora Amador Chapter "Volcanic Hazard Assessment of Poás (Costa Rica) Based on the 1834, 1910, 1953–1955 and 2017 Historical Eruptions"). Since the active vent is located only 20 km from the second largest city of Costa Rica, Alajuela, where the main international airport is located, and just 30 km from the capital, San José, Poás represents a significant hazard for ash falling and ash clouds, which may affect air traffic, and lahars and acid rain. Additionally, since 1980, the Costa Rican Institute for Electricity (ICE) and private companies have developed hydroelectric projects in the northern side of Poás edifice, taking advantage of the high mean annual precipitation of the area (3,000–6,000 mm), and the steep slopes (between 25° and 30°) of the Poás volcano flanks.

3 Field and Analytical Methods

The geologic map of Poás volcano presented in this chapter (Fig. 2) is an upgrade of the one presented in Ruiz et al. (2010). Herein we completed the geological mapping and introduced more details on several areas that were not included in previous works, using LiDAR data and high-resolution orthophotos. In particular, the updated areas include the northwest sector of the Congo cone, the zone of Bajos del Toro, and the eastern flank of the volcano, where the road from Vara Blanca to San Miguel (Road 126) is located. Fieldwork was carried out in 2009 and 2013, taking advantage of the newly exposed outcrops after the landslides produced by the Cinchona earthquake (Ruiz et al. Chapter "Coseismic Landslide Susceptibility Analysis Using LiDAR Data PGA Attenuation and GIS: The Case of Poás Volcano, Costa Rica, Central America").

The complete geological map of Poás volcano (Fig. 2) is built on a digital elevation model (DEM) background based on two sources: the topographic maps to scale 1:50,000 of Poás, Barva, Río Cuarto, Quesada and Naranjo from the National Geographic Institute of Costa Rica (NGI), 520 km^2 of LiDAR data from Poás, and parts of Barva and Platanar volcanoes. The LiDAR data and orthophotos were obtained during an airplane flight in April 2009 by the Spanish and Costa Rican companies STEREO-CARTO and AERODIVA, respectively, with an ALS50-II LEICA system, using a resolution of 3–4 points/m^2, to create a DEM with an x and y resolution of 50 cm and a z resolution of 15 cm (Ruiz et al. 2014). Differences in altitude between the images and the benchmarks of the topographic maps are less than 11 cm. The high quality of these data has an unprecedented resolution, which allowed us to identify volcanic features not recognizable by standard photogrammetric techniques. These images were processed using the commercial software packages Quick Terrain Modeler, SURFER 10.0 and GLOBAL MAPPER 13.0.

Several geologic studies were carried out in the northern part of area by the Costa Rican Institute of Electricity (ICE) to provide geologic characterization for hydropower projects. More than 155 borehole profiles from ICE were located on the NE and NW flanks of Poás volcano edifice, and were used in this chapter to identify the thickness of some of the lava flows from different volcanic units and members. In addition to the cores, information from a 6 km horizontal tunnel from the Cariblanco hydroelectrical project was also used. Cores and walls from the Cariblanco and Toro III tunnels were logged and sampled in detail for petrographic and geochemical characterization.

Furthermore, a geochronology database inclusive of several new $^{40}Ar/^{39}Ar$ and calibrated ^{14}C ages is here presented. The new $^{40}Ar/^{39}Ar$ age determinations were obtained on matrix separates at the Noble Gas Laboratory of Rutgers University, following the method described in Carr et al. (2007). These data were used to construct a new chronostratigraphic column of the Poás units (Fig. 3). Samples from each geologic

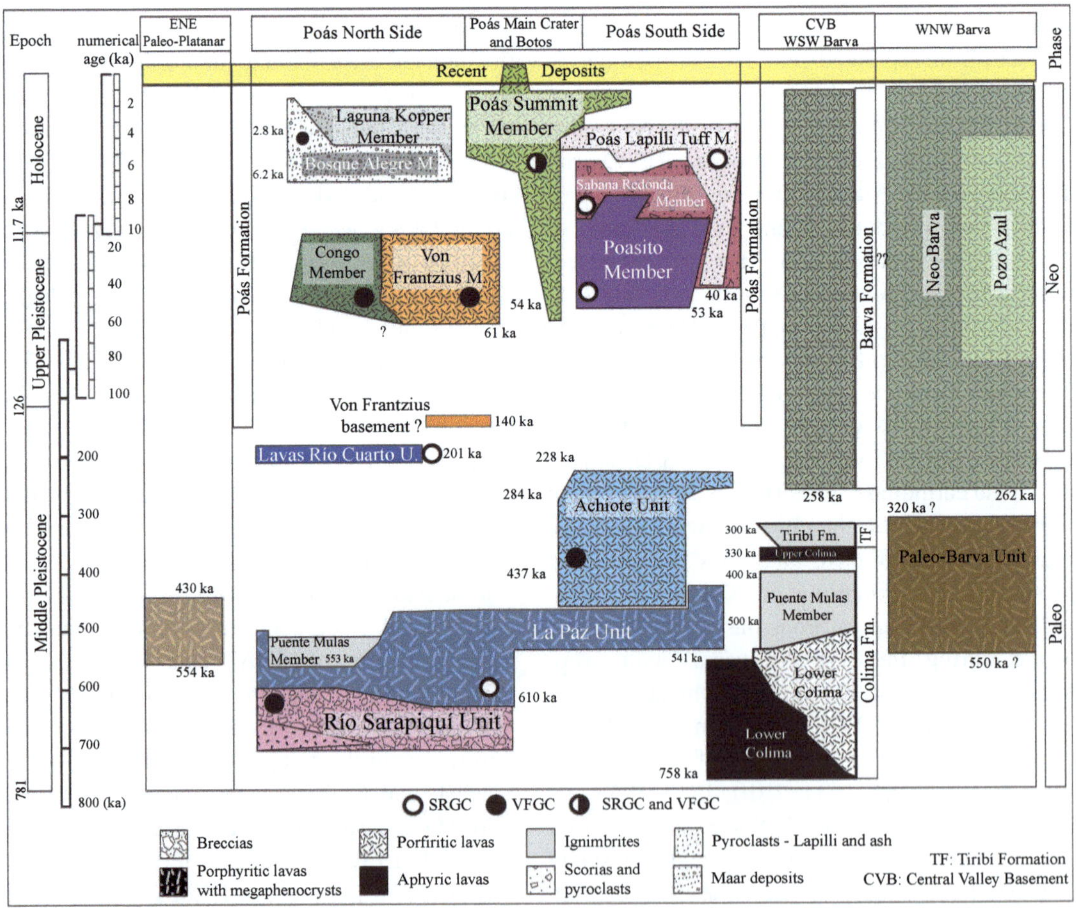

Fig. 3 Chronostratigraphic scheme of the Poás volcano products units. The main and subordinated lithologies are presented according to their geographical position relative to the Main Crater. Sabana Redonda Geochemical Component (SRGC), Von Frantzius Geochemical Component (VFGC). Colors of the units as in Fig. 2

member of Poás volcano were petrographically described. The resulting microscope photos and description are reported in the supplementary material (Table S2 and Fig. S3).

The geochemical database compiled for this chapter includes a total of 155 analyses of major and trace elements, as well as 27 new XRF determinations, and previous compilations by Kussmaul and Sprechmann (1982), Kussmaul et al. (1982) Kussmaul (1988) as well as those of McBirney and Williams (1965), Krushensky and Escalante (1967), Prosser (1983), Tournon (1984), Alvarado (1985), Paniagua (1985), Prosser and Carr (1987), Cigolini et al. (1991), Malavassi (1991), Soto (1999), Patino et al. (2000), Carr (2002), Gazel and Ruiz (2005), Carr et al. (2007), and Ruiz et al. (2010). The new geochemical data were obtained at the geochemistry laboratory of the Department of Geological Sciences (Michigan State University) by XRF and ICP-MS for major and trace elements, respectively. Details on precision, accuracy and used internal and international standards can be found in Hannah et al. (2002). The major and trace elements diagrams were made using IGPET 2012. The geochemical, geochronological and petrographical databases compiled for this chapter are available in the supplementary material.

3.1 New Formal Description of Poás Formation and Poás Stratigraphy

In this section, we summarize the evolution of the terminology used, while details on each unit from Poás volcano are provided below. Over the years the stratigraphy Poás was defined by a sequence of units, overlying a folded, faulted and tilted sedimentary and volcanic Neogene sequence, including deposits that present similar appearance, behavior and geochemical signature. Units were defined based on their stratigraphic sequence without any knowledge about their ages and relationship to the Central Volcanic Range. The considerable recent improvements in the age constraints and regional ties call for a revision of the nomenclature used.

The Colima (lower unit) and Tiribí (upper unit) formations were introduced on the basis of a limited number of outcrops and the geological description of numerous drilling cores from water wells in the Central Valley (Fernández 1969; Echandi 1981). It was not until recently that these formations were dated and introduced into the volcanic stratigraphy of the CVR (i.e., Pérez et al. 2006; Alvarado and Gans 2012). Barva Formation, overlying the Tiribí Formation, was first introduced in the 1980s because of its similarity to the rocks that host the Barva aquifer (Echandi 1981; BGS-SENARA 1985; Protti 1986). With time the understanding of the CVR stratigraphic greatly improved, allowing a better definition of the Barva Formation that now includes only mapped members formed during the Neo-stage of Barva volcano (i.e., Arredondo and Soto 2006). Similarly, the name Poás Andesites Formation, previously proposed by Kussmaul and Sprechmann (1982) for the whole CVR, using Poás volcano as stratotype, was shown to be impractical. Moreover, some confusion still exists since Echandi (1981) used the same name for those rocks associated with Poás. Similarly to Barva, the name Poás aquifer was suggested for Poás rocks by Pérez (2001) and Pérez et al. (2002). Soto (2005) reviewed this hypothesis following the same reasoning that was used for Barva, suggesting that the Poás Formation should only include those units related to Neo-Poás.

Accordingly, we subdivided the stratigraphy of Poás into Colima Formation, Tiribí Formation, and Poás Formation. Where the Poás Formation includes all the best-known and mapped units, which can be defined with the status of members. These members mainly consist of lava bodies, complex lava fields and some interlayered and minor interstratified pyroclastic flows, with a common vent source, which could span over several thousands of years. The newly defined members include: Poasito, Sabana Redonda, Von Frantzius, Congo, Bosque Alegre, Laguna Kopper, Poás Summit and Poás Lapilli Tuff. Units older than those members, which can be grouped geographically and chronologically, but whose sources can poorly be constrained or even unknown, were called as units (i.e., Río Cuarto Lavas, Achiote, La Paz, and Río Sarapiquí).

4 Results

4.1 Geology and Stratigraphy of Poás Stratovolcano

Poás stratovolcano is located between the Platanar and Barva volcanic edifices (Fig. 1c). Based on the location of its units, their ages and the history of other volcanoes from the CVR, it is inferred that it was formed by the superimposition of volcanic rocks during at least three principal chronological stages (Proto, Paleo and Neo Poás) occurring over almost one million years (Soto 1994, 1999; Carr et al. 2007; Ruiz et al. 2010; Alvarado and Gans 2012). Here, we present a description of the geologic units that build up the volcano that also includes some units related to fissural activity near the Central Valley and Barva volcano. Because the Proto Poás stage rocks are not clearly shown stratigraphically in the area, we focused on the lithology of the last two constructive stages of the volcano. Nevertheless, it is possible that the Sarapiquí Unit could be part of Proto Poás, and even parts of the Lower Colima Member may represent the volcanism of the oldest stage of the stratovolcanoes of the Central Volcanic Range. In Fig. 3, a stratigraphic sketch is shown where the vents of the various volcanic units and members are located according to their geographic position relative to the central Main Crater. The geologic sketch map (Fig. 2) does not include the more recent pyroclastic deposits that cover the volcano in order to facilitate the understanding of the underlying units and members and shows as clearly as possible their geographic distribution.

4.1.1 Colima Formation

Originally Williams (1952) named this unit as Intracanyon. Fernández (1969) introduced the name of Colima Formation that is currently in use. This formation corresponds to the local basement of the southern part of the study area (Fig. 3) and is overlain by the Tiribí Formation ignimbrite from Barva Volcano. The Colima Formation is composed by three members: (a) Lower Colima, (b) Puente de Mulas and, (c) Upper Colima. On the basis of cores from water wells drilled in the southern area of Poás (Campos et al. 2004), the average thickness of this formation under the Poás volcano units is $\sim$100 m (Lower Colima: ca. 50 m, Puente de Mulas: ca. 20 m and Upper Colima: ca. 30 m). The Lower Colima Member is mainly composed by porphyritic lavas with phenocrysts of plagioclase, augite, hypersthene, magnetite and some olivine in a matrix with intersertal texture. The intermediate member, Puente de Mulas, is a sequence of ignimbrites (at least two) outcropping together with a series of tuffs and lake sediments in the rivers Tacares and Prendas (Borgia et al. 1990). The Upper Colima lavas present an aphyric texture with only 4% of phenocrysts in a flow matrix with elongated vesicles. Radiometric $^{40}Ar/^{39}Ar$ dating from Marshall and Idleman (1999), Gans et al. (2002), Marshall et al. (2003), Gans et al. (2003), Pérez et al. (2006), and Alvarado and Gans (2012) produced ages of 758–535 ka for Lower Colima, 505 ka for one of the units of Puente de Mulas and 331 ka for Upper Colima (Table 1).

For this chapter, we sampled and dated an aphyric lava located in the southern flank of Poás at the base of an old quarry next to the Route No. 107, nearby the Alajuela thrust fault scarp (Fig. 2). Because of the texture, stratigraphic position and age (541 $\pm$ 4 ka; Table 1) the sample (PO-28/7/09-1) was related to the Lower Colima Member.

Based on petrographical and geochemical data, Soto (1999), Soto et al. (2008), and Ruiz et al. (2010) proposed that an ignimbrite outcropping on the northern side of Poás could be related to the Tiribí Formation, whose age (326 $\pm$ 8 ka) was obtained by Pérez et al. (2006) (Table 1). However, in this chapter we present a new $^{40}Ar/^{39}Ar$ age of 553 $\pm$ 9 ka for the ignimbrite located on the northern side of Poás (Table 1), suggesting that this ignimbrite is actually one of the units of Puente de Mulas Member and does not belong to the Tiribí Formation. This ignimbrite is a few meters thick and

Table 1 Ages for the Poás Volcano units (n.a. = not available, mtx = matrix, plat = plateau age, iso = isochron age), samples ACH-01 and ATN-58 from Ruiz et al. (2010) were revised for this study

Sample ID	Lithology	Geologic unit/member	Location	Age (ka)	Reference
n.a.	Carbonized material, on pyroclastic deposit	Bosque Alegre (Pata de Gallo Maar)	n.a.	2.79 ± 0.07–0.02 (^{14}C)	Malavassi et al. (1990), revised by Alvarado and Salani (2004, 2009). Reinterpreted by Alvarado et al. (2011)
n.a.	n.a.	Bosque Alegre (Hule maar)	n.a.	5.92 ± 0.25 (^{14}C)	Calibrated in this study (Calib Rev 7.0.4-Reime et al. 2013), revised from Melson et al. (1988), Alvarado and Salani (2004, 2009) and Alvarado et al. (2011)
n.a.	n.a.	Bosque Alegre (Hule maar)	Barros creek quarry, 474472 E-1138354 N	6.11 ± 0.16 (^{14}C)	Calibrated in this study (Calib Rev 7.0.4-Reime et al. 2013), revised from Soto (1999)
n.a.	n.a.	Poás Summit (Botos)	n.a.	8.33 ± 0.70 (^{14}C)	Malavassi et al. (1990), revised by Ruiz et al. (2010)
n.a.	n.a.	Poás Summit (Botos)	n.a.	10.89 ± 0.30 (^{14}C)	Malavassi et al. (1990), revised by Ruiz et al. (2010)
n.a.	n.a.	Poás Summit (Botos)	n.a.	11.36 ± 0.25 (^{14}C)	Malavassi et al. (1990), revised by Ruiz et al. (2010)
n.a.	n.a.	Congo	n.a.	35.6 ± 0.6 (^{14}C)	Malavassi et al. (1990), revised by Ruiz et al. (2010), Van Der Plicht et al. (2004)
n.a.	n.a.	Poás Lapilli Tuff	n.a.	40–3.3 ± 1 (^{14}C)	Prosser and Carr (1987), revised by Soto (1999)
SR-12	Charcoal	Sabana Redonda	Outcrop east from Sabana Redonda Town, 477083.7 E-1118572.4 N	40.04 ± 2.4–1.37 (^{14}C)	Ruiz et al. (2010)
CR-014B	Porphyritic basaltic andesite	Von Frantzius	Lava flow 25 m-thick, Ángel river and hill, 480818.8 E-1134320.1 N	41 ± 2 (^{40}Ar/^{39}Ar: mtx)	Alvarado and Gans (2012)
CR-095	Andesite	Von Frantzius	Drill core, upper section (Junction of Segundo and Toro rivers), 469421.8 E-1132540.0 N	42 ± 4 (^{40}Ar/^{39}Ar: mtx)	Alvarado and Gans (2012)
VF-01	Wood on lahars	Von Frantzius	NE side of Poás volcano, lahars below lavas from Von Fratzius cone	>46 (^{14}C)	Ruiz et al. (2010)
SR-28	Basaltic andesite	Poasito	Outcrop in Poás river, S from Sabana Redonda town, 476330.2 E-1116884.1 N	53 ± 40 (^{40}Ar/^{39}Ar: mtx, iso)	This study
PO-27/7/09-1	Andesite	Poás Summit (Botos)	Lava flow, outcrop on Pulga stream, 481740.3 E-1127127.7 N	42 ± 8 (^{40}Ar/^{39}Ar: mtx, plat)	This study

(continued)

Table 1 (continued)

Sample ID	Lithology	Geologic unit/member	Location	Age (ka)	Reference
PO-08/01/09-01	Andesite	Poás Summit (Botos)	Lava flow, outcrop on Pulga stream, 480590.4 E-1126178.9 N	54 ± 28 ($^{40}Ar/^{39}Ar$: mtx, plat)	This study, revised from Ruiz et al. (2010)
CR-013	Banded aphyric andesite	Von Frantzius	100 m from La Paz Grande bridge, 481991.9 E-1128677.3 N	61 ± 2 ($^{40}Ar/^{39}Ar$: mtx)	Alvarado and Gans (2012)
CR-097	Andesite	Von Frantzius	Drill core, middle section (Junction of Segundo and Toro rivers), 469421.8 E-1132540.0 N	140 ± 50 ($^{40}Ar/^{39}Ar$: mtx)	Alvarado and Gans (2012)
CR-PO-02-29	Basaltic andesite (aphyric)	Río Cuarto Lavas	Old quarry, near the W border of Río Cuarto maar, 475783.2 E-1144912.6 N	201 ± 30 ($^{40}Ar/^{39}Ar$: mtx, plat)	Carr et al. (2007)
CR-PO-02-28	Basaltic andesite (aphyric)	Río Cuarto Lavas	Old quarry 1.5 km WNE from Río Cuarto maar, 474374.1 E-1145742.0 N	201 ± 30 ($^{40}Ar/^{39}Ar$: mtx, plat)	Carr et al. (2007)
ACH-01	Andesite	Achiote	Achiote river outcrop, 473986.0 E-1122336.1 N	228 ± 34 ($^{40}Ar/^{39}Ar$: mtx, plat)	This study, revised from Ruiz et al. (2010)
PO-28/7/09-4	Andesite	Achiote	Outcrop next to the road, W from Poás river, 474178.8 E-1115336.4 N	320 ± 13 ($^{40}Ar/^{39}Ar$: mtx, plat)	This study
CR-053-1.2	Ignimbrite	Tiribí Formation	Bridge over Poás river, Route 1, after Manolos junction, 465869.7 E-1105695.7 N	336 ± 6, 326 ± 8 ($^{40}Ar/^{39}Ar$: plag)	Pérez et al. (2006)
ATN-58	Basaltic andesite	Achiote	San Juan river outcrop, 467201.3 E-1120227.2 N	437 ± 18 ($^{40}Ar/^{39}Ar$: mtx)	This study, revised from Ruiz et al. (2010)
PO-28/7/09-2	Basaltic andesite (porphyritic)	La Paz	Old quarry next to Route 107, 473255.2 E-1112079.7 N	480 ± 17 ($^{40}Ar/^{39}Ar$: mtx, plat)	This study
PO-28/7/09-3	Basaltic andesite (porphyritic)	La Paz	Poás river outcrop, 473310.6 E-1112868.8 N	506 ± 8 ($^{40}Ar/^{39}Ar$: mtx, plat)	This study
CR-100	Andesite	La Paz	Drill core, Base section (Junction of Segundo and Toro rivers), 469421.8 E-1132540.0 N	516 ± 12 ($^{40}Ar/^{39}Ar$: mtx)	Alvarado and Gans (2012)
CR-141	Andesite	La Paz	Drill core H.P.C. 28 TUN, Upper section, 481073.8 E-1133320.9 N	520 ± 20, 515 ± 10 ($^{40}Ar/^{39}Ar$: mtx)	Alvarado and Gans (2012)

(continued)

Table 1 (continued)

Sample ID	Lithology	Geologic unit/member	Location	Age (ka)	Reference
CR-144	Andesite	La Paz	Drill core H.P.C. 28 TUN, Depth 77.5 m (=975 m a.s.l.), 481073.8 E-1133320.9 N	517 ± 10, 490 ± 20 ($^{40}Ar/^{39}Ar$: mtx)	Alvarado and Gans (2012)
CR-012	Basaltic andesite (porphyritic)	La Paz	Montaña Azul hill, near La Paz waterfall, 482419.1 E-1128895.8 N	527 ± 6 ($^{40}Ar/^{39}Ar$: mtx)	Alvarado and Gans (2012)
PO-28/7/09-1	Aphyric lava	Lower member-like (Colima Formation)	Old quarry next to Route 107, 473255.2 E-1112079.7 N	541 ± 4 ($^{40}Ar/^{39}Ar$: mtx, plat)	This study
PO-19	Ignimbrite	Puente de Mulas Member-like (Colima Formation)	Junction of María Aguilar and Sarapiquí rivers, 480202.9 E-1139128.4 N	553 ± 9 ($^{40}Ar/^{39}Ar$: mtx, plat)	This study
CR-153	Andesite (block in volcanic breccia)	La Paz	Drill core H.P.C. 28 TUN, Depth 182 m (=871 m a.s.l.), 481073.8 E-1133320.9 N	588 ± 12, 585 ± 16 ($^{40}Ar/^{39}Ar$: mtx)	Alvarado and Gans (2012)
CR-146	Andesite	La Paz	Drill core H.P.C. 28 TUN, Depth 111.3 m (=939 m a.s.l.), 481073.8 E-1133320.9 N	607 ± 20 ($^{40}Ar/^{39}Ar$: mtx)	Alvarado and Gans (2012)
CR-148	Andesite	La Paz	Drill core H.P.C. 28 TUN, Depth 138.8 m (=913 m a.s.l.), 481073.8 E-1133320.9 N	610 ± 36, 590 ± 20 ($^{40}Ar/^{39}Ar$: mtx)	Alvarado and Gans (2012)

outcrops exclusively at the intersection of the María Aguilar and Sarapiquí rivers, about 1 km away from the San Miguel fault scarp (Fig. 3; according to boulders seen into the canyon of Sarapiquí river, it is likely cropping out somewhere in the canyon). In the stratigraphy of the northern side of Poás, Puente de Mulas ignimbrite overlies the Rio Sarapiquí Unit and some of the lavas from La Paz unit (Fig. 2).

Because of geomorphological observations, their lateral extension and the different chemical composition with respect to other CVR stratovolcanoes, which is very similar to that of Tiribí Formation, Kussmaul (1988) interpreted the Colima Formation lavas (lower and upper members) as the result of fissural activity along a NE-SW trend. As Lower Colima is not cropping out but in few sites along the Virilla canyon, south of Poás massif, it is still matter of debate where the multiple sources was located. The same situation stands for the Puente de Mulas Member ignimbrites. It can only be affirmed that some ignimbrites are distributed on both sides of the paleo-Cordillera Central (i.e., Caribbean and Pacific slopes). Following the distribution observed in water boreholes, we hypothesized that the deposits related to the Upper Colima Member were effused from a series of fissures on the present southern flank of Poás and Barva volcanoes, running WNW, and located near the towns of San Pedro de Poás (southern flank of Poás massif), and Santa Bárbara (southern flank of Barva massif).

4.1.2 Tiribí Formation

Originally named the Glowing Avalanche Deposits by Williams (1952), this formation presents a basal fall pumice layer (maximum observed thickness of 3 m) overlaid by an ignimbrite deposit with different welding facies. This formation is outcropping mainly in the southern part of the study area. Thanks to the reinterpretation of some deposits discussed in the previous section, it is believed that so far no outcrops of this unit were found in the northern flank of Poás edifice. In the southern sector, the Tiribí Formation is found in quarries and river valleys, e.g. Rosales and Poás rivers, and several places close to the Alajuela fault scarp (Echandi 1981; Borgia et al. 1990; Hannah et al. 2002; Campos et al. 2004; Pérez et al. 2006; Alvarado and Gans 2012).

In the southern sector of Poás, Tiribí Formation overlies the Colima Formation (Fig. 2). It appears to be underneath the La Paz Unit, although this observation might be caused by stratigraphic repetition due to the activity along the Alajuela reverse fault. Additionally, because this is a flow deposit it is also possible that it preferentially traveled through river canyons depositing in lateral contact between two units of older age.

The thickness of this formation in the southern side of the Poás volcano reaches 40 m (Campos et al. 2004). The radiometric ages for this ignimbrite are 326 ± 8 ka (Table 1; Pérez et al. 2006). Pérez (2000) interpreted this unit as generated by a powerful eruption that caused the subsequent development of Barva's volcano major caldera.

4.1.3 Río Sarapiquí Unit

This unit only crops out in the Sarapiquí river canyon and its tributaries (Soto 1999). It forms the local basement on the NE side of the Poás volcano. It is underlying the La Paz Unit and its base is not cropping out, thus only a minimum thickness of 50 m can be estimated. It mostly consists of breccias and ash-flow tuffs, with epiclastic lenses and subordinate lavas interdigitated between them. Based on stratigraphic correlations, Ruiz et al. (2010) speculated a minimum age of 600–700 ka (Fig. 2). There are no evidences that can indicate the location of its source. However, we hypothesized that its vent could be set near the N-S oriented volcano-tectonic cortical fracture on the Poás edifice as the rest of the units of this volcano.

4.1.4 La Paz Unit

Although the La Paz Unit is mainly found in the NE flank of Poás volcano, some patches of this unit were also found in the northwestern side as well as in the canyon of the river Poás in the southern flank of the volcano (Fig. 2). This unit overlies unconformably the Río Sarapiquí Unit and is overlain by products emitted from Río Cuarto Lavas Unit and the Congo and Von Frantzius cones, in the northern slope of the Poás edifice. In the southern flank of the Poás volcano, most lavas from Achiote Unit overlie it, although some of them could be contemporaneous.

Based on the study of drill cores obtained by ICE´s hydro-electrical projects, it was estimated that this unit mainly consists of several (at least seven) andesitic lava flows interdigitated with breccias and tuffs. The best outcrops for this unit are the result of a series of landslides developed along the slopes of Sarapiquí river canyon and Route No. 126. These lavas have a typical porphyritic texture with megaphenocrysts (2–3 cm) of plagioclase (Ruiz et al. 2010). Two new $^{40}Ar/^{39}Ar$ plateau ages for this unit: 480 ± 17 ka (PO-28/7/09-2) and 506 ± 8 ka (PO-28/7/09-3) (Tables 1 and 2) were determined (this chapter). At the surface, fresh outcrops of this unit are difficult to locate because the plagioclase phenocrysts and the glassy matrix are easily weathered in tropical environments. Thus, an effort was made to collect the samples dip enough to ensure freshness. Because of its texture and age span (600 and 480 ka; Gans et al. 2003; Ruiz et al. 2010; Alvarado and Gans 2012; this chapter), these lavas underwent high rates of weathering, which caused the formation of residual soils, reducing considerably their geotechnical conditions and ultimately affecting the geomorphology of the area by making the slopes more susceptible to sliding. Morphologically, this unit is similar to the Paleo-Barva Unit, presenting uneven slopes with angles between 30° and 60°. These slopes are covered by tuffs and weathered lapilli tuffs with thickness between 5 and 40 m, showing deeply eroded river valleys that are mainly truncated by right lateral faults. A maximum thickness of ~260 m was measured for this unit, which consists of lavas, tuffs and breccias. There are no evidences to locate the La Paz Unit vent. However, based on the distribution of its effusive products, we hypothesized that it could be positioned close to where the Main Crater and also influenced by the N-S running volcano-tectonic cortical fracture on Poás edifice.

4.1.5 Achiote Unit

The Achiote Unit consists of several lava flows, which only outcrops in the southern part of the Poás edifice. These flows present heterogeneous textures and can reach a thickness of ~110 m. They could be contemporaneous with the La Paz Unit and in some sectors they seem to overlie it (south of the study area). The much younger Poasito and Sabana Redonda Members lie on top of the Achiote Unit, which in some areas is covered by material from the Poás Lapilli Tuff Unit (see below) and/or residual soils (Campos et al. 2004; Montes 2007). Ruiz et al. (2010) constrained the age of these lavas at 540–200 ka, although they are likely younger (480–200 ka) according to Ruiz et al. (2010). Two lava flows from this unit were dated by $^{40}Ar/^{39}Ar$, whose plateau ages were of 320 ± 37 ka (PO-28/7/09-4) and 228 ± 34 ka (ACH-01) (Tables 1 and 2). The geomorphology of this unit is similar to the La Paz Unit, with uneven slopes between 30° and 60°, although its river valleys are less truncated and arranged in a sub-parallel to parallel drainage system. Similarly to La Paz Unit, we suggested a source location in the central part of the study area. This vent could have been the same as La Paz, or a different one, though located nearby. Based on our $^{40}Ar/^{39}Ar$ ages, lava flows from La Paz and Achiote units (PO-28/7/09-2 480 ± 17 and

Table 2 New $^{40}Ar/^{39}Ar$ dating results, the plateau and isochron diagrams are available in the supplementary material

Sample ID	Material	Total fusion age (ka)	% Rad	% ^{39}Ar in plateau	Plateau age (ka)	MSWD	Isochron age of plateau steps (ka)	$^{40}Ar/^{36}Ar$ intercept	MSWD	Selection criteria
SR-28	Matrix	144 ± 21	1.5	52.3	0.117 ± 12	13.9	**53 ± 40**	301.6 ± 2.2	1.636	II
PO-27/7/09-1	Matrix	29 ± 19	1.1	93.8	**42 ± 8**	1.2	58 ± 11	295.9 ± 1.8	0.963	I
ACH-01	Matrix	−213 ± 52	5.6	88.9	**228 ± 34**	0.4	236 ± 63	297.4 ± 10.5	0.537	I
PO-28/7/09-4	Matrix	298 ± 37	3.7	83.9	**320 ± 13**	0.4	331 ± 23	297.9 ± 1.3	0.431	I
ATN-58	Matrix	336 ± 23	7.2	92	**437 ± 18**	0.6	504 ± 53	292.1 ± 5.0	0.347	I
PO-28/7/09-2	Matrix	−1340 ± 46	20.7	63.6	**480 ± 17**	1.3	155 ± 110	316 ± 12.9	0.839	I
PO-28/7/09-3	Matrix	485 ± 18	13.8	79.9	**506 ± 8**	0.5	519 ± 23	296.7 ± 3.0	0.486	I
PO-28/7/09-1	Matrix	543 ± 06	42	63.6	**541 ± 4**	1.1	547 ± 07	290.5 ± 7.7	1.186	I
PO-19	Matrix	593 ± 13	12.8	43.9	**553 ± 09**	0.8	532 ± 10	304.61 ± 8.2	0.935	I

The bold ages are considered the most reliable ages for any given sample. Selection criteria used in the selection of the analysis are: (I) the plateau age is the best estimate of the sample age given that the $^{40}Ar/^{36}Ar$ intercept is atmospheric and is of higher precision than the isochron age; (II) given the evidence for a non-atmospheric "trapped" component we used the inverse isochron age

ATN-58 437 ± 18) are statistically undistinguishable in age and can overlap, meaning that these vents were likely be active at the same time.

4.1.6 Río Cuarto Lavas Unit

The main outcrops of this unit are found in the vicinity of Río Cuarto town that is located in the northern sector of the study area. This unit consists of a lava field that extends beyond the north face of the San Miguel fault scarp with a slight downward slope (3°–5°) to the north and with a parallel drainage system (Fig. 2). Stratigraphically, this unit overlies the La Paz Unit and underlies the Congo and Laguna Kopper Members (Fig. 3). This lava field has a minimum thickness of ~15 m. It consists of basaltic-andesitic lavas with typical aphyric and fluidal textures. Carr et al. (2007) dated this unit by $^{40}Ar/^{39}Ar$ at 201 ± 30 ka (Table 1). Based on the petrographic textures of these lavas and the geomorphologically features of the area, we suggested that they were likely related to fissure volcanism associated with the volcano-tectonic fracture of Poás volcano.

4.1.7 Poasito Member

This member is only present in the southern sector of the Poás volcano, with the main outcrops observable in the river canyons near Poasito town (Fig. 2). It consists of a minimum of 80 m massive lava flows with aphyric and fluidal textures (Gazel and Ruiz 2005). These rest unconformably on the Achiote Unit, congruent with or in some sectors overlying the Sabana Redonda Member (Fig. 3). Based on stratigraphic correlations Ruiz et al. (2010) estimated an age between 40 and 25 ka. In this chapter we support this interpretation with a new $^{40}Ar/^{39}Ar$ isochron age of 53 ± 40 ka (SR-28) (Tables 1 and 2). Since it is mostly covered by the Poás Lapilli Tuff Member and/or residual soils, where in some areas they reach up to 7 m, this unit presents a smooth topography with a parallel to sub-parallel drainage system and slope angles between 10° and 30°. The petrographic and rheological features of this unit suggest an origin related to fissure volcanism (Prosser and Carr 1987; Gazel and Ruiz 2005 and this chapter).

4.1.8 Sabana Redonda Member

The products of Sabana Redonda Member are located only in the southern sector of the Poás volcano. They correspond to the deposits (i.e. bombs and blocks in lapilli matrix, with lava agglomerates, and pseudo-lava flows from accumulation of bombs in the base) from at least six pyroclastic unnamed cones located between Poasito and Sabana Redonda towns (Fig. 2). Some of these cones were exploited as quarries where the best outcrops of this unit were found. Additionally, there are some outcrops in road cuts on the slopes of Route No. 146. Most of the original morphology of these cinder cones is now disappeared and even with the LiDAR data they can difficulty by recognized. The deposits from this member are overlying the Achiote Unit and Poasito Member, and it is underlying the Poás Lapilli Tuff Member deposits (Fig. 3). The pyroclastic deposits from these cones have a thickness that can reach ~60 m near the emission centers (Gazel and Ruiz 2005). The ages range between 40 and 10 ka (Ruiz et al. 2010). The origin of these six cinder cones is related to extensional movement of the N-S aligned volcano-tectonic fracture (Gazel and Ruiz 2005; Ruiz et al. 2010; this chapter).

4.1.9 Von Frantzius Member

The source of this unit is a composite cone located to the north of the actual active crater of Poás volcano (Fig. 2), the Von Frantzius Member is mainly made up by lava flows with breccias, epiclasts and pyroclasts on the top, for a total maximum thickness of 70 m while residual soils rarely are 5 m thick. Stratigraphically it lies above the La Paz Unit and probably Río Cuarto Lavas, it is contemporaneous with part of the Congo Member and it is below the deposits from Bosque Alegre Member (Fig. 3). The outcropping lavas from this member have an age from 41 to 10 ka (Gans et al. 2003; Ruiz et al. 2010). An $^{40}Ar/^{39}Ar$ age (140 ± 50 ka) was obtained by Alvarado and Gans (2012) for a drill core sample located on the northwest flank of the Von Frantzius cone and here interpreted as related to the base of this member. Alternatively, this sample could be part of another unidentified unit not outcropping because completely covered. The Von Frantzius cone shows a semi-radial drainage system with smooth slopes that present angles mainly between 30° and 60°.

4.1.10 Congo Member

The source of this member is a composite cone located in the northern side of the study area, between the Von Frantzius cone and the San Miguel fault (Fig. 2). Stratigraphically, the products of this cone underlie the Von Frantzius Member although the uppermost Congo lavas could be contemporaneous (Fig. 3). The Congo Member consists of lavas, epiclasts and pyroclastic flows. It has a minimum thickness of ~60 m and in some areas is only covered by ~5 m of residual soils. According to Ruiz et al. (2010) the age of this member ranges between 10 and 40 ka. The Congo cone is dissected by two main landslide scarps to the NNW and NNE and does not have a well-defined crater (Fig. 2). After witnessing the erosive effects that the Cinchona Earthquake had on this cone (Alvarado 2010; Ruiz et al. 2014, Chapter "Coseismic Landslide Susceptibility Analysis Using LiDAR Data PGA Attenuation and GIS: The Case of Poás Volcano, Costa Rica, Central America"), it is likely that these gullies formed due to repeated erosive episodes linked to similar seismic events over the last few thousand years. The geomorphology of the Congo cone is similar to that of Von Frantzius, presenting a semi-radial drainage system with smooth slopes of angles between 30° and 60°.

4.1.11 Poás Summit Member

The Poás Summit Member is located in the central zone of the study area. It includes the products from the Main Crater and the Botos Cone (Fig. 2). The lava flows and pyroclastic materials from the Main Crater are exposed in the crater walls and the outcrops extend westward. The lava flows from the Botos sub-unit extend eastward and overlie a thick set (~10 m) of pyroclasts in the Pulga stream valley on the eastern flank of the volcano. Based on ^{14}C and ^{40}Ar/^{39}Ar dates by Ruiz et al. (2010), the age of this member ranges from 56 to 8 ka for the Botos lavas and it is proposed that the lavas from the Main Crater could be contemporaneous or younger than the lavas from Botos. A new ^{40}Ar/^{39}Ar age from another lava flow from Botos (PO-27/7/09-1) was measured. It was collected further to the east and topographically below the one dated by Ruiz et al. (2010). Our new age determination produced a coherent age of 42 ± 8 ka (Tables 1 and 2). The slopes of the Poás Summit Member present a big range with angles been >15° in some areas (south of Botos cone) and close to 60° in river valleys and the Main Crater walls.

4.1.12 Poás Lapilli Tuff Member

The pyroclastic products of this member were defined by Prosser and Carr (1987) and later analyzed by Campos et al. (2004), Gazel and Ruiz (2005), and Montes (2007). It extends from the summit of the Poás volcano to the southwest, down to Grecia, which is located 15 km from the Main Crater. It consists of a juvenile lapilli tuff with a thickness of >7 m. In the sector of Sabana Redonda the Poás Lapilli Tuff crops out in road cuts, which are now better exposed following the Cinchona Earthquake of January 2009. The tuff and lapilli are light grey when fresh and turn into a range of brown and purple when weathered. Based on its geochemistry and regional distribution Gazel and Ruiz (2005) related this member to the Botos cone. Prosser and Carr (1987) have assigned an age close to 40 ka based on ^{14}C dating of older products (Table 1). Based on the relations with Sabana Redonda and Poasito Members, it was inferred that the upper materials of this deposits are much younger. This member is presented as a layer of dots in the geologic sketch map (Fig. 2) in order to show more clearly the extent of the lavas from Achiote Unit and Poasito Member on the south sector of the study area (Fig. 2), an isopleths map (Φ max) of this member deposits can be found in the supplementary material (Fig. S1).

4.1.13 Bosque Alegre Member

The Bosque Alegre Member is located north of the Congo cone, in the northern side of the study area (Figs. 2 and 4). It comprises the eruptive products of the Hule and Pata de Gallo maars.

The explosion crater walls of Hule have steep slopes (~60°) whereas those of the two small cones within the Hule crater and the Pata de Gallo maar have less steep slopes (~40°). There are three lava flows, with gentle slopes (<5°) that are associated with the small cones within the Hule maar (Salani and Alvarado 2010; Soto et al. 2010; Alvarado et al. 2011). Parametric data from these maars are presented in Table 3. Stratigraphically, Bosque Alegre Member

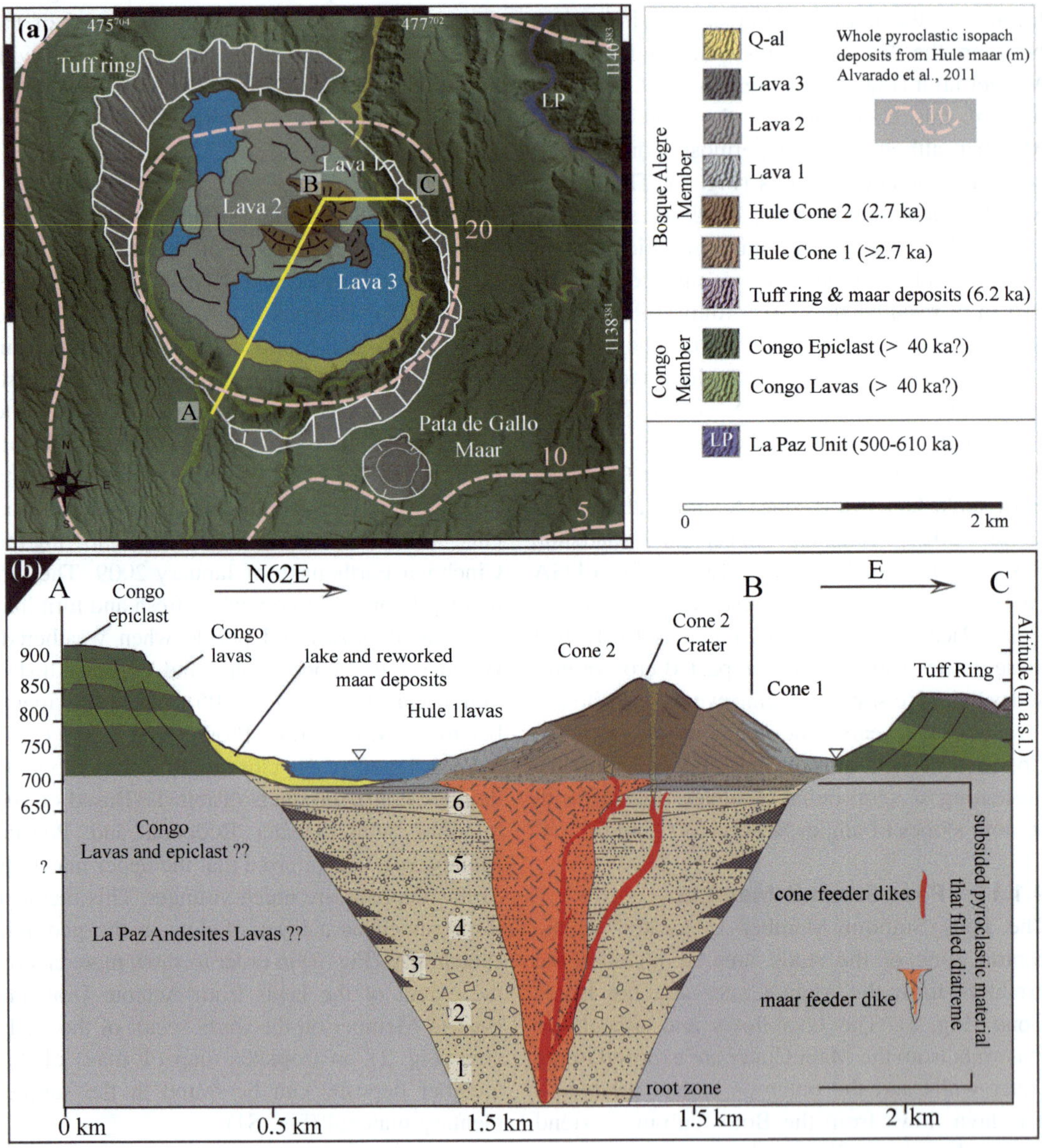

Fig. 4 **a** Interpretation of the geological features for the Hule maar over DEM based on LiDAR data with a resolution of 3–4 points/m^2. Isopachs data from Alvarado et al. (2011). **b** Schematic geological cross-section of the Hule maar and its possible internal structures based on Kurszlaukis and Fülop (2013). The stratigraphy includes: (1) resedimentation zone by collapse of the original maar and tuff ring; (2) coarse collapse material; (3) tuff rings that formed during the formation of the maar; (4) sedimentation of lapilli and medium size material; (5) sedimentation of ash and fine materials; and (6) sedimentation of lacustrine material related to the formation of the maar

overlies the Congo and Von Frantzius Members (Fig. 3). The tephra products of this unit are mainly pyroclastic surges, flow and fall deposits outside the maar (Alvarado et al. 2011). Within the maar, the two cones produced at least three basaltic lava fields (Fig. 4a). Hule was formed 6.0 ka ago and Pata de Gallo 2.8 ka ago while Alvarado et al. (2011) assigned the intra-maar products to 1.7 or 0.7 ka In Fig. 4b a schematic cross-section of internal structure of the Hule maar based on the morphology obtained by LiDAR imaging data and the models proposed by Kurszlaukis and Fülop (2013) for this kind of volcanic structures is drawn.

4.1.14 Laguna Kopper Member

The Laguna Kopper Member is located in the northernmost part of the study area (Figs. 2 and 5). It is genetically related to the maar known as Laguna Río Cuarto, which presents walls with steep slopes ($\sim 60°$). The fall deposits of this member appear in a narrow axis with an EW direction from the maar over 2 km with a variable thickness (no more than 15 m) that quickly pinches out disappearing. Locally, they overlie unconformably the Río Cuarto Lavas Unit (Figs. 2 and 5). Parametric data from this maar are presented in Table 3. The deposits are composed by the alternation of lithics, pumices and pyroclasts. Alvarado et al. (2011) estimated an age of 3–4 ka for this maar.

4.1.15 Recent Deposits

We included mostly fluvial and unconsolidated epiclastics. These sedimentary deposits occur in several sectors of the study area, although mainly located on the northern side, where they likely represent the distal facies of Congo and the foothills from the San Miguel fault scarp. These deposits also occur in the zone of Bajos del Toro, bordering the Poás massif to the west. The lithology is related to fluvial deposits with lahars and colluvial material, reaching tens of meters in thickness. These deposits were not included in the map to facilitate the understanding of the geology of the basement.

5 Geologic History of Poás Volcano

In the last 700–600 ka, Poás volcano has risen over materials that came from the proto-cordillera, aphyric lavas from fissure eruptions and ignimbrite layers from the Barva volcano. Different episodes of effusive, explosive, and erosive activity interspersed over time during the development of this composite volcano. Each one of the volcanic units and members that make up the edifice has distinct characteristics and diverse origins. Some units, e.g. the Río Cuarto lavas, and members, e.g. Poasito and Sabana Redonda, have origins related to extension processes along the volcano-tectonic fracture in the southern part of

Table 3 Morphometric parameters from Poás volcano maars revised from Umaña (1993), Haberyan and Horn (1999), Alvarado et al. (2011), and Ruiz et al. (2014)

Maar	Laguna Hule	Pata de Gallo	Río Cuarto
Shape	Subcircular	Circular	Circular
Major axis (km)	2.3	0.4	847
Minor axis (km)	1.8	0.38	758
Total area (km^2)	3.5	0.14	0.45
Total perimeter total (km)	7.2	1.32	2.51
Maximum altitude (m a.s.l.)	978	990	412
Minimum altitude (m a.s.l.)	777	889	362
Internal slopes (°)	27-76	25-79	Oct-70
Lake area (km^2)	0.46		0.33
Lake depth (m)	26.5		66
Surface water temperature (°C)	22.2–26.5	No water	24.6–29.9

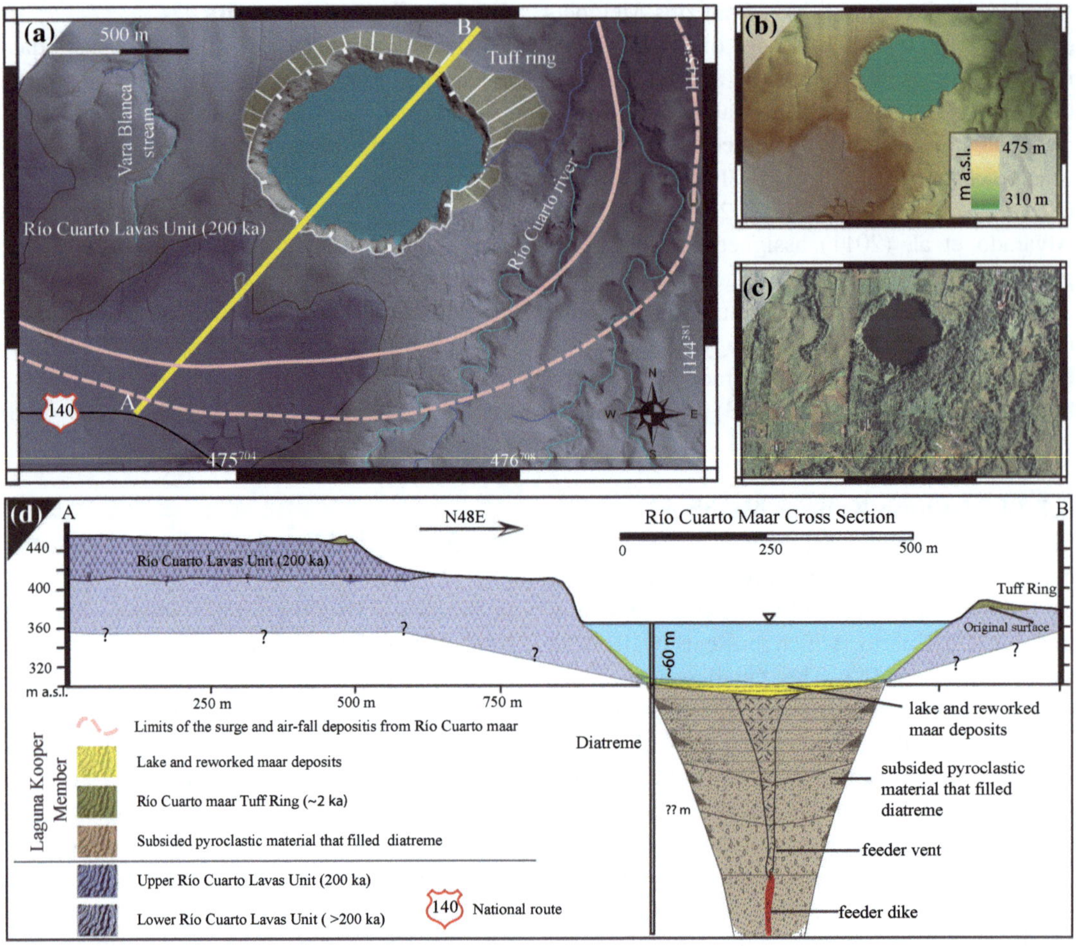

Fig. 5 **a** Interpretation of the geological features from Laguna Kooper Unit over DEM based on LiDAR data; **b** topographic features related to Río Cuarto maar; **c** aerial photo of Río Cuarto maar; **d** Geologic profile and interpretation of the Río Cuarto maar internal portion

Poás (Fig. 6). Other units, such as La Paz and Achiote as well as the Poás Summit Member, are products mainly related to effusive activity from a common central vent or vents that erupted at different times as typical for composite volcanoes. On the northern flank of the Poás volcano edifice, over the same volcano-tectonic fracture, two stratocones are located: Von Frantzius and Congo, which grew from subsequent effusive and explosive phases (Fig. 6). During thousands of years all the erupted units were constantly eroded, with high erosive peaks likely associated with seismic activity from active faults running along the flanks of the volcano. Finally, on the northern side of the study area, three explosive craters formed as maars and deposited the members of Bosque Alegre and Kopper, which pierced through the previous lava surfaces leaving behind their explosive materials (Fig. 6). The genesis of the Laguna Hule maar likely occurred in subsequent stages: first we identify an "aperture phase" due to a series of phreatic explosions that cut through the lavas from La Paz and Congo. Subsequently, a "strombolian phase" took place, during which the activity continued with phreatomagmatic (sub-plinian?) explosions, depositing pumice flows that buried charcoal dated at 6.1 ka. After the maar formed, the tuff

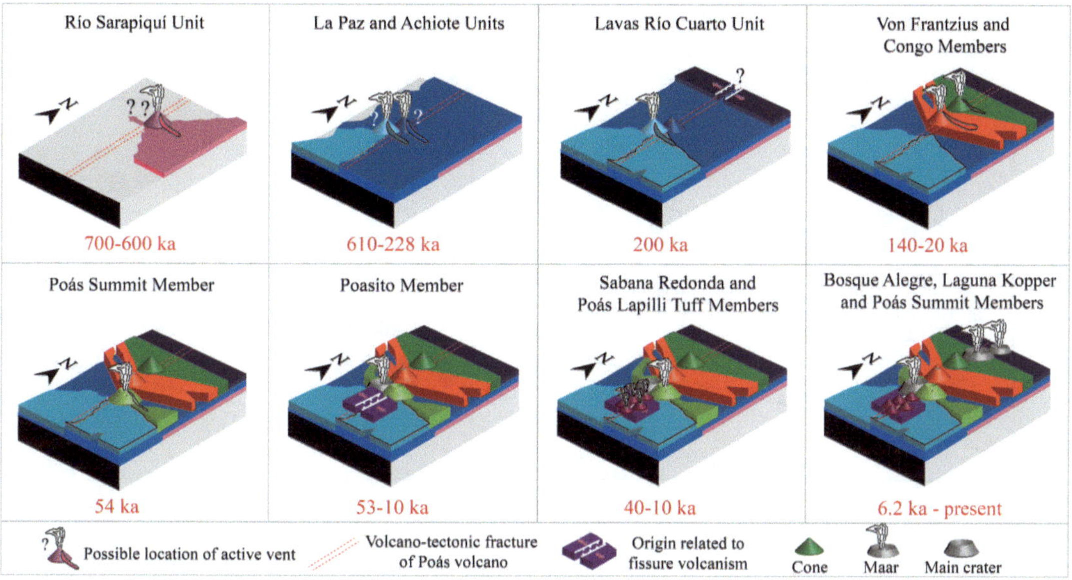

Fig. 6 Simplified evolution sketch of Poás volcano edifice. It only includes the units and members whose origin was related to Poás volcano

ring of a basaltic magma built the two pyroclastic cones by a strombolian eruption and simultaneously lava erupted creating small lava flows at about 2.8 ka (Alvarado et al. 2011). These flows and the erupted material divided the original lake of the maar into three different water surfaces, named Hule, Congo and an unnamed one.

The last growing episode of the Poás volcano is related to the intra-crater dome in the Main Crater that occurred in 1953 (Fig. 6). The total volume of the volcanic units that form the present volcano edifice represents a minimum volume since erosive episodes linked to landslides and accelerated by earthquakes have occurred throughout the volcano history. The Congo Member is a good example for this phenomenon as two main landslide (NNW and NNE) scarps dissected its cone due to the Cinchona earthquake and previous events.

6 Poás Volcano Geochemistry

The most common rock types, based on the SiO_2 versus K_2O diagram, which is commonly used in volcanic rocks saturated in silica from island arcs, are basaltic-andesites with variable potassium content and andesites (Fig. 7). The lavas from the Poás volcano show a sequence that spans from basalts to dacites, with each volcanic unit presenting variable degrees of magmatic evolution. The products from La Paz Unit, Poás Summit Main Crater and Botos sub-units present the complete range of variation from basalts to dacites, while other units such as the Río Cuarto Lavas, Poasito and Von Frantzius Members the basaltic end-member in their products is laking (Fig. 7).

Due to its geochemical variability and origin, the Poás Summit Member products are divided in two sub-units: Poás Summit Main Crater and Botos. The products from Botos are further subdivided into two geochemical sub-units. The Main Crater sub-unit has no basalts and ranges from basaltic andesites to dacites. Unlike Botos lavas, those from the Main Crater are all part of the low-K Calc-alkaline magma series (Fig. 7) and present a unimodal distribution of TiO_2, with all values having $TiO_2 < 1\%$ (Fig. 8). The geochemical behavior of the Botos sub-unit is widely variable with compositions that are from basalts to dacites and represent both the high- and low-K

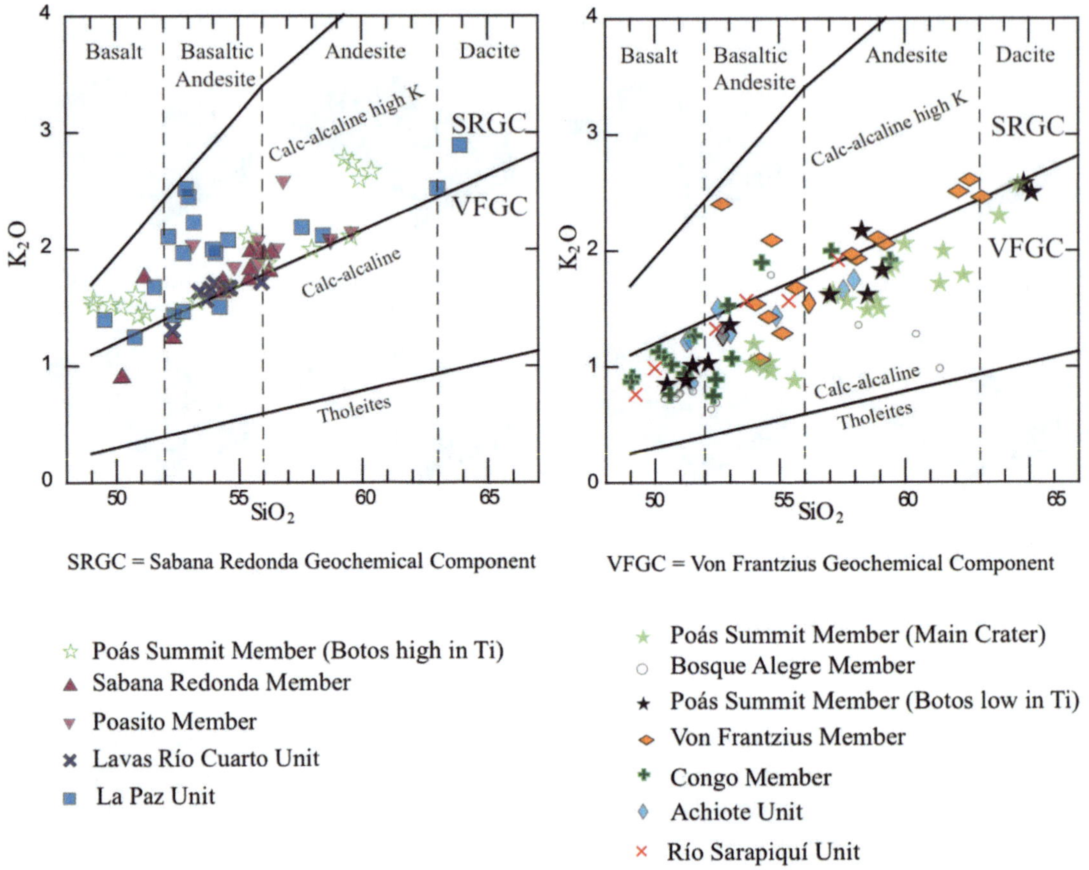

Fig. 7 Rock classification diagrams for SRGC and VFGC lithologic units of Poás volcano based on Peccerillo and Taylor (1976). Major elements (oxides) in wt%

series (Fig. 7). The Botos sub-unit is also very variable in TiO_2 and MgO. Due to these differences, this unit was further divided in Poás Summit Botos (high-Ti) and Poás Summit Botos (low-Ti). The samples that have high MgO contents, MgO > 7%, have low SiO_2 contents (<52%), which makes them true basalts. The Poás Summit Botos sub-unit (high-Ti) belongs to the SRGC (Sabana Redonda Geochemical Component) whereas the Poás Summit Botos (low-Ti) belongs to the VFGC (Von Frantzius Geochemical Component).

Following the geochemical classification identified for some units by Gazel and Ruiz (2005) and later extended to the rest of the volcanic units of the Poás volcano by Ruiz et al. (2010), we have assigned each volcanic unit and member to one of the two characteristic geochemical components of the Poás volcano: SRGC and VFGC (see Figs. 7 and 8).

The SRGC and the VFGC are distinct because the magmas that produced the former were the most enriched magmatic component of the Poás volcano. The SRGC matches the high-K Calc-alkaline magma series. It has values of TiO_2 > 1% and higher values than the VFGC for other major oxides (e.g. P_2O_5) and trace elements (e.g. Zr and Ba). The SRGC is relatively enriched in HFSE and several incompatible elements. The VFGC matches the low-K Calc-alkaline magma series. It has values of TiO_2 < 0.8% and has HFSE contents. In addition to the well-defined geochemical components, there are some intermediate lavas (TiO_2 0.8–1%) that are interpreted as a mix of the two geochemical components (Gazel and Ruiz 2005; Ruiz et al. 2010). The most

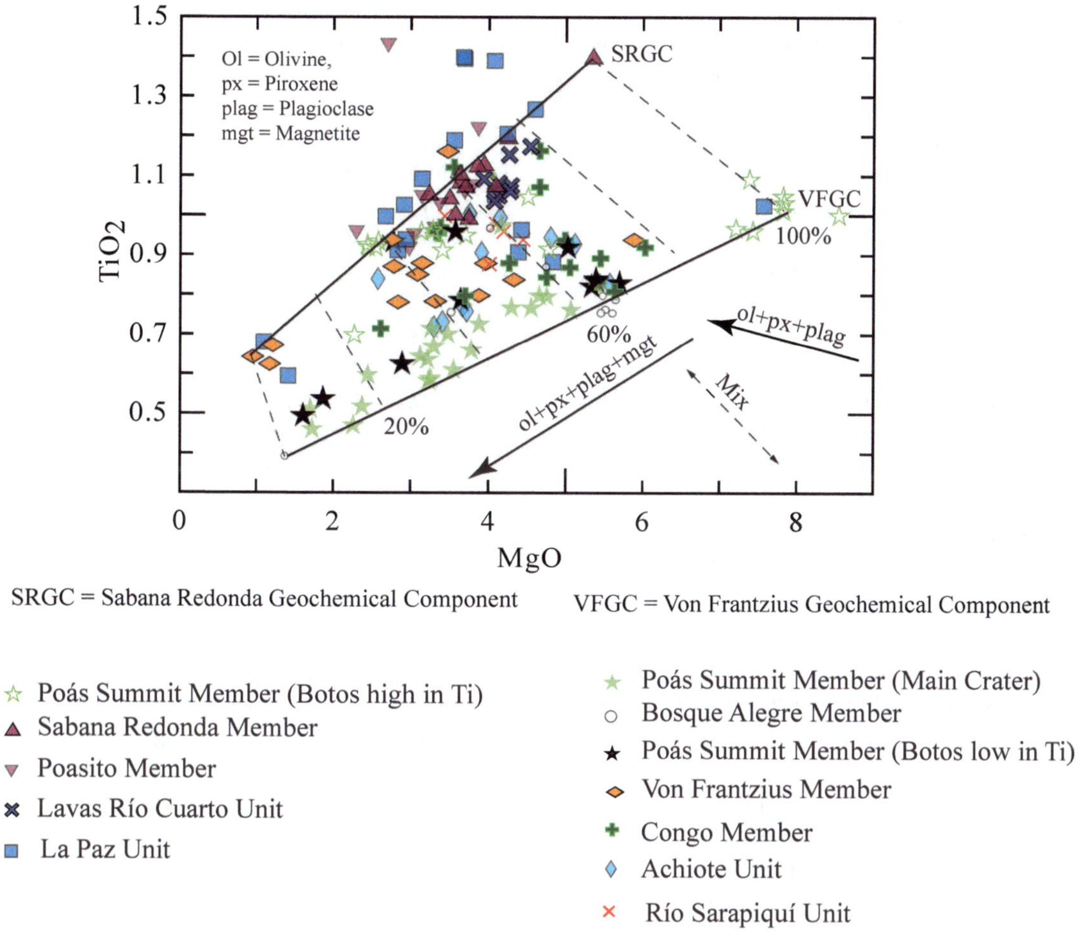

Fig. 8 TiO_2 versus MgO (in wt%) variation diagram showing the effects of crystal fractionation. Two independent magma series are present: one high-TiO_2 (SRGC) and the other relatively depleted in TiO_2 (VFGC). The vectors show the direction of chemical change expected at different stages due to crystal fractionation of the observed mineral phases (see supplementary material for thin sections description)

important geochemical ranges for major oxides and some trace elements used in the diagrams for each unit are presented in Tables 4 and 5.

Two magma series are observed due to variations in terms of K_2O content: (1) Calc-alkaline magma series, and (2) a high-K Calc-alkaline magma series (Fig. 7). The units of La Paz, Lavas Río Cuarto, and Poasito, Sabana Redonda and Poás Summit (Botos high-Ti) Members belong to the high-K Calc-alkaline series while the volcanic units of Río Sarapiquí, Achiote, Congo, Von Frantzius, Poás Summit (Botos low-Ti), Bosque Alegre and Poás Summit Main Crater are part to the Calc-alkaline series (Fig. 7). Previous studies (Prosser and Carr 1987; Patino et al. 2000; Carr 2002) suggested that some Poás lavas showed a trachyte tendency. Plots derived from our geochemical database confirm that the samples from the La Paz, Poasito and Sabana Redonda are relatively higher in the total alkali content than the rest of the units due to higher Na_2O values. This behavior is due to fractional crystallization where the more evolved specimens present a trachyte tendency. In particular, the La Paz Unit samples show a steep Na_2O trend at low degrees of fractionation (e.g. $SiO_2 < 58$ wt%). With increasing fractionation ($SiO_2 > 58$ wt%) a change in slope is observed, which is consistent with a transition

Table 4 Geochemical ranges for major oxides (in %) and some trace (in ppm) elements for units related to the Von Frantzius Geochemical Component (VFGC). The complete data set can be found in the supplementary material

Geologic unit	SiO_2	Al_2O_3	MgO	K_2O	Na_2O	TiO_2	P_2O_5	Sr	Rb	Zr	Ba	Age range
Poás Summit Main Crater	54.3–66.1	16.2–19.4	1.7–5.1	0.9–2.6	2.3–3.7	0.5–0.8	0.12–0.20	428–697	18–71	76–177	472–976	<1 ka
Bosque Alegre	51.3–62.6	16.8–20.9	1.4–5.7	0.6–1.8	2.2–3.9	0.4–1.1	0.12–0.44	510–789	13–50	54–145	374–933	6–2 ka
Poás Summit (Botos low-Ti)	51.8–65.7	16.9–19.1	1.6–5.7	0.9–2.5	2.2–3.5	0.5–0.9	0.12–0.23	451–615	16–48	70–156	480–1243	54–8 ka
Von Frantzius	54.3–63.7	17.2–19.6	1.0–5.9	1.1–2.7	2.6–4.1	0.6–0.9	0.15–0.44	499–804	20–73	85–206	572–1059	40–10? ka
Cerro Congo	49.6–59.8	14.7–21.0	2.6–6.0	0.8–2.2	2.4–3.3	0.7–1.1	0.16–0.30	552–746	13–40	57–129	427–662	40–10? ka
Achiote	51.3–59.6	16.9–21.0	2.6–5.6	0.9–1.8	2.6–3.4	0.8–1.0	0.17–0.27	471–651	13–37	75–145	455–829	0.54–0.20? Ma
Río Sarapiquí	51.4–58.9	17.4–20.7	3.4–4.4	1.0–2.0	2.9–3.2	0.8–1.0	0.20–0.23					>0.6 Ma

Table 5 Geochemical ranges for major oxides (in %) and some trace (in ppm) elements for units related to the Sabana Redonda Geochemical Component (SRGC). The complete data set can be found in the supplementary material

Geologic unit	SiO_2	Al_2O_3	MgO	K_2O	Na_2O	TiO_2	P_2O_5	Sr	Rb	Zr	Ba	Age range
Poás Summit (Botos high-Ti)	50.0–61.0	15.9–18.5	2.3–8.5	1.4–2.9	2.5–3.5	0.9–1.1	0.20–0.45	362–741	24–79	102–210	598–1010	54–8 ka
Poás Lapilli Tuff	53.4–55.3	17.4–18.4	3.9–4.5	1.1–1.5	2.7–2.8	0.9–1.0	0.12–0.23	536–559	22–44	106–136	494–610	>40?–10 ka
Sabana Redonda	52.1–56.9	16.6–18.6	3.2–5.4	0.9–2.0	2.7–3.3	1.0–1.4	0.21–0.70	508–582	16–57	143–161	657–781	40–10? ka
Poasito	51.9–60.3	15.9–17.9	2.3–3.9	1.9–2.6	3.0–3.8	0.9–1.4	0.23–0.61	500–713	42–67	139–259	702–1591	40?–25? ka
Lavas Río Cuarto	52.9–55.2	16.7–18.2	3.9–4.5	1.3–1.7	2.5–3.1	1.0–1.2	0.24–0.28	496–541	36–47	125–145	509–777	0.2–0.15? Ma
La Paz	51.5–63.5	16.7–20.1	1.4–4.8	1.3–2.5	2.5–4.0	0.6–1.2	0.19–0.47	487–733	36–58	131–193	635–854	0.6–0.5 Ma

of Na_2O from highly incompatible to higher compatibility (Fig. 9). We model this change in behavior as derived from fractionation of plagioclase with higher Na content. This is consistent with the presence of relatively large plagioclase phenocrysts in these samples (Fig. 9).

The two magma series differ in several geochemical components and cannot be related to each other by crystal fractionation. The variation diagram (Fig. 8) shows the effects of crystal fractionation. It is evident that there are two independent fractionation series: one high-TiO_2 (SRGC) and the other relatively depleted in TiO_2 (VFGC). The vectors show the direction of chemical change expected at different stages due to crystal fractionation of the observed mineral phases. The La Paz Unit has indications of the presence of fractional crystallization of olivine,

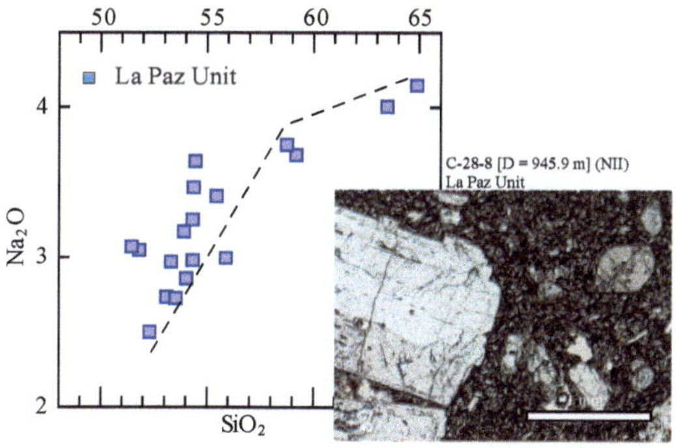

Fig. 9 SiO_2 versus Na_2O diagram for the La Paz Unit samples and petrographic thin section where typical large plagioclase phenocrysts present in these lavas are visible

pyroxene, plagioclase and magnetite. The products of Poás Summit (Main Crater) present the same fractional crystallization evolution but with lower levels of TiO_2 (Fig. 8). Additionally, there are indications of differentiation and mixing processes. Major elements from the analyzed rocks show a good general inverse correlation with MgO: the higher the concentration of SiO_2, Na_2O and K_2O the lower that of MgO (Fig. 8). On the other side, Fe_2O_3, CaO and TiO_2 show a positive correlation, as typical for incompatible oxides (Fig. 10). The crystal fractionation of the two different magma series is clearer when looking at TiO_2, Fe_2O_3 variation diagrams (Fig. 10).

Some magmatic vents erupted both the geochemical components while others were erupting only one of them (i.e. Sabana Redonda and Poasito). Therefore, these magmatic components can be either clearly separated or shared the same vent and eventually mix each other, suggesting that some of the magma chambers were separated while others served to mix these magmatic components to generate intermediate compositions (Fig. 10). This magmatic process is not exclusive for Poás as a similar geochemical behavior was attributed to the Irazú volcano (Alvarado et al. 2006).

MgO was used as a differentiation index to construct the Poás variation diagrams in Figs. 10 and 11. Zr and Nb present a negative tendency, as a result of their incompatible nature (Fig. 11). Ni and Cr have a positive correlation with MgO, highlighting their compatible behavior (Fig. 11). In both major (Fig. 10) and trace (Fig. 11) elements diagrams it is possible to observe how the HKCA basaltic andesites from the Poás Summit (Botos high-Ti) form a different group with respect to the rest of the geochemical units of the Poás volcano, displaying the highest MgO content among Poás products.

Based on geochemical observations and the regional distribution of the Poás Lapilli Tuff Unit deposits it is possible to associate them with the Botos crater. The eruption of the Poás Lapilli Tuff should have occurred during the time when this vent was erupting magmatic products related to the high-Ti phase. Furthermore, Gazel and Ruiz (2005) mentioned that the materials of Poás Lapilli Tuff Unit have major elements compositions similar to the geochemistry of the Sabana Redonda Unit: TiO_2 (0.9–1%), Fe_2O_3 (9–10%) and CaO (<6.5%), which were used to define the SRGC.

Central Costa Rica has an OIB-like mantle source (Feigenson et al. 2004; Hoernle et al. 2008; Gazel et al. 2009). This is the case for both VFGC and SRGC. The latter requires a relatively lower degree of partial melting, probably produced primarily by a decompression melting. Decompression melting was also suggested to be the mechanism of magma generation of other rocks with high contents of MgO, TiO_2 and Nb in some of the Quaternary Central American volcanic arc (i.e. Cameron et al. 2002). On the contrary, the VFGC lavas are related to magmas primarily produced by flux melting, which is typical in subduction zones (i.e. Cameron et al. 2002). This difference may be related to the extension within the Poás volcano-tectonic fracture (Gazel and Ruiz 2005; Ruiz et al. 2010; this chapter).

Isotopic data of $^{207}Pb/^{204}Pb$ and $^{206}Pb/^{204}Pb$ from literature (Feigenson et al. 2004) indicated that products from the Poás volcano plot on a mixing line between the DMM and HIMU isotopic reservoirs (Fig. 12; Feigenson et al. 2004). In fact, the Galapagos hotspot plots closely to the HIMU reservoir. Feigenson et al. (2004) interpreted the HIMU contribution observed in the Central Costa Rican volcanic front lavas as a

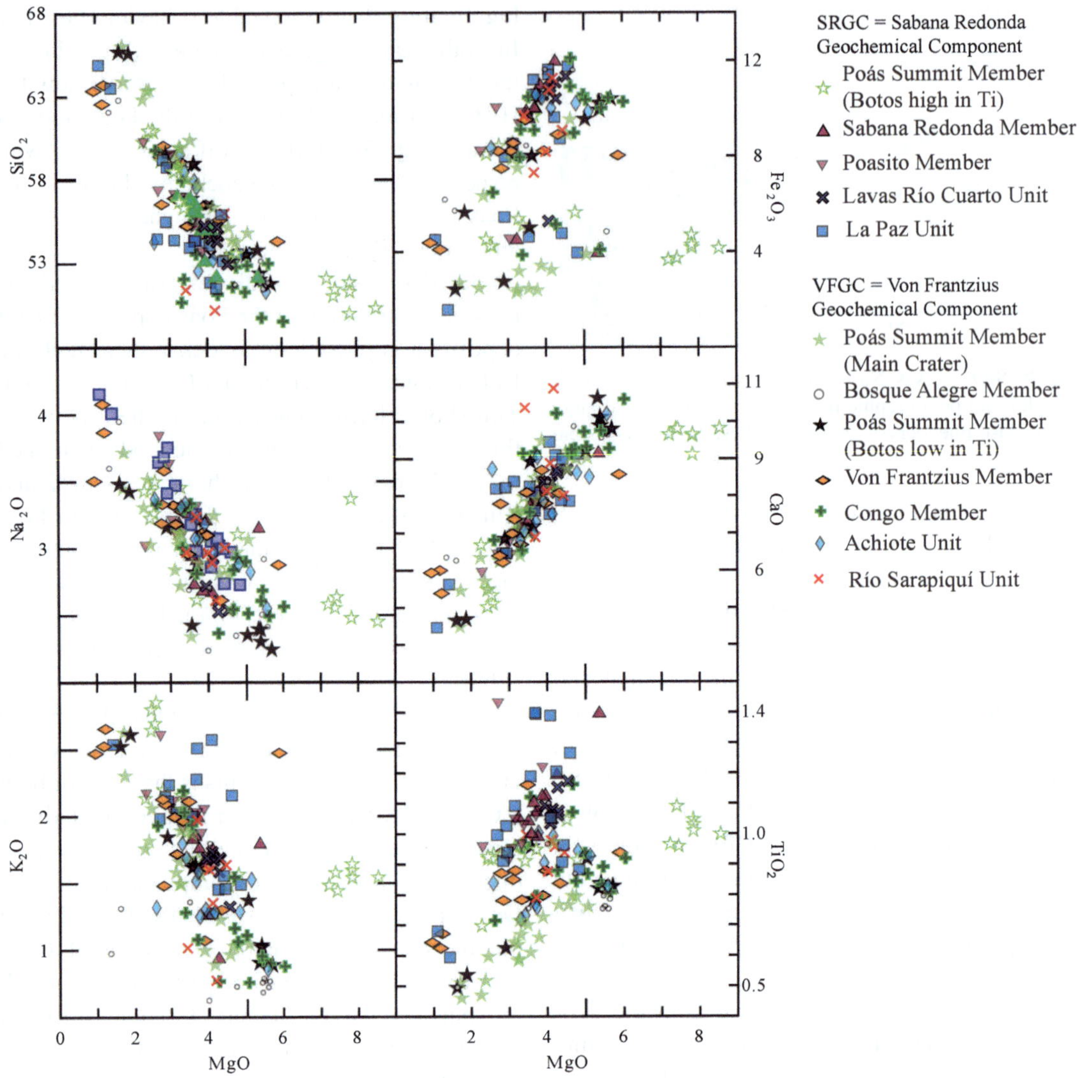

Fig. 10 Variation diagrams for major elements using MgO (in wt%) as differentiation index of the geochemical components for Poás volcano

consequence of contributions from the Galapagos plume-influenced mantle (see also Gazel et al. 2009, 2011).

6.1 Geochemical Evolution of Poás Volcano

Distinct range of variation of key major elements (TiO_2, P_2O_5) and trace elements (Zr and Ba) through time (Fig. 13) are observed and interpreted here to highlight the geochemical evolution of the Poás Volcano during the last 600 ka. The geochemical data for each unit are here assigned to their age averages while the exact sample by sample values can be found in Table 1. The reported units are consistent with the ones shown in the geologic map (Fig. 2) and chronostratigraphic column of the Poás volcano (Fig. 3). Although the hiatuses present between

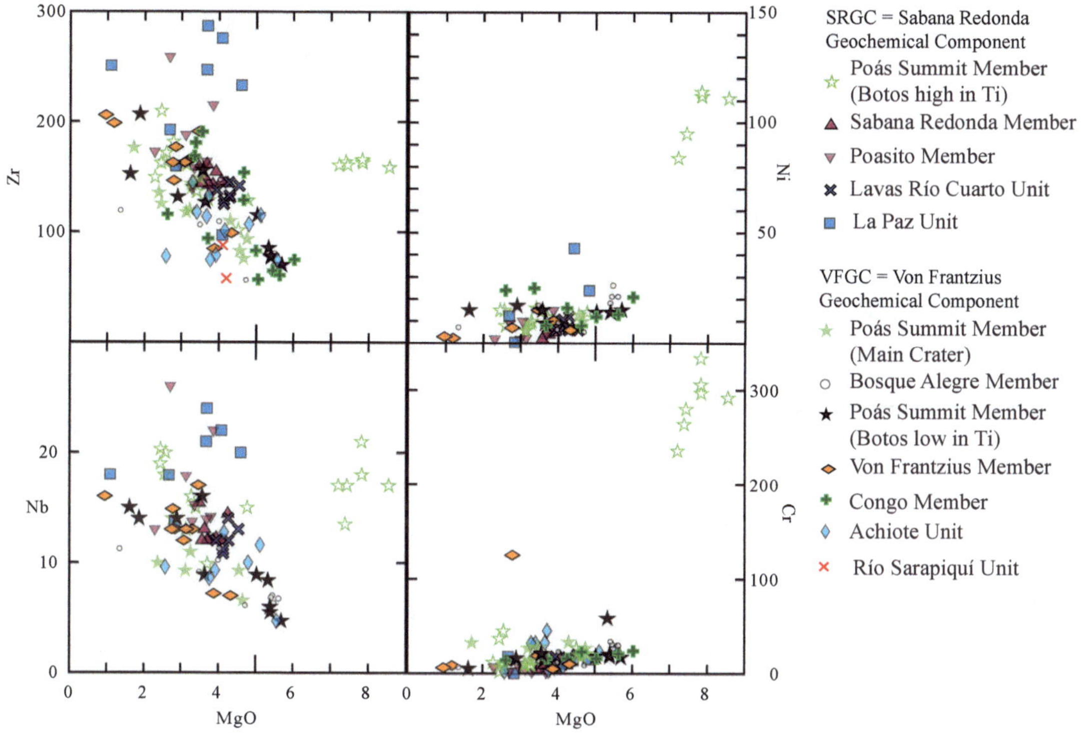

Fig. 11 Variation diagrams for trace elements using MgO (in wt%) as differentiation index of the geochemical components for Poás volcano

units are exaggerated by plotting only average ages, they represent periods of time during which the deposited units were eroded or periods of time when no activity occurred. Similar hiatuses were identified in other volcanic edifices in the Costa Rica like Irazú and Turrialba volcanoes (Carr et al. 2007; Alvarado and Gans 2012).

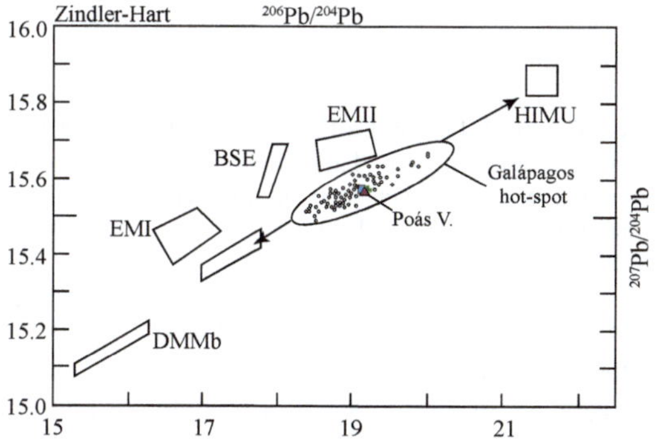

Fig. 12 $^{206}Pb/^{204}Pb$ versus $^{207}Pb/^{204}Pb$ isotopic plot (Zindler and Hart 1986) of Poás volcano rock samples. The isotopic distribution can be used to define relative magma source. Modified from Feigenson et al. (2004) and Gazel et al. (2009, 2011). DMMb: Depleted Mantle; EMI: Enriched Mantle I; EMII: Enriched Mantle II; BSE: Bulk Silicate Earth; HIMU: high μ mantle

The main observations derived from the age-geochemistry diagrams are: (1) there are units with similar age ranges that present different geochemical behavior; (2) there are units with different age ranges but similar geochemical behavior. Some units consistently have the same characteristic through time and can be correlated to the geochemical components previously described above (e.g. La Paz, Río Cuarto Lavas, Sabana Redonda and Botos are consistently high-Ti; and Río Sarapiquí, Achiote, Von Frantzius, Botos and Poás Summit Main Crater display low-Ti features). In other cases some units differ among themselves in the same time span (e.g. Poás Summit Botos high- and low-Ti units).

As discussed above, lavas from Poás volcano are distinguished into two magmatic components

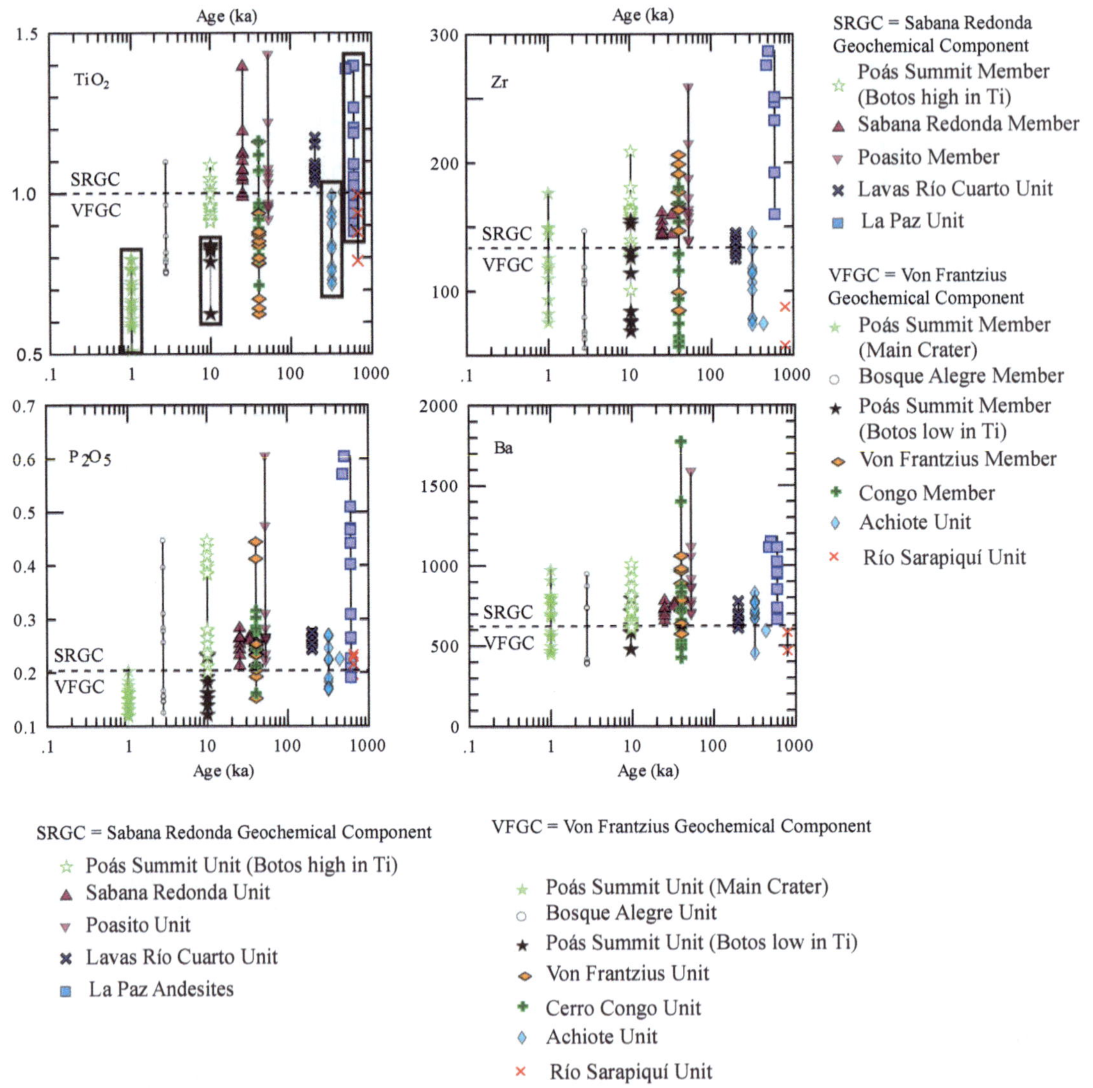

Fig. 13 Geochemical evolution of the Poás volcano. TiO_2, P_2O_5 are in wt%, while Zr and Ba are in ppm. The x-axis is logarithmic to be able to graph the ages of the units with a range from almost 600 ka and to less than 2 ka. The black box on La Paz, Achiote, Botos and Main Crater samples in the TiO_2 diagram is to show the chemical behavior over time of products that could be related to the same vent location

since the Paleo-Cordillera phase. Assuming a similar central location vent for some of the units of Poás (Fig. 6), there is an overall decrease of TiO_2 and P_2O_5 through time (i.e. from the oldest unit: La Paz, followed by Achiote and finishing with the Poás Summit, including a sub unit of Botos). However, there are also units with relatively high values for these elements according to their age (e.g. Poasito and Sabana Redonda). However, these lavas are known to derive from a different vent location related to extensional movement of the N-S aligned volcano-tectonic fracture at Poás, thus confirming the presence of two different geochemical components that are independent and that can be erupted from different vents (e.g. Sabana Redonda and Poás Summit Main Crater), as well as the same vent (e.g. Botos high- and low-TiO_2) (Fig. 13).

The fact that these magmatic components on some occasions are clearly separated while in other events they share and mix in the same vents suggests that the magmatic chambers was either separated or served to mix these magmatic components to generate intermediate compositions (Figs. 8 and 9). This magmatic process is not exclusive for Poás and similar behavior was observed at Irazú (Alvarado et al. 2006).

7 Discussion and Summary

Poás volcano is a relative young complex stratovolcano, its the edifice was formed by different stratocones, lava fields, cinder cones and explosion craters. All these volcanic features are related and located close to a volcano-tectonic fracture that runs N-S throughout the apparatus, as clearly evidenced by the alignment of all the volcanic structures along this direction. It is likely that the maximum tensional stress for the Central Volcanic Range is responsible for the generation of a perpendicular extensional movement where the Poás volcano-tectonic fracture could fit, although more work focusing on this topic is needed.

A new stratigraphic nomenclature for Poás volcano was proposed. This approach groups all the volcanic products that are younger than 200 ka (Neo-Poás temporal phase), and that can be easily mapped and have their vents location related to the Poás volcano edifice in a new formation called Poás Formation. Subsequently the former volcanic units from Neo-Poás temporal phase in Ruiz et al. (2010) are now referred to as members. These new members are: Von Frantzius, Congo, Poás Summit, Poasito, Sabana Redonda, Poás Lapilli Tuff, Bosque Alegre and Laguna Kopper. The volcanic products related to the Poás edifice and are older than 200 ka (Paleo-Poás temporal phase), maintain their designation of units, since some of their features (i.e. vent location and total extension) are unknown. These units are: Río Sarapiquí, La Paz, Achiote and Lavas Río Cuarto. The basement of Poás volcano on the southern sector of the study area was identified as part of the Colima and Tiribí Formations. Additionally, the Puente Mulas Member from Colima Formation was found interdigitated with the Paleo-Poás temporal phase units in the northern sector of the Poás edifice. No outcrops of the Tiribí Formation were found in the northern flank of Poás thus far. However, we do not exclude a priori the possibility that some exposure might be found in the future on the northern side of Barva volcano.

Ruiz et al. (2010) presented the very first complete geologic map of the Poás volcano. The updated version shown in this chapter was mainly based on LiDAR topographic data for most of the volcano edifice (Fig. 2) together with the stratigraphy of Poás volcano developed by using new geological, petrographical, geochronological and geochemical analyses carried out on the Poás units between 2009 and 2013.

The volcanic products of Poás volcano have unique characteristics and diverse origins. Some lavas with fluid textures have origins related to extension processes in the volcano-tectonic fracture in both the northern and southern sectors (i.e. Río Cuarto Lavas and Poasito). Sabana Redonda cinder cones could also be related to this fracture in the southern flank of Poás. Other units, e.g. La Paz, Achiote and Poás Summit Member, are products of a mainly effusive activity from a common central vent or vents that erupted at different times as typically occurs in composite volcanoes. On the northern flank of the Poás volcano edifice, over the same volcano-tectonic fracture, two stratocones Von Frantzius and Congo are located. During thousands of years all the cones and lavas were constantly eroded and the high erosive peaks were likely associated with seismic activity from active faults running along the flanks of the volcano. The northern side of the study area has three explosive craters formed as maars and deposited the members of Bosque Alegre and Laguna Kopper, which pierced through the previous lava surfaces leaving their explosive materials around them.

Geochemically, the lavas from the Poás volcano show the presence of a fractional crystallization sequence that span from basalts to dacites, with each volcanic unit presenting variable degrees of magmatic evolution. The presence of more evolved lavas indicates that there have been periods of quiescence where the original magma had time to evolve and probably was discharged before more basic magmas refilled the magma chamber. Thus, these eruptions were probably gas rich and violent at the beginning and became less explosive through time.

The volcanic units that form the Poás volcano have ages younger than 600 ka. Throughout its volcanic evolution the lavas erupted show the presence of two geochemical components: The Sabana Redonda Geochemical Component ($TiO_2 > 1\%$) and the Von Frantzius Geochemical Component ($TiO_2 < 1\%$). The former is interpreted to be related to processes that require a relatively lower degree of partial melting, produced primarily by a decompression melting that may be related to the extension generated within the Poás volcano-tectonic fracture, while the latter represents magmas produced primarily by flux melting related to subduction.

The SRGC and VFGC magmatic activities are observed as being derived from clearly separated vents as well as sharing the same vent. This observation suggests that in some occasions the magma chambers were separated and in others they had a key role since they likely mixed these magmatic components to generate intermediate compositions.

Acknowledgements We would like to thank Dr. Tyrone Rooney from the Department of Geological Sciences at Michigan State University for the new geochemical analyses presented in this chapter. Carolina Suárez and Amalia Gutiérrez from Lanamme UCR and several people from the Department of Earth and Planetary Sciences at Rutgers University and ICE who collaborated during this work. We thank Gianfilippo de Astis and Marinella Laurenzi for their insightful reviews, which greatly improved this manuscript.

References

Alvarado GE (1985) Mapa geológico para el P.H. Toro II. In: Piedra J, Vega OC, Alvarado G, Lezama G (eds) Informe geológico-geotectónico de avance a la prefactibilidad. P.H. Toro II, vol 1–2, vol 1, 156 p and Appendix A and B (In Spanish)

Alvarado GE (2009) Los volcanes de Costa Rica: Geología, historia, riqueza natural y su gente, vol 32, 3rd edn, Ed. Univ. Estatal a Distancia, Costa Rica, p 335 (In Spanish)

Alvarado GE (2010) Aspectos geohídricos y sedimentológicos de los flujos de lodo asociados al terremoto de Cinchona (Mw 6.2) del 8 de enero del 2009, Costa Rica. Rev Geol Am Cent 43:67–96 (in Spanish with English abstract)

Alvarado GE, Carr MJ (1993) The Platanar-Aguas Zarcas volcanic centers, Costa Rica: spatial-temporal association of Quaternary cal-alkaline and alkaline volcanism. Bull Volcanol 55:443–453

Alvarado GE, Climent A (1985) Informe sismológico para la etapa de avance a la prefactibilidad. In: Piedra J, Vega OC, Alvarado G, Lezama G (eds) Proyecto Hidroeléctrico Toro II, ICE, San José, 46 p (Internal report) (In Spanish)

Alvarado GE, Gans PB (2012) Síntesis geocronológica de magmatismo, metamorfísmo y metalogenia de Costa Rica, América Central. Rev Geol Am Central 46:7–122 (In Spanish with English abstract)

Alvarado GE, Salani FM (2004) (Tefroestratigrafía (40000–2000 a.P.) en el sector Caribe de los volcanes Barva, Congo y Hule, Cordillera Central, Costa Rica. Rev Geol Am Cent 30:59–72 (In Spanish with English abstract)

Alvarado GE, Carr MJ, Turrin DB, Swicher III CC, Schmincke H-U, Hudnut KW (2006) Recent volcanic history of Irazú volcano, Costa Rica: alternation and mixing of two magma batches, and pervasive mixing. In: Rose WI, Bluth GJS, Carr MJ, Ewert J, Patino LC, Vallance J (eds) Volcanic hazards in Central America, Geological Society of America Special Paper, vol 412, pp 259–276

Alvarado GE, Soto GJ, Salani FM, Ruiz P, Hurtado L (2011) The formation and evolution of Hule and Río Cuarto maars, Costa Rica. J Volcanol Geotherm Res 201:342–356

Arredondo SG, Soto G (2006) Edad de las lavas del Miembro Los Bambinos y sumario cronoestratigráfico de la Formación Barva, Costa Rica. Rev Geol Am Central 34–35:59–71 (In Spanish with English abstract)

BGS-SENARA (1985) Mapa hidrogeológico del Valle Central de Costa Rica. 1:50000-E.S.R. Limited, Inglaterra (In Spanish)

Borgia A, Burr J, Montero W, Morales LD, Alvarado GE (1990) Fault propagation folds induced by gravitational failure and slumping of the Central Costa Rica Volcanic Range: implications for large terrestrial and Martian volcanic edifices. J Geophys Res 95:14357–14382

Cameron BI, Walter JA, Carr MJ, Patino LC, Matias O, Feigenson MD (2002) Flux versus decompression melting at stratovolcanoes in southeastern Guatemala. J Volcanol Geotherm Res 119:21–50

Campos LA, Castro L, Gazel E, Montes N, Murillo S, Ramírez S, Ruiz P, Sequeira M (2004) Geología, geomorfología, amenazas naturales del Cantón de Poás, Alajuela, Univ Costa Rica, San José, p 60 (Report) (In Spanish)

Carr MJ (2002) CAGeochem database. http://www.rci.rutgers.edu/~carr/
Carr MJ, Feigenson MD, Bennett EA (1990) Incompatible element and isotopic evidence for tectonic control of source mixing and melt extraction along the Central American arc. Contrib Mineral Petrol 105:369–380
Carr MJ, Alvarado GE, Bolge L, Linsay F, Milidakis K, Turrin B, Feigenson M, Swisher CC III (2007) Element fluxes from the volcanic front of Costa Rica and Nicaragua. Geochem Geophys Geosyst 8: Q06001. https://doi.org/10.1029/2006GC001396
Casertano L, Borgia A, Cigolini C (1983) El volcán Poás, Costa Rica: Cronología y características de la actividad. Geofís Int 22:215–236
Cigolini C, Kudo AM, Brookins DG, Ward D (1991) The petrology of Poás Volcano lavas: basalt-andesite relationship and their petrogenesis within the magmatic arc of Costa Rica. J Volcanol Geotherm Res 48:367–384
de Moor JM, Aiuppa A, Pacheco J, Avard G, Kern C, Liuzzo M, Martínez M, Giudice G, Fisher TP (2016) Short-period volcanic gas precursors to phreatic eruptions: Insights from Poás Volcano, Costa Rica. Earth Planet Sci Lett 442:218–227
DeMets C, Gordon RG, Argus DF (2010) Geologically current plate motions. Geophys J Intern 181:1–80
Echandi E (1981) Unidades volcánicas de la vertiente N de la cuenca del río Virilla. Univ Costa Rica, San José (M.Sc. Thesis), 123 p
Feigenson M, Carr MJ, Maharaj SV, Bolge L (2004) Lead isotope composition of Central American volcanoes: influence of the Galapagos plume. Geochem Geophys Geosyst 5:Q06001. https://doi.org/10.1029/2003GC000621
Fernández M (1969) Las unidades hidrogeológicas y los manantiales de la vertiente norte de la cuenca del río Virilla. Investigaciones de aguas subterráneas en Costa Rica. Technical Report 27, 56 p (In Spanish)
Gans PB, Macmillan I, Alvarado GE, Pérez W, Sigarán C (2002) Neogene evolution of the Costa Rican Arc. Geological Society of America 2002 Annual Meeting, Denver Annual Meeting, 27–30 Oct 2002
Gans PB, Macmillan I, Alvarado GE, Pérez W, Sigarán C (2003) Neogene evolution of the Costa Rican arc and development of the Cordillera Central. Geological Society of America, Cordilleran Section, 99th Annual meeting, 1–3 Apr, Puerto Vallarta), vol 4, p 74
Gazel E, Ruiz P (2005) Los conos piroclásticos de Sabana Redonda: Componente magmático enriquecido del volcán Poás, Costa Rica. Rev Geol Am Central 33:45–60
Gazel E, Carr MJ, Hoernle K, Feigenson MD, Szymanski D, Hauff F, Van Den Bogaard P (2009) Galapagos-OIB signature in southern Central America: Mantle refertilization by arc-hot spot interaction. Geochem Geophys Geosyst 10. https://doi.org/10.1029/2008gc002246
Gazel E, Hoernle K, Carr MJ, Herzberg C, Saginor I, Van Den Bogaard P, Feigenson M, Swisher C (2011) Plume-subduction interaction in southern Central America: mantle upwelling and slap melting. Lithos 12:117–134
Gazel E, Hayes JL, Hoernle K, Kelemen P, Everson E, Holbrook WS, Hauff F, van den Bogaard P, Vance EA, Chu S, Calvert AJ, Carr MJ, Yogodzinski GM (2015) Continental crust generated in oceanic arcs. Nat Geosci 8:321–327
Goss AR, Kay SM (2006) Steep REE patterns and enriched Pbisotopes in southern Central American arc magmas: evidence for forearc subduction erosion? Geochem Geophys Geosyst 7(111):B09206. https://doi.org/10.1029/2005GC001163
Haberyan KA, Horn SP (1999) Chemical and physical characteristics of seven volcanic lakes in Costa Rica. Brenesia 51:85–95
Hannah RS, Vogel TA, Patino LC, Alvarado GE, Pérez W, Smith DR (2002) Origin of silicic volcanic rocks in Central Costa Rica: a study of a chemically variable ash-flow sheet in the Tiribi Tuff. Bull Volcanol 64:117–133
Herrstrom EA, Reagan MK, Morris JD (1995) Variations in lava composition associated with flow of asthenosphere beneath southern Central America. Geology 23:617–620
Hoernle K, Abt DL, Fisher KM, Nichols H, Hauff F, Abears GA, Bogaard P, Van Den Bogaard P, Heydolph K, Alvarado G, Protti M, Stauch W (2008) Arc parallel flow in the mantle wedge beneath Costa Rica and Nicaragua. Nature 451:1094–1097
Hofmann AW (2003) Sampling mantle heterogeneity through oceanic Basalts: isotopes and trace Elements. Carlson RW (ed) Treatise on geochemistry, vol 2, pp 61–101
Huapaya S, Rojas V (2012) Mapa Geológico de la Hoja Naranjo (3346-III), Escala 1:50000, Dirección de Geología y Minas, San José, Costa Rica (In Spanish)
Krushensky RD, Escalante G (1967) Activity of Irazú and Poás volcanoes, Costa Rica, November 1964/July 1965. Bull Volcanol 31:75–84
Kurszlaukis S, Fülop A (2013) Factors controlling the internal facies architecture of maar-diatreme volcanoes. Bull Volcanol 75. https://doi.org/10.1007/s00445-013-0761-y
Kussmaul S (1988) Comparación petrológica entre el piso del valle central y la cordillera central de Costa Rica. Ciencia y Tecnología 12:109–116
Kussmaul S, Sprechmann P (1982) Estratigrafia de Costa Rica (América Central) II: Unidades litoestratigráficas ígneas. Proceedings 5th Latin-American Congress of Geology, Buenos Aires, vol I, pp 73–79 (In Spanish)
Kussmaul S, Paniagua S, Gaínza J (1982) Recopilación, clasificación e interpretación petroquímica de las rocas ígneas de Costa Rica. Report IGN, vol 1982, no 2, pp 17–79 (In Spanish)
Malavassi E (1991) Magma sources and crustal processes at the terminus of the Central American Volcanic Front. University of Santa Cruz, California (Ph.D. Thesis), 435 p
Malavassi E, Gill JB, Trimble D (1990) Nuevas dataciones radiométricas del alineamiento volcánico de

Poás (Costa Rica): Contribución a la evaluación de peligros volcánicos. Geological Congress of Central America, 19–23 Nov 1990, San José, Costa Rica

Marshall JS, Idleman BD (1999). $^{40}Ar/^{39}Ar$ age constraints on quaternary landscape evolution of the central volcanic arc and Orotina debris fan, Costa Rica. In: Proceedings Geological Society of America, Annual Meeting

Marshall JS, Idleman BD, Gardner TW, Fisher DM (2003) Landscape evolution within a retreating volcanic arc, Costa Rica, Central America. Geology 31:419–422

Mcbirney W, Williams H (1965) Volcanic history of Nicaragua. Univ Calif Publ Geol Sci 55:1–65

Melson W, Sáenz R, Barquero J, Fernández J (1988) Edad relativa de las erupciones del Cerro Congo y Laguna Hule. Univ Nacional, Heredia, Bol Vulcanol 19:8–10

Montero W, Soto GJ, Alvarado GE, Rojas W (2010) División del deslizamiento tectónico y transtensión dentro del macizo del volcán Poás, Costa Rica. Rev Geol Am Central 43:13–36

Montes N (2007) Clasificación de los suelos a partir de sus propiedades físicas, mecánicas e hidráulicas y su relación con el potencial de infiltración, en el sector occidental del cantón de Grecia, Alajuela. Univ. Costa Rica, San José (B.Sc. Thesis) 97 p (In Spanish)

Mora-Amador R, Rouwet D, González G, Vargas P, Ramírez C (Chapter 11) Volcanic hazard assessment of Poás (Costa Rica) based on the the 1834, 1910, 1953–1955 and 2017 historical eruptions. In: Tassi F, Mora-Amador R, Vaselli O (eds) Poás volcano (Costa Rica): the pulsing heart of Central America Volcanic Zone. Springer, Heidelberg

Paniagua S (1985) Geoquímica de los elementos traza de las vulcanitas del Cenozoico superior de la región central de Costa Rica. Rev Geol Am Central 2:33–62

Patino L, Carr MJ, Feigenson MD (2000) Local and regional variations in Central America arc lavas controlled by variations in subducted sediment input. Contrib Mineral Petrol 138:265–283

Peccerillo A, Taylor SR (1976) Geochemistry of Eocene calc-alkaline volcanic rocks from the Kastamonu area, northern Turkey. Contrib Mineral Petrol 58:63–81

Peraldo G, Montero W (1999) Sismología histórica de América Central. Instituto Panamá Geography and History, México 347 p (In Spanish)

Pérez W (2000) Vulcanología y petroquímica del evento explosivo del Pleistoceno Medio (0.33 Ma) del Valle Central, Costa Rica. Universidad de Costa Rica, San José (B.Sc. Thesis), 170 p

Pérez JW (2001) Hidrogeología del área oeste del Valle Central. Informe Técnico, SENARA, San José, Costa Rica (In Spanish)

Pérez JW, Hirata R, Reynolds J (2002) Estudio de la hidrogeología del área oeste del Valle Central, Costa Rica, utilizando isótopos ambientales. In: Reynolds J (ed) Manejo integrado de aguas subterráneas. Un reto para el fururo. UNED, pp 203–214

Pérez W, Alvarado GE, Gans P (2006) The 322 ka Tiribí Tuff: stratigraphy, geochronology and mechanisms of deposition of the largest and most recent ignimbrite in the Central Valley, Costa Rica. Bull Volcanol 69:25–40

Plank T, Langmuir CH (1993) Tracing trace elements form sediment input to volcanic output at subduction zones. Nature 362:739–741

Prosser JT (1983) The geology of Poás Volcano, Costa Rica. Dartmouth College, Hanover, New Hampshire (M.Sc. Thesis), 165 p

Prosser JT, Carr MJ (1987) Poás Volcano, Costa Rica: geology of the summit region and spatial and temporal variations among the most recent lavas. J Volcanol Geotherm Res 33:131–146

Protti R (1986) Geología del flanco sur del volcán Barva. Univ Nac Bol. Vulcanol. Univ Nac Costa Rica 17: 23–31

Ranero CR, Villaseñor A, Phipps Morgan J, Weinrebe W (2005) Relationship between bend-faulting at trenches and intermediate-depth seismicity. Geochem Geophys Geosyst 6:Q12002. https://doi.org/10.1029/2005GC000997

Red Sismológica Nacional (RSN: ICE-UCR) (2009) El terremoto de Cinchona del jueves 8 de enero de 2009. Rev Geol Am Cent 40:91–95

Reime PJ, Bard E, Bayliss A, Beck JW, Blackwell PG, Bronk Ramsey C, Buck CE, Cheng H, Eduards RL, Friedrich M, Grootes PM, Guilderson TP, Haflidason H, Hajdas I, Hatté C, Heaton TJ, Hogg AG, Hughen KA, Kaiser KF, Kromer B, Manning SW, Niu M, Reimer RW, Richards DA, Scott EM, Southon JR, Turney CSM, van der Plich J (2013) IntCal13 and MARINE 13 radiocarbon age calibration curves 0–50000 years calBP. Radiocarbon 55. https://doi.org/10.2458/azu_jsrc.5516947

Rojas L (1993) Estudio geológico-geotécnico de un sector del P. H. Laguna Hule. Univ Costa Rica, San José (Report) (In Spanish) 42 p

Rudnick RL, Gao S (2003) Composition of the Continental Crust. In: Rudnick RL, Holland HD, Turekian KK (eds) Treatise on geochemistry—the crust, pp 1–64

Ruiz P, Gazel E, Alvarado GE, Carr MJ, Soto GJ (2010) Caracterización geoquímica y petrográfica de las unidades geológicas del macizo del volcán Poás, Costa Rica. Rev Geol Am Cent 43:37–66

Ruiz P, Garro JF, Soto GJ (2014) El uso de imágenes lídar en Costa Rica: casos de estudio aplicados en geología, ingeniería y arqueología. Rev Geol Am Cent 51: 7–31

Ruiz P, Carr MJ, Alvarado GE, Soto G, Mana S, Feigenson MD, Sáenz LF (2016) Coseismic landslide susceptibility analyses using LiDAR data PGA attenuations and GIS: the case of Poás volcano, Costa Rica, Central America (This volume)

Ruiz P, Carr MJ, Alvarado GE, Soto G, Mana S, Feigenson MD, Sáenz LF (Chapter 4) Coseismic landslide susceptibility analyses using LiDAR data PGA attenuations and GIS: the case of Poás volcano, Costa Rica, Central America. In: Tassi F, Mora-Amador R, Vaselli O (eds) Poás volcano (Costa

Rica): The pulsing heart of Central America Volcanic Zone. Springer, Heidelberg

Russo RM, Silver PG (1994) Trench-parallel flow beneath the Nazca plate from seismic anisotropy. Science 263:1105–1111

Ryan JG, Chauvel C (2014) The subduction-zone filter and the impact of recycled materials on the evolution of the mantle. In: Holland HD, Turekian KK (eds) Treatise in geochemistry, vol 3, pp 479–508

Rymer H, Cassidy J, Locke CA, Barboza MV, Barquero J, Brenes J, Van Der Laat R (2000) Geophysical studies of recent 15-year eruptive cycle at Poás Volcano, Costa Rica. J Volcanol Geotherm Res 97:425–442

Rymer H, Locke AA, Borgia A, Martinez M, Brenes J, Van Der Laat R, Willians J (2009) Long-term fluctuations in volcanic activity: implications for future environmental impact. Terra Nova 21:304–309

Saginor I, Gazel E, Carr MJ, Swisher CC, Turrin B (2011) New Pliocene–Pleistocene $^{40}Ar/^{39}Ar$ ages fill in temporal gaps in the Nicaraguan volcanic record: J Volcanol Geotherm Res 202:143–152

Saginor I, Gazel E, Condie C, Carr MJ (2013) Evolution of geochemical variations along the Central American volcanic front. Geochem Geophys Geosyst 14:4504–4522

Salani MF, Alvarado GE (2010) El maar poligenético de Hule (Costa Rica). Revisión de su estratigrafía y edades. Rev Geol Am Cent 43:97–118

Soto GJ (1994) Volcanología Física. In: Denyer P, Kussmaul S (eds) Atlas Geológico del Gran Área Metropolitana. Ed Tecnol Costa Rica, pp 131–146

Soto GJ (1999) Geología Regional de la Hoja Poás (1:50 000) Instituto Costarricense de Electricidad: 1 hoja. In: Alvarado GE, Madrigal LA (eds) Estudio Geológico-Geotécnico de Avance a la Factibilidad del P.H. Laguna Hule, San José (Internal Report), ICE 30 p (In Spanish)

Soto GJ (ed) (2005) Geología del cantón de Poás y estudios adicionales. + apéndices. FUNDEVI, Universidad de Costa Rica (Final Report), 198 p

Soto GJ, Alvarado GE (1989) Procesos volcánicos asociados con el agua subterránea. El caso de los volcanes Arenal y Poás, Costa Rica. In: Proceedings 3rd National Congress on Groundwater Resources, Nov 1989, San José, pp 249–261

Soto GJ, Arredondo SG (2007) Chronostratigraphic summary of Barva Formation (Costa Rica). Abstract volume, Workshop to integrate Subduction Factory and Seismogenetic Zone Studies in Central America, 18–22 June 2007, Heredia, Costa Rica, 103 p

Soto G, Pérez W, Arredondo S (2008) ¿Cuán extensa y voluminosa es la Formación Tiribi, Costa Rica? Nuevos hallazgos y reinterpretaciones. 9th Geological Congress of Central America, San José, July 2008 (In Spanish)

Soto GJ, Alvarado GE, Salani FM, Ruiz P, Hurtado De Mendoza L (2010) The Hule and Río Cuarto maars, Chapter One: The beginning. IAVCEI Comm Volcan Lakes. 7th Workshop on Volcanic Lakes Costa Rica, 10–19 Mar 2010, p 37

Thorpe CA, Locke GC, Brown PW (1981) Magma chamber below Poás volcano, Costa Rica. J Geol Soc 138:367–373

Tournon J (1984) Magmatismes du Mesozoique a l'actuel en Amérique Centrale: l'exemple de Costa Rica, des ophiolites aux andésites. Mém Sc Terre Univ Pierre et Marie Curie, París, 84–49, 335 p

Umaña G (1993) The planktonic community of Laguna Hule, Costa Rica. Rev Biol Trop 41(3):499–507

Van Der Plicht J, Beck JW, Bard E, Baillie MGL, Blackwell PG, Buck CE, Friedrich M, Guilderson TP, Hughen KA, Kromer B, Mccormac FG, Bronk Ramsey C, Reimer PJ, Reimer RW, Remmele S, Richards DA, Southon JR, Stuiver M, Weyhenmeyer CE (2004) NotCal04-Comparasion/Calibration ^{14}C Records 26-50 calkyr BP. Radiocarbon 46:1225–1238

Vannucchi P, Mason JP (Chapter 1) Overview of the tectonics and geodynamics of Costa Rica. In: Tassi F, Mora-Amador R, Vaselli O (eds) Poás volcano (Costa Rica): The pulsing heart of Central America Volcanic Zone. Springer, Heidelberg

Vogel TA, Patino LC, Alvarado GE, Gans PB (2004) Silicic ignimbrites within the Costa Rican volcanic front: evidence for the formation of continental crust. Earth Planet Sci Lett 226:149–159

Williams H (1952) Volcanic history of the Meseta Central Occidental, Costa Rica. Univ Calif Publ Geol Sci 29:145–180

Zindler A, Hart SR (1986) Chemical geodynamics. Ann Rev Earth Planet Sci 14:493–571

The Extraordinary Sulfur Volcanism of Poás from 1828 to 2018

Raúl Alberto Mora Amador, Dmitri Rouwet, Priscilla Vargas and Clive Oppenheimer

Abstract

This chapter is arguably the most complete compilation of sulfur volcanism of any given volcano on Earth: Poás. Sulfur volcanism at Poás is described in historical literature since 1828, and in scientific literature since the 1960's. We first classify the various manifestations of sulfur volcanism at crater lake bearing volcanoes (subaerial and sub-lacustrine sulfur pools, sulfur spherules, flows, cones/*hornitos*, and sweat, and pyroclastic and burning sulfur), based on work by Japanese pioneers in the early 1900s. Their first observations and models have passed the test of time and still stand as theories today. Comparing the sulfur volcanism at Poás with that in other (55) volcanoes, it is honest to say that only White Island (New Zealand) and Kawah Ijen (Indonesia) are the only ones comparable with Poás, being the most dynamic of them all.

Keywords

Poás volcano · Laguna caliente · Sulfur volcanism · Sub-lacustrine sulfur pool · Sub-aerial sulfur pool · Sulfur flow · Spherules · Sulfur cones · Volcanic lake

R. A. Mora Amador (✉)
Centro de Investigaciones en Cienca Geológicas CICG, Escuela Centroamericana de Geología, PREVENTEC, Universidad de Costa Rica, San José, Costa Rica
e-mail: raulvolcanes@yahoo.com.mx

R. A. Mora Amador · P. Vargas
Laboratorio de Ecología Urbana, Universidad Estatal a Distancia, San José, Costa Rica

D. Rouwet
Istituto Nazionale di Geofisica e Vulcanologia, Sezione di Bologna, Bologna, Italy

C. Oppenheimer
Department of Geography, University of Cambridge, Downing Place, Cambridge CB2 3EN, UK

1 Introduction

After lava and water, sulfur is the most common liquid eruptive product of volcanoes on Earth. Sulfur is also an abundant component of volcanic products on the Jovian moon, Io, often described as the most volcanically-active body in the Solar System. "Wet volcanoes" promote the formation of elemental sulfur (S°) due to the absorption of SO_2- and H_2S-rich gases in their aquifers or lakes, whereas open-conduit volcanoes (with a magma-air interface) typically discharge gaseous SO_2 and H_2S to the atmosphere. The presence of water is hence a strong catalyst for the assorted manifestations of sulfur volcanism. As such, crater lakes that efficiently tap degassing magmas are ideal settings to observe this important element in its various morphologies and physical states. Acting as calorimeters and chemical condensers, they also represent "windows" into

F. Tassi et al. (eds.), *Poás Volcano*, Active Volcanoes of the World,
https://doi.org/10.1007/978-3-319-02156-0_3

subsurface plumbing systems and magmatic processes. Further, the extreme conditions of crater lakes, hyper-saline and hyper-acidic brines, in which S° and related species flourish are often regarded as analogs of the environments where early life thrived. Crater lakes represent windows into the underlying magmatic-hydrothermal systems, invaluable to shed light into magmatic dynamics of the underlying threatening volcano. Hot crater lakes often display a turquoise color, which arises from the suspension of sulfur colloids. In the absence of these colloids, crater lakes are often green (Ohsawa et al. 2010; Christenson et al. 2016; Murphy et al. 2018). Grey-colored lakes indicate the mixing of lake water with sediments, which often accompanies periods of unrest (Christenson et al. 2016; Rouwet et al. 2016).

Poás volcano (Fig. 1), Costa Rica, hosts one of the most dynamic crater lakes on Earth, Laguna Caliente. Since 1828, Poás has experienced four main magmatic eruptive episodes (1834, 1910, 1953–1955; April 2017-present; Mora Amador et al. Chapter "Volcanic Hazard Assessment of Poás (Costa Rica) Based on the 1834, 1910, 1953–1955 and 2017 Historical Eruptions"), which have often been preceded by phases of phreatic eruptions. Since 1978, three unrest phases, accompanied by phreatic eruptions, have occurred: March 1978 to August 1980; May 1986 to May 1996; and March 2005 to present (Brantley et al. 1987; Brown et al. 1989; Rowe et al. 1992a, b; Martínez et al. 2000; Rymer et al. 2000, 2009; Fischer et al. 2015; de Moor et al. 2016; Rouwet et al. 2016, Rouwet et al. Chapter "39 Years of Geochemical Monitoring of Laguna Caliente Crater Lake, Poás:

Fig. 1 Aerial photograph of Poás and Laguna Caliente. The diameter of the lake is ≈300 m (until April 2017). On the lake surface mats of sulfur spherules are observed, generated by multiple phreatic eruptions. Southeast of Laguna Caliente the 1953 dome/pyroclastic deposit degasses vigorously with precipitation of sulfur. The white arrow indicates the former site of the Fumarola Naranja

Patterns from the Past as Keys for the Future"). During the 1986–1996 eruption cycle, the Laguna Caliente dried out several times, revealing spectacular sulfur volcanism. Oppenheimer and Stevenson (1989) and Oppenheimer (1992) described this unique activity in detail (see also https://www.youtube.com/watch?v=TUvkqsldfcQ for video footage of these eruptions). In April 2017, phreatic activity transitioned to phreatomagmatic activity, leading once again to the disappearance of Laguna Caliente (re-appearing in January 2018), and the almost total destruction of the 1953–1955 dome/ pyroclastic deposit. During the most recent eruptive phase (since 2006), sulfur volcanism has been observed several times. Here, we review and discuss these more recent episodes of sulfur volcanism at Poás volcano in the context of a longer period of observations, stretching back to 1828 (Table 1).

To gather information on manifestations of sulfur volcanism prior to 1978 when more systematic observations are available, we have drawn on historical reports from the local press and from some pioneering expeditions (see Mora Amador et al. Chapter "Volcanic Hazard Assessment of Poás (Costa Rica) Based on the 1834, 1910,

Table 1 Chronology of sulfur volcanism observed at Poás from 1828 to 2018

Date	Manifestations	Source(s)
1828	Burning sulfur, sulfur cone	Von Frantzius (1861)
1834	Sulfur pyroclastic	Oersted (1863)
1906	Burning sulfur	Leiva (1906)
March 1860	Molten sulfur and sulfur cones	Von Frantzius (1861)
January 1910	Molten sulfur, pouring sulfur	Rudín (1910)
1925	Sulfur spherules	Sapper (1925)
December 1964	Rocks sweat sulfur	Krushensky and Escalante (1967)
December 1965	Molten sulfur and spherules	Bohnenberger et al. (1966)
1967	Molten sulfur	Taylor (1967)
1977	Sulfur pyroclastics, rocks sweat sulfur and sulfur pool	Bennett and Raccichini (1978)
January 1978	Burning sulfur, pyroclastic sulfur, rocks sweat sulfur, sulfur spherules and sulfur pool	Francis et al. (1980)
September 1978	Burning sulfur, pyroclastic sulfur, rocks sweat sulfur, sulfur spherules and sulfur pool	Francis et al. (1980)
November 1978	Ash and mud	SEAN 03:11
February 1979	Burning sulfur, pyroclastic sulfur, rocks sweat sulfur, sulfur spherules and sulfur pool	Francis et al. (1980)
March 1979	Burning sulfur, pyroclastic sulfur, rocks sweat sulfur, sulfur spherules and sulfur pool	Francis et al. (1980)
May 1981	Molten sulfur and spherules	SEAN 06:05
January 1988	Precipitates of sulfur	SEAN 13:01
Apr-88	Sulfur pyroclastics and sulfur spherules	SEAN 13:04
May 1988	Sulfur spray, pyroclastic sulfur and spherules	Red Sismológica Nacional, Costa Rica
June 1988	Pyroclastic sulfur and sulfur spherules	SEAN 13:06
July 1988	Pyroclastic sulfur and sulfur spherules	SEAN 13:07
January 1989	Pyroclastic sulfur and sulfur spherules	SEAN 14:01

(continued)

Table 1 (continued)

Date	Manifestations	Source(s)
February 1989	Sulfur cones	SEAN 14:02
March 1989	Sulfur cones and pyroclastic sulfur	SEAN 14:03
Apr-89	Molten sulfur, pyroclastic sulfur and sulfur cones	SEAN 14:04
May 1989	Burning sulfur, sulfur cone	Red Sismológica Nacional, Costa Rica
June 1989	Burning sulfur, pyroclastic sulfur, sulfur cones and spherules	SEAN 14:06
July 1989	Sulfur cones, molten sulfur, pyroclastic sulfur and spherules	SEAN 14:07
August 1989	Sulfur flows, molten sulfur, sulfur cones and spherules	SEAN 14:08
September 1989	Sulfur cones, castle like forms, and spherules	SEAN 14:09
October 1989	Sulfur flow, sulfur cones and sulfur pool	SEAN 14:10
January 1990	Sulfur pool	BGVN 15:01
February 1990	Sulfur spherules	BGVN 15:02
March 1990	Burning sulfur and pyroclastic sulfur	Red Sismológica Nacional, Costa Rica
Apr-90	Sulfur pool, sulfur cones, burning sulfur, pyroclastic sulfur and sulfur sprays	BGVN 15:04
May 1990	Sulfur cones, sulfur precipitates and sulfur spray	BGVN 15:05
June 1990	Sulfur cones, sulfur flow, pyroclastic sulfur and sulfur precipitates	BGVN 15:06
July 1990	Spherules, sulfur flow and sulfur pool	BGVN 15:07
August 1990	Sulfur precipitates	BGVN 15:08
September 1990	Sulfur spray and sulfur cones	BGVN 15:09
October 1990	Sulfur pool	BGVN 15:10
November 1990	Sprays, sulfur cones	BGVN 15:11
January 1991	Spherules	BGVN 16:01
March 1991	Burning sulfur and pyroclastic sulfur	BGVN 16:03
Apr-91	Burning sulfur	BGVN 16:04
October 1991	Spherules	BGVN 16:10
November 1991	Spherules	BGVN 16:11
December 1991	spherules	Red Sismológica Nacional, Costa Rica
Apr-92	Sulfur deposits	Red Sismológica Nacional, Costa Rica
June 1992	Burning sulfur, sulfur cones	BGVN 17:06
August 1992	Sulfur pool	BGVN 17:08
September 1992	Pyroclastic sulfur	BGVN 17:09
November 1992	Molten sulfur	BGVN 17:11
January 1993	Pyroclastic sulfur?	BGVN 18:01
February 1993	Sulfur springs (?) and spherules	BGVN 18:02
March-April 1993	Spherules	BGVN 18:03
May 1993	Sulfur deposits	Red Sismológica Nacional, Costa Rica
July 1993	Sulfur pool (?), sulfur cone	Red Sismológica Nacional, Costa Rica

(continued)

Table 1 (continued)

Date	Manifestations	Source(s)
September 1993	Spherules	Red Sismológica Nacional, Costa Rica
October 1993	Spherules	Red Sismológica Nacional, Costa Rica
November 1993	Spherules	Red Sismológica Nacional, Costa Rica
January 1994	Spherules	BGVN 19:01
May 1994	Spray and pyroclastic sulfur	Red Sismológica Nacional, Costa Rica
July 1994	Burning sulfur	BGVN 19:07
September 1994	Sulfur pyroclastic and spherules	BGVN 19:09
March 1995	Spherules	BGVN 20:03
May 1995	Sulfur deposits	Red Sismológica Nacional, Costa Rica
July 1995	Spherules	BGVN 20:07
May 1996	Spherules	BGVN 21:05
March 1998	Spherules	BGVN 23:03
October 1998	Spherules	This study
February–May 1999	Spherules	This study
January–February 2000	Spherules	This study
February–March 2001	Spherules	This study
October–November 2001	Spherules	This study
March 2002	Spherules	This study
June 2002	Spherules	This study
September 2002	Spherules	This study
May 2003	Spherules	This study
September 2003	Spherules	This study
October–November 2003	Spherules	This study
January 2004	Spherules	This study
Apr-04	Spherules	This study
March–April 2005	Spherules	This study
May 2005	Sulfur flow, spherules and pyroclastic sulfur	This study
June–December 2005	Spherules	This study
January 2006–September 2007	Spherules, pyroclastic sulfur, molten sulfur	This study
October 2007–August 2009	Spherules	This study
September 2009	Spherules and burning sulfur	This study

(continued)

Table 1 (continued)

Date	Manifestations	Source(s)
October 2009–March 2012	Spherules	This study
Apr–12	Spherules and rocks sweat sulfur	This study
May 2012–October 2015	Spherules	This study
November 2015	Molten sulfur and spherules	This study
December 2015–July 2017	Spherules	This study
August–December 2017	Sulfur cone, pyroclastic sulfur	Mora Amador et al. (this issue)
January–February 2018	Spherules, molten sulfur	Mora Amador et al. (this issue)

1953–1955 and 2017 Historical Eruptions" for details of methodology). We compare our findings and interpretations with reports of sulfur volcanism for other volcanoes, including cases with and without crater lakes.

2 Poás Volcano

Here, we summarize the geological and tectonic setting of the volcano. Further background to the geology of Poás is part of a north-trending system of composite cones (Von Frantzius, Botos), cinder cones (Sabana Redonda cones, Bosque Alegre cone), small stratovolcanoes (Congo), and maars (Hule, Los Ángeles and Río Cuarto) (see Ruiz et al. Chapter "Geochemical and Geochronological Characterisation of the Poas Stratovolcano Stratigraphy"). It belongs to the Quaternary Central America volcanic front, which is associated with the subduction of the Cocos plate beneath the Caribbean plate.

Streamflow into the Laguna Caliente crater lake is ephemeral and episodic. The age of the lake is not well established. It is younger than the Botos cone (8300 yr B.P., calibrated age after Prosser and Carr 1987), but it appears to have been present for several centuries based on the stratigraphy of the crater interior (Casertano et al. 1983; Cigolini et al. 1991), and also according to Amerindian legends (Zeledón 2007). Preliminary findings from radiocarbon dating of organic matter in lacustrine deposits suggest a calibrated age of at least 890 ± 40 yr B.P. (Mora-Amador et al. 2004).

The lake deposits consist of alternating fine- and coarse-grained laminations, containing either elemental sulfur, or poorly crystallized silica precipitates and gypsum (Brantley et al. 1987). Disseminated sulfur, and sulfur "pipes", veins and tubes, are present within the lake sediments (Prosser and Carr 1987). The sediments were precipitated by the disproportionation of H_2S in the lake water. Documented activity at Poás dates back to 1828. Mora Amador et al. (Chapter "Volcanic Hazard Assessment of Poás (Costa Rica) Based on the 1834, 1910, 1953–1955 and 2017 Historical Eruptions") report details on the four historical eruptions of Poás. Molten sulfur lakes, tephra, and cones at Poás were described by Raccichini and Bennet (1977), Bennett and Raccichini (1978), Francis et al. (1980), Oppenheimer and Stevenson (1989), and Oppenheimer (1992), supplemented by numerous reports in the SEAN (Scientific Event Alert Network) and GVN (Global Volcanism Program) bulletins. Elemental sulfur occurs in crater lake environments as banded sediments exposed on the lakeshore, tephra and fumarolic sublimates, and in an inferred subaqueous molten body on the crater floor. A similar alternation between sulfur and clay layers was described at Tateyama volcano, Japan (Kusakabe et al. 1986), where it provided an estimate of the minimum sedimentation rate.

3 Physical-Chemical Properties of Sulfur

The occurrence of elemental sulfur at volcanoes is an indicator of elevated temperature magmatic degassing ($T_{gas\text{-}vapor} > 300$ °C), following the reactions (Giggenbach 1987):

$$3SO_2 + 7H_2 = H_2S + 2S^{\circ} + 6H_2O \quad (1)$$

or (Mizutani and Sugiura 1966):

$$SO_2 + 2H_2S = 3S^{\circ} + 2H_2O \quad (2)$$

For lower temperature gases (T < 300 °C), SO_2 will be converted into H_2S through kinetically slow processes in equilibrium with the Fe^{2+}–Fe^{3+} mineral-bearing rocks, without the formation of elemental sulfur (i.e. geothermal buffer by Giggenbach 1987). Thus, the presence of elemental sulfur at a volcano points to an origin by magmatic degassing rather than "steam-heating" by H_2S-rich vapor boiled off from a hydrothermal system. When SO_2–H_2S rich gases intercept water in their rise towards the surface, elemental sulfur can become supersaturated in the water through various reactions (see Delmelle and Bernard 2015, and references therein, for further details):

$$3SO_2 + 3H_2O = 2SO_4^{2-} + H_2O + S^{\circ} + 4H^+ \quad (3)$$

$$SO_4^{2-} + 3H_2S + 2H^+ = 4S^{\circ} + 4H_2O \quad (4)$$

Acid crater lakes receiving direct inputs of magmatic fluids provide ideal sites for elemental sulfur to be manifested in its various forms (see Sect. 4). From reactions (3) and (4) it is clear that the absorption of SO_2 generates acidity, to extreme pH values near 0. Sulfur chemistry becomes even more complex if we consider the role of thiosulfate ($S_2O_3^{2-}$) and polythionates ($S_xO_6^{2-}$, with x = 4, 5, 6), indicative of the relative proportions of SO_2 and H_2S present (Takano and Watanuki 1990; Takano et al. 1994a), and the third most abundant sulfur species after SO_4^{2-} and HSO_4^-, following the reactions (Delmelle and Bernard 2015):

$$S_2O_3^{2-} + 2H_2S + 2H^+ = 4S^{\circ} + 3H_2O \quad (5)$$

$$S_xO_6^{2-} + (3x-7)HSO_3^- = (2x-3)SO_4^{2-} + (2x-4)S^{\circ} + (x-1)H^+ + (x-3)H_2O \quad (6)$$

$$S_xO_6^{2-} + H_2S = 2S_2O_3^{2-} + (x-2)S^{\circ} + 2H^+ \quad (7)$$

Evaporation from hot crater lakes increases salinity and acidity, promoting further S speciation and elemental S° formation. The complexity of sulfur chemistry in volcanic acid aqueous environments is beyond the scope of the present study but it is ripe for further research (e.g. Delmelle and Bernard 2015; Scolamacchia and Cronin 2016). What is important to note is that elemental sulfur is ubiquitous at Poás' Laguna Caliente crater lake.

Next, we highlight some of the basic physical properties of elemental sulfur. Pure sulfur melts at 116–119 °C (Fig. 2); this temperature can be reached during water boiling in a pressurized hydrothermal system (e.g. with the hydrostatic loading of a deep crater lake). Reaching this threshold temperature, vugs and vents can be cleared out by the higher-T gas, generating bodies of liquid sulfur. Due to the higher density of liquid sulfur (1.819 g/cm^3) with respect to water (even dense brines), liquid sulfur will reside at the crater lake bottom. When the temperature rises up to ~159 °C, the viscosity of liquid sulfur increases by four orders of magnitude due to polymerization into cyclic octatomic S_8 (Oppenheimer 1992; Takano et al. 1994a; Scolamacchia and Cronin 2016, and references therein) (Fig. 2). It is thought that this acts to reduce permeability in magmatic-hydrothermal systems resulting in further elevated temperatures and explosive (phreatic) eruptions (Hurst et al. 1991; Takano et al. 1994a; Christenson et al. 2010; Rouwet and Morrissey 2015; Rouwet et al. 2016). Impurities in the liquid elemental sulfur, ubiquitous in magmatic-hydrothermal systems (e.g. volatilized metals such as As, Au, Bi, Cd, Cu, Mo, Ni, Pb, Se, Sn, Te, Tl, and W, as well as Fe, pyrite, H_2S, halogens, organics from CH_4) can modify the color of S°, and perturb the

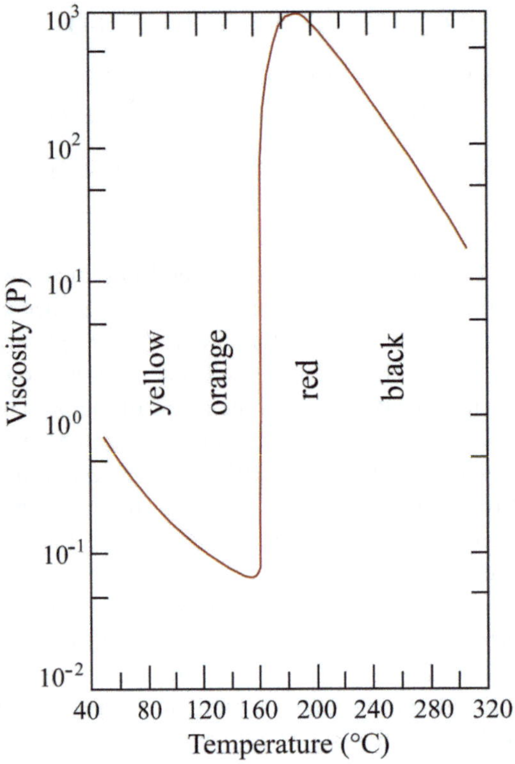

Fig. 2 Relationship between viscosity and temperature of sulfur, and associated color changes. Note the sudden increase in viscosity by four orders of magnitude near 160 °C. (modified from MacKnight and Tobolsky 1965)

melting-viscosity thresholds in the temperature domain 116–160 °C (Scolamacchia and Cronin 2016). Above 200 °C liquid sulfur turns dark-red in color (Fig. 2). Sulfur burns near 400 °C, oxidizing S° to SO_2 accompanied by a characteristic blue flame (e.g. at Kawah Ijen, Java), whereas it boils at 444 °C (Oppenheimer 1992).

4 Sulfur Volcanism Recognized Since the Early 1900s

Sulfur is omnipresent in many volcanic systems, though relatively few volcanoes present active sulfur volcanism, i.e. with eruptions of liquid sulfur. This suggests that certain physical-chemical conditions, and circumstances of the heating-cooling history of the sulfur, are required to promote sulfur volcanism.

4.1 Sulfur Spherules and "Sweat" Testify Sub-lacustrine Sulfur Pools

Where sulfur volcanism is observed, the variety in eruptive manifestations and associated products is wide and spectacular. Since the late nineteenth-early twentieth centuries, crater lake pioneers provided the first description of sulfur spherules and debated on their origin. Kawasaki (1903) was the first to infer that the presence of such spherules pointed to the existence of a molten sulfur pool at the bottom of Yugama crater lake (Kusatsu-Shirane volcano, Japan). He found the bubbly sulfur spherules on the lake shore (Fig. 3) noting

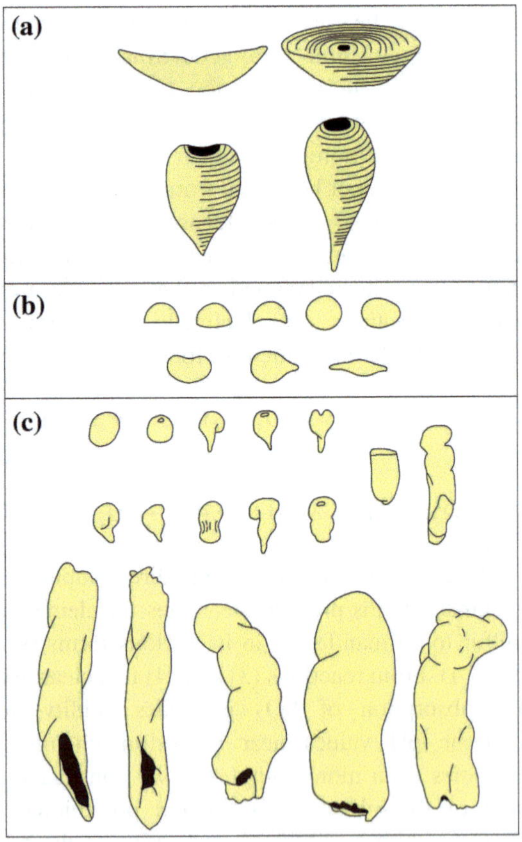

Fig. 3 Reproduction of sketches of sulfur spherules from pioneering studies in the early 1900s, by: **a** Kawasaki (1903), **b** Ōinouye (1916), and **c** Ohashi (1919). In these three studies the authors suggested that the spherules originated from molten sulfur pools at the bottom of the acidic crater lakes. The same structures and morphologies were described for Laguna Caliente, Poás

that: "*The spherules are minute pieces of sulfur, detached in molten state from the hot liquid mass on the lake bottom by the action of gas bubbles, which the sulfur completely envelopes and are thus thrown up just to the surface of the water. The perforation is due to the escape of gases confined in the sulfur shell*".

Kawasaki (1903) explained the presence of sulfur at Yugama by (i) sublimation due to gaseous reactions between and SO_2 and H_2S, and sublimed sulfur covered highly altered rocks, (ii) precipitation and (iii) as molten sulfur. He termed sulfur that formed in acidic hot spring waters "precipitated sulfur". In such hot springs, H_2S was oxidized to sulfuric acid upon contact with air creating a high acidity. Precipitated sulfur was very fine and "muddy", and often stratified at the bottom. Before 1897, molten sulfur from crater lake walls was observed and collected. Kawasaki (1903) also observed dish-shaped sulfur, suggesting that the temperature was higher than the melting point at the moment of its formation. These interpretations remain consistent with present day understanding of the formation of such materials.

Ōinouye (1916) described "sulfur grains" at Ponto Lake, Kunashiri volcano (Kuril Islands, Russia), which was at that time part of Japan: "*The greater part of the Sulfur is dark gray in color, but some is yellow. The grains are hemispherical, oval, kidney-, fig-, or spindle-shaped, and they are usually about 0.2–3.0 mm in diameter. The grains are not solid, but hollow, and the cell walls, which are usually rough on the outside on account of a coating of impurities and of very minute sulfur particles, are so thin that they are very easily broken.*"

Three years later, Ohashi (1919), followed Kawasaki (1903) in suggesting that sulfur spherules originate from sulfur pools at the bottom of Kusatsu-Shirane's hot-acid crater lake. Ohashi (1919) presented a sophisticated classification system of the shapes, sizes and colors of those spherules (Fig. 3): "*Three different forms are at once distinguished by observers:—namely,* (1) *hollow spherules,* (2) *solid grains and* (3) *tubes*".

More recently, Takano et al. (1994a) correlated changes in the shape and size of sulfur spherules with changes in volcanic activity, describing the following stages:

- I: absence of sulfur spherules at the lake surface due to feeble fumarolic activity at the lake bottom, with very low SO_2 and H_2S contents in the gas phase;
- II: presence of black hollow sulfur spherules and mainly broken yellow sulfur spherules at the lake surface. The fumarolic temperature at the lake bottom is sufficient (>116 °C) to melt sulfur and develop a molten pool at the lake bottom;
- III: presence of mud plumes, black (pyrite rich) and yellow sulfur spherules with "tails" at the lake surface, testifying to increased viscosity and hence a higher fumarolic temperature (>159 °C) at the lake bottom;
- IV: vent clearing phreatic or geyser-like eruptions, ejecting the sulfur pool (temporarily) from the eruptive vent.

Karpov et al. (1996) extracted a 60 kg black block of sulfur from the top of the sulfur pool at the bottom of Bannoe Lake, Uzon Caldera (Kamchatka, Russia). This observation is suggestive of the Stage III black spherules according to the Takano et al. (1994a) scheme. Similar findings resulted from an earlier study by Luke (1950) that involved drilling into sulfur pools at White Island (New Zealand).

During phreatic eruptions, highly altered rocks are launched together with muddy sediments from the lake basin up to several tens of meters from the vent. On the outside of these altered ballistics, solid sulfur bubbles are the relict of "sweating" of liquid sulfur out of the pores. Deposits with such sulfur structures were recognized after the September 2007 Ruapehu eruption (New Zealand, Christenson et al. 2010). These authors explained this manifestation by flash boiling of interstitial pore water after the pressure drop during expulsion from the lake. Alternatively, or contemporaneously, sulfur was present as a liquid ($T > 116$ °C) originating from the molten sulfur pool mixed with the altered material at the lake bottom, and then solidified during the expulsion when the ballistics pass through the air (sudden T drop). The evidence for liquid sulfur demonstrates that the temperature of the ballistics remained above 116 °C up to impact, and again indicates the presence of a molten sulfur pool at the lake bottom.

Other authors have described sulfur pools from crater lake volcanoes, e.g. Murozomi et al. (1966), Giggenbach (1974), Bennett and Raccichini (1978), Ossaka et al. (1980), Xu et al. (1988), Christenson (1994), Delmelle and Bernard (1994), Pasternack and Varekamp (1994) and Delpino and Bermúdez (1996a, b).

4.2 Sub-aerial Sulfur Pools and Eruptive Products of Sulfur Volcanism

Sub-aerial sulfur pools can form in depressions inside fumarolic fields, or be revealed at the lake bed following desiccation of the crater lake itself. Outlet temperatures of the discharging gases should be above around 116 °C to keep the sulfur in the liquid state (Fig. 2). Such pools may be associated with sulfur flows, sulfur cones and *hornitos*, and pyroclastic sulfur in various forms (Pelé hair, sulfur "sweat").

Watanabe (1940) was the first to report the emplacement of a sulfur flow at Shiretoko-Iozan volcano, Japan. This flow reached a length of 1400 m. Subsequent reports describing sulfur flows include those of Luke (1950), Kirsanov et al. (1964), Skinner (1970), Colony and Nordlie (1973), Naranjo (1985), White et al. (1988), Imai and Geshi (1999), Harris et al. (2000), Mora-Amador and Ramírez (2008) and Masubuchi (2012). Section 7 and Table 2 provide a detailed overview.

Pyroclastic sulfur can be associated with sulfur cones and *hornitos*, sometimes associated with crater lake desiccation. At several volcanoes (including Copahue and Poás) sulfur pyroclasts have been expelled from shallow lakes during progressive desiccation. If temperatures of sub-aerial sulfur pools exceed around 400 °C, sulfur combustion can occur.

Table 2 Compilation of global sulfur volcanism.

Volcano, Country	Volcano type	Tectonic context	1	2	3	4	5	6	7	References
Poás, Costa Rica	Stratovolcano	Subduction	x	x	x	x	x	x	x	Von Frantzius (1861), Leiva (1906), Krushensky and Escalante (1967), Bennett and Raccichini (1978), Oppenheimer and Stevenson (1989), Mora-Amador and Ramírez (2008); this study
Kawah Ijen, Indonesia	Stratovolcano	Subduction	x	x	x	x	x		x	Delmelle and Bernard (1994), Delmelle et al. (1996, 2000), Africano et al. (2007), Jonguk Kim et al. (2011); BGVN 24:9
White Island, New Zealand	Stratovolcano	Subduction	x	x	x	x		x		Luke (1950), Hamilton and Baumgart (1959), von Hochstetter (1864); SEAN 14:10; BGVN 23:10; BGVN 23:11
Rincón de la Vieja, Costa Rica	Complex	Subduction	x	x	x	x		x		BGVN 17:5; BGVN 17:9; BGVN 19:1; BGVN 20:4; Tassi et al. (2005, 2009), RM-A personal observations (2015)

(continued)

Table 2 (continued)

Volcano, Country	Volcano type	Tectonic context	1	2	3	4	5	6	7	References
Golovnin, Russia	Caldera	Subduction	x	x	x	x		x		Naboko (1958, 1959), Averyanoy et al. (1972), Markhinin and Stratula (1977), Karpov et al. (1996), Taran, personal communication (2010)
Cinder Pool Yellowstone, USA	Caldera	hotspot	x	x	x			x		White et al. (1988), Xu et al. (1988, 2000) Grimes et al. (1999)
Copahue, Argentina	Stratovolcano	Subduction	x	x			x	x		Delpino and Bermúdez (1996a, b), Varekamp et al. (2001); BGVN 17:7; BGVN 38:9, Caselli et al. (2016), Agusto et al. (2017)
Vulcano, Italy	Stratovolcano	Subduction			x	x			x	Harris et al. (2000, 2004); SEAN 14:10; BGVN 20:04
Sierra Negra, Ecuador	Shield volcano	Hotspot			x	x		x		Colony and Nordlie (1973), Theilig (1982)
Kuttara-Noboribetsu, Japan	Stratovolcano	Subduction	x	x	x					Kawasaki (1903), Ohashi (1919), Murozomi et al. (1966), Delmelle and Bernard (2015)
NW Rota-1, USA	Submarine	Subduction	x	x			x			Butterfield (2006); BGVN 29:03
Waiotapu, New Zealand	Caldera	Subduction	x	x				x		Grimes et al. (1999)
Ebeko, Russia	Stratovolcano	Subduction			x	x			x	Kirsanov et al. (1964), Menyailov et al. (1985)
Kudryavy, Russia	Somma volcano	Subduction			x	x			x	Korzhinskii et al. (1996), Taran, personal communication (2010); BGVN 24:10
Irazú, Costa Rica	Stratovolcano	Subduction	x	x		x				Bullard (1956); SEAN 7:11
El Chichón, Mexico	Dome complex	Subduction	x	x	x					BGVN 17:6; SEAN 7:10
Mauna Loa, USA	Shield volcano	Hotspot			x	x				Skinner (1970), Theilig (1982)
Turrialba, Costa Rica	Stratovolcano	Subduction			x	x				BGVN 32:08
Momotombo, Nicaragua	Stratovolcano	Subduction			x	x				BGVN 22:07
Santa Ana, El Salvador	Stratovolcano	Subduction	x	x						BGVN 26:4
Loloru, Papua New Guinea	Pyroclastic shield	Subduction					x	x		SEAN 13:4
Karymsky, Russia	Stratovolcano	Subduction	x	x						Takano et al. (1995)
Bannoe Uzon, Russia	Caldera	Subduction	x	x						Takano et al. (1994b), Karpov and Fazlullin (1995)
Kusatsu-Shirane, Japan	Stratovolcano	Subduction	x	x						Kawasaki (1903), Ōinouye (1916), Ohashi (1919), Takano et al. (1994a)

(continued)

Table 2 (continued)

Volcano, Country	Volcano type	Tectonic context	1	2	3	4	5	6	7	References
Taupo-Rotokawa, New Zealand	Caldera	Subduction	x	x						Krupp and Seward (1987)
Aso, Japan	Caldera	Subduction	x	x						Shinohara et al. (2015)
Tokachidake, Japan	Stratovolcano	Subduction			x				x	Ishikawa et al (1971); SEAN 10:7
Maly Semiachik, Russia	Stratovolcano	Subduction	x	x						Takano et al. (1994b)
Ilopango, El Salvador	Caldera	Subduction	x	x						López et al. (2004)
Satsuma-Iwojima, Japan	Caldera	Subduction				x			x	Mori et al. (2002)
Tateyama, Japan	Stratovolcano	Subduction				x			x	Kouno (1988), Masubuchi (2012)
Yumoto Oyunuma-Niseko, Japan	Stratovolcano	Subduction	x	x						Imai and Geshi (1999)
Kawah Putih, Indonesia	Stratovolcano	Subduction	x	x						Sriwana et al. (2000)
Shiretoko-Iozan, Japan	Stratovolcano	Subduction	x			x				Watanabe (1940)
Ruapehu, New Zealand	Stratovolcano	Subduction	x	x						Giggenbach (1974), Christenson (1994)
Kelimutu, Indonesia	Complex	Subduction	x	x						Pasternack and Varekamp (1994), Murphy et al. (2018), BGVN 20:06
Zao, Japan	Complex	Subduction		x						Toraishi and Tominaga (1940)
Kelud, Indonesia	Stratovolcano	Subduction			x					Takano (1987)
Stromboli, Italy	Stratovolcano	Subduction			x					BGVN 16:9
Morne Plat Pays, Dominica	Stratovolcano	Subduction							x	BGVN 19:5
Galeras, Colombia	Complex	Subduction							x	BGVN 15:9
Concepción, Nicaragua	Stratovolcano	Subduction						x		BGVN 18:03
Emmons Lake, USA	Caldera	Subduction						x		BGVN 15:9
Mendeleev, Japan	Stratovolcano	Subduction				x				Ōinouye (1916)
Gorely, Russia	Caldera	Subduction	x							Gavrilenko et al. (2008)
Kirishima, Japan	Shield volcano	Subduction	x			x				SEAN 04:06
Yakeyama, Japan	Stratovolcano	Subduction		x						Ohashi (1919)
Lastarria, Chile	Stratovolcano	Subduction	x							Naranjo (1985, 1988)
Bayo, Chile	Complex volcano	Subduction	x							Naranjo (1988)
Ohyunuma, Japan	Stratovolcano	Subduction	x							Uzumasa et al. (1959), Inoue and Aoki (2000)

(continued)

Table 2 (continued)

Volcano, Country	Volcano type	Tectonic context	1	2	3	4	5	6	7	References
Ambae, Vanuatu	Shield volcano	Subduction	x							Bani et al. (2009)
Guagua Pichincha, Ecuador	Stratovolcano	Subduction				x				Beate (2006) personal communication
Taftan, Iran	Stratovolcano	Hotspot				x				Shakeri et al. (2008)
Fogo, Cape Verde	Stratovolcano	Hotspot				x				RM-A personal observations (2011)
Dempo, Indonesia	Stratovolcano	Subduction					x			Delmelle and Bernard (2015)

Numbers denote various sulfur manifestations: 1. Sublacustrine pool, 2. Sulfur spherules, 3. Subaerial sulfur pool, 4. Sulfur flow, 5. Pyroclastic sulfur, 6. Sulfur cone, 7. Burning sulfur

5 Sulfur Volcanism at Poás

Poás has exhibited several phases of sulfur volcanism during the period investigated in this study (1828–2018). In addition, the repeated desiccation of the crater lake, and relative ease of access makes Poás an ideal location to study this peculiar activity. Below, we (i) describe the numerous manifestations of sulfur volcanism observed at Poás, and (ii) present conceptual models for the generation of the specific features.

5.1 The 1987–1990 Laguna Caliente Dry-Out

In June 1987, Poás resumed phreatic activity after seven years of quiescence (since September 1980). The temperature of Laguna Caliente and fumaroles increased, giving way to steady lake dry-out. A series of reports (SEAN Bulletins), summarized below, described the subsequent changes at Laguna Caliente. Figure 4 shows how the lake started to shrink, as well the appearance of structures inside the active crater such as terraces, active faults, and several sulfur manifestations: sulfur pools, cones, "castles", flows and burning sulfur. Sketches in Fig. 4 were adapted on the basis of the descriptions by Gerardo Soto and one of us (CO).

The April 1988 SEAN Bulletin reports: "*strong phreatic explosions and deposition of acidic mud and sulfur, with the emission of a vapor plume with an estimated height of 2 km. Around the hot lake there was a spray of pale yellow sulfur pyroclastics, kidney-shaped or "grease-like" spots, with mean diameters of 0.5–1 mm and maximum lengths of 2 cm. The shapes of the pyroclastics indicated that the sulfur was molten when ejected. The temperature of the lake was 65 °C and its pH was 0.45 on 11 April. A fumarole on the remnant of the 1953–1955 [dome] had a temperature of 298 °C on its upper part, while 587 °C was registered W of it*" (Fig. 4).

The June 1988 SEAN Bulletin reports: "Geyser-like phreatic activity was continuing, ejecting muddy plumes every 2–3 min. Plume heights were predominantly less than 15 m, with maximum observed heights of 80 m. The level of the hot lake continued to descend, revealing a wide littoral crescent formed by terraces of precipitated sediment. These sediments were thixotropic, making measurements of lake level and temperature difficult" (Fig. 4).

In February 1989 it was reported: "*the crater lake had continued to shrink, and by the end of February, the cumulative decrease in depth was estimated at 30 m. The lake was only 10–15 m deep and its diameter was 100–125 m. Seven desiccation terraces surrounded the lake, which was colored gray by solutes and suspended mud. The lake contained multiple centers of convective bubbling. Mud plumes were rarely ejected, some with small blocks that fell back into the lake. On*

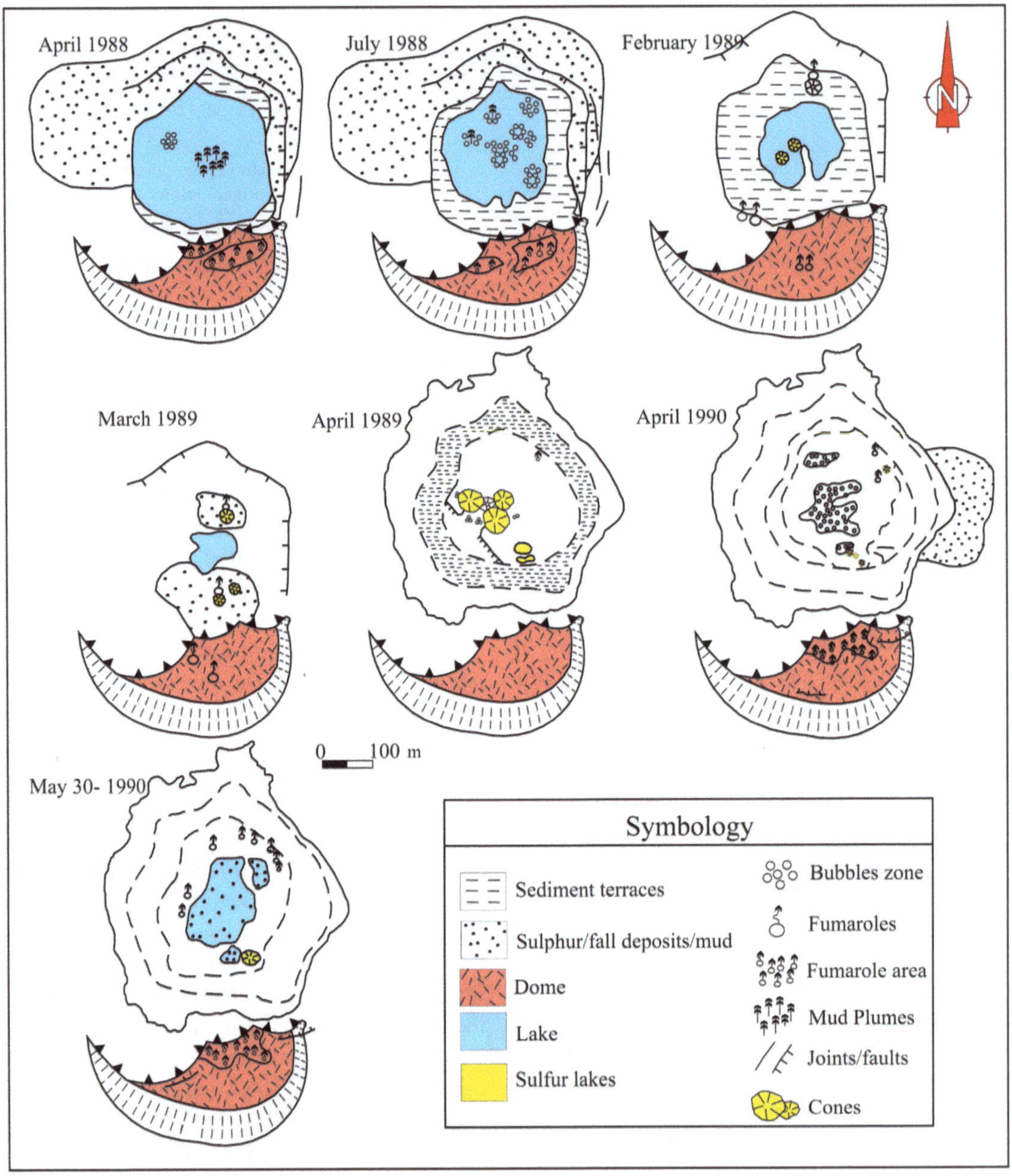

Fig. 4 Sketches of the April 1988-May 1990 desiccation of Laguna Caliente and manifestations of sulfur volcanism revealed at the bottom of the former lake basin (modified from SEAN Bulletins)

the N shore and in the south-center of the lake there were small mud/sulfur cones with fumaroles" (Fig. 4).

For March 1989, SEAN Bulletin reports: "*the level of the crater lake continued to descend, dropping another 2 m in March (for a cumulative decline of about 32 m since early 1987). Lake water was colored muddy gray by suspended sediment, and convective bubbling was continuous. At the beginning of March, movement of rafts of floating sediment illustrated convective trajectories within the lake. Rafts were not seen in the middle of the month, although there was some thick floating sediment. By the end of March, no floating material was evident. The temperature of the lake water averaged 82 °C. Small sulfur cones had formed on sedimentary terraces at the SE edge of the lake, constantly ejecting sprays of mud and pyroclastic sulfur that carpeted the surrounding area. Some of the cones were topped by chimney-like structures of ~1 m high, made of sulfur. Two small sulfur cones had also developed on the lake's NE edge, venting gases with a mean temperature of 98 °C*" (Fig. 4).

In April 1989: "*Until mid-April, thermal activity remained similar to that observed in March, with boiling mud springs and vigorous fumaroles in the crater lake, which has been shrinking since early 1987. Two ponds of molten sulfur (115 °C) have persisted since 16 March at the former site of small sulfur and mud cones 50 m SE of the center of the inner crater. Small pyroclastic sulfur cones surrounded the lakes, collapsing occasionally. On 12 April, the crater lake was convecting vigorously, but shallow areas were visible. The lake level dropped about 2 m during the following week, and by 19 April only a few small mud pools remained. On 19 April, small bursts of gas and mud that contained sulfur particles emerged through the mud surface to heights of about 10 m, rarely to 25–30 m. Steaming was continuous. During observations on 23 April, a thick white plume coalesced from numerous vents, two of which were discharging a mixture of white condensed steam and yellow sulfur. Dark cypressoid plumes were emitted every few seconds. At least one vent continuously discharged fine dark material. Most of the ejecta fell within 10–20 m of the vents, building cones to about 10 m height with funnel-shaped craters up to 5 m in diameter. The ejecta appeared dry and included blocks more than 20 cm across. Radiant temperatures of dark plumes were only about 80 °C as measured from about 150 m away. Activity occasionally reached a level at which at least one of the six or more phreatic vents was erupting at a given time*" (Fig. 4).

Subsequently, Laguna Caliente reformed. We highlight the descriptions provided in the April 1990 SEAN Bulletin: "*By the beginning of April, lake level had dropped 4 m since December 1989, shrinking to a small pool in the center of the former lake, where minor phreatic eruptions occurred. The principal fumarole in this area had a vent 2–3 m in diameter. Sprays of very pure yellow sulfur were emitted, as were bluish SO_2 and water vapor. The gases were expelled under high pressure with a jet aircraft sound, and burned with yellow-orange flames. At the beginning of the month, temperature (190 °C with an infrared thermometer), sound level, and pressure were less than on the 18th, when activity was stronger with temperatures to 793 ° C. The continuous ejection of material also built other cones, a few meters high and rich in pyroclastic sulfur, which periodically collapsed and recycled their contents*" (Fig. 4).

In May 1990 the SEAN Bulletin records "*strong gas emissions but phreatic activity decreases; small lake and mud/sulfur cones in crater. May activity was less intense than in April, but similar to that of previous months. Strong gas emissions occurred from fumaroles in the crater and there were at least two mud eruptions that reached heights of several hundred meters, one at roughly 0800 on 4 May, the other on 8 May. [...] The crater lake had shrunk substantially, to a small feature, several centimeters deep, that fluctuated slightly with changes in rainfall recharge. Lake temperatures were >72 °C (measured by infrared thermometer) and small hot springs around the lake had temperatures of 85 °C. Small cold springs from the E wall of the crater had pH values of 1.5. Fumaroles on the 1953–55 dome emitted gas (91 ± 1 °C), primarily water vapor, and precipitated sulfur*" (Fig. 4).

Until the most recent (April-July 2017) peak activity at Poás, leading to the destruction of the 1953 dome, and consequent desiccation of Laguna Caliente, with vigorous sulfur volcanism in its former lake basin, the 1987–1990 activity of Poás was the most spectacular manifestations of sulfur volcanism recorded at Poás.

5.2 Rise and Demise of the Fumarola Naranja (2000–2009)

Since early 2000, the fumarolic activity that until then had been concentrated in the western portion of the crater of Poás ceased and migrated towards the eastern and northeastern sector of the active crater (Vaselli et al. 2003; Mora-Amador et al. 2004; Vaselli et al. Chapter "The Last Eighteen Years (1998–2014) of Fumarolic Degassing at the Poás Volcano (Costa Rica) and Renewal Activity"). During the following months, new fumaroles emerged. During the second half of 2000, substantial fumarolic activity resumed in the northern sector of the crater, located in a narrow valley. This site was named "Fumarola Naranja" for the orange-yellow appearance of sulfur mineral sublimates. Some months later, a cylindrical structure, of 1.3 m height by 0.35 m in diameter, was formed emanating gases at near boiling temperature.

During 2001, the Fumarola Naranja exceeded boiling temperatures, and a gas plume that could be observed from the "mirador" (tourist viewpoint located on the southern crater rim). In March 2002, small boiling pools formed in the Fumarola Naranja area, exposing grey mud at pH of zero. From this site, explosions occurred that coated the narrow walls of the rill with grey acid mud.

In September 2002, degassing increased broadly across the crater area, to such a degree that sulfur odors could be perceived at the Tourist Centre (approximately 1 km southwest of the crater rim). At Fumarola Naranja, the emissions were so vigorous as to be deafening. Emission columns reached tens of meters, and a crust of sulfur deposits accumulated. Meanwhile, the mud explosions from the boiling pools continued, expelling centimeter-size rocks.

In 2003 and 2004, the Fumarola Naranja transformed into a fumarolic field with higher temperatures and feeding a steady plume (Fig. 5), comparable to the main plume emanating from the dome. Temperatures varied within a range of 15 °C, possibly indicating the buffering effect of the hydrothermal system. Inside the cleft, small cones (1 × 0.5 m) were formed, similar to the small sulfur volcanoes that formed in April 1989 within the desiccated Laguna Caliente basin (Sect. 5.1, Oppenheimer and Stevenson 1989). These "fumarolic cones" exhibited vigorous degassing at elevated pressure (Fig. 5).

Small streams of near-zero pH thermal waters descended towards Laguna Caliente from the fumarolic field. Occasionally, these waters were grey and rich in sediment, at other times they were greenish and opaque. These thermal springs carried floating sulfur spherules. Only rarely have such sulfur spherules exiting thermal springs been previously reported (Watanabe 1940; Xu et al. 1988). Ephemeral small boiling pools and mud pots appeared. Towards Laguna Caliente, an alluvial fan built up, containing decimeter-sized rocks and agglomerates of sulfur crystals up to 40 cm × 30 cm in size. These fragments were transported down-slope during intense rain events. Small explosions plastered the walls of the valley with mud. Between March and November 2003, yellow and orange sulfur minerals were deposited. Reflecting the high temperature ($T > 150$ °C) inside the fumarolic cone, the sulfur turned orange (Fig. 2).

Fumarolic activity increased in 2005, with temperatures reaching 200–230 °C by March 2005 (Fig. 6). A 3 m × 3 m *hornito*-like structure (Fig. 6a), sourced a 200-m-high plume. Inside the structure, small scarlet-red molten sulfur pools formed (Fig. 6f). Pelé's hair and tears were erupted from the *hornito*'s mouth (Fig. 6c–f). During a gas sampling campaign in April 2007, sulfur eruptions coated the mask and clothing of field scientists with molten sulfur lapilli.

Fresh relicts of an 83-m-long sulfur flow were observed after its emplacement during the third week of May 2005 (Fig. 7). The volume was estimated at 12.5 m^3. Its predominantly yellow color suggested that it was deposited at relatively low temperature and viscosity (113–160 °C and $\sim 10^{-1}$ poise; Fig. 7). The flow preserved sulfur "cascades" (Fig. 7a), and its textures resembled those typical of basaltic lava flows on Hawaii, with a pāhoehoe-like morphology closer to the vent (Fig. 7b–e) and decimeter-size levées

Fig. 5 Fumarola Naranja developed by mid-2000, emerging as the highest-temperature emission spot within the active crater of Poás. Dynamic changes were observed with an evolution from boiling pools, to boiling mud pools that generated small explosions, substantial sulfur deposition, with the formation of a sulfur *hornito* of 3 m × 3 m. This *hornito* sourced an 83-m-long sulfur flow in May 2005, and multiple eruptions of sulfur pyroclasts

and lobes (Fig. 7b–e; Mora-Amador and Ramírez 2008), and a texture more reminiscent of ‘a’ā and blocky lava towards the flow toe. The flow deposit revealed evidence of an initial major flow and subsequent smaller flows that travelled 10–15 m from Fumarola Naranja. Minor sulfur flows up to a few meters in length had exited from vents on the valley walls (Fig. 7c). As recognized elsewhere (e.g., Watanabe 1940; Naranjo 1985), the sulfur flow had incorporated rocky debris from the substrate (Fig. 7f).

Between May 2005 and March 2006, great quantities of floating sulfur spherules were seen at the surface of Laguna Caliente, culminating in a series of phreatic eruptions in March 2006 after more than 10 years of quiescence (de Moor et al. 2016; Rouwet et al. 2016; Rouwet et al. Chapter “39 Years of Geochemical Monitoring of Laguna Caliente Crater Lake, Poás: Patterns from the Past as Keys for the Future”). This indicates that sulfur volcanism was also active in the sub-lacustrine setting beneath Laguna Caliente.

By July 2008, fumarolic activity at Fumarola Naranja had decreased drastically, with much lower emission temperatures. On January 8, 2009, the M_w 6.2 Cinchona tectonic earthquake, whose epicenter was 6 km from Poás, caused landslides and rock falls inside the crater, blocking the relicts of the *hornito* and covering large portions of the sulfur flow. Later, strong rainfall and resulting erosion scoured out the bottom of the Fumarola Naranja valley leaving few remnants of the sulfur.

Fig. 6 **a** The *hornito* formed in 2005 at Fumarola Naranja. **b** Sulfur "dreads" similar to Pelé's hair. **c** Structures formed by Pelé's hair of more than a meter in dimension, named "Pelé's hair piece". **d** Two meters from the Fumarole Naranja *hornito* sulfur lapilli deposited in layers of up to 2 cm thickness. **e** Sulfur lapilli on the walls of the valley of Fumarola Naranja, known as Pelé sulfur tears. **f** Inside the *hornito* sulfur pools formed; the scarlet red color suggests temperatures above 200 °C

5.3 Tracking Unrest Using Sulfur Spherules (2000–2018)

The three types of sulfur spherules, as described by Ohashi (1919) (Fig. 3c), were observed at Laguna Caliente, prior to and during the most recent (phreatic) eruption cycle. The various morphologies of sulfur spherules in relation to the four stages of lake activity of Takano et al. (1994a) for the six years prior to the onset of the eruption cycle are summarized in Fig. 8. The water temperature of Laguna Caliente and

Fig. 7 **a**, **b** and **c** Several cascades along the May 2005 sulfur flow path. Note the different lobes and pahoehoe textures. **d** Typical levée of 10–15 cm width surrounded by other sulfur flows with its respective lobes. **e** Detail of the lobes and its fractures. Note the superposition of different flows. The intense yellow droplets are Pelé's tears. **f** At the distal portions of the sulfur flow (near Laguna Caliente), a section of the sulfur flow shows how andesitic rocks were incorporated along the flow path, while the color of sulfur was yellowish to grey-colored

Fumarola Naranja is also shown. Figure 9 applies the model for phreatic eruptions of Takano et al. (1994a) to the most recent eruption cycle at Poás. Increases in crater lake and fumarole temperatures in 2001 were anticipated by the appearance of yellow sulfur spherules on the lake surface (Stage II, Fig. 8). During the period 2001–2004, the presence of tailed yellow sulfur spherules (Stage II, Figs. 8 and 9) characterized typical background activity for Poás rather than signifying

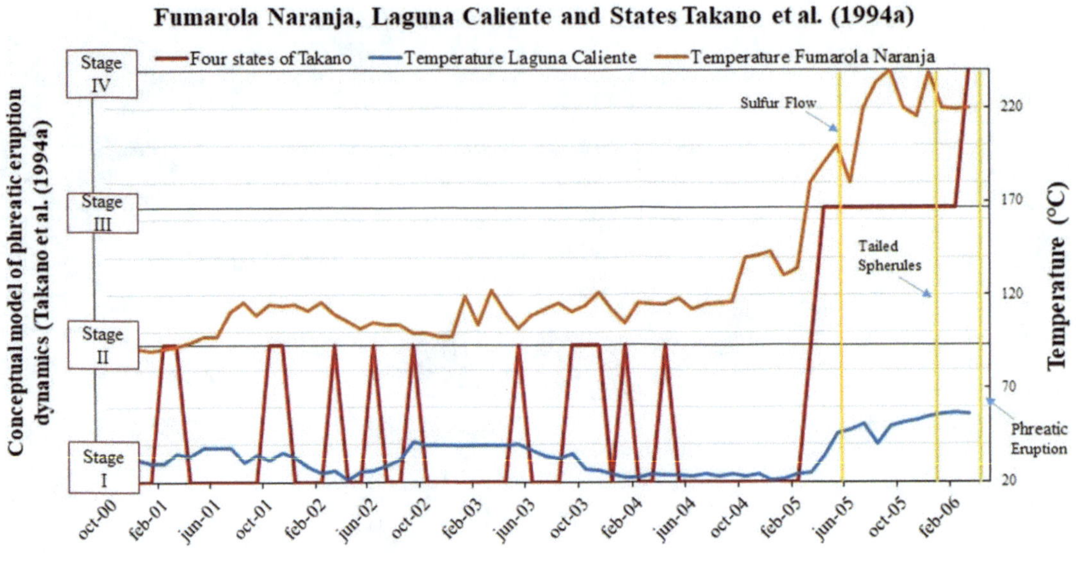

Fig. 8 Temporal variations of the temperature of Fumarola Naranja and Laguna Caliente, and the four activity stages of the Takano et al. (1994a), based on the type of sulfur spherules. The vertical yellow lines indicate key manifestations, from left to right: (1) the sulfur flow from Fumarole Naranja and black spherules on Laguna Caliente (May 2005), (2) the appearance of tailed sulfur spherules on Laguna Caliente (December 2005), (3) the onset of the phreatic eruption cycle (March 2006), after more than 10 years of quiescence

unrest (as it might elsewhere, for example at Kawah Ijen, Tonini et al. 2016). The heating of Fumarola Naranja from mid-2004 to mid-2005, and the contemporaneous cooling of Laguna Caliente point to sealing of fluid pathways beneath Laguna Caliente, and a reorganization of the shallow magmatic-hydrothermal system with major heat and mass dissipation through fumaroles. This implies that the lake water cooled and pressurized a magmatic-hydrothermal degassing system beneath the lake, presumably similar as the Fumarola Naranja system, in order to keep sulfur immobile and hence sealing the lake bottom for mass and heat.

When the fumarole temperature increased above 200 °C, tailed spherules appeared on Laguna Caliente (Stage III, Figs. 8 and 9), suggesting the clearing of the sub-lacustrine vents. Takano et al. (1994a) explained that when the temperature of the molten sulfur pools at the crater lake bottom rises above 150 °C, spherules are subject to physical changes related to the increase in viscosity of molten sulfur (Sect. 4.1). Stage III spherules are considered precursory to phreatic eruptions. At Poás, tailed spherules could be observed on December 25, 2005, after over a decade-long period of crater lake quiescence (Fig. 10). Tailed sulfur spherules are hence the surface expression of sub-lacustrine molten sulfur remobilization, whereas we interpret the Naranja sulfur flow, that developed some months earlier, as the subaerial expression of the same process: vent clearing in advance of phreatic eruptions. In fact, phreatic eruptive activity resumed (Stage IV, Fig. 8) on March 26, 2006 and continued until April 2017, before culminating in even more violent phreatomagmatic and magmatic activity (April 2017-ongoing), followed by complete lake desiccation.

The most recent lake dry-out after the April 2017 phreatomagmatic eruptions and the observation of sulfur volcanism, as reported by one of us (CO) in the late 1980s (Sect. 5.1), prompts us to propose a Stage V to extend the Takano et al. (1994a, b) model (Fig. 9). We believe that the presence of the crater lake is necessary pre-requisite for the generation of sulfur volcanism in the desiccated lake basin.

Hundreds of phreatic eruptions were recorded during the period March 2006-April 2017,

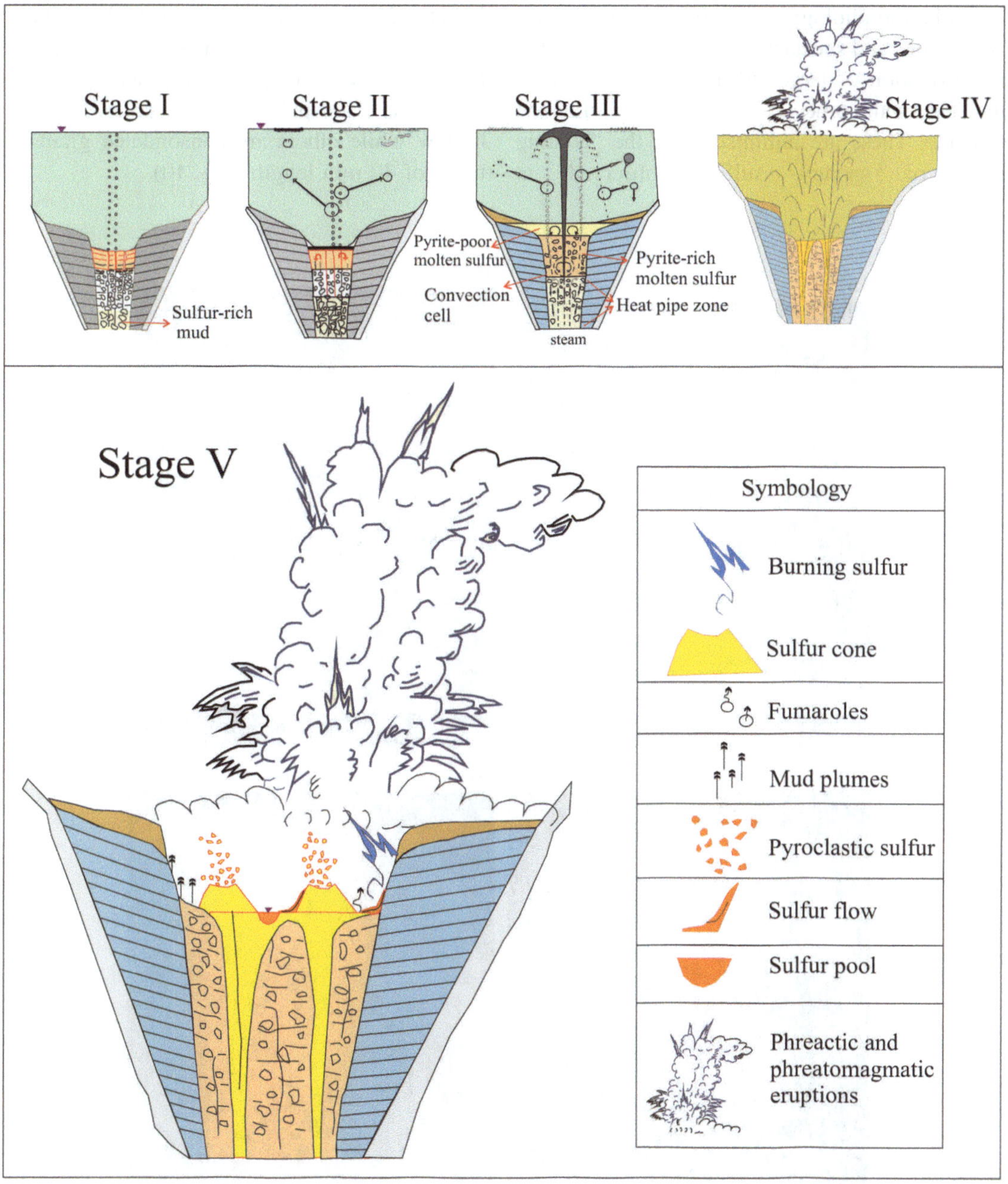

Fig. 9 Model for precursory signals of phreatic eruptions based on the appearance and morphology of sulfur spherules on acidic crater lakes (modified from Takano et al. (1994a) for Kusatsu-Shirane. The same four stages were recognized for Poás. We recognized an additional Stage V associated with the complete desiccation of Laguna Caliente, phreatic eruptions and potentially culminating in phreatomagmatic activity, as occurred in 1953–1965, 1989, 1994 and, currently, in 2017

accompanied by the production of abundant sulfur spherules in various forms. Sulfur spherules and grey mud generally rose concentrically outwards from the main vent in the middle of the lake (Ramírez et al. 2011, 2013). The timeline of spherule generation along with details of spherule morphology are reported in Fig. 11. Anomalous and spectacular spherule bursts led to the formation of large sulfur "mats" that accumulated with time at the shore of Laguna Caliente. These are reminiscent of the "floating mats of spongy sulfur" observed at Kusatsu-Shirane and described by Ohashi (1919). The most abundant sulfur spherules at Laguna Caliente range in size between 2 and 10 mm, and are pale yellow in color. Some are hollow while others are consolidated grains, or tubes of 40 mm length (Fig. 10).

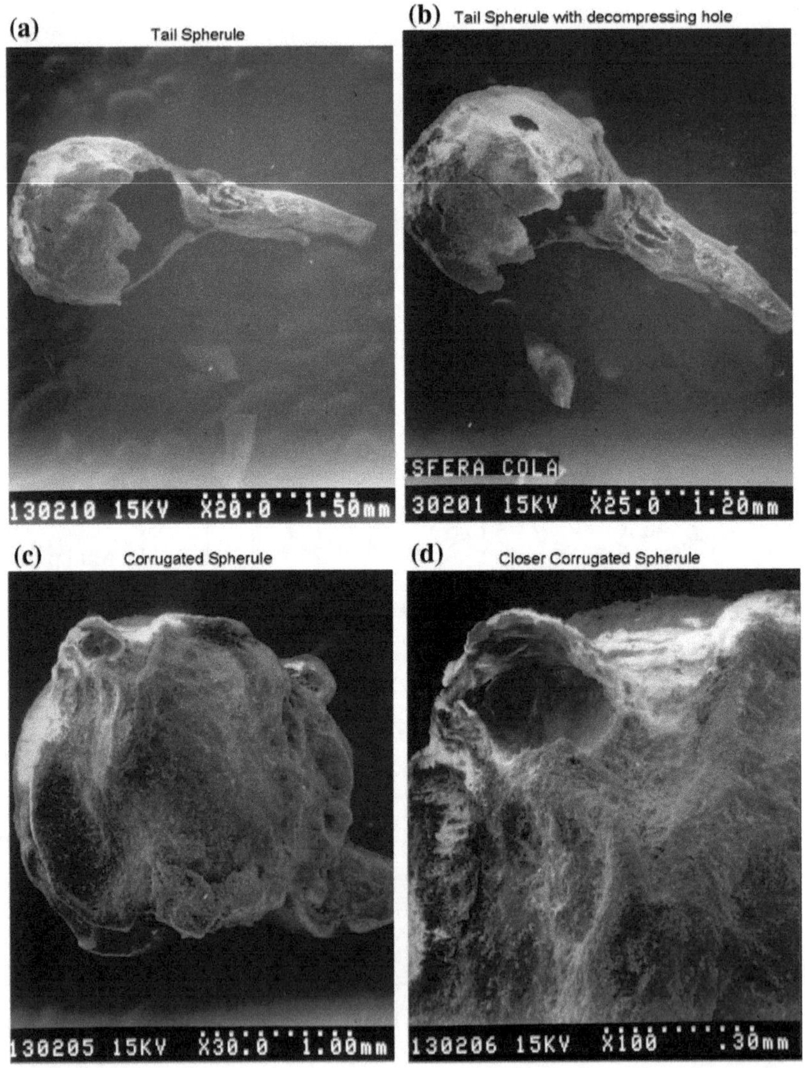

Fig. 10 SEM images of sulfur spherules sampled at Laguna Caliente on 25 and 26 December 2005 (a, b, c and d), when tailed spherules were observed for the first time since the 1990s, and prior to the onset of phreatic activity (March 2006). Tailed sulfur spherules suggest a temperature of the molten sulfur pool at the bottom of Laguna Caliente near 160 °C (Stage III, following Takano et al. 1994a), recognized as a precursory signal for phreatic eruptions. Some spherules have smooth surfaces, others have more "wavy" surfaces, probably pointing to higher viscosity sulfur for the latter ones

6 Poás Compared with Other Volcanoes

We have striven here to develop a nomenclature for sulfur volcanism in the context of its origin in sub-lacustrine and sub-aerial molten sulfur pools. The presence of a crater lake facilitates sulfur volcanism. The range of exotic eruptive products at Poás observed over the past three decades is shown in Fig. 12, and summarized in Table 1 for the period 1828–2018. We believe that this record of observations of sulfur volcanism at

Fig. 11 Reports of sulfur spherules at Laguna Caliente back at least to the 19th century. Between 2002 and 2017 spectacular sulfur slicks were observed at the lake surface, often precursory to phreatic eruptions. **a** Mats of tailed sulfur spherules at the ends of December 2005. **b** A phreatic eruption of September 2006 left mud and black sulfur at the surface. **c** Wave action accumulates sulfur spherules at the shores of Laguna Caliente (October 2006). **d** Sulfur accumulated near the Dome. **e** Color changes at Laguna Caliente are common, and on some occasions the presence of sulfur spherules accompanies a turquoise color. **f** Sulfur mats covering areas with a diameter of up to 15 m (January 2011). **g** Before and after phreatic eruptions, sulfur spherules with an intense yellow color are commonly observed. **h** Prior to a phreatic eruptions, convection cells with a diameter of up to 30 m accompany sulfur rise (April 2012). **i** On some occasions sulfur output occurs through the Dome fumaroles that is subsequently fed to the lake (May 2013). **j** Spectacular sulfur manifestations on 4 May 2014, just after a phreatic eruption. **k** Sulfur “sponges” with a thickness of up to 30 cm accumulate near the shore of Laguna Caliente (September 2015)

Poás is unique in respect of the 190 year time span of documented activity, and the wide range of manifestations: from the early nineteenth century, when Miguel Alfaro described sulfur cones and sulfur combustion in the crater of Poás, up to the observations of black sulfur spherules on the shores of Laguna Caliente, anticipating the phreatomagmatic eruptions of April 12–22, 2017.

To provide a broader context, Table 2 documents 55 volcanoes where sulfur volcanism was described. Not all of these host crater lakes but

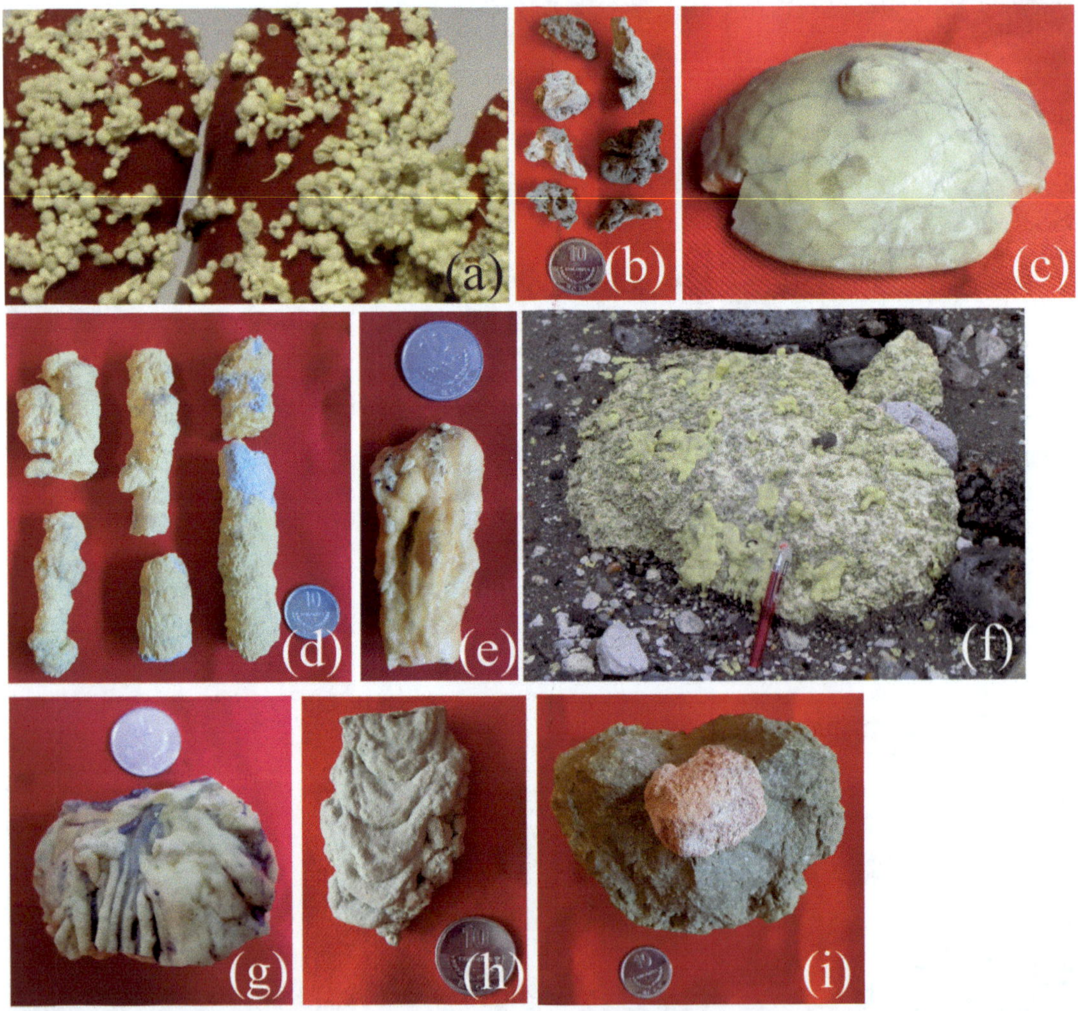

Fig. 12 Compilation of different sulfur products at Poás. **a** Hollow and tailed sulfur spherules sampled in December 2005 at the surface of Laguna Caliente. **b** Green sulfur lapilli from the eruption of pyroclasts as described by Bennett and Raccichini (1978) and Francis et al. (1980). **c** Sulfur bomb collected on 27 July 1988 from the shores of Laguna Caliente when phreatic eruption were most intense, prior to complete lake dry-out (sample, courtesy of Geol. Héctor Flores Albertazzi). **d** Tube-shaped spherules, originating from the molten sulfur pool below Laguna Caliente, sampled in September 2007. **e** Sulfur stalactites formed at the northern part of the 1953 Dome. **f** Hydrothermalized rocks with "sulfur sweat", originating from the bottom of the lake ejected during phreatic eruptions. The rock was found 350 m from the emission centre at the Playon area of Poás. **g** Small cluster of a mini-cascade structure from the May 2005 Fumarole Naranja sulfur flow in the northern inner wall of the crater of Poás. **h** A typical pahoehoe lobe from the May 2005 Fumarole Naranja sulfur flow. **i** A concretion of rock inside a green sulfur pyroclast, emitted during a phreatic eruption from Laguna Caliente in the late-1980s

few, if any other volcanoes for which we have found evidence, show the variety in sulfur volcanism seen at Poás. Most comparable to Poás are other subduction zone volcanoes with well-developed magmatic-hydrothermal systems in the form of crater lakes or thermal pools, including Kawah Ijen (Indonesia), White Island (New Zealand), Cinder Pool (USA), Copahue (Argentina), Rincón de la Vieja (Costa Rica), Kusatsu-Shirane (Japan), and Goryachee (Russia).

Most similar to the Poás' sulfur volcanism are the manifestations seen at Kawah Ijen (Indonesia) and White Island. Kawah Ijen hosts the largest acidic crater lake on Earth (27×10^6 m^3, 170 m deep). It requires a prodigious degassing of acidic species (SO_2–H_2S, HCl, HF) to keep this amount of water heated and of such low pH. Kawah Ijen is renowned for its artisanal sulfur mining, 400 °C fumaroles, sulfur spherules and spectacular blue-flamy sulfur combustion (Delmelle and Bernard 1994, 2015), though clearly some of the sulfur volcanism manifestations are triggered or modified by the mining operations. Recent activity at White Island (Christenson et al. 2017) includes the migration of fumaroles and vents in the environs of the hot acid crater lake.

7 Summary

Active crater lakes promote formation of near-surface native sulfur magmas that fuel a range of eruptive styles and products. The lake water acts to maintain the underlying magmatic-hydrothermal system close to the melting temperatures of sulfur (116–119 °C). Meanwhile, the acid aqueous environment promotes disproportionation of SO_2. If temperature rises beyond around 159 °C, activity can transition due to the 10,000 fold increase in viscosity of liquid sulfur. The narrow range of environmental conditions conducive to generation of sulfur magmas limits the manifestation of sulfur volcanism on Earth. Other acid crater lakes including Kawah Ijen and White Island are among the most similar in terms of sulfur volcanism, to Poás, reinforcing the association of sulfur volcanism with acid crater lakes.

Nevertheless, sulfur volcanism at Poás is gas-driven, not melt driven. The difference with magmatic eruptions (e.g. Strombolian activity) is that the gas/vapor is not directly sourced from the molten sulfur, but from fumarolic degassing from the magmatic-hydrothermal system. The presence of sulfur crystals provides evidence for this, since sulfur pyroclasts (accretionary lapilli, and single crystals) are ejected during sulfur cone formation, or *hornitos* built-up from a spray of sulfur upon the passage of gas through molten sulfur pools.

Two of the major perturbations in stratospheric sulfate aerosol observed in past decades resulted from the eruptions of El Chichón (Mexico) in 1982, and Pinatubo (the Philippines) in 1991 (Rampino and Self 1984; Kirchner et al. 1999; Krueger et al. 2008). It has been suggested that a significant component of the sulfur budgets of these eruptions was sourced by pyrolysis of sulfur that had been sequestered over prolonged periods of time within the upper regions of the magmatic-hydrothermal envelope (Oppenheimer 1996). Notably, both volcanoes hosted crater lakes following these eruptions, and geothermal explorations had previously identified vigorous magmatic-hydrothermal systems. Poás provides a remarkable opportunity to observe the process by which such large quantities of sulfur may be generated and stored in the shallow parts of a volcanic edifice.

NOTE: Figure 13 shows the most recent manifestations of sulfur volcanism during the phreatomagmatic eruptive activity and partial dome destruction of April 2017 at Poás.

Fig. 13 In April 2017, Poás entered its fourth historical magmatic eruption. After the phreatomagmatic eruptions on April 12, 14 and 22, 2017, Laguna Caliente disappeared due to enhanced evaporation. **a** Photograph of the dome/pyroclastic deposit a day after (April 13) the first phreatomagmatic eruption. The 1953 dome was fractured and intense evaporation was observed at Laguna Caliente. Convection cells originating from sulfur pools distribute black sulfur mats, testifying high temperature degassing. **b** In July 2017 Laguna Caliente dried out entirely, and a pyroclastic cone was formed at the former lake bottom, from which brown ashes (left) were released, whereas a small sulfur cone released yellow-tinted sulfur-loaded gases (right). **c** Apart from the sulfur-enriched exhalations, sulfur tephra (ash and lapilli) was ejected. The small sulfur cone is formed and destroyed, a common dynamics during such stages of activity. Its location coincides with the location of the former convection cells at the former Laguna Caliente, and hence finds its origin in the former molten sulfur pools. **d** Formation of sulfur *hornitos* of 0.5 m height in January 2018. **e** In January 2018 Laguna Caliente reformed. Lake water temperature was 40 °C in February 2018, with a pH near 1. Vigorous exhalation activity is observed. The picture was taken from the eastern terraces

References

Africano F, van Bergen M, Bernard A, Mason PRD, Martínez M (2007) Trace element contents of sulfur spherules in acid crater lakes: signals of volcanic activity. Geophysical Res Abstracts 9:07883

Agusto MR, Caselli A, Daga R, Varekamp J, Trinelli A, Afonso M, Vélez ML, Euillades P, Guevara S (2017) The crater lake of Copahue volcano (Argentina): geochemical and thermal changes between 1995 and 2015. Geol Soc, London, Special Public 437:107–130

Averyanoy IP, Zhizhin DP, Tyutrin II (1972) Potential for discovered of native sulfur deposits in caldera volcanoes of the Kuril Islands. Internal Geol Rev 14 (8):880–886

Bani P, Oppenheimer C, Varekamp JC, Quinou T, Lardy M, Carn S (2009) Remarkable geochemical changes and degassing at Voui crater lake, Ambae volcano, Vanuatu. J Volcanol Geoth Res 188:347–357

Bennett FD, Raccichini SM (1978) Subaqueous Sulphur sulphur lake in Volcan Poás. Nature 271:342–344

Bohnenberger O, Bengoeche A, Dóndoli C, Marroquín A (1966) Report on active volcanoes in Central America during 1957 to 1965. Segunda Reunión Geólogos, Guatemala, Central America, p 35

Brantley SL, Borgia A, Rowe G, Fernández JF, Reynolds JR (1987) Poás volcano crater lake acts a condenser for acid metal-rich brine. Nature 330:470–472

Brown G, Rymer H, Dowden J, Kapadia P, Stevenson D, Barquero J, Morales LD (1989) Energy Budget analysis for Poás Crater lake: implications for predicting volcanic activity. Nature 339:370–373

Bullard F (1956) La actividad en Costa Rica y Nicaragua en 1954. In Vargas C (1979) Volcán Poás Ed UNED, San José 109–118

Butterfield DA (2006) Role of Magmatic fluids in hydrothermal vents on volcanic arcs. NOAA Ocean Explorer Submarine Ring of Fire 2006 Vent Chemistry 1–6

Caselli A, Agusto M, Velez ML, Forte P, Bengoa C, Daga R, Albite JM, Capaccioni B (2016) The 2012 Eruption. In: Tassi F et al. (eds) Copahue Volcano, Active Volcanoes of the World.https://doi.org/10.1007/978-3-662-48005-2_4

Casertano L, Borgia A, Cigolini C (1983) El Volcán Poás, Costa Rica: Cronologia y características de la actividad. Mex Geofis Intern 3:215–236

Christenson BW (1994) Convection and stratification in Ruapehu Crater lake, New Zealand: implications for lake Nyos-type gas release eruptions. Geochem J 28:185–198

Christenson BW, Németh K, Rouwet D, Tassi F, Vandemeulebrouck J, Varekamp JC (2016) Volcanic Lakes. In: Rouwet D, Christenson BW, Tassi F, Vandemeulebrouck J (eds) Volcanic lakes. Springer, Heidelberg, pp 1–20. https://doi.org/10.1007/978-3-642-36833-2_1

Christenson BW, Reyes AG, Young R, Moebis A, Sherburn S, Cole-Baker J, Britten K (2010) Cyclic processes and factors leading to phreatic eruption events: insights from the 25 September 2007 eruption through Ruapehu Crater Lake, New Zealand. J Volcanol Geotherm Res 191:15–32

Christenson BW, White S, Britten K, Scott BJ (2017) Hydrological evolution and chemical structure of a hyper-acidic spring-lake system on Whakaari/White Island, NZ. J Volcanol Geotherm Res 346: 180–211. https://doi.org/10.1016/j.jvolgeores.2017.06.017

Cigolini C, Kudo A, Brookins DG, Wardo D (1991) The petrology of Poás Volcano lavas: basalt-andesite relationship and petrogenesis with the magmatic arc of Costa Rica. J. Volcanol Geotherm Res 48:367–384

Colony WE, Nordlie BE (1973) Liquid sulfur at Volcan Azufre. Galapagos Islands Econ Geol 68:371–380

de Moor JM, Aiuppa A, Pacheco J, Avard G, Kern C, Liuzzo M, Martínez M, Giudice G, Fischer TP (2016) Short-period volcanic gas precursors to phreatic eruptions: insights from Poás Volcano, Costa Rica. J Volcanol Geotherm Res 442:218–227

Delmelle P, Bernard A (1994) Geochemistry, mineralogy and chemical modeling of the acid crater lake of Kawah Ijen Volcano, Indonesia. Geochim Cosmochim Acta 58:2445–2460

Delmelle P, Bernard A (2015) The remarkable chemistry of sulfur in volcanic acid crater lakes: a scientific tribute to Bokuichiro Takano and Minoru Kusakabe. In: Rouwet D, Christenson BW, Tassi F, Vandemeulebrouck J (eds) Volcanic lakes. Springer, Heidelberg, pp 239–260. https://doi.org/10.1007/978-3-642-36833-2_10

Delmelle P, Bernard A, Kusakabe M, Fischer TP, Takano B (2000) Geochemistry of the magmatic-hydrothermal system of Kawah Ijen volcano, East Java Indonesia. J Volcanol Geotherm Res 97(1–4):31–53

Delmelle P, Bernard A, Kusakabe M, Fisher T, Takano B (1996) Isotropic geochemistry of the hydrothermal system of Kawah Ijen, Indonesia. In: Proceedings chapman conference, Crater Lake, Oregon, p 29

Delpino D.H, Bermúdez A.M (1996) Eruptions of pyroclastic sulfur at crater lake of Copahue volcano, Argentina. Newsletter IAVCEI Commissio n on Volcanic Lakes 8: 23

Delpino DH, Bermúdez AM (1996b) Eruptions of pyroclastic sulfur at crater lake of Copahue volcano, Argentina. Newlett IAVCEI Comm Volcanic Lakes 8:23

Fischer TP, Ramírez C, Mora-Amador R, Hilton DR, Barnes JD, Sharp ZD, Le Brun M, de Moor JM, Barry PH, Furi E, Shaw AM (2015) Temporal variations in fumarole gas chemistre at Poás volcano, Costa Rica. J Volcanol Geotherm Res 294:56–70

Francis PW, Thorpe RS, Brown GC, Glasscock J (1980) Pyroclastic sulphur eruption at Poás volcano Costa Rica. Nature 283:754–756

Gavrilenko G, Melnikov D, Ovsyannikov A (2008) Current state the thermal Lake into active crater Gorely (Kamchatka): materials of Russian scientific conference "100 years of the Kamchatka expedition of the Russian geographical society 1908–1910." (Petropavlovsk-Kamchatsky, 22–27 Sept 2008) (In Russian) Petropavlovsk, 2009, p 267

Giggenbach WF (1974) The chemistry of crater lake, Mt Ruapehu (New Zealand) during and after the 1971 active period. NZJ Sci 17:33–45

Giggenbach WF (1987) Redox processes governing the chemistry of fumarolic gas discharges from White Island, New Zealand. Appl Geochem 2:143–161

Grimes S, Rickard D, Browne P, Simmons S, Jull T (1999) Sub-aqueous sulfur volcanoes at Waiotapu, New Zealand. Geothermics 28:729–738

Hamilton WM, Baumgart IL (1959) White Island. Department of Scientific and Industrial Research Bulletin 127: 84

Harris A, Carniel R, Patrick M, Dehn J (2004) the sulfur flow fields of the Fossa di vulcano. Bull Volcanol 66:749–759

Harris AJL, Sherman SB, Wright R (2000) Discovery of self-combusting volcanic sulfur flows. Geology 28:415–418

Hurst AW, Bibby HM, Scott BJ, McGuinness MJ (1991) The heat source of Ruapehu crater lake: deductions from the energy and mass balances. J Volcanol Geotherm Res 46:1–20

Imai A, Geshi N (1999) Spinifex texture of native sulfur flow eruptions at Shiretoko—Iwosan volcano, Hokkaido, Japan. Res Geol 49: 99–104

Inoue A, Aoki M (2000) Mineralogy of Ohyunuma explosion crater lake, Horkaido, Japan. Part 1: geochemistry, hydrology, and bulk mineralogy. Clay Sci 11:147–168

Ishikawa T, Yokoyama I, Katsui Y, Kasahara M (1971) Tokachi-dake, its volcanic geology, history of eruption, present state of activity and prevention of disaster. Committee for Sapporo committee for prevention of disasters of Hokkaido, Sapporo, pp 136 (In Japanese with English abstract)

Karpov GA, Fazlullin SM (1995) The creation and prolonged existence of the zone of molten native sulfur at the bottom of thermal lake within the volcanogenic-hydrothermal system (Uzon caldera, Kamchatka). In: Kharaka YK, Chudaev OV (eds) Water-rock interaction 8, A.A. Balkema, Rotterdam, pp 307–310

Karpov GA, Fazlullin SM, Nadeznaya TB (1996) Liquid sulfur at the bottom of a thermal lake in the Uzon Caldera, Kamchatka. Volc Seism 18:171–186

Kawasaki H (1903) Genesis of sulphur on the volcano Shirane, in Japanese. J Geol Soc Tokyo 10:425–428

Kervyn M, Ernst GGJ, Klaudius J, Keller J, Kervyn F, Mattsson HB, Belton F, Mbede E, Jacobs P (2008) Voluminous lava flows at Oldoinyo Lengai in 2006: chronology of events and insights into the shallow magmatic system. Bull Volcanol 70:1069–1086

Kim J, Lee KY, Kim JH (2011) Metal-bearing molten sulfur collected from a submarine volcano: implications for vapor transport of metals in seafloor hydrothermal systems. Geology 39:351–354

Kirchner I, Stenchikov G, Graf H-F, Robock A, Antuna J (1999) Climate model simulation of winter warming and summer cooling following the 1991 Mount Pinatubo volcanic eruption. J Geophys Res 104:19039–19055

Kirsanov, IT, Serafimova, EK, Sidorov, SS, Trubenko, VF, Farberov, AI, Fedorchenko, VI. and Shilov, VN (1964) The eruption of Ebeko volcano in March-April, 1963. Bull Volcanol. St., 36:66-72 (in Russian)

Korzhinskii, MA, Tkachenko, SI, Bulgakov, RF, Shmulovich, KI (1996). Condensate compositions and native metals in sublimates of high-temperature gas streams of Kudryavyi Volcano, Iturup Island, Kuril Islands. Geochem Intern—GeoKhimia 34: 1057–1064

Kouno T (1988) "Hoguchi" Sulfur from the Jigokudani Valley, Tateyama Volcano. Geological reports of Shimane University, 7: 91–97 (in Japanese)

Krueger A, Krotkov N, Carn S (2008) El Chichón: the genesis of volcanic sulfur dioxide monitoring from space. J Volcanol Geotherm Res 175:408–414

Krupp RE, Seward TM (1987) The Rotokawa geothermal system, New Zealand: an active epithermal gold-deposing environment. Ecom Geol 81:1109–1129

Krushensky R, Escalante G (1967) Activity of Irazú and Poás volcanoes, Costa Rica, November 1964—July 1965. Bull Volcanol 31:75–84

Kusakabe M, Hayashi N, Kobayashi T (1986) Genetic environments of the banded sulfur sediments at the Tatyeama Volcano, Japan. J Geophys Res 91:12159–12166

Leiva E (1906) Una excursión al volcán Poás. In: Vargas C, (1979) Volcán Poás Ed UNED, San José, pp 69–70

López DL, Ransom L, Pérez N, Hernández P, Monterrosa J (2004) Dynamics of diffuse degassing at Ilopango Caldera, El Salvador. In: Rose WI, Bommer JJ, López DL, Carr MJ and Major JJ (eds) Natural hazards in El Salvador, Geol Soc Am, Special paper no. 375: 191–202

Luke J (1950) History in White Island. Dept Sci Ind Res Bull 127:14–24

MacKnight WJ, Tobolsky AV (1965) Properties of polymetric sulfur. In: Meyer B (ed) Elemental sulfur, chemistry and physics. Interscience Publ, New York, pp 95–107

Markhinin EK (1977) Stratula DS (1977) Gidritermy Kurilskix ostrovov (Hydrotherms of the Kurile Islands). Nauka, Moscow, p 212

Markhinin EK, Stratula DS (1977) Hydrothermal Systems of Kurile Islands. Nauka, Moscow, p 212

Martínez M, Fernández E, Valdés J, Barboza V, Van der Laat R, Duarte E, Malavassi E, Sandoval L, Barquero J, Marino T (2000) Chemical evolution and activity of the active crater lake of Poás volcano, Costa

Rica, 1993–1997. J Volcanol Geotherm Res 97:127–141
Masubuchi Y (2012) Field occurrence and petrological characteristics of sulfur lava flow formed in may 2010 in Jigokudani, Tateyama Volcano. Volcanol Soc Japan 2012 Fall Meeting. https://doi.org/10.18940/vsj.2012.0_153
Menyailov IA, Nikitina LP, Shapar VN (1985) Results of geochemical monitoring of the activity of Ebeko volcano (Kurile Islands) used for eruption prediction. J Geodyn Res 3:259–274
Mizutani Y, Sugiura T (1966) The chemical equilibrium of the $2H_2S + SO_2 = 3S + 2H_2O$ reaction in solfataras of the Nasudake volcano. Bull Chem Soc Japan 39:2411–2414
Mora-Amador R, Ramírez C (2008) Sulfur flows at Poás volcano, Costa Rica. IAVCEI general assembly 2008, Reykjavík Iceland:85
Mora-Amador R, Ramírez C, Fernández M (2004) La actividad de los volcanes de la cordillera central, Costa Rica, entre 1998–2002. Rev Geol Amer Central 30:189–197
Mora-Amador R, Rouwet D, González G, Ramírez C (Chapter 11). Volcanic hazard assessment of Poás (Costa Rica) based on the major historical eruptions of 1834, 1910, 1953–1955 and 2017–2018. In: Tassi F, Mora-Amador R, Vaselli O (eds) Poás volcano (Costa Rica): the pulsing heart of Central America Volcanic Zone. Springer, Heidelberg (Germany)
Mori T, Sato M, Shimoike Y, Notsu K (2002) High SiF_4/HF ratio detected in Satsuma-Iwojima volcano's plume by remote FT-IR observation. Earth Planet Space 54:249–256
Murozomi M, Abiko T, Nakamura S (1966) Geochemical investigation of the Noboribetsu Oyunuma explosión crater lake. Bull Volcanol Soc Japan 11:1–16
Murphy S, Wright R, Rouwet D (2018) Color and temperature of the crater lakes at Kelimutu volcano through time. Bull Volcanol 80:2. https://doi.org/10.1007/200445-017-1172-2
Naboko SI (1958) Formation of lake sulfuro n Golvnin volcano. Byul Vulkanol Stantisiy AN SSSR 27:43–50
Naboko SI (1959) Volcanic exhalations and the products of their reactions as exemplified by Kamchatka-Kuriles volcanoes. Bull Volcanol 20:121–136
Naranjo JA (1985) Sulphur flows at Lastarria volcano in the North Chilean Andes. Nature 313:778–780
Naranjo JA (1988) Coladas de azufre de los volcanes Lastarria y Bayo en el Norte de Chile: Reología, génesis e importancia en geología planetaria. Rev Geol Chile 15:3–11
Oersted AS (1863) L'Amerique Centrale: Recherches sur sa flore et sa Géographie Physique. Imprenta Blanco Luna, Copenhague
Ohashi R (1919) On the peculiar Sulphur spherules produced in a crater lake of the Volcano Shirane, in the province of Kozuke, central Japan. J Akita Min Coll 1:1–10
Ohsawa S, Saito T, Yoshikawa S, Mawatari H, Yamada m, Amita K, Takamatsu N, Sudo Y, Kagiyama T (2010) Color change of lake water at the active crater lake of Aso volcano, Yudamari, Japan: is it in response to change in water quality induced by volcanic activity? Limnology 11: 207–215
Ōinouye Y (1916) A peculiar process of sulphur deposition. J Geol 24:806–808
Oppenheimer C (1992) Sulfur eruptions at Volcán Poás, Costa Rica. J Volcanol Geotherm Res 49:1–21
Oppenheimer C, Pyle DM, Barclay J (eds) Special Memory of the Geol Soc London (UK), Special Issue on: "volcanic degassing" 213: 247–262
Oppenheimer C, Stevenson D (1989) Liquid sulfur lakes at Poás Volcano. Nature 342:790–793
Oppenheimer C (1996) On the role of hydrothermal systems in the transfer of volcanic sulfur to the atmosphere. Geophys Res Lett 23(16). https://doi.org/10.1029/96gl02061
Ossaka J, Hirabayashi J, Nomura T, Ossaka T, Hayashi T, Masuda Y (1980) On the molten sulfur at crater lake Yugama Kusatsu-Shirane volcano. Bull Volcanol Soc Japan 25:309 (in Japanese)
Pasternack GB, Varekamp JC (1994) The geochemistry of the Keli Mutu crater lake, Flores, Indonesia. Geochem J 28:243–262
Prosser JT, Carr MJ (1987) Poás volcano, Costa Rica: geology of the summit region and spatial and temporal variations among the most recent lavas. J Volcanol Geoth Res 33:131–146
Raccichini S, Bennett FD (1977) Nuevos aspectos de las erupciones del volcán Poás. Rev Geogr Amér Central 5–6:37–53
Ramírez CJ, Mora-Amador R, González G (2011) Field thermal imaging monitoring of fumaroles, crater lakes and mud pools, at Costa Rican volcanoes. 11th gas workshop. IAVCEI commission on volcanic gases, Kamchatka, Russia, 40
Ramírez CJ, Mora-Amador R, González G, Alpizar Y (2013) Applications of infrared cameras at Costa Rican volcanoes, crater lakes and thermal features. 8th workshop on volcanic lakes. IAVCEI commission on volcanic lakes, Japan:6
Rampino MR, Self S (1984) Sulphur-rich volcanic eruptions and stratospheric aerosols. Nature 310. https://doi.org/10.1038/310677a0
Rouwet D, Mora-Amador R, Ramírez CJ, González G, Inguaggiato S (2016) Dynamic fluid recycling at Laguna Caliente (Poás, Costa Rica) before and during the 2006—ongoing phreatic eruption cycle (2005–10). In: Ohba T, Capaccioni B, Caudron C (eds) Geochemistry and Geophysics of active volcanic lakes. Geol Soc London, Special Publication 437: 73-96. https://doi.org/10.1144/sp437.11
Rouwet D, Mora-Amador R, Sandri L, Ramírez C, González G, Pecoraino G, Capaccioni B (Chapter 9). 39 years of geochemical monitoring of Laguna Caliente crater lake, Poás: patterns from the past as keys for the future. In: Tassi F, Vaselli O,

Mora-Amador R (eds) Poás volcano (Costa Rica): the pulsing heart of central America volcanic zone. Springer, Heidelberg (Germany)

Rouwet D, Morrisey MM (2015) Mechanisms of crater lake breaching eruptions. In: Rouwet D, Christenson BW, Tassi F, Vandemeulebrouck J (eds) Volcanic lakes. Springer, Heidelberg:73–91. https://doi.org/10.1007/978-3-642-36833-2_3

Rowe GL, Ohsawa S, Takano B, Brantley SL, Fernández JF, Barquero J (1992a) Using crater lake chemistry to predict volcanic activity at Poás Volcano, Costa Rica. Bull Volcanol 54:494–503

Rowe GL, Brantley SL, Fernández M, Fernández JF, Borgia A, Barquero J (1992b) Fluid volcano interaction in an active stratovolcano: the crater lake system of Poás Volcano, Costa Rica. J Volcanol Geotherm Res 49:23–51

Rudín J, Alfaro A, Michaud G, Rudín A (1910) Gran erupción de ceniza del volcán Poás. In Vargas C (1979) Volcán Poás Ed UNED, San José:75–82

Ruiz P, Mana S, Gazel E, Soto GJ, Carr MJ, Alvarado GE (Chapter 2). Geochemical and geochronological characterisation of the Poas stratovolcano stratigraphy. The case of Poás volcano, Costa Rica. In: Tassi F, Mora-Amador R, Vaselli O, (eds) Poás volcano (Costa Rica): The pulsing heart of Central America Volcanic Zone. Springer, Heidelberg (Germany)

Rymer H, Cassidy J, Locke CA, Barboza M, Barquero J, Brenes J, Van der Laat R (2000) Geophysical studies of the recent 15-year eruptive cycle at Poás volcano, Costa Rica. J Volcanol Geotherm Res 97:425–442

Sapper K (1925) Los volcanes de América Central.Halle (Saale) Verlang ron Max Niemeyer

Scolamacchia T, Cronin SJ (2016) Idiosyncrasies of volcanic sulfur viscosity and the triggering of unheralded volcanic eruptions. Front Earth Sci 4:24

Shakeri A, Moore F, Kompani-Zare M (2008) Geochemistry of the thermal springs of Mount Taftan, southeastern Iran. J Volcanol Geotherm Res 178:829–836

Shinohara H, Yoshikawa S, Miyabuchi Y (2015) Degassing activity of a volcanic crater lake: Volcanic plume measurements at the Yudamari crater lake, Aso volcano, Japan. In: Rouwet D, Christenson BW, Tassi F, Vandemeulebrouck J (eds) Volcanic lakes. Springer, Heidelberg (Germany)

Skinner BJ (1970) A sulfur lava flow on Mauna Loa. Pac Sci 24:144–145

Sriwana T, van Bergen MJ, Varekamp JC, Sumarti S, Takano B, van Os BJH, Leng MJ (2000) Geochemistry of the acid Kawah Putih lake, Patuha Volcano, West Java, Indonesia. J Volcanol Geotherm Res 97:77–104

Takano B (1987) Correlation of volcanic activity with sulfur oxyanion speciation in a crater lake. Science 235:1633–1635

Takano B, Ohsawa S, Glover RB (1994a) Surveillance of Ruapehu crater lake, New Zealand, by aqueous polythionates. J Volcanol Geotherm Res 178:145–168

Takano B, Saitoh H, Takano E (1994b) Geochemical implications of subaqueous molten sulfur at Yugama crater lake, Kusatsu-Shirane volcano, Japan. Geochem J 28:199–216

Takano B, Watanuki K (1990) Monitoring of volcanic eruptions at Yugama crater lake by aqueous sulfur oxyanions. J Volcanol Geotherm Res 40:71–87

Takano B, Matsuo M, Suzuki K (1995) Bathymetric and chemical investigation of crater lake at Maly Semiachik, Volcano, Kamchatka. Proceedings 8th Inter Symp Water- Rock Interaction, Balkema, Rotterdam:47–49

Tassi F, Vaselli O, Capaccioni B, Giolito C, Duarte E, Fernández E, Minissale A, Magro G (2005) The hydrothermal-volcanic system of Rincón de la Vieja volcano (Costa Rica): a combined (inorganic and organic) geochemical approach to understanding the origin of the fluid discharges and its possible application to volcanic surveillance. J Volcanol Geotherm Res 148:315–333

Tassi F, Vaselli O, Fernández E, Duarte E, Martínez M, Delgado-Huertas A, Bergamaschi F (2009) Morphological and geochemical features of crater lakes in Costa Rica: an overview. J Limnol 68(2):193–205

Taylor H (1967) Croquis del fondo del volcán Poás. Inedito

Theilig E (1982) A primer on Sulfur for the Planetary Geologist. NASA contractor report 3594, p 42

Tonini R, Sandri L, Rouwet D, Caudron C, Marzocchi W (2016) A new Bayesian Event Tree tool to track an quantify volcanic unrest and its application to Kawah Ijen volcano. Geochem Geophys Geosyst 17. https://doi.org/10.1002/2016gc006327

Toraishi S, Tominaga H (1940) Sulfur spherules in crater lake, Okama, Zao Volcano. Kagaku. Science 10: 3–4

Toraishi S, Tominaga H (1940b) Sulfur spherules in crater lake, Okama, Zao Volcano. Kagaku (Science) 10:3–4 (in Japanese)

Uzumasa Y, Nasu Y, Seo T (1959) Chemical studies on hot springs in Niseko district, southwest of Hokkaido. Nippon Kaseko Zasshi 80:992–995 (in Japanese)

Varekamp JC, Ouimette AP, Herman SW, Bermúdez A, Delpino D (2001) Hydrothermal element fluxes from Copahue, Argentina: a "beehive" volcano in turmoil. Geology 29:1059–1062

Vaselli O, Tassi F, Fischer TP, Tardani D, Fernandez Soto E, Duarte E, Martinez M, De Moor MJ, Bini G (Chapter 10). The last eighteen years (1998–2015) of fumarolic activity at the Poás volcano (Costa Rica) and the renewed activity. In: Tassi F, Mora-Amador R, Vaselli O, (eds) Poás volcano (Costa Rica): the pulsing heart of Central America Volcanic Zone. Springer, Heidelberg (Germany)

Vaselli O, Tassi F, Montegrossi G, Duarte E, Fernández E, Bergamaschi F (2003) Fumarole migration and fluid chemistry at Poás volcano (Costa Rica) from 1998 to 2001. In: Oppenheimer C, Pyle DM, Barkley J (eds) Volcanic degassing, vol 213. Geol Soc London, Special Publications, pp 247–262. https://doi.org/10.1144/gsl.sp.2003.213.01.15

Von Frantzius A (1861) Aporte al Conocimiento de los Volcanes de Costa Rica, Escalamiento al Volcán Poás,

marzo 1860–1979.In: Vargas C (1979) Volcán Poás Ed UNED, San José, pp 11–34
von Hochstetter F (1864) Lecture on the Geology of the Province of Auckland. In: von Hochstwttern F, Petermann A (eds) The geology of New Zealand: in explanation of the geographical and topografical atlas (Novara Expedition) T. Delattre, Auckland, pp 113
Watanabe T (1940) Eruptions of molten sulphursulphur from the Siretoko-Iosan Volcano, Hokkaido, Japan. Japan J Geol Geogr 17:289–310
White DE, Hutchinson RA, Keith TEC (1988) The geology and Remarkable Thermal Activity of Norris Geyser Basin, Yellowstone National Park. U.S. Government Printing Office, Washington, pp 42–44
Xu Y, Schoonen MAA, Nordstrom DK, Cunningham KM, Ball JW (1988) Sulfur geochemistry of hydrothermal waters in Yellowstone National Park: I. The origin of thiosulfate in hot spring waters. Geochim Cosmochim Acta 62(23–24): 3729–3743
Xu Y, Schoonen MAA, Nordstrom DK, Cunningham KM, Ball JW (2000) Sulfur geochemistry of hydrothermal waters in Yellowstone National Park, Wyoming, USA. II. Formation and decomposition of thiosulfate and polythionate in Cinter Pool. J Volcanol Geotherm Res 97(1–4): 407–423
Zeledón E (2007) Leyendas Costarricenses. EUNA, pp 65–66

Scientific Event Alert Network and Bulletin of the Global Volcanism Network

Global Volcanism Program, 1978. Report on Poás (Costa Rica). In: Squires, D (ed) Scientific Event Alert Network Bulletin, 3: 11. Smithsonian Institution. https://dx.doi.org/10.5479/si.GVP.SEAN197811-345040
Global Volcanism Program (1979) Report on Kirishimayama (Japan). In: Squires D (ed) Scientific event alert network Bulletin 4: 6 Smithsonian Institution. https://dx.doi.org/10.5479/si.GVP.SEAN197906-282090
Global Volcanism Program (1981) Report on Poás (Costa Rica). In: McClelland L (ed) Scientific event alert network Bulletin 6: 5. Smithsonian Institution. https://dx.doi.org/10.5479/si.GVP.SEAN198105-345040
Global Volcanism Program (1982) Report on El Chichon (Mexico). In: McClelland L (ed) Scientific event alert network Bulletin, 7: 10, Smithsonian Institution. https://dx.doi.org/10.5479/si.GVP.SEAN198210-341120
Global Volcanism Program 1982. Report on Irazu (Costa Rica). In: McClelland L (ed) Scientific event alert network Bulletin, 7: 11, Smithsonian Institution. https://dx.doi.org/10.5479/si.GVP.SEAN198211-345060
Global Volcanism Program (1985) Report on Tokachidake (Japan). In: McClelland L (ed) Scientific event alert network Bulletin, 10: 7. Smithsonian Institution. https://dx.doi.org/10.5479/si.GVP.SEAN198507-285050
Global Volcanism Program (1988) Report on Poás (Costa Rica). In: McClelland L (ed) Scientific event alert network Bulletin, 13: 1. Smithsonian Institution. https://dx.doi.org/10.5479/si.GVP.SEAN198801-345040
Global Volcanism Program (1988) Report on Poás (Costa Rica). In: McClelland L (ed) Scientific event alert network Bulletin, 13: 4. Smithsonian Institution. https://dx.doi.org/10.5479/si.GVP.SEAN198804-345040
Global Volcanism Program (1988) Report on Loloru (Papua New Guinea). In: McClelland L (ed) Scientific event alert network Bulletin, 13: 4. Smithsonian Institution. https://dx.doi.org/10.5479/si.GVP.SEAN198804-255030
Global Volcanism Program (1988) Report on Poás (Costa Rica). In: McClelland L (ed) Scientific event alert network Bulletin, 13: 6. Smithsonian Institution. https://dx.doi.org/10.5479/si.GVP.SEAN198806-345040
Global Volcanism Program (1988) Report on Poás (Costa Rica). In: McClelland L (ed) Scientific event alert network Bulletin, 13: 7. Smithsonian Institution. https://dx.doi.org/10.5479/si.GVP.SEAN198807-345040
Global Volcanism Program (1989) Report on Poás (Costa Rica). In: McClelland L (ed) Scientific Event Alert Network Bulletin, 14: 1. Smithsonian Institution. https://dx.doi.org/10.5479/si.GVP.SEAN198901-345040
Global Volcanism Program (1989) Report on Poás (Costa Rica). In: McClelland L (ed) Scientific event alert network Bulletin, 14:2. Smithsonian Institution. https://dx.doi.org/10.5479/si.GVP.SEAN198902-345040
Global Volcanism Program (1989) Report on Poás (Costa Rica). In: McClelland L (ed) Scientific event alert network Bulletin, 14: 3, Smithsonian Institution. https://dx.doi.org/10.5479/si.GVP.SEAN198903-345040
Global Volcanism Program (1989) Report on Poás (Costa Rica). In: McClelland L (ed) Scientific event alert network Bulletin, 14: 4. Smithsonian Institution. https://dx.doi.org/10.5479/si.GVP.SEAN198904-345040
Global Volcanism Program, 1989. Report on Poás (Costa Rica). In: McClelland L (ed) Scientific event alert network Bulletin, 14: 6. Smithsonian Institution. https://dx.doi.org/10.5479/si.GVP.SEAN198906-345040
Global Volcanism Program (1989) Report on Poás (Costa Rica). In: McClelland L (ed) Scientific event alert network Bulletin, 14: 7. Smithsonian Institution.

https://dx.doi.org/10.5479/si.GVP.SEAN198907-345040
Global Volcanism Program (1989) Report on Poás (Costa Rica). In: McClelland L (ed) Scientific Event Alert Network Bulletin, 14:8. Smithsonian Institution. https://dx.doi.org/10.5479/si.GVP.SEAN198908-345040
Global Volcanism Program (1989) Report on Poás (Costa Rica). In: McClelland L (ed) Scientific event alert network Bulletin, 14: 9. Smithsonian Institution. https://dx.doi.org/10.5479/si.GVP.SEAN198909-345040
Global Volcanism Program (1989) Report on Poás (Costa Rica). In: McClelland L (ed) Scientific event alert network Bulletin, 14: 10. Smithsonian Institution. https://dx.doi.org/10.5479/si.GVP.SEAN198910-345040
Global Volcanism Program (1989) Report on Vulcano (Italy). In: McClelland L (ed) Scientific event alert network Bulletin, 14: 10. Smithsonian Institution. https://dx.doi.org/10.5479/si.GVP.SEAN198910-211050
Global Volcanism Program (1990) Report on Poás (Costa Rica). In: McClelland L (ed) Bulletin of the global volcanism network, 15: 1. Smithsonian Institution. https://dx.doi.org/10.5479/si.GVP.BGVN199001-345040
Global Volcanism Program (1990) Report on Poás (Costa Rica). In: McClelland L (ed) Bulletin of the global volcanism network, 15: 2. Smithsonian Institution. https://dx.doi.org/10.5479/si.GVP.BGVN199002-345040
Global Volcanism Program (1990) Report on Poás (Costa Rica). In: McClelland L (ed) Bulletin of the global volcanism network, 15: 4. Smithsonian Institution. https://dx.doi.org/10.5479/si.GVP.BGVN199004-345040
Global Volcanism Program (1990) Report on Poás (Costa Rica). In: McClelland L (ed) Bulletin of the global volcanism network, 15: 5. Smithsonian Institution. https://dx.doi.org/10.5479/si.GVP.BGVN199005-345040
Global Volcanism Program (1990) Report on Poás (Costa Rica). In: McClelland L (ed) Bulletin of the global volcanism network, 15: 6. Smithsonian Institution. https://dx.doi.org/10.5479/si.GVP.BGVN199006-345040
Global Volcanism Program (1990) Report on Poás (Costa Rica). In: McClelland L (ed) Bulletin of the global volcanism network, 15: 7. Smithsonian Institution. https://dx.doi.org/10.5479/si.GVP.BGVN199007-345040
Global Volcanism Program (1990) Report on Poás (Costa Rica). In: McClelland L (ed) Bulletin of the global volcanism network, 15: 8. Smithsonian Institution. https://dx.doi.org/10.5479/si.GVP.BGVN199008-345040
Global Volcanism Program, 1990. Report on Emmons Lake (United States). In: McClelland L (ed) Bulletin of the global volcanism network, 15: 9. Smithsonian Institution. https://dx.doi.org/10.5479/si.GVP.BGVN199009-312020
Global Volcanism Program (1990) Report on Galeras (Colombia). In: McClelland L (ed) Bulletin of the global volcanism network, 15: 9. Smithsonian Institution. https://dx.doi.org/10.5479/si.GVP.BGVN199009-351080
Global Volcanism Program, 1990. Report on Poás (Costa Rica). In: McClelland L (ed) Bulletin of the global volcanism network, 15: 9. Smithsonian Institution. https://dx.doi.org/10.5479/si.GVP.BGVN199009-345040
Global Volcanism Program, 1990. Report on Poás (Costa Rica). In: McClelland L (ed) Bulletin of the global volcanism network, 15: 10. Smithsonian Institution. https://dx.doi.org/10.5479/si.GVP.BGVN199010-345040
Global Volcanism Program (1990) Report on Poás (Costa Rica). In: McClelland L (ed) Bulletin of the global volcanism network, 15: 11. Smithsonian Institution. https://dx.doi.org/10.5479/si.GVP.BGVN199011-345040
Global Volcanism Program (1991) Report on Poás (Costa Rica). In: McClelland L (ed) Bulletin of the global volcanism network, 16: 1. Smithsonian Institution. https://dx.doi.org/10.5479/si.GVP.BGVN199101-345040
Global Volcanism Program (1991) Report on Poás (Costa Rica). In: McClelland L (ed) Bulletin of the global volcanism network, 16: 3. Smithsonian Institution. https://dx.doi.org/10.5479/si.GVP.BGVN199103-345040
Global Volcanism Program (1991) Report on Poás (Costa Rica). In: McClelland L (ed) Bulletin of the global volcanism network, 16: 4, Smithsonian Institution. https://dx.doi.org/10.5479/si.GVP.BGVN199104-345040
Global Volcanism Program (1991) Report on Stromboli (Italy). In: McClelland L (ed) Bulletin of the global volcanism network, 16: 9. Smithsonian Institution. https://dx.doi.org/10.5479/si.GVP.BGVN199109-211040
Global Volcanism Program (1991) Report on Poás (Costa Rica). In: McClelland L (ed) Bulletin of the global volcanism network, 16: 10. Smithsonian Institution. https://dx.doi.org/10.5479/si.GVP.BGVN199110-345040
Global Volcanism Program, 1991. Report on Poás (Costa Rica). In: McClelland L (ed) Bulletin of the global volcanism network, 16: 11. Smithsonian Institution. https://dx.doi.org/10.5479/si.GVP.BGVN199111-345040
Global Volcanism Program (1992) Report on Rincon de la Vieja (Costa Rica). In: McClelland L (ed) Bulletin of the global volcanism network, 17: 5, Smithsonian Institution. https://dx.doi.org/10.5479/si.GVP.BGVN199205-345020
Global Volcanism Program, 1992. Report on El Chichon (Mexico). In: McClelland L (ed) Bulletin of the global volcanism network 17: 6, Smithsonian Institution.

https://dx.doi.org/10.5479/si.GVP.BGVN199206-341120
Global Volcanism Program (1992) Report on Poás (Costa Rica). In: McClelland L (ed) Bulletin of the global volcanism network, 17: 6, Smithsonian Institution. https://dx.doi.org/10.5479/si.GVP.BGVN199206-345040
Global Volcanism Program (1992) Report on Copahue (Chile-Argentina). In: McClelland L (ed) Bulletin of the global volcanism network, 17: 7. Smithsonian Institution. https://dx.doi.org/10.5479/si.GVP.BGVN199207-357090
Global Volcanism Program (1992) Report on Poás (Costa Rica). In: McClelland L (ed) Bulletin of the global volcanism network, 17: 8, Smithsonian Institution. https://dx.doi.org/10.5479/si.GVP.BGVN199208-345040
Global Volcanism Program, 1992. Report on Poás (Costa Rica). In: McClelland L (ed) Bulletin of the global volcanism network, 17: 9, Smithsonian Institution. https://dx.doi.org/10.5479/si.GVP.BGVN199209-345040
Global Volcanism Program (1992) Report on Rincon de la Vieja (Costa Rica). In: McClelland L (ed) Bulletin of the global volcanism network, 17: 9, Smithsonian Institution. https://dx.doi.org/10.5479/si.GVP.BGVN199209-345020
Global Volcanism Program (1992) Report on Poás (Costa Rica). In: McClelland L (ed) Bulletin of the global volcanism network, 17: 11, Smithsonian Institution. https://dx.doi.org/10.5479/si.GVP.BGVN199211-345040
Global Volcanism Program (1993) Report on Poás (Costa Rica). In: McClelland L (ed) Bulletin of the global volcanism network, 18: 1, Smithsonian Institution. https://dx.doi.org/10.5479/si.GVP.BGVN199301-345040
Global Volcanism Program (1993) Report on Poás (Costa Rica). In: Venzke E (ed) Bulletin of the global volcanism network, 18: 2, Smithsonian Institution. https://dx.doi.org/10.5479/si.GVP.BGVN199302-345040
Global Volcanism Program (1993) Report on concepcion (Nicaragua). In: Venzke E (ed) Bulletin of the global volcanism network, 18: 3, Smithsonian Institution. https://dx.doi.org/10.5479/si.GVP.BGVN199303-344120
Global Volcanism Program (1993) Report on Poás (Costa Rica). In: Venzke E (ed) Bulletin of the global volcanism network, 18: 3, Smithsonian Institution. https://dx.doi.org/10.5479/si.GVP.BGVN199303-345040
Global Volcanism Program (1993) Report on Poás (Costa Rica). In: Venzke E (ed) Bulletin of the global volcanism network, 18: 4, Smithsonian Institution. https://dx.doi.org/10.5479/si.GVP.BGVN199304-345040
Global Volcanism Program (1994) Report on Poás (Costa Rica). In: Wunderman R (ed) Bulletin of the global volcanism network, 19: 1, Smithsonian Institution. https://dx.doi.org/10.5479/si.GVP.BGVN199401-34504
Global Volcanism Program (1994) Report on Rincon de la Vieja (Costa Rica). In: Wunderman R (ed) Bulletin of the global volcanism network, 19: 1, Smithsonian Institution. https://dx.doi.org/10.5479/si.GVP.BGVN199401-345020
Global Volcanism Program (1994) Report on Morne Plat Pays (Dominica). In: Wunderman R (ed) Bulletin of the global volcanism network, 19: 5, Smithsonian Institution. https://dx.doi.org/10.5479/si.GVP.BGVN199405-360110
Global Volcanism Program (1994) Report on Poás (Costa Rica). In: Wunderman R (ed) Bulletin of the global volcanism network 19: 7, Smithsonian Institution. https://dx.doi.org/10.5479/si.GVP.BGVN199407-345040
Global Volcanism Program (1994) Report on Poás (Costa Rica). In: Venzke E (ed) Bulletin of the global volcanism network, 19: 9, Smithsonian Institution. https://dx.doi.org/10.5479/si.GVP.BGVN199409-345040
Global Volcanism Program, 1995. Report on Poás (Costa Rica). In: Wunderman R (ed) Bulletin of the global volcanism network, 20: 3, Smithsonian Institution. https://dx.doi.org/10.5479/si.GVP.BGVN199503-345040
Global Volcanism Program (1995) Report on Rincon de la Vieja (Costa Rica). In: Wunderman R (ed) Bulletin of the global volcanism network, 20: 4, Smithsonian Institution. https://dx.doi.org/10.5479/si.GVP.BGVN199504-345020
Global Volcanism Program (1995) Report on Vulcano (Italy). In: Wunderman R (ed) Bulletin of the global volcanism network, 20: 4, Smithsonian Institution. https://dx.doi.org/10.5479/si.GVP.BGVN199504-211050
Global Volcanism Program (1995) Report on Kelimutu (Indonesia). In: Wunderman R (ed) Bulletin of the global volcanism network, 20: 6, Smithsonian Institution. https://dx.doi.org/10.5479/si.GVP.BGVN199506-264140
Global Volcanism Program (1995) Report on Poás (Costa Rica). In: Wunderman R (ed) Bulletin of the global volcanism network, 20: 7, Smithsonian Institution. https://dx.doi.org/10.5479/si.GVP.BGVN199507-345040
Global Volcanism Program (1996) Report on Poás (Costa Rica). In: Wunderman R (ed) Bulletin of the Global Volcanism Network, 21: 5, Smithsonian Institution. https://dx.doi.org/10.5479/si.GVP.BGVN199605-345040
Global Volcanism Program (1997) Report on Momotombo (Nicaragua). In: Wunderman R (ed) Bulletin of the global volcanism network, 22: 7, Smithsonian Institution. https://dx.doi.org/10.5479/si.GVP.BGVN199707-344090
Global Volcanism Program (1998) Report on Poás (Costa Rica). In: Wunderman R (ed) Bulletin of the global volcanism network, 23: 3, Smithsonian Institution.

https://dx.doi.org/10.5479/si.GVP.BGVN199803-345040
Global Volcanism Program (1998) Report on White Island (New Zealand). In: Wunderman R (ed) Bulletin of the global volcanism network, 23: 10, Smithsonian Institution. https://dx.doi.org/10.5479/si.GVP.BGVN199810-241040
Global Volcanism Program (1998) Report on White Island (New Zealand). In: Wunderman R (ed) Bulletin of the global volcanism network, 23: 11, Smithsonian Institution. https://dx.doi.org/10.5479/si.GVP.BGVN199811-241040
Global Volcanism Program (1999) Report on Moyoro-dake [Medvezhia] (Japan—administered by Russia). In: Wunderman R (ed) Bulletin of the Global Volcanism Network, 24: 10, Smithsonian Institution. https://dx.doi.org/10.5479/si.GVP.BGVN199910-290100
Global Volcanism Program (2001) Report on Santa Ana (El Salvador). In: Wunderman R (ed) Bulletin of the global volcanism network, 26: 4. Smithsonian Institution. https://dx.doi.org/10.5479/si.GVP.BGVN200104-343020
Global Volcanism Program (2004) Report on NW Rota-1 (United States). In: Wunderman R (ed) Bulletin of the Global Volcanism Network, 29: 3, Smithsonian Institution. https://dx.doi.org/10.5479/si.GVP.BGVN200403-284211
Global Volcanism Program (2007) Report on Turrialba (Costa Rica). In: Wunderman R (ed) Bulletin of the global volcanism network, 32: 8, Smithsonian Institution. https://dx.doi.org/10.5479/si.GVP.BGVN200708-345070
Global Volcanism Program (2013) Report on Copahue (Chile-Argentina). In: Wunderman R (ed) Bulletin of the global volcanism network, 38: 9, Smithsonian Institution. https://dx.doi.org/10.5479/si.GVP.BGVN201309-357090

Red Sismológica Nacional

Red Sismológica Nacional (UCR-ICE) May (1988) Boletín Sismológico y Volcanológico
Red Sismológica Nacional (UCR-ICE) May (1989) Boletín Sismológico y Volcanológico
Red Sismológica Nacional (UCR-ICE) March (1990) Boletín Sismológico y Volcanológico
Red Sismológica Nacional (UCR-ICE) Dec (1991) Boletín Sismológico y Volcanológico
Red Sismológica Nacional (UCR-ICE) Apr (1992) Boletín Sismológico y Volcanológico
Red Sismológica Nacional (UCR-ICE) May (1993) Boletín Sismológico y Volcanológico
Red Sismológica Nacional (UCR-ICE) July (1993) Boletín Sismológico y Volcanológico
Red Sismológica Nacional (UCR-ICE) Sept (1993) Boletín Sismológico y Volcanológico
Red Sismológica Nacional (UCR-ICE) Oct (1993) Boletín Sismológico y Volcanológico
Red Sismológica Nacional (UCR-ICE) Nov (1993) Boletín Sismológico y Volcanológico
Red Sismológica Nacional (UCR-ICE) May (1994) Boletín Sismológico y Volcanológico
Red Sismológica Nacional (UCR-ICE) May (1995) Boletín Sismológico y Volcanológico

Coseismic Landslide Susceptibility Analysis Using LiDAR Data PGA Attenuation and GIS: The Case of Poás Volcano, Costa Rica, Central America

Paulo Ruiz, Michael J. Carr, Guillermo E. Alvarado, Gerardo J. Soto, Sara Mana, Mark D. Feigenson and Luis F. Sáenz

Abstract

A landslide susceptibility model for Poás volcano was created in response to the most recent event that triggered landslides in the area (the M_w 6.2 Cinchona earthquake, which occurred on the 8th of January, 2009). This earthquake was the sixth event related to destructive landslides in the last 250 yr in this area and it severely affected important infrastructures. This chapter refers to a study, which consisted of three phases, as follows: (1) creation of a post-Cinchona earthquake landslide catalog, which was done manually based on a set of high resolution orthophotos and LiDAR data and it includes 4,846 landslides; (2) a landslide susceptibility model, based on the Mora-Vahrson-Mora method, the data from our landslide inventory, and a new modeling of earthquake triggering indicators based on the attenuation of the peak ground acceleration of the event, and (3) an evaluation of the methodology used, which for the Cinchona case resulted in an overlap of the actual landslides and the higher susceptibility zones of ~97%. Based on our new methodology, four landslide susceptibility models were simulated: the Cinchona earthquake, the M_w 5.5 Sarchí earthquake 1912, and two hypothetical earthquakes: one on the Angel fault (M_w 6.0) and the other one on the San Miguel fault (M_w 7.0). The Toro and Sarapiquí river canyons, the non-vegetated corridor located west from the main crater of Poás and the areas where the La Paz Andesites Unit are located are always the zones with the highest susceptibility to slide values. Meanwhile, the northern part of the study area, where the Río Cuarto Lavas unit outcrops, always presented the lowest susceptibility values due to both the low slope angles and weathering level of its rocks.

Keywords

Landslide susceptibility · Coseismic landslides · Cinchona earthquake · Costa Rica · Poás volcano

P. Ruiz (✉)
Laboratorio de Materiales y Modelos Estructurales de la Universidad de Costa Rica (LANAMME-UCR), San Pedro, Costa Rica
e-mail: paulo.ruizcubillo@ucr.ac.cr

M. J. Carr · M. D. Feigenson
Department of Earth & Planetary Sciences, Rutgers University, New Brunswick, NJ, USA

G. E. Alvarado · L. F. Sáenz
Instituto Costarricense de Electricidad (ICE), San Jose, Costa Rica

G. J. Soto
Terra Cognita Consultants S.A, San Jose, Costa Rica

S. Mana
Salem State University, Salem, USA

F. Tassi et al. (eds.), *Poás Volcano*, Active Volcanoes of the World,
https://doi.org/10.1007/978-3-319-02156-0_4

1 Introduction

Volcanic edifices in Central America are exposed to high mean annual precipitation (3,000–6,000 mm/yr), intense weathering, hydrothermal alteration, elevated erosion rates, and fluctuating temperatures (18–30 °C), typical of tropical environments. These factors make them extremely susceptible to mass wasting events, which have produced damage and high death tolls in the past. The most common triggers of landslides in the area are: heavy rain (e.g., Casitas volcano Nicaragua by the Hurricane Mitch in 1998, described by Kerle and van Wyk de Vries 2001), earthquakes $M_w > 5.5$ (e.g., Guatemala 1976; El Salvador 2001; Costa Rica 2009: Bommer and Rodríguez 2002, Evans and Bent 2004, and this chapter) or a combination of both. Moreover, the region is known to have a disproportionally high number (at least an order of magnitude) of landslides triggered by earthquakes when compared with other regions worldwide (Keefer 1984; Rodríguez et al. 1999; Bommer and Rodríguez 2002 and this chapter). Central America is a highly populated area with ~42 million people, and many of the major cities (e.g., Guatemala City, Tegucigalpa, San Salvador, Managua and San José) are located near active volcanoes and/or active seismic zones (Small and Naumann 2001, Fig. 1a).

Since a large number of inhabitants live near areas at risk of coseismic landslides, the study of this phenomenon based on high quality data is important to facilitate effective hazard mitigation strategies. This chapter focuses on the coseismic landslides generated during the M_w 6.2 earthquake of January 8, 2009 in Costa Rica, called the Cinchona earthquake (Fig. 1b, c). This is an excellent opportunity to study in detail one of the most recent coseismic landslide events in the region, due to the collection of a set of high quality airborne light detection and ranging (LiDAR) data and orthophotos, obtained just three months after the earthquake, together with field work immediately after the event. Our work is pioneering in the region and in the application of this technology to study coseismic landslides on the flanks of an active volcano.

Previous works on the coseismic landslides triggered by the Cinchona earthquake include: (a) an aerial mapping of the landslides that occurred within the walls of the main active crater of Poás volcano (GVN 2009), (b) the geologic description of the area affected by landslides on route 126 on the eastern side of Poás volcano (Méndez et al. 2009), (c) the effect of the earthquake on the route 126 slopes near Poás volcano based on a geotechnical approach (Laporte 2009a, b), and (d) a sedimentological account of the debris flows related to this seismic event, including a preliminary map of the landslides in the Poás massif (Alvarado 2010).

Due to the location of several hydropower projects around the area that account for 7.2% of the total electricity production of the country, the Costa Rican Institute for Electricity (ICE) performed several seismological and earthquake hazard studies before the occurrence of the Cinchona event. However, no previous studies taking into account the possibility of coseismic landslides occurring in the area were carried out. After the effects of the Cinchona earthquake (see below), the necessity of a study of this kind became evident.

The civil structures related to the hydroelectric projects that are not underground and can be affected by coseismic landslides in the study area include: dams and water reservoirs, power plants, surge tanks, water pipes and power lines. The other civil structure of importance located on Poás volcano is the national route 126. It is one of the routes that connects the Central Valley with the Caribbean region and is located in one of the major tourist attractions of the country.

The main objective of this chapter is to analyze the susceptibility to slide of the Poás area. By correlating the main factors that determine the occurrence of landslides and our observations from the events generated by the 2009 earthquake. The principal factors that influenced the generation of landslides studied here are: lithology, weathering degree, slope angle, precipitation rates, ground moisture, and attenuation of the peak ground acceleration. The correlations found here, together with the historical data from past

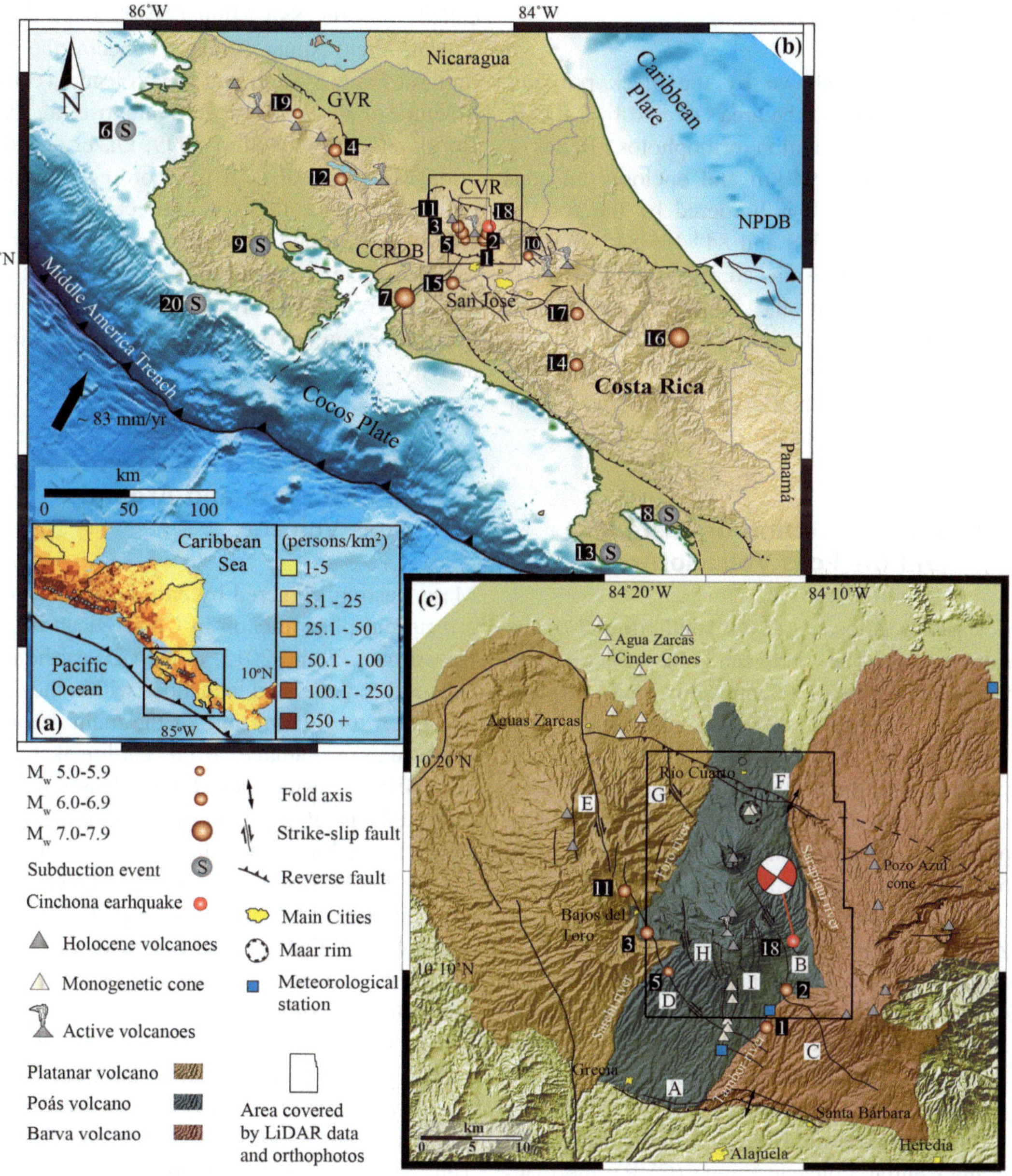

Fig. 1 **a** Central American Volcanic arc and its population distribution (CIESIN 2005). **b** Digital elevation map of Costa Rica, volcanic fronts and its tectonic setting. Location of earthquakes associated with coseismic landslides in the last 250 yr, numbers 1–20 correspond to event code from Table 2. The black rectangle denotes a close-up to western Central Volcanic Range and the study area (see Fig. 1c). Bathymetry is from Ranero et al. (2005). **c** Cinchona earthquake location with Harvard (MIT) Body-Wave moment tensor solution, the letters A-I correspond to the fault code in Table 1. North Panamá Deformed Belt (NPDB), Central Costa Rica Deformed Belt (CCRDB)

earthquakes and the most recent neotectonic studies near Poás volcano (Montero et al. 2010), are applied to estimate areas and volumes affected in past events and to determine which localities would be the most affected by landslides in the case of future earthquakes.

1.1 Study Area

The study area, shown by the black polygon (Fig. 1c), represents the area covered by LiDAR data and high-resolution orthophotos. It has a total perimeter of 94 km and encloses an area (A_{ts}) of 519 km^2. It is bordered by the coordinates 10° 7′–10° 21′N and 84° 18′–84° 07′W. It covers the mesoseismic area where most landslides triggered by the Cinchona earthquake occurred, including the entire north flank and part of the southern flank of the Poás volcano, the northwest slope of the Barva volcano and part of the northeast flank of the Platanar volcano.

1.2 Tectonic Setting of Costa Rica and the Poás Volcano Area

Costa Rica is located at the southern end of the subduction zone between the Cocos and Caribbean plates (Vannucchi and Mason, Chapter "Overview of the Tectonics and Geodynamics of Costa Rica"). The former is subducting to the northeast underneath the latter at a rate of ~83 mm-yr^{-1} in the northwestern part of Costa Rica (DeMets et al. 2010, Fig. 1a, b). The majority of earthquakes in Costa Rica are associated with subduction, the Panama Fracture Zone (PFZ), the North Panama Deformed Belt (NPDB) and crustal faults located in the interior of the country (Fig. 1b). The latter are historically responsible for generating heavy damages and induced landslides (Mora and Mora 1994; Climent et al. 2009).

Platanar, Poás and Barva volcanoes are located in the Central Volcanic Range (CVR), the magmatic products of the subduction, which is partly cut by a belt of neotectonic faults named Central Costa Rica Deformed Belt (CCRDB) (Fig. 1b, c), defined and described by Marshall et al. (2000), Montero (2001) and Montero et al. (2010). In this area, most of the tectonic structures are right lateral faults with a northwest strike and left lateral faults with northeast strike. Both systems present vertical components as well.

The most important tectonic structures near Poás volcano are the faults mapped in Fig. 1c, called Alajuela, San Miguel, Carbonera, Ángel, Volcán Viejo-Aguas Zarcas, Venecia, the Poás Summit Fault System and the Volcano-Tectonic Fracture of Poás (VTFP) (Alvarado et al. 1988; Borgia et al. 1990; Soto 1999; Gazel and Ruiz 2005; Montero et al. 2010). According to Montero et al. (2010), the faults that affect Poás volcano could be controlled by an interaction between the regional tectonic stress and volcanic processes (Fig. 1b, c). In Table 1, we present a summary of the main features of these faults and their historical activity.

1.3 Historical Coseismic Landslides in Costa Rica and Around Poás Volcano

Mass wasting triggered by earthquakes has been a recurrent phenomenon in Costa Rican history. Between 1772 and 2012, at least 20 earthquakes ($M_w > 5.5$) generated landslides within the Costa Rican territory, being one of the main causes of infrastructure damage from earthquakes and, since 1950, causing more human losses than the direct effects of the earthquake shaking (Mora 1985, 1989; Mora and Mora 1994; Peraldo and Rojas 2000; Bommer and Rodríguez 2002; Climent et al. 2009; and this chapter, Fig. 1b, c, and Table 2). Only five of these 20 earthquakes that have induced landslides in Costa Rica were originated by subduction, while the remaining events were generated by upper crustal faults resulting into a ratio similar to the one obtained for all the Central America region by Bommer and Rodríguez (2002) (Fig. 1b, c and Table 2). Ten of these shallow (depth < 23 km) earthquakes occurred near volcanic centers, three in the Guanacaste Volcanic Range (GVR), affecting Miravalles and Tenorio volcanoes, and seven in the CVR, generating landslides on the Platanar, Poás, Barva and Irazú volcanoes (Fig. 1b). Historically, the volcanic massif of Poás volcano has been the most affected by coseismic landslides, since four of these landslides generating earthquakes occurred within the edifice of the volcano and two nearby. The Cinchona earthquake is the most recent case (Fig. 1c).

Table 1 Summary of the main features of tectonic faults and their historical activity around the Poás volcano

Fault code (*)	Fault name	Location	Fault type	Strike angle	Length (km)	Geomorphology expression	Historical and recent seismic activity	References
A	Alajuela	S flank of Poás volcano, from Grecia to Santa Bárbara	Thrust propagation fault bend fold	E-W and WNW	~20	Anticline and synclinal folds in the front scarp, tectonic valleys	1772 earthquake?	(b and e)
B	Ángel	E and NE flanks of Poás volcano, W flank of Barva volcano	Dextral-small normal component	N25W, N70W, N30W, N-S, N50W	>20	Right lateral displacement of drainage divide (Angel River), slopes changes, linear valleys, sag ponds, drainage offset	Fraijanes Earthquake 1888, Cinchona Earthquake 2009	(a, c and e)
C	Sabanilla	W and SW flanks of Poás volcano, from NW of Bajos de Toro to N of Alajuela	Strike-slip fault (dextral)	NW	>25	Right lateral displacement of drainage systems, linear valleys, fault berms	Toro Amarillo Earthquake 1911, Sarchí Earthquake 1912	(e)
D	Carbonera	W flank of Poás volcano	Strike-slip fault (dextral)	NNW	>6.5	Fault scarps, right lateral displacement of streams	Seismic swarm, March 1990, January 1997	(a and e)
E	Volcán Viejo-Aguas Zarcas	NE flank of Platanar-Viejo volcano	Strike-slip fault (dextral)	N10W	>25	Linear valley in the Aguas Zarcas river	Bajos del Toro Earthquake 1955, seismic swarm April-May 1998	(a and e)
F	San Miguel	N flank of Poás volcano	Bifurcated reverse propagation fault bend fold	N70W	~15	Straight fault scarp related with an anticline fold in the front and a synclinal fold in the back	Potentially active (CO_2 and CH_4 emissions and Sarapiqui and Toro river displacements	(b, c and e)
G	Venecia	SW of Platanar-Viejo volcano	Strike-slip fault (dextral)	N20W	~8	Alignment NW	Seismic swarm September 1989?	(a)

(continued)

Table 1 (continued)

Fault code (*)	Fault name	Location	Fault type	Strike angle	Length (km)	Geomorphology expression	Historical and recent seismic activity	References
H	Poás summit faults system	E and W side of Main and Botos craters on Poás volcano	Normal arched	N-S, NNW-SSE	~2–3	Faceted scarps, seudo-caldera, volcanic graven, rivers and streams are parallel to these faults	Seismic activity after Cinchona earthquake?	(a, c and e)
I	Volcano tectonic fracture of Poás	S flank of Poás volcano, near Sabana Redonda	Normal	N-S	~5	Depression (graben) N-S limited in the E and W by linear scarps	Seismic activity after Cinchona earthquake?	(d and e)

(*) Fault code corresponds to letter code used in (Figs. 1c and 3a). References: (a) Alvarado et al. (1988), (b) Borgia et al. (1990), (c) Soto (1999), (d) Gazel and Ruiz (2005) and (e) Montero et al. (2010)

Table 2 Earthquakes in Costa Rica resulting in coseismic landslides in the last 250 yr

Event code (*)	Earthquake name/location	Date	Epicenter		Depth	Magnitude	I_o	Death toll	Ref.
			Lat N	Lon W	(km)	Mw			
1(Po)	Barva	15-Feb-1772	10.12	−84.2	10	6	VIII	–	f, h, j, k
2(Po)	Fraijanes	30-Dec-1888	10.15	−84.18	15	6	VIII	5	a, j, k
3(Po)	Bajos del Toro	29-Aug-1911	10.2	−84.3	10	6	VIII	3	a, i, j, k
4	Guatuso	10-Oct-1911	10.6	−84.92	10	6.5	VIII	6	i, j
5(Po)	Sarchí	12-Jun-1912	10.17	−84.28	18	5.5	VII	–	a, i, j, k
6(S)	Papagayo	27-Feb-16	10.7	−85.99	50	7.3	IX	–	i, j
7(Po)	Orotina	04-Mar-24	9.83	−84.56	15	7	IX	–	h, i, j
8(S)	Golfito	5-Dec-1941	8.7	−83.2	28	7.4	IX	–	j, k
9(S)	Samara	05-Oct-1950	10.1	−85.3	55	7.8	VIII	–	j, k
10	Patillos	30-Dec-1952	10.05	−83.92	10	6.2	VIII	21	g, j, k
11(Po)	Bajos del Toro	01-Set-1955	10.23	−84.32	3	6.1	VIII	10	a, j, k, l
12	Tilaran	14-Apr-1973	10.45	−84.9	10	6.5	IX	23	f, j, k
13(S)	Golfito	3-Apr-1983	8.5	−83.5	34	7.3	VIII	–	h, j, k
14	División-Buvis	03-Jul-1983	9.49	−83.67	14	6.2	VIII	–	b, j, k
15	Piedras Negras	May–Dec 1990	9.91	−84.31	4	6	VIII	–	c, j, k
16	Limón	22-Apr-1991	9.63	−83.15	23	7.7	X	–	e, j, k
17	Pejibaye swarm	1013-Jul-1993	9.75	−83.67	14	6	VIII	–	d, j, k
18(Po)	Cinchona	8-Jan-09	10.19	−84.18	4.6	6.2	IX	30	h, k, l
19	Upala	July-2011	10.79	−85.11	10	5.5	VI	–	This study
20	Sámara	Sept-2012	9.69	−85.62	15.4	7.6	VII		This study

(*) Event code corresponds to the letter code used in (Figs. 1b, c), (S): subduction event, (Po): event that produced coseismic landslides near or on the Poás volcano. I_0 epicentral modified Mercalli Intensity. References: (a) Alvarado et al. (1988), (b) Boschini et al. (1988), (c) Montero et al. (1991), (d) Barquero and Peraldo (1993), (e) Mora and Mora (1994), (f) Peraldo and Montero (1994), (g) Montero and Alvarado (1995), (h) Montero (1999), (i) Peraldo and Montero (1999) (1, 7), (j) Bommer and Rodríguez (2002), (k) Climent et al. (2009) and (l) Montero et al. (2010)

1.4 The Cinchona Earthquake and Its Impact

The Cinchona earthquake occurred on January 8, 2009 at 13:21:34 local time (UTC-6 h) with a magnitude of M_w 6.2 (Fernández-Arce and Mora Amador, Chapter "Seismicity of Poás Volcano, Costa Rica"). It was located only 6.5 km east from the active main crater of Poás volcano, with a hypocentral depth of 4.5 km (RSN: ICE-UCR 2009, Fig. 1c.). The maximum intensity reported was IX (Modified Mercalli Intensity scale, MMI) near the epicenter (Fig. 2). However, intensities of VIII-VII were dominant in the mesoseismic area (Fig. 2). The peak ground acceleration (PGA) measured was (0.67 g) at a hypocentral distance of 15 km (Fig. 2b). However, field observations suggested that values close to 1 g were reached in places near the epicenter (RSN: ICE UCR 2009, Rojas et al. 2009; Climent and Moya 2009).

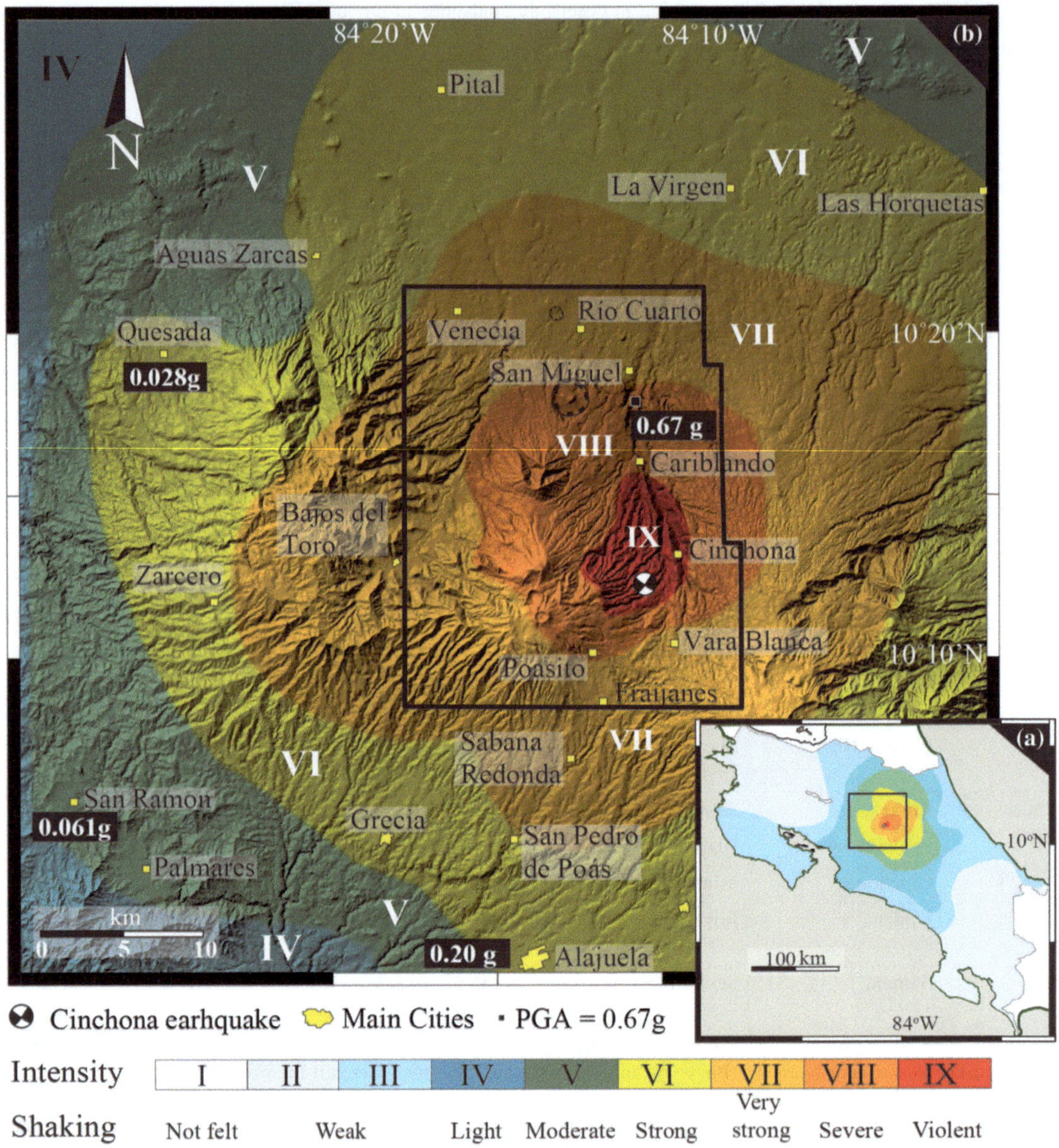

Fig. 2 **a** Modified Mercalli Intensity map from the Cinchona earthquake based on RSN: ICE-UCR (2009), Climent and Moya (2009) and Montero et al. (2010). Peak ground accelerations (PGA) in the horizontal axis measured in the nearest stations to the epicenter based on Climent and Moya (2009)

Based on the location and focal mechanism (Fig. 1c), Rojas et al. (2009) determined that the origin of the event was the rupture of 72 km^2 of the Ángel dextral strike-slip fault. No superficial rupture was found near the fault trace. However, this earthquake triggered thousands of landslides all around Poás, Barva and Platanar volcanoes.

The effects caused by the coseismic landslides included environmental damage (mainly destruction of primary tropical rainforest, soil erosion and silting of rivers), damage to housing, public services, at least two hydro-electrical power plants (Cariblanco and Ángel), road network (specially on route 126), agriculture, dairy cattle, tourism and casualties (29 out of 30 were caused by landslides) (Méndez et al. 2009 and this chapter). The total losses in infrastructure from the earthquake were >US$ 100 million (RSN: ICE-UCR 2009).

1.5 Geology and Geomorphology Setting of the Affected Region

In this section, the main features of the three volcanic edifices (Poás, Barva and Platanar) that were affected by the Cinchona coseismic landslides are summarized. In Fig. 3a we present a geological map for the study area with the location of each coseismic landslide studied for our catalog. In Table 3 the most important geologic and geomorphologic features of each volcanic unit that was affected by the coseismic landslides are summarized. More stratigraphic details, geochemical and geochronological data of these volcanoes and their units can be found in Prosser and Carr (1987), Alvarado and Carr (1993), Soto (1999), Gazel and Ruiz (2005), Alvarado et al. (2009), Ruiz et al. (2010), Ruiz et al. (Chapter "Geochemical and Geochronological Characterisation of the Poas Stratovolcano Stratigraphy").

1.5.1 Platanar Volcano

Platanar volcano (dormant) is located in the northwestern end of the CVR, about 15 km northwest of the active crater of Poás volcano (Fig. 1c). It is formed by at least one eroded caldera (Chocosuela) infilled by two young (<0.4 Ma) volcanic centers: Platanar and Porvenir volcanoes (Alvarado 2009). Alvarado and Carr (1993) recognized for these volcanoes eight volcanic units, and their products correspond to basaltic and andesitic lavas, pyroclastic flows, volcanic breccias, lahars and volcanic alluvium. In addition, on the northern flank of the Platanar volcano, a group of nine Quaternary monogenetic cones called Aguas Zarcas Cinder Cones are located along a NNW trend extending ~10 km (Fig. 1c).

1.5.2 Barva Volcano

The summit of this dormant andesitic composite complex is located 10 km southeast from the active crater of the Poás volcano (Fig. 1c). It has at least a dozen eruptive centers on its summit and several satellite cones on its northern and southern flanks (Soto 1999; Alvarado 2009). Similar to Poás, and other volcanoes from the CVR, Barva was subdivided into two extensive units that define the last two temporal phases: Paleo (800–240 ka) and Neo-Barva (<240 ka) (Soto 1999). The products from the Paleo-Barva temporal phase are grouped into a volcanic unit called the Paleo-Barva Unit, while the products from the Neo-Barva temporal stage are divided in two units: Neo-Barva and Pozo Azul. Segments of these three units are located within the study area and were affected by coseismic landslides (Fig. 3). More volcanic units were mapped on Barva (i.e. Arredondo and Soto 2006), but here we are focusing only on those located within the study area.

1.5.3 Poás Volcano

Poás volcano is one of the five currently active volcanoes in Costa Rica (Fig. 1b) and is located between the Platanar and the Barva volcanic centers (Fig. 1c). It was formed by the stacking of volcanic rocks during at least three principal stages (Proto-Paleo-Neo), occurring over almost one million years (Soto 1994, 1999; Ruiz et al. 2010).

Poás volcano is bounded by the canyons of Sarapiquí and Toro rivers to the NE and to NW, respectively (Fig. 1c). These rivers present narrow (~400 m wide) and profound (~250 m high) canyons with internal slopes that vary from high (45°–60°) to extremely high (75°–89°) slopes. On the southern flank, the rivers that mark the limit of the volcano are Sarchí in the southwest and Tambor in the southeast, though their river valleys are neither as deep nor as wide as their counterparts on the northern flank (Fig. 1c).

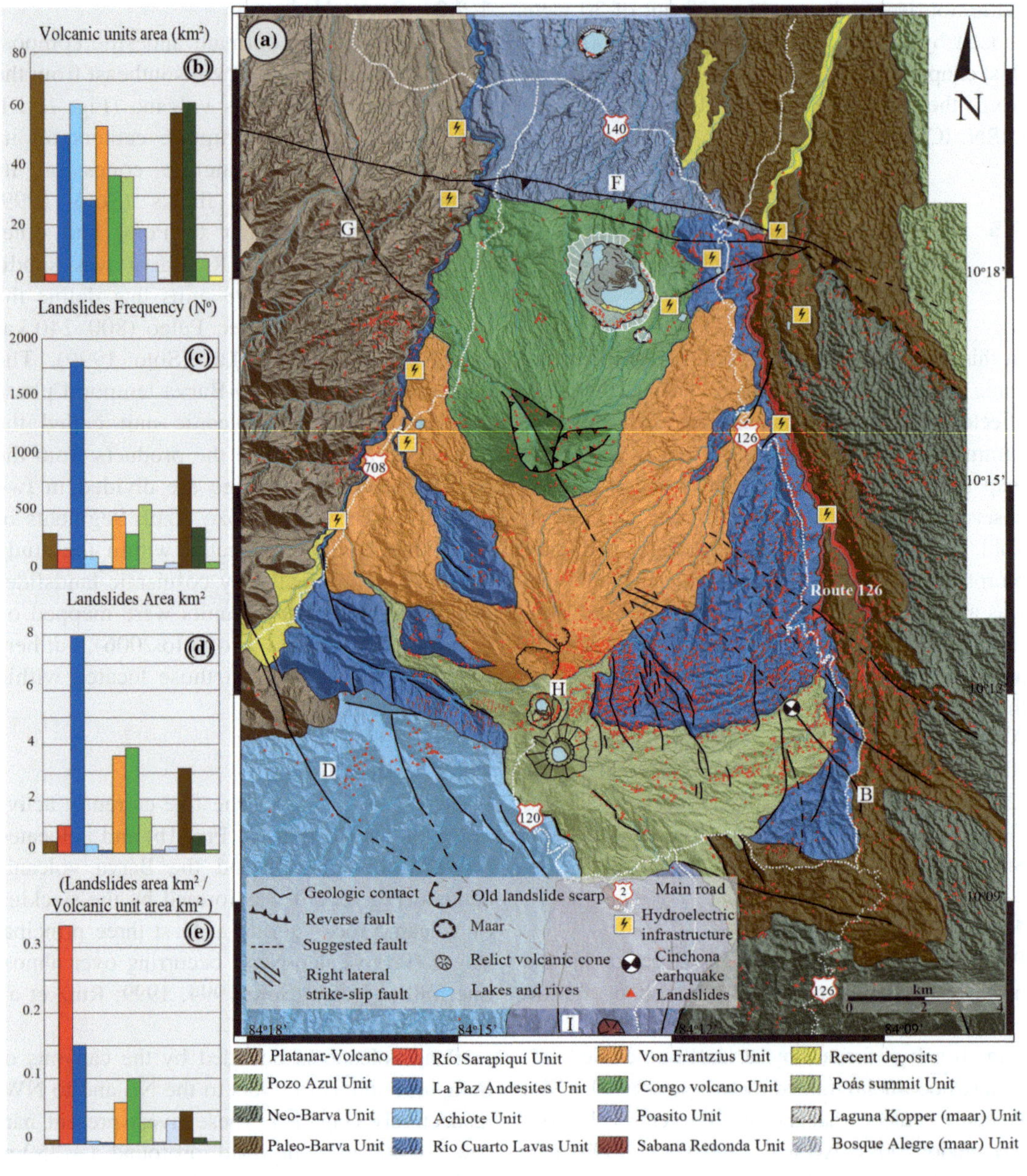

Fig. 3 **a** Geologic map of Poás volcano and location of coseismic landslides from Cinchona earthquake. Letters A-I correspond to fault code in Table 1. **b** Area in km^2 from each volcanic unit of the study area. **c** Number of coseismic landslides per volcanic unit. **d** Stripped area (km^2) from coseismic landslides per volcanic unit. **e** Stripped area from landslides per volcanic unit divided by the area of each volcanic unit

1.6 Orographic Regions and Climatic Conditions Around Poás Volcano

The territory of Costa Rica is cut by a magmatic range that divides the country into two orographic regions: the northeastern region (Caribbean slope) and the southwestern region (Pacific slope). These regions have significant differences in aspects that might play a significant role in the generation of landslides such as: mean temperature, type of vegetation cover,

Table 3 Geology summary of the volcanic units that were affected by the induced landslides from the Cinchona earthquake

Volcano	Unit	Age and temporal phase (Paleo or Neo)	Lithology	Slope angle	Geomorphology	Weathering	Drainage system	Soil type and thickness
Platanar V.	Viejo	~1 Ma (Paleo-Platanar)	Lavas, volcanic breccias and pyroclastic flows	30°–60°	Uneven slopes, deeply eroded river valleys truncated by faults	Very badly weathered	Sub-parallel	Residual volcanic (~5 m)
Barva	Paleo-Barva	0.8–0.2 Ma (Paleo-Barva)	Porphyritic lava flows, breccias and subordinated tuffs	30°–60°	Uneven slopes, deeply eroded river valleys truncated by strike-slip faults	Very badly weathered	Sub-parallel	Tuff and weather lapilli (5–25 m)
Barva	Neo-Barva	<0.26 Ma (Neo-Barva)	Lava flows, tuffs and epiclasts	10°–30°	Smooth surface given by a pumiceous flow 10 m thick	Slightly to moderately weathered	Parallel to sub-parallel	Pumiceous flow and residual soil (~5 m)
Barva	Pozo Azul	<0.2 Ma (Neo-Barva)	Lava flows	10°–30°	Shield morphology given by the lava flow and smoothed by pumice flow	Moderately weathered	Semi radial	Residual volcanic (~2 m)
Poás	Río Sarapiquí	0.6–0.7 Ma (Paleo-Poás)	Breccias and ash-flow tuffs, with epiclastic lenses and subordinate lavas	> 50°	Sarapiquí river canyon	Moderately to badly weathered	Parallel	Residual volcanic <2 m
Poás	La Paz Andesites	0.6–0.5 Ma (Paleo-Poás)	Several (at least 7) andesitic lava flows with a characteristic porphyritic texture with megaphenocrysts of plagioclase (2–3cm)	30°–60°	Uneven slopes, deeply eroded river valleys truncated by right lateral faults	Very badly weathered	Sub-parallel	Tuffs and weathered lapilli tuffs with thickness between 5 and 40 m
Poás	Achiote	0.5–0.2 Ma (Paleo-Poás)	Lava flows	30°–60°	Uneven slopes with river valleys are less truncated than La Paz Andesites U.	Moderately weathered	Parallel to sub-parallel	Material from the Lapilli Tuff Unit (see below) and/or residual soils > 2m
Poás	Río Cuarto Lavas	0.2 Ma (Paleo-Poás)	Lavas	3°–5°	Relatively flat lava field with a slight downward slope to the north	Unweathered to slightly weathered	Parallel	Residual soil and material from Laguna Río Cuarto Maar (see below) < 8m

(continued)

Table 3 (continued)

Volcano	Unit	Age and temporal phase (Paleo or Neo)	Lithology	Slope angle	Geomorphology	Weathering	Drainage system	Soil type and thickness
Poás	Von Frantzius	0.06–0.01 Ma (Neo-Poás)	Lava flows with breccias, epiclasts and pyroclasts on the top	30°–60°	Volcanic cone with smooth slopes	Unweathered to slightly weathered	Semi radial	Residual soils rarely surpass 5 m
Poás	Congo volcano	0.06–0.01 Ma (Neo-Poás)	Lavas, epiclasts and pyroclastic flows	30°–80°	Volcanic cone which does not have a well defined crater; instead it is open in two main previous landslide scarps to the NNW and NNE	Slightly to moderately weathered	Semi radial	Residual volcanic soils ~5 m
Poás	Poás Summit	0.054–present (Neo-Poás)	Lavas and pyroclasts	15°–60°	Main crater and Botos crater	Slightly to moderately weathered	Radial to semi radial	Residual soil and thick set (~10 m) of pyroclasts in the Pulga stream valley
Poás	Poasito	0.05–0.025 Ma (Neo-Poás)	Massive lava flows	10°–30°	Smooth surface	Unweathered to slightly weathered	Parallel to sub-parallel	Residual volcanic soils and material from the Poás Lapilli tuff unit >7 m
Poás	Bosque Alegre	0.0062–0.0028 Ma (Neo-Poás)	Lavas (inner cones)but mostly pyroclasts	40°–60°	Explosion crater	Moderately to badly weathered	Radial into the maar	Residual volcanic soils and pyroclasts ~20 m
Poás	Laguna Kooper	0.0003–0.0004 Ma (Neo-Poás)	Pyroclastic material	~60°	Explosion crater	Moderately to badly weathered	Radial into the maar	Residual soils and pyroclastic material <15 m
Poás	Poás Lapilli tuff	<0.04 Ma (Neo-Poás)	Lapilli tuff	10°–30°	Smooth topography	Moderately to badly weathered	Parallel to sub-parallel	Residual soils and pyroclastic material <7 m

Table 4 Distribution of landslides and their area based on the orographic regions of the study area approach and precipitation data from meteorological stations near Poás volcano

Orographic regions	Pacific region		Caribbean region	
Region area (km^2), (%)	181.65	35	337.25	65
Landslides frequency (No), (%)	251	5	4595	95
Landslides area (km^2), (%)	0.6	2.7	21.38	97.3
(Landslides frequency No/region area km^2)	1.38		13.62	
(Landslides area km^2/region area km^2)	0.003		0.06	
Precipitation one year previous the earthquake (mm)	4796[a], 5042[b]		3994[c]	
Precipitation one month previous the earthquake (mm)	13[a], 93[b]		435[c], 233[d]	
Precipitation one day previous the earthquake (mm)	0[a], 0.9[b]		6.4[c], 1.5[d]	

[a]San Rafael meteorological station (15 km southern from epicenter), [b]Fraijanes meteorological station (10 km from epicenter), [c]La Selva de Sarapiquí meteorological station (25 km northern from epicenter) and [d]Poás meteorological station (located near the summit of the volcano and 6 km western of the epicenter, this station does not have complete the data of mm of rain one year before the earthquake)

hydrology and climate, including precipitation rates (PNUD-IMN-MINAET 2009). Since the volcanoes of the CVR are part of this orographic division, their slopes are differently affected depending on which side they are located. In general, the Platanar, Barva and Poás volcanoes are located in a tropical climate zone with annual mean temperature variations of 8 °C and mean temperature of 19 °C (Peel et al. 2007). The study area is affected by two climate zones based on the Köppen climate classification: (1) from the Poás summit to Río Cuarto, that corresponds to the northern part of the study area (Caribbean slope) presents a warm humid climate zone (A_f) meaning that there is variable precipitation every month and no dry season with a mean annual precipitation between 4,000 and 6,000 mm (Peel et al. 2007; IMN 2008a, b) and (2) from the Poás summit to Alajuela city; the southern part of the study area (Pacific slope) is characterized by a tropical rain forest climate in spite of a short dry season in a monsoon type cycle zone (A_m) with a mean annual precipitation between 3,000 and 4,000 mm (Peel et al. 2007; IMN 2008a, b). The rainfall season in the Pacific slope extends from May to November. In Table 4 we present the rainfall data from four meteorological stations near the study area one year prior to the Cinchona earthquake. Two of them were located in the Pacific slope of Poás volcano, one near Poás summit and one in the Caribbean slope.

1.7 Land Cover and Soil Type

Based on high-resolution aerial photos from 2009 we defined and measured five different types of land cover in the study area. These types of land cover are: primary tropical rain forest, secondary tropical rain and/or gallery forest, crops and/or farming areas, no vegetation cover areas, and water bodies (Fig. 4).

Most of the primary tropical rain forests is located within the limits of the Poás Volcano National Park (Fig. 4). The secondary forested areas correspond to zones naturally recovered after being used for grazing, other zones are teak forests for industrial uses, private reforestation areas and gallery forest in the river and stream valleys. The most important activities in the crop and farming areas are pastures for dairy cattle, plantations of coffee, sugarcane and cinchona (this last plant, the source of quinine, gave the name to the town that was destroyed by the earthquake and subsequently gave its name to this event). The zones without vegetation cover are located primarily in the main active crater of Poás volcano and in a corridor

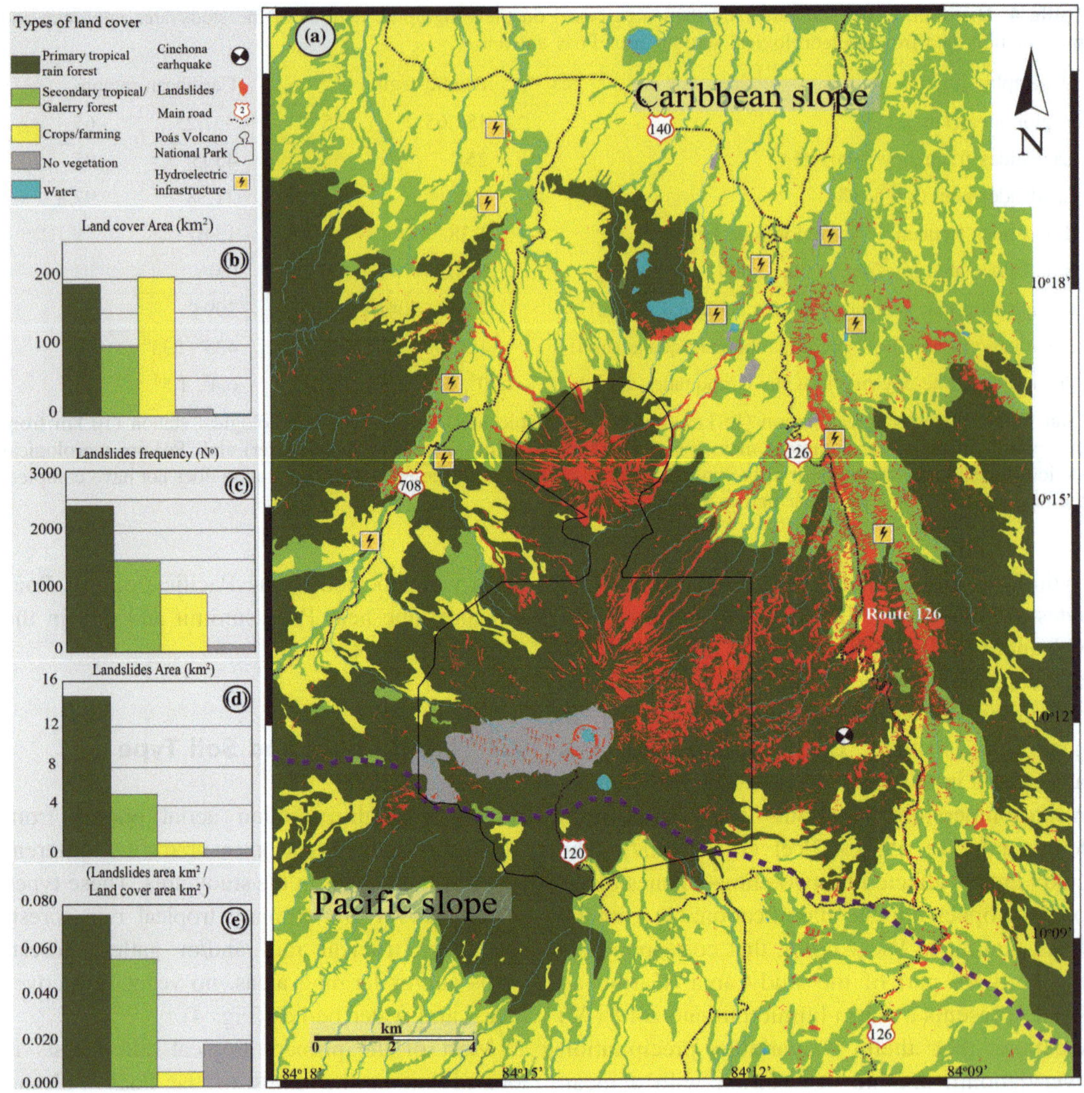

Fig. 4 **a** Land cover and slope failures map. **b** Area in km^2 from each land cover type. **c** Number of coseismic landslides per land cover type. **d** Stripped area (km^2) from coseismic landslides per land cover type. **e** Stripped area from landslides per land cover type divided by the area of each land cover type. The purple line marks the limit between the northern side of CVR (Caribbean slope) and the southern side (Pacific slope)

4 km long and 2 km wide west from it that, due to the prevailing wind direction, has been exposed through years to acid rain and acid gas emissions from the fumarole vents located inside the main crater. Also, in the northern sector of the study area there are rock quarries with lavas completely exposed. The water bodies mostly correspond to crater lakes (Poás main crater, Botos, Hule and Laguna Río Cuarto) and artificial reservoirs from the hydroelectric projects within the zone (Table 5, Fig. 4).

The study area comprises soils (andisols), which have a high content of allophane as the main product of the decomposition of volcanic ash in wet areas (PNUD-IMN-MINAET 2009). The allophone is a very unstable Al-silicate clay mineraloid that gives to andisols special features such as a well-defined structure, which facilitates

Table 5 Distribution of landslides and their areas based on land use approach

Land use	Land use area		Landslides frequency		Landslides area		Density	
	(km^2)	(%)	(No)	(%)	(km^2)	(%)	(Landslides frequency No/land use area km^2)	(Landslides area km^2/land use area km^2)
Primary tropical rain forest	197.2	38	2381	49	14.84	67.5	12.07	0.075
Secondary tropical rain and/or gallery forest	103.8	20	1494	31	5.72	26	14.39	0.055
Crops and/or farming areas	207.6	40	892	892	1.23	5.7	4.3	0.006
No vegetation	9.6	1.9	79	18	0.17	0.8	8.23	0.018
Water	0.8	0.2	NA	NA	NA	NA	NA	NA

a good drainage, but at the same time it tends to retain a lot of the humidity. They have a low bulk density, low plasticity and low cohesion. Close to the volcanic craters, the andisols tend to have a texture that is sand-rich and coarse, while in the intermediate zones its texture is silt-rich and in the lower altitude zones, its texture is mostly clay.

In situ studies made after the Cinchona earthquake in different slopes of the study area showed soils with high humidity contents (as high as 82–108%), and Atterberg liquid limit values (LL) close to the unit (71–143%) (Laporte 2009a, b).

2 Methodology

The methodology used in this work to build a coseismic landslide susceptibility model for the study area is a combination of heuristic and statistic methods divided in three steps, as follows: (a) construction of a landslide catalog for the most recent coseismic landslides in the area, (b) application of a heuristic model based on the Mora-Vahrson approach (Mora et al. 1992; Mora and Vahrson 1993, 1994) with some modifications and (c) evaluation of the methodology used.

2.1 Coseismic Landslide Catalog for the Cinchona Earthquake

The set of high resolution LiDAR data and orthophotos to create our coseismic landslide catalog were acquired during an airborne survey in April 2009 by STEREOCARTO S.L. with an ALS50-II Leica system. The resolution of these LiDAR data is three points per m^2, which were used to create a digital elevation model (DEM) with a resolution of 50 cm in the x- and y-axes, and 15 cm the z-axis.

Using the existing geological information of the area (Alvarado and Carr 1993; Soto 1999; Ruiz et al. 2010), field observations after the earthquake and the new high resolution DEM created with the LiDAR data, we built a variety of thematic maps (land use, slope, geology, and temporal phase of the volcanic units), each one with its corresponding histograms to show correlations between the theme and the occurrence and areas striped by the coseismic landslides from the Cinchona earthquake. Our results were normalized based on the % of the total study area that every parameter (land cover, slope, geology, and others) presented.

The new DEM allowed us to refine some geological contacts and interpretations previously presented in the geologic map by Ruiz et al. (2010). Based on field observations of the new outcrops produced by the landslides, we assigned a level of weathering (unweathered, slightly, moderate, badly, very badly and extremely weathered) to each volcanic unit (Table 3). Due to their geomorphological characteristics, the Bosque Alegre and Laguna Kooper units are very similar and because most of the landslides occurred within the crater walls of these two maars, for this chapter we grouped these units into a single unit called, maar unit (Fig. 3).

To map the coseismic landslides from the Cinchona earthquake and create our catalog, the aerial photos taken during the acquisition of the LiDAR data were overlapped to the DEM and then each landslide was drawn over these composite images using Geographic Information Systems (GIS) tools. To facilitate the mapping of the landslides, we use photos that had a false-color, i.e. a combination of near infrared, red and green light in which the vegetation is bright red, while areas stripped by the landslides appear in the brown and dark gray ranges. The use of this false-color was more convenient than the true-color images in which it was sometimes difficult to spot the limits of the landslides. Using commercial software packages (Global Mapper 13.0, Surfer 10.0, Rockworks 14.0 and Arc Map 10) to process the images, we obtained different parameters (geographical location, slope angle, bearing of the flow, length, maximum and minimum heights, perimeter and area) directly for each landslide.

The areas measured from each mapped landslide correspond to the zone disturbed by each event. In this chapter we considered the disturbed zone as the sum of the detachment and deposition areas. Using the mapped landslides we created a slope failure map that was superposed on some of the thematic maps previously mentioned to better understand the relationship between each factor and the distribution and size of the landslides.

The geographical location (latitude and longitude) for each landslide was assigned from the midpoint of the crown of each feature. The bearing of each flow was also measured from this same point. Comparing each landslide's geographical location and the thematic maps, each event was cataloged based on their position according to: orographic region (Pacific slope or Caribbean slope), land cover, volcanic edifice (Platanar, Barva or Poás), geologic unit, temporal volcanic phase (Paleo or Neo-phase) and slope angle ranges.

Based on the aerial photos, oblique pictures and fieldwork each landslide was identified and classified using the classification of Varnes (1978) and Skinner and Porter (1992). We distinguished rock fall, slide/earth flows, slump, and debris flows, according to their type of movement and rock, soil or a combination of both according to the type of material that was removed. We also identified whether the material removed entered directly into the drainage system (rivers and streams).

Since the study area does not have any previous LiDAR data, the measurements of the removed volume by the coseismic landslides (V_{ls}) could not be obtained directly from the actual images. Instead, we used Eq. (1) from Parker et al. (2011) and the parameters for global relationships ($\alpha = 0.146$ and $\gamma = 1.332 \pm 0.005$) defined by Larsen et al. (2010) to obtain an estimation of the volume for the Cinchona event:

$$V_{1\varepsilon} = \sum_{i}^{n} \alpha A_i^{\gamma} \quad (1)$$

We compared some of our parametric results (area, volume and landslides magnitude) with the coseismic events from the world catalog of Malamud et al. (2004). Our volume results for this comparison were obtained using Eq. (2) from Hovius et al. (1997) with $\varepsilon = 0.005 \pm 0.002$ to be in agreement with the method used by Malamud et al. (2004).

$$V = \varepsilon A_L^{1.50} \quad (2)$$

Since the LiDAR data and orthophotos were taken three months after the earthquake, they were not adequate to spot landslides in areas without vegetation cover due to the lack of noticeable differences between the failure slopes and the surroundings slopes. The region affected by this situation was the main crater of the Poás volcano and the corridor located western from it (Fig. 4). Most of the landslides mapped in the main crater where done via aerial and oblique pictures taken days after the earthquake, some reported in GVN (2009). Based on the number and size of these intra-crater landslides, the steep slopes >45° and the proximity to the epicenter ~8 km we estimated a number of ~93 landslides by km^2 for the non-vegetated corridor. It is also possible that we missed mapping some landslides in various river valleys and especially in the Toro river canyon because of the steep slopes and tree shadows in the aerial pictures. Therefore, we estimated that our coseismic landslide database has an error of about 200 events, this number of small events could account for about 0.32 km^2 of the total area removed by the landslides in this chapter. This value was added to the sum of the areal error that each landslide presented from our measurements (0.5% of the landslide area from each event). The error obtained for the volume was also calculated for each landslide based on the equation used (1 or 2) and the value estimated for the landslides that we miss mapping.

To compare our results with other worldwide coseismic catalogs we used the Malamud et al. (2004) data and plot (total volume of landslides triggered versus earthquake moment magnitude). Although this is a log-log plot it has a reasonably good power law dependence of the total landslide volume on the earthquake's moment magnitude. We added information to each earthquake in this plot to evaluate the geological (rock type), geophysical (earthquake type and depth) and climatic (mean annual rainfall rate) conditions in order to understand better the differences in the volumes obtained between events of similar moment magnitude.

2.2 Coseismic Landslide Susceptibility Model

The coseismic landslide susceptibility model for Poás volcano is essentially based on the methodology proposed for the slope instability hazard method (Mora et al. 1992; Mora and Vahrson 1993, 1994). However, our model presents some differences from this method, as follows: (a) we use slope angle ranges instead of ranges for topographic gradient by unit area as proposed by Mora et al. (2002), (b) the lithological susceptibility (S_l) was determined based on statistical data of the two temporal phase units (see below) and not geotechnical data, being this statistical approach in agreement with the van Westen and Soeters (2000) methodology, (c) we use only one possible trigger mechanism (earthquakes), since our model is applied only for coseismic landslides and not landslides triggered by excess of rain, (d) for the trigger event, we used weighted values derived from a formula for peak ground acceleration attenuation (PGAa) for crustal earthquakes (depth < 25 km) in Costa Rica, instead of values derived only from the maximum Modified Mercalli Intensity. Therefore, our method defines the slope susceptibility to slide (H) as the product of the slope's intrinsic susceptibility (Susc) and the earthquake trigger mechanism (Earthq Trig) (Eq. 3).

$$\mathrm{H} = (\mathrm{Susc}) * (\mathrm{Earthq\ Trig}) \tag{3}$$

where the slope's intrinsic susceptibility is defined as the product of the lithological susceptibility (S_l), the slope angle susceptibility (S_s) and the ground moisture susceptibility (S_h) (Eq. 4).

$$\mathrm{Susc} = (S_l * S_s * S_h) \tag{4}$$

The lithological susceptibility (S_l) quantifies the influence of the different geologic units in the landslide generation. To obtain the weight values for this factor, we first grouped the units based on their temporal phases (Paleo and Neo), and their weathering level (Table 3). Although the Rio Cuarto Lavas Unit belongs to the Paleo temporal phase units, because its weathering level is nil to slightly weathered for the lithological susceptibility analysis, we used it as if it belonged to the Neo-phase units. Once all the units are grouped, and then based on the frequency and density of coseismic landslides from the Cinchona inventory for each temporal phase unit we obtained a value for their lithological susceptibility using Eqs. (5)–(9).

$$S_l = \frac{d1}{d2} + \frac{f1}{f2} \quad (5)$$

where,

$$d1 = \frac{\text{Area stripped by landslides for each temporal phase unit}}{\text{Area of each temporal phase unit}} \quad (6)$$

$$d2 = \frac{\text{Total area stripped from landslides } (21.98\ \text{km}^2)}{\text{Total study area } (519\ \text{km}^2)} \quad (7)$$

$$f1 = \frac{\text{Number of landslides in each temporal phase unit}}{\text{Area of each temporal phase unit}} \quad (8)$$

$$f2 = \frac{\text{Total number of landslides } (4846)}{\text{Total study area } (519\ \text{km}^2)} \quad (9)$$

To obtain the values for the factor slope angle susceptibility S_s we built a slope angle map from the DEM based on the LiDAR data, similar to the one created for the landslide inventory. The classification used to create this slope angle map is based on van Zuidam (1986) and it has seven angle ranges (0°, 0.1–4°, 4–8°, 8–16°, 16–35°, 35–55° and >55°), to which weights from 0 to 6 were assigned based on Mora et al. (2002) classification.

The value of the S_h represents the prevalent ground moisture in the study area and was derived from a simple hydrologic balance based on the methodology used by Mora et al. (1992). Our hydrologic balance was computed by using the rainfall data from seven meteorological stations located within the study area for a time period between 1959 and 2002 (Paniagua and Soto 1986; ICE 2008 and this chapter). To obtain the weights for S_h we analyzed the data (mm/month) from each station and a monthly value from 0 to 2 was assigned respectively (<125 mm = 0, 125–250 mm = 1 and >250 mm = 2). Afterwards, the sum of the 12 monthly values for each station was associated to a S_h weight (0–5), following the classification from Mora et al. (1992) respectively (0–4 = 1, 5–9 = 2, 10–14 = 3, 15–19 = 4 and 20–24 = 5).

The calculation of the earthquake trigger factor is derived from a process that begins with the calculation of the peak ground acceleration attenuation (PGAa) of the earthquake that generated the landslides, the Cinchona earthquake in this case. For this step, we used the peak ground acceleration attenuation equation from Schmidt (2010) for crustal earthquakes in Costa Rica (Eq. 10), where the cb variables are constant (cb1 = 0.15454, cb2 = 0.48743, cb3 = 1.03269, cb4 = 3.83891, cb5 = 0.21489 and cb6 = 0.11115), Mw = the earthquake moment magnitude, d = hypocentral distance, and the values for s and h depend on the type of soil where the formula is applied. Based on Schmidt (2010), the soil classification that corresponds to our study area presents homogeneous values of s = 1 and h = 0:

$$\begin{aligned} \log_{10}(\text{PGAa}) = {} & \text{cb1} + \text{cb2} * (\text{M}_\text{w}) \\ & + \text{cb3} * \log_{10}\sqrt{(\text{d}^2 + \text{cb4}^2)} \\ & + \text{s} * \text{cb5} + \text{h} * \text{cb6} \end{aligned} \quad (10)$$

The PGAa values obtained from Eq. (10) for each point of the grid of the study area were grouped into the ranges proposed for each grade of the MMI scale following Wald et al. (1999) and Linkimer (2008). Finally, to transform the PGA associated value ranges into the weight values proposed by Mora et al. (1992) for the Earthq Trig, we used an equation obtained by us for this chapter from an empirical regression with a logarithmic best fit:

$$\text{Earthq Trig} = 1.2597 \ln(\text{PGAa}) - 1.2517 \quad (11)$$

The values of the slope susceptibility to slide obtained from the application of Eq. (3) were grouped in five different ranges. These ranges were obtained by dividing in five equal intervals the value obtained by multiplying the maximum value of each of the intrinsic susceptibility factors and the trigger event for the study area. Then, to each range we assigned a classification of susceptibility: very low, low, moderate, high and very high. These same values and classification ranges obtained are used for all the models tested in the study area and can be applied only to this zone. According to Mora et al. (1992), we can expect to have landslides in zones with a hazard susceptibility level of moderate, high and very high.

2.3 Evaluation of the Methodology

The evaluation of the methodology is a measurement of the overlapping degree between the coseismic landslides from Cinchona earthquake catalog and the zones from the susceptibility to slide map that have H values of moderate, high and very high. The number of coseismic landslides occurred within these ranges was divided by the total number of landslides from the catalog. To obtain a relationship based on landslide density, we divided the total area stripped by the coseismic landslides by the total area covered by the moderate, high and very high hazard zones.

3 Data and Results

In this section, we first report the results from our post-Cinchona earthquake landslide inventory. The total number of landslides (N_{Lt}) measured here was 4,846, if we include the number of landslides missed during the mapping we obtain a total number of 4,946 $\pm$ 100 events. However our distribution results are based only in the sampled landslides. A simple statistical approach was used to study the collected data of landslides from the LiDAR data. Our analysis includes the landslide frequency and landslide density within the areas of the analyzed factors (orographic regions, type of land cover, volcanic units, temporal phase paleo-neo, slope angle and distance from the epicenter). In a second section of our coseismic inventory we present the morphometric parameters obtained including type of landslides, area, and volume to later compare the Cinchona results to data from other world earthquakes that generated landslides.

In separate sub-sections, the results of the landslide susceptibility to slide model of the study area and the evaluation of our methodology by recreating the Cinchona earthquake are presented. Finally, we discuss the results of testing our model in three scenarios. Firstly, with a simulation of a historical event in the southwest of the study area and secondly, by simulating two hypothetical earthquakes, one located in the northwest sector of the study area and the other one in the northeast side of Poás volcano.

3.1 Post-Cinchona Earthquake Landslide Inventory

3.1.1 Landslide Distribution by Orographic Regions (Caribbean-Pacific)

Although the epicenter of the Cinchona earthquake is located $\sim$4 km northern from the border line between the Pacific slope and the Caribbean slope of Poás volcano (Fig. 4a), most of the landslides (95%) are located in the Caribbean side (Table 4 and Fig. 4). This irregular distribution could be explained by different factors such as geology, slope angle, and soil saturation between these two orographic regions. In the following sections we analyze in detail all these factors, while here we focus on the differences in terms of precipitation rates between the two slopes, using it as an indirect measurement of soil saturation in the study area.

During November and December of 2008, Costa Rica was directly affected by at least seven atmospheric events (five cold fronts, one low pressure system and one trade wind system) and was indirectly affected by the Paloma Hurricane (IMN 2008a, b). These events generated precipitation rates above average for these months, in three regions of the country (Northern, Caribbean and Central) (IMN-CR 2008a, b). Based on the rainfall data of the stations close to the study area (Table 4), historical accounts since 1961 (PNUD-IMN-MINAET 2009), and additional information from IMN (2008a, b), we determined that the surplus of rain for November and December 2008 in the study area was between 15 and 50%. This surplus was more obvious in the Pacific slope and near the summit of Poás volcano than in the Caribbean slope, because normally by November-December the Pacific slope of the Poás volcano is already influenced by dry season conditions, while the Caribbean side usually experiences its maximum amount of rain during the year.

By January 8, 2009 when the Cinchona earthquake occurred, the Pacific slope was already showing the normal conditions of the beginning of the dry season, while the Caribbean slope was experiencing the typical high amounts of rain during that time of the year. The total accumulated rain measured in La Selva de Sarapiquí meteorological station (~25 km northward of the epicenter, Fig. 1c, Caribbean slope), 12 months previous to the Cinchona earthquake, was 3,994 mm. The same station recorded 435 mm of rain one month before the event (Table 4). Meanwhile, the meteorological station located in San Rafael de Poás (~15 km southward of the epicenter Fig. 1c, Pacific slope) recorded 4,796 mm for the year previous the earthquake but only 13 mm for the month before the event (Table 4). Therefore, the level of the ground water table was shallower near the summit and especially in the Caribbean slope. On the Caribbean side a high ground water table could have influenced the slopes failures. However, on the Pacific slope this was unlikely.

3.1.2 Landslide Distribution by Type of Land Cover

The majority and the most extensive landslides occurred in areas that were covered by primary tropical rain forest, this vegetation being the second most common land cove in the study are (Fig. 4). Most of this land cover is located on the Caribbean slope and near the epicenter of the earthquake (Table 5 and Fig. 4). We also noted that the density of landslides that occurred in zones where the rocks were completely exposed is less reliable because (a) we were not able to map all the landslides that occurred in these zones and (b) their size was relatively small (Table 5 and Fig. 4).

Soils in areas covered by tropical rain forest act as sponges and retain high levels of moisture (Bonell et al. 1981). Because the epicentral area had a considerable amount of rain during the whole year and especially in the two months previous to the earthquake, the dense canopy favored the generation of landslides by retaining more moisture in the soils. Other cases of widespread stripping of saturated superficial materials and jungle cover from steep slopes by coseismic landslides in other humid tropical areas are: New Guinea 1935 and 1970 (Marshall 1937; Pain 1972), Panama 1976 (Garwood et al. 1979) and Ecuador 1987 (Schuster et al. 1996). It is well known that forest and vegetation cover can prevent landslides and erosion, however due the high peak ground acceleration near the epicenter (0.67 to about 1 g), the expected protective effect of the forest cover was not observed, at least close to the epicenter (0–5 km).

3.1.3 Landslide Distribution by Volcanic Edifice

Volcanoes from CVR affected by the Cinchona landslides were Poás, Platanar and Barva. Because the epicenter was located on the northern flank of the Poás volcano, the edifice of this volcano was the most affected, presenting the majority and biggest landslides of the study area (Table 6 and Fig. 3). The Barva volcano was the second most affected and the Platanar volcano was the least affected, having the fewest and smallest landslides (Table 6 and Fig. 3). Although the study area encloses more area of Poás volcano and only portions of the Platanar and Barva volcanoes (Table 6 and Fig. 3), based on our field observations the occurrence of landslides on these last two edifices outside the limits of the study area was reduced and mainly located on steep slope zones near their summits.

3.1.4 Landslide Distribution by Volcanic Unit

Based on the detailed geologic map of the study area (Fig. 3), a more careful analysis of the incidence and area affected by the landslides

Table 6 Distribution of landslides and their areas based on volcanic edifice approach

Volcano	Volcano area		Landslides frequency		Landslides area		Density	
	(km^2)	(%)	(No)	(%)	(km^2)	(%)	(Landslides frequency No/volcano area km^2)	(Landslides area km^2/volcano area km^2)
Platanar V.	83.04	16	306	6	0.38	1.8	3.68	0.005
Poás V.	285.45	55	3322	69	18	82	11.64	0.063
Barva V.	150.51	29	1218	25	3.6	16.2	8.09	0.024

based on which volcanic unit they occurred was conducted. From Poás, Platanar and Barva, a total of 13 volcanic units had landslides (Table 7 and Fig. 3). From N_{Lt}, the three volcanic units with higher landslide frequency are: La Paz Andesites Unit, Paleo-Barva Unit and Poás Summit Unit. Since the epicenter of the earthquake was located almost between the La Paz Andesites and the Poás Summit units (Fig. 3), the high frequency of landslides on these units is understandable. Furthermore, other factors like steep slopes (see below), high weathering level, hydrothermal alteration and the presence of several active faults on the La Paz Andesites unit (Fig. 3) could also play an important role in the high frequency of landslides on this unit. The Paleo-Barva Unit is located close to the epicenter as well and it has a weathering level similar to La Paz Andesites unit, which may also explain its high landslide frequency (Table 7, Fig. 3).

The three volcanic units that presented the most extensive landslides are: La Paz Andesites, Congo volcano and Von Frantzius units (Table 7, Fig. 3). The differences between the frequency and the areal approaches can be explained by the fact that in the Congo volcano and the Von Frantzius units there were more events of the type known as debris flows than in Paleo Barva and Poás Summit units. Debris flow events tend to expose more area than the less complex slide/earth flows or slumps events because they travel longer distances.

Although the Platanar unit had a significant number of landslides they were relatively small and their contribution in the total area affected by landslides is relatively small, especially if we compare it to the Congo volcano Unit that presented a similar frequency of landslides but had landslides that were much bigger and stripped about nine times more area (Table 7, Fig. 3).

3.1.5 Landslide Distribution by Volcanic Temporal Phase Units (Paleo-Neo)

Because of the stratigraphic control we had on the Platanar, Barva and especially on the Poás volcanic units based on Ruiz et al. (2010) and Ruiz et al. (Chapter "Geochemical and Geochronological Characterisation of the Poas Stratovolcano Stratigraphy"), we were able to compare the occurrence of landslides from two major temporal units (Paleo-Stage units: older than 0.2 My or Neo-Stage units: younger than 0.2 My) (Table 8, Fig. 5). This approach was done because of the significant differences in geomorphology and weathering level between these age groups (Fig. 5, Table 3). From N_{Lt}, 65% occurred in the Paleo temporal stage units, while the remaining 35% occurred in the Neo temporal stage units (Table 8, Fig. 5a–e). However, some of the landslides that occurred in the Neo stage units were relatively big and the area between the Paleo and Neo stage units was not as disproportionate as the frequency of landslides (Table 8, Fig. 5a–e).

Table 7 Landslides distribution and their areas based on volcanic units approach; NA not analyzed

Unit	Volcanic units area		Landslides frequency		Landslides area		Density	
	(km^2)	(%)	(No)	(%)	(km^2)	(%)	(Landslides frequency No/Volcanic unit area km^2)	(Landslides area km^2/Volcanic unit area km^2)
Platanar	74.84	14.42	306	6	0.38	1.73	4.09	0.005
Río Sarapiquí	2.85	0.55	129	3	0.95	4.32	45.19	0.333
La Paz Andesites	53.04	10.22	1756	35.8	7.98	36.3	33.11	0.15
Achiote	64.36	12.4	93	2	0.26	1.18	1.45	0.004
Río Cuarto Lavas	29.58	5.7	22	0.1	0.02	0.08	0.74	0.001
Von Frantzius	56.21	10.83	448	9	3.53	16.06	7.97	0.063
Congo volcano	38.41	7.4	288	6	3.82	17.37	7.5	0.099
Poás Summit	38.2	7.36	517	11	1.25	5.69	13.53	0.033
Poasito	19.46	3.75	14	0.1	0.01	0.05	0.72	0.001
Maar units	5.71	1.1	55	1	0.2	0.9	9.63	0.035
Sabana Redonda	0.67	0.13	NA	NA	NA	NA	NA	NA
Paleo-Barva	61.03	11.76	851	18	3.03	13.78	13.94	0.05
Neo-Barva	64.36	12.4	343	7	0.54	2.45	5.33	0.008
Pozo Azul	8.3	1.6	24	1	0.02	0.09	2.89	0.002
Recent deposits	1.97	0.38	NA	NA	NA	NA	NA	NA

Table 8 Landslides distribution and their areas based on volcanic temporal phases units (Paleo and/or Neo)

Temporal volcanic phase unit	Temporal volcanic phase unit area		Landslides frequency		Landslides area		Density	
	(km^2)	(%)	(No)	(%)	(km^2)	(%)	(Landslides frequency No/temporal volcanic phase unit area km^2)	(Landslides area km^2/temporal volcanic phase unit area km^2)
Platanar	74.84	14.42	306	6.3	0.38	1.73	4.09	0.005
Paleo-Poás	149.73	28.85	1997	41.2	9.21	41.88	13.34	0.061
Paleo-Barva	60.98	11.75	851	17.56	3.03	13.77	13.95	0.05
Neo-Barva	160.58	30.94	1325	27.34	8.81	40.08	8.25	0.055
Neo-Poás	72.87	14.04	367	7.58	0.56	2.54	5.04	0.008
Paleo-unit (Platanar, Poás and Barva)	285.55	55.02	3154	65.08	12.61	57.38	11.05	0.044
Neo-units (Poás and Barva)	233.45	44.98	1692	34.92	9.37	42.62	7.25	0.04

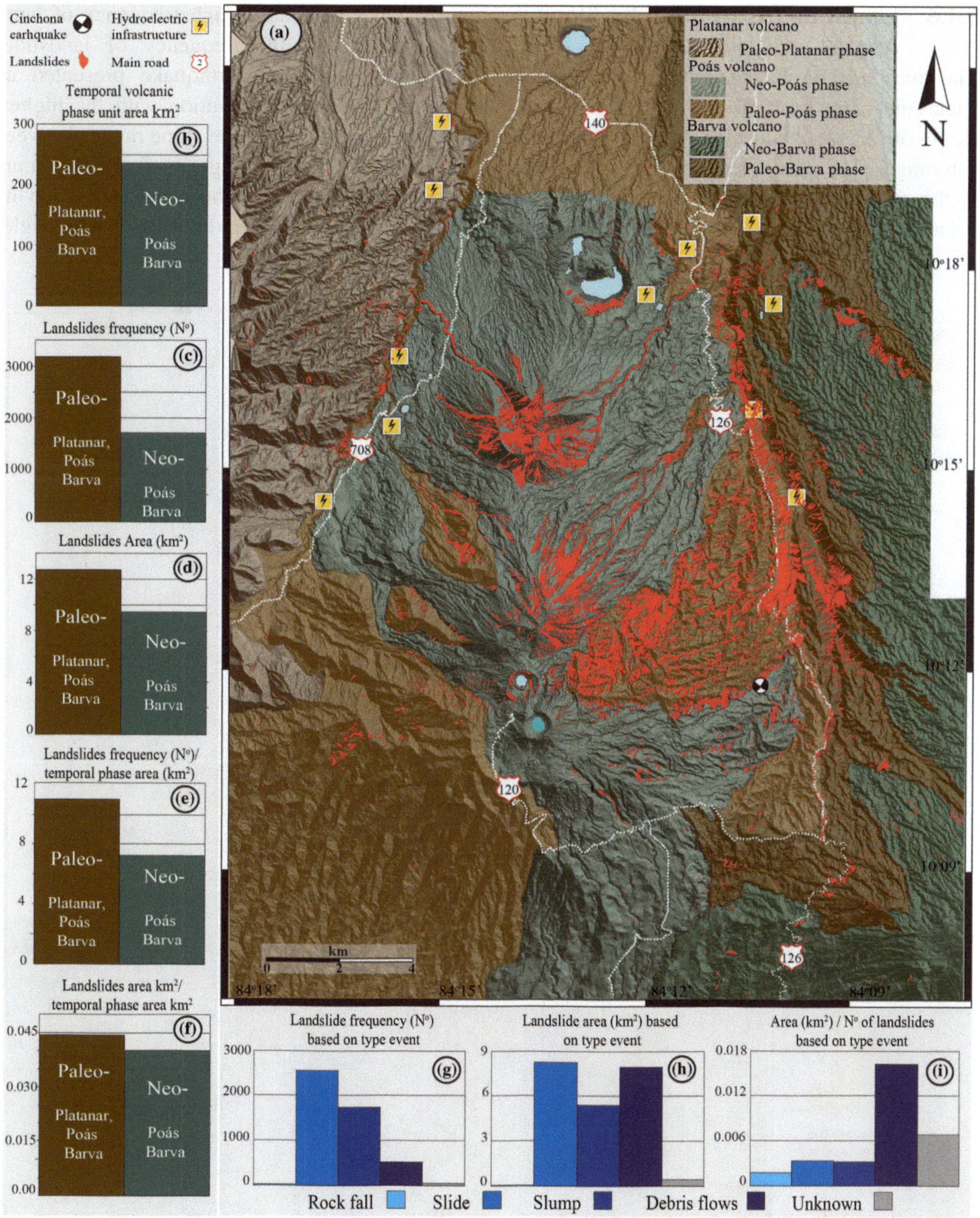

Fig. 5 a Occurrence of landslides based on Paleo or Neo temporal phase units. Brown column in histograms (**b–f**) represent the sum of the data from the Paleo Temporal units from Platanar, Poás and Barva volcanoes, the green column reported the data from the Neo temporal units for Poás and Barva volcanoes. Histograms (**g–f**) show the frequency and area striped based on the type of landslide

3.1.6 Landslide Distribution by Slope Angle

The study area is characterized by having slope angles that range from 0 to 90° (Fig. 6a). By using ranges of 15°, we obtained six different slope angle sub-ranges. The area covered by these sub-ranges in the study zone decreases as the slope angle values increase (Table 9, Fig. 6a). As expected, steep slopes generate a high frequency of landslides. However, the frequency of landslides shown by Cinchona earthquake presented an apparent Gaussian distribution with the highest number of landslides in the slope ranges between (30° and 59°) (Table 9, Fig. 6b, c). This apparent Gaussian distribution is the result of not having large enough areas covered by slopes with angles

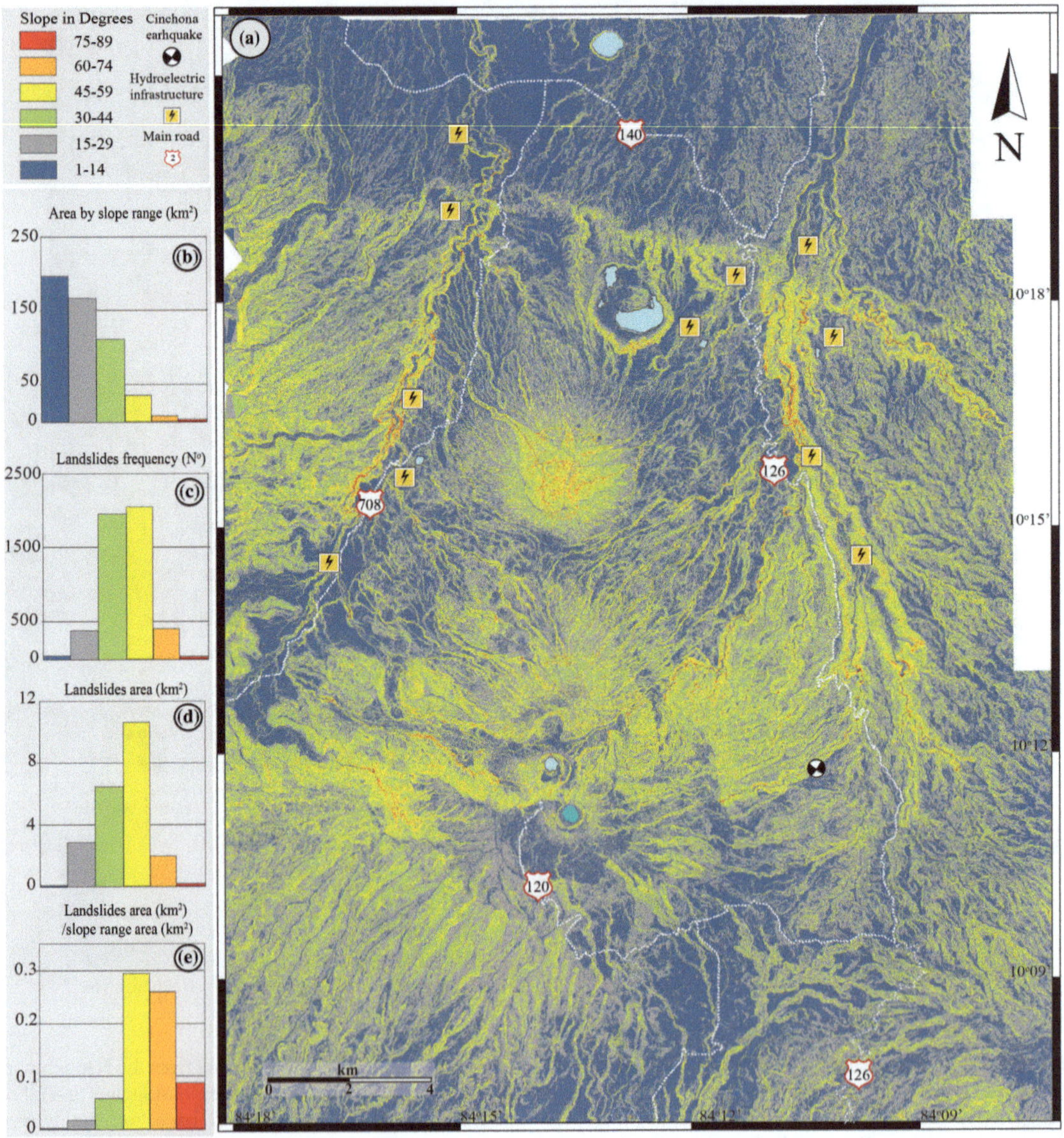

Fig. 6 **a** Slope angle map. **b** Area (km^2) covered by slope range. **c** Number of coseismic landslides per slope range. **d** Stripped area (km^2) from coseismic landslides per slope range. **e** Stripped area from landslides per slope range divided by the area covered by each slope range

Table 9 Landslides distribution based on slope

Slope angle range (°)	Slope angle range area		Landslides frequency		Landslides area		Density	
	(km^2)	(%)	(No)	(%)	(km^2)	(%)	(Landslides frequency No/slope range area km^2)	(Landslides area km^2/slope range area km^2)
Jan-14	196.55	37.87	42	0.87	0.04	0.17	0.21	0.0002
15–29	166.7	32.12	373	7.7	2.84	12.92	2.24	0.017
30–44	110.72	21.33	1956	40.36	6.42	29.21	17.67	0.058
45–59	36.08	6.95	2058	42.47	10.62	48.32	57.05	0.294
60–74	7.46	1.44	390	8.05	1.93	8.78	52.25	0.259
75–89	1.49	0.29	27	0.56	0.13	0.6	18.09	0.088

>60°. This makes the estimated landslide frequency for these slope bins less reliable. Otherwise, the landslide frequency strongly increases with slope. Meanwhile, the small areas of landslides that occurred in slopes with angles <30° is understandable since these low angle slopes are less susceptible to slide and are located far from the epicenter. There is a relationship between the landslides that occurred in slopes with angles >75° and the type of material that they involved, since most of them removed only rock (see below).

3.1.7 Types of Landslides and Types of Material that They Involved

From the N_{Lt}, 53% presented translational movement (slides/earth flows), 35% were rotational events (slumps), 10% corresponded to debris flows and only 0.4% were recognized as rock fall events (Fig. 5f–h and Table 10). This last type of landslides were located as expected in areas where the bedrock is completely exposed and where the slopes are very steep >60°. The remaining 1.6% of landslides could not be recognized because they occurred next to roads that were stabilized by machinery just weeks after the earthquake before the LiDAR data and orthophotos were taken.

We determined that most of the landslides involved displacements of andisols. Besides the typical characteristics of these soils previously described, the soils also included recent (brown tuffs) and older (orange and yellowish tuffs) ashes. In some cases a combination of soils and rocks also occurred in the displacement of material. Alvarado (2010) reported that some of the debris flows included lava blocks (mostly boulders) and mega clasts of ignimbrite blocks as big as 5 m × 4 m × 9 m. Only a few coseismic landslides removed rock and they were located within the walls of the main crater of Poás volcano and in some areas of the Toro river canyon where columnar lavas are exposed in cliffs as high as ~50 m and with slopes angles >75°.

Table 10 Landslides distribution base on event type

Landslide type	Number of event by type	Area of events by type (km^2)	Area/No
Rock fall	17	0.03	0.0017
Slide	2555	8.22	0.0032
Slump	1725	5.38	0.0031
Debris flow	486	7.93	0.0163
Unknown	63	0.43	0.0069

3.1.8 Distance to the Epicenter

From N_{Lt} about 49% are located within a radius of 5 km from the epicenter of the earthquake (Fig. 7a). The attenuation of landslide density in five kilometer windows from the epicenter exhibits a logarithmic trend from the epicenter to as far as ~21 km (Fig. 7a, b). Outside the study area, small and isolated landslides occurred as far as 35 km from the epicenter in places with steep slopes, primarily in river valleys and near the summits of the Barva and Platanar volcanoes.

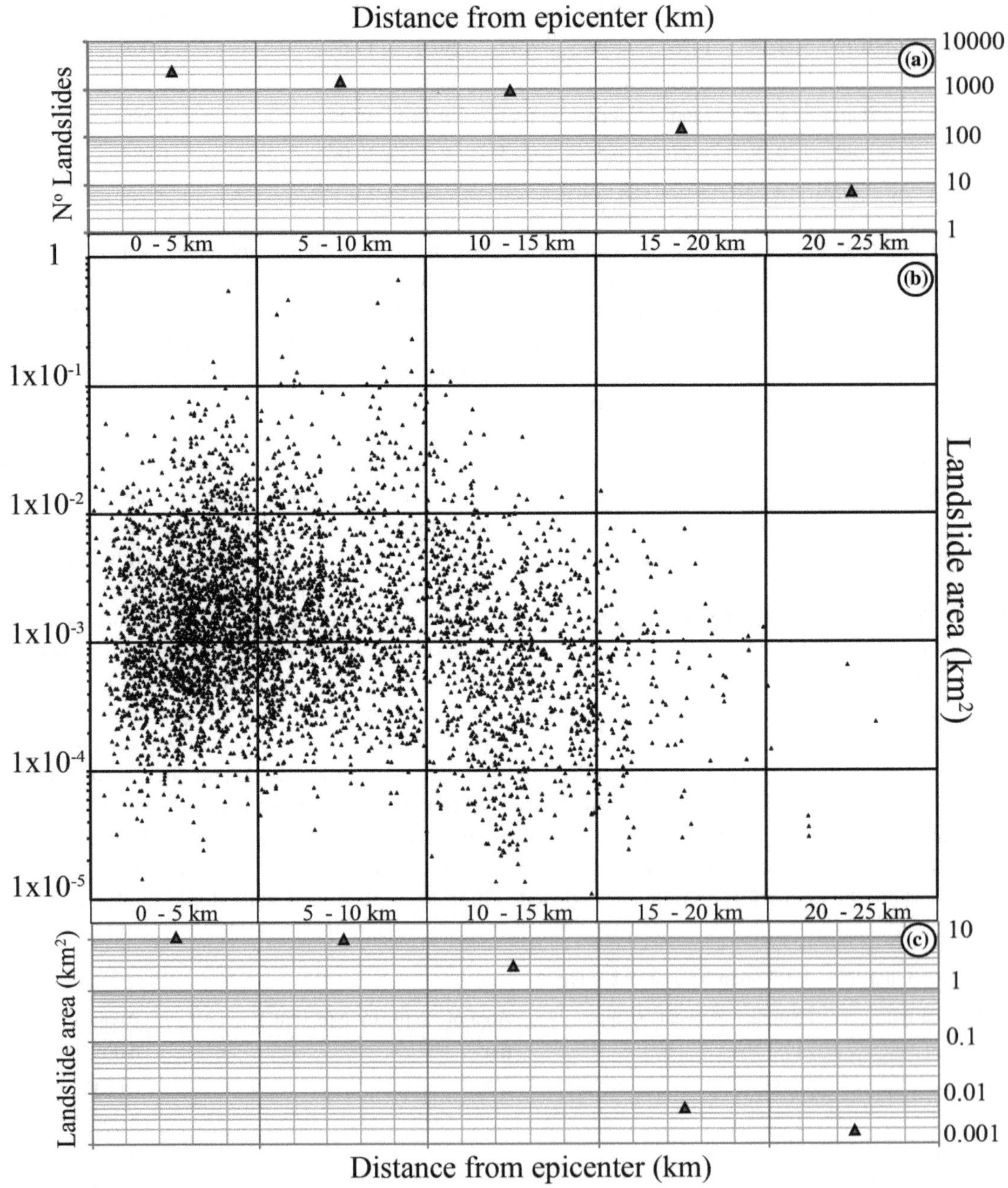

Fig. 7 **a** Number of coseismic landslides by distance from the Cinchona earthquake epicenter in 5 km windows. **b**, **c** Area of coseismic landslides by distance from the Cinchona earthquake epicenter

3.1.9 Area and Volume Removed from the Coseismic Landslides

The total disturbed area from the landslides that we measured (A_{Lt}) in the study area was 22.14 ± 0.27 km^2, an area equivalent to ~2,020 soccer fields with the maximum official dimensions. The disturbed area most likely was removed in less than 8.2 s, which was the duration of the strong movement of the earthquake in the mesoseismic area (Climent and Moya 2009), resulting in an average of 2.7 km^2 per second.

Single landslide areas ranges from 10.91 m^2 to more than 671,700 m^2 with an average of 4,500 m^2. As shown in Figs. 4, 5, 6 and 7c, the largest landslides did not occur close to the epicenter, but were actually located between 4 and 18 km from it. Most of these large landslides were debris flows that occurred in the Congo volcano and Von Frantzius cones (Neo-temporal Units). Debris flows accounted for 36% of the total slope failure area measured in this chapter (Table 10 and Fig. 5f–h.). The Congo volcano and Von Frantzius cones are characterized by steep slopes between (30° and 60°) covered by a thin layer of volcanic soils no greater than five meters with fresh lava flows underneath (Table 3). These lava flows worked as sliding surfaces for the debris flows that traveled for several kilometers and affected large areas.

We estimate that the landslides from Cinchona earthquake produced ~0.24 ± 0.0013–0.39 ± 0.1017 km^3 of erodible material (depending on Eq. (1) or (2) used to calculate the volume). By analyzing the aerial pictures, we also determined either the material removed from the landslides entered directly into the drainage system (rivers and streams) or it was deposited on the volcano's flanks without direct access to river valleys or streams. From N_{Lt} only ~20% of the landslides did not enter directly the drainage system, leaving almost all these materials meters away from the source area. Since the size of most of these landslides was small they only account from a volume between 0.047 and 0.0087 km^3. The remaining 80% of the landslides occurred in the river valleys walls and/or in the river headwaters, which have direct access to the streams. Alluvial processes will eventually carry these materials out from the volcanoes and transport them into different basins.

However, this does not mean that all the material removed by the landslides that entered into the drainage system was transported outside the volcano. An exception could be the debris flows, but not even these events rapidly removed all the mobilized material. Alvarado (2010) estimated a minimum volume between 2.5×10^{-3} and 3.5×10^{-3} km^3 of material that was removed outside the volcanic system through the debris flows (lahars) that occurred minutes and days after the earthquake.

Using a volume of 0.3 km^3 per event, a 50 yr return period for earthquakes (Mw > 5.5–6) and 20 km length of arc segment, we calculated a mass flux from coseismic mass wasting of ~300 ± 150 km^3/km/Myr, a rate comparable to estimates of magma flux at arc volcanic systems (Holbrook et al. 1999); Clift and Vannucchi 2004; Carr et al. 2007). The same recurrence interval yields an average erosion rate due to landslides of 0.19–0.75 mm/yr given an area of 519 km^2. However, estimations of the amount of material that remained in the volcanic edifice are still needed.

3.1.10 Comparison of the Cinchona Event with Other Coseismic Worldwide Events

Using the equations for landslide event magnitude (m_L), proposed by Malamud et al. (2004), which is based on (N_{Lt}) or (A_{LT}), values for m_L of 3.68 and 3.85, respectively, for the Cinchona earthquake were obtained. By comparing our (m_L) results with an analogous event (Mammonth Lakes, CA, USA earthquake, 1980), which had the same moment magnitude (M_w 6.2), a similar focal mechanism solution (right lateral slip) and focal depth (8 km) (Julian and Sipkin 1985; Malamud et al. 2004), we realized that the Cinchona event result was very similar to the Mammoth Lakes event which had a N_{Lt} of 5,253 (Malamud et al. 2004). However, if the m_L is compared with the total area affected by the landslides, our result significantly differs because the Cinchona earthquake removed a total landslide area at least one order of magnitude bigger than the Mammoth Lakes event. The same amount of difference is noticeable between these two events in earthquake moment magnitude versus volume removed binary plot (Fig. 8). With additional information included in the original plot of Malamud et al. (2004), we found that there are significant differences in the mean annual precipitation rates between the region of Poás volcano (3,000 and 6,000 mm) and Mammonth Lakes (1,000 and 2,500 mm) that could help to explain both the area and volume differences. These rainfall rates at Poás were sufficient to produce a high level of moisture in the soils that, as observed in this landslide catalog, facilitated the generation of debris flows events which contributed to about 36% of the total area stripped in the Cinchona event (Table 10). Meanwhile, the occurrence of this type of landslides was not considered normal in higher latitude zones with lower rainfall levels. We also suggest that the intense weathering and thick regolith soils at volcanic edifices in tropical regions could help to explain why the Poás landslides removed more material than the Mammonth Lakes event.

By using the difference between the expected volume removed for a 6.2 M_W magnitude earthquake (best fit line) and the results obtained for the Cinchona earthquake (Fig. 8), we defined a range where the other five historic events near Poás volcano from Table 2 may be located. Based on what we observed in this study, we differentiated the historical earthquakes that had the same moment magnitude using their differences in location (Pacific or Caribbean slope) and time of the year when they occurred. The events located in the north and central region of the Poás volcano present higher frequency, density and volume removed by the landslides than those located in the southern flanks and that occurred between January and May, i.e. during the dry season.

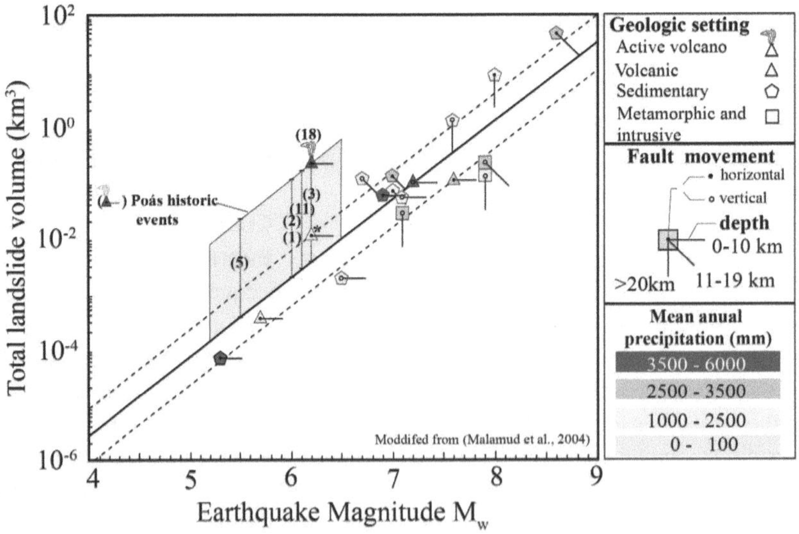

Fig. 8 Comparison of Cinchona event with coseismic world catalog from Malamud et al. (2004). Bracketed numbers correspond to events near Poás volcano reported in Table 2, the shade area is the range between the maximum and minimum estimated volume removed from these events. Error lines for these events are based on the difference between the Cinchona event and the average obtained from Malamud et al. (2004) (*) Mammoth Lakes earthquake (5-25-1980) data from Julian and Sipkin (1985)

3.2 Coseismic Landslides Susceptibility Model for the Poás Volcano

The results and analyses of the slope's intrinsic susceptibility factors (S_l, S_s and S_h,), the earthquake trigger event (Earthq Trig) and the susceptibility to slide (H) are presented in this sub section. These results are presented in maps and tables within the maps of Figs. 9 and 10 where the ranges and weight values used for each factor in our susceptibility model are shown. The maps were created using a grid of ~ 1100 pixels per km^2 in which each pixel presents the values obtained from each factor from Eqs. (3) to (11), to fill the spaces between pixels, a triangulation with linear interpolation was applied for each map.

3.2.1 Lithological Susceptibility (S_l)

Because of the differences in frequency and density of coseismic landslides observed in the Cinchona event between the Paleo and Neo phase units and described in the previous sub-section units, we selected these two temporal phase units to obtain the lithological susceptibility factor for the study area. According to a statistical approach following Eqs. (5–9), we obtained that S_l for the Paleo-phase units is 2.22 while for the Neo-phase units this factor is 1.72 (Fig. 9a). The first weight value covers 49.35% of the study area while the second one refers to 50.65%.

3.2.2 Slope Angle Susceptibility (S_s)

The results for this factor were directly obtained by firstly transforming the altitude values from the LiDAR based DEM into the slope angle ranges suggested by Van Zuidam (1986). Then, each slope angle value was grouped and transformed into the S_s weights values from 0 to 6 based on Mora et al. (2002) classification (Fig. 9b). From the slope angle susceptibility map (Fig. 9b) we observed that the distribution for each class of the range of values for Ss is: 0 = 0.05%, 1 = 6.30%, 2 = 10.90%, 3 = 21.22%, 4 = 40.18%, 5 = 19.03% and 6 = 2.32%.

3.2.3 Ground Moisture Susceptibility (S_h)

Based on the simple hydrologic balance executed for this chapter following the methodology proposed by Mora et al. (1992), we observed that our study area presents only two zones for ground moisture susceptibility, which are $S_h = 4$ and $S_h = 5$. The distribution of the zones affected by these two ranges is not proportionate since about 97% of the study area presents values for $S_h = 5$ (Fig. 9c). As previously mentioned, the Caribbean side of the CVR presents higher rainfall rates than the Pacific side. The Fraijanes meteorological station (FRA) located on the Pacific slope (Fig. 9c and Table 11) is the only station out of the seven considered in this study that showed significant differences between the Pacific and the Caribbean rainfall rates based on our hydrologic balance. This station has a short dry season between January and April and the amount of rain in this time period never exceeded 150 mm/month, meanwhile the rest of the stations had values between 150 and >250 mm/month during the entire year (Table 11).

3.2.4 Earthquake Trigger (Earthq Trig)

The result for this factor was obtained using the methodology proposed in this chapter (Eqs. 10 and 11). The main idea of this new methodology consists in using data from the earthquake (location, depth, moment magnitude) that generated the landslides and transforming those data into weight values for the earthquake trigger factor. The main difference from previous susceptibility models is that the final result is not a unique value, based on the maximum intensity for the study area in the last 100 years. In contrast, our results have a range of values that attenuate from the source of the event following a similar trend described by the attenuation of the peak ground acceleration values, which were calculated with the data from the earthquake.

Using the Cinchona earthquake data, $M_w = 6.2$ and depth = 4.5 km in Eq. (10), we obtained a maximum value for the PGAa = 392.92 cm/s^2 that corresponds to the earthquake epicenter

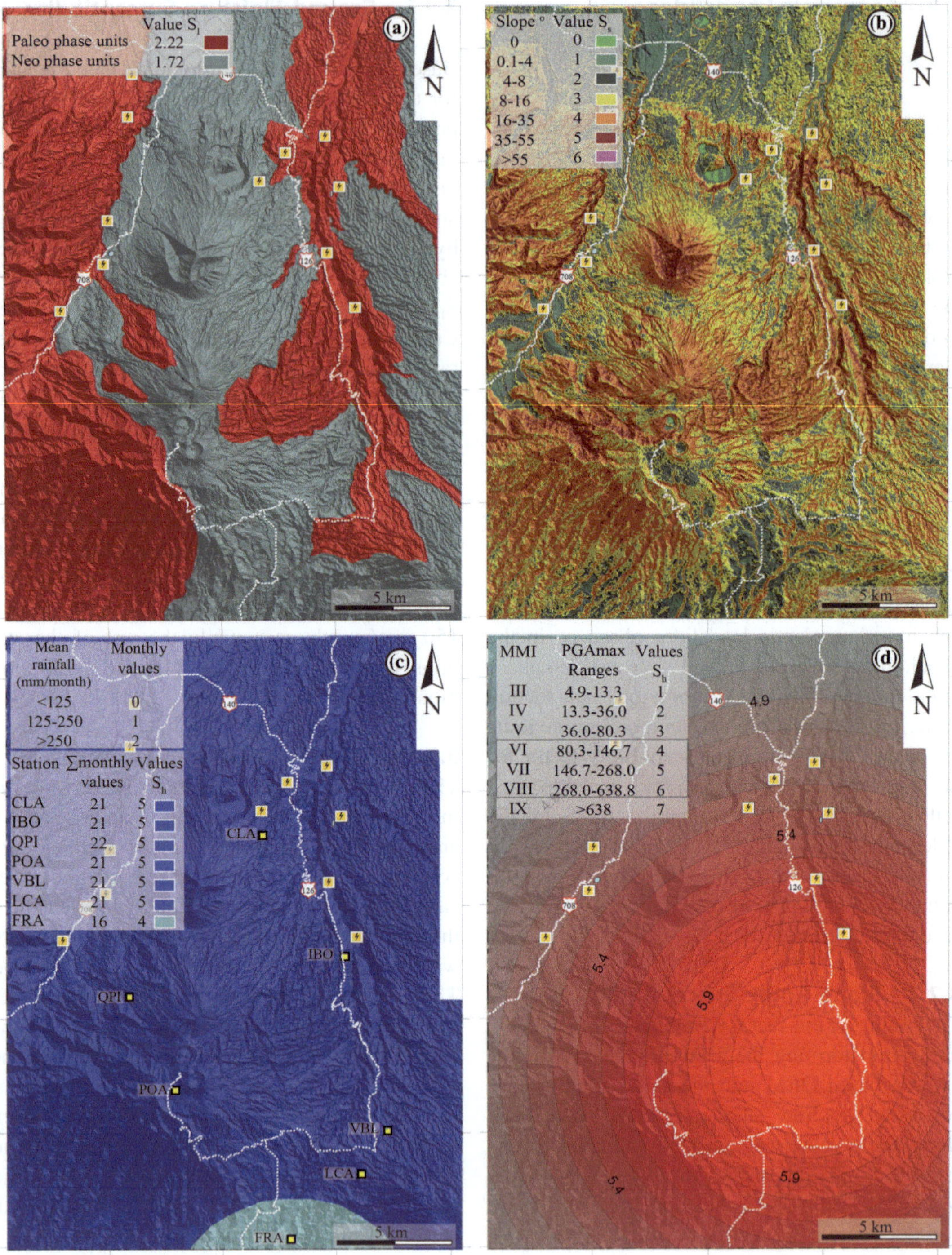

Fig. 9 **a** Lithological susceptibility map. **b** Slope angle susceptibility map. **c** Ground moisture susceptibility map. **d** Cinchona earthquake trigger event map

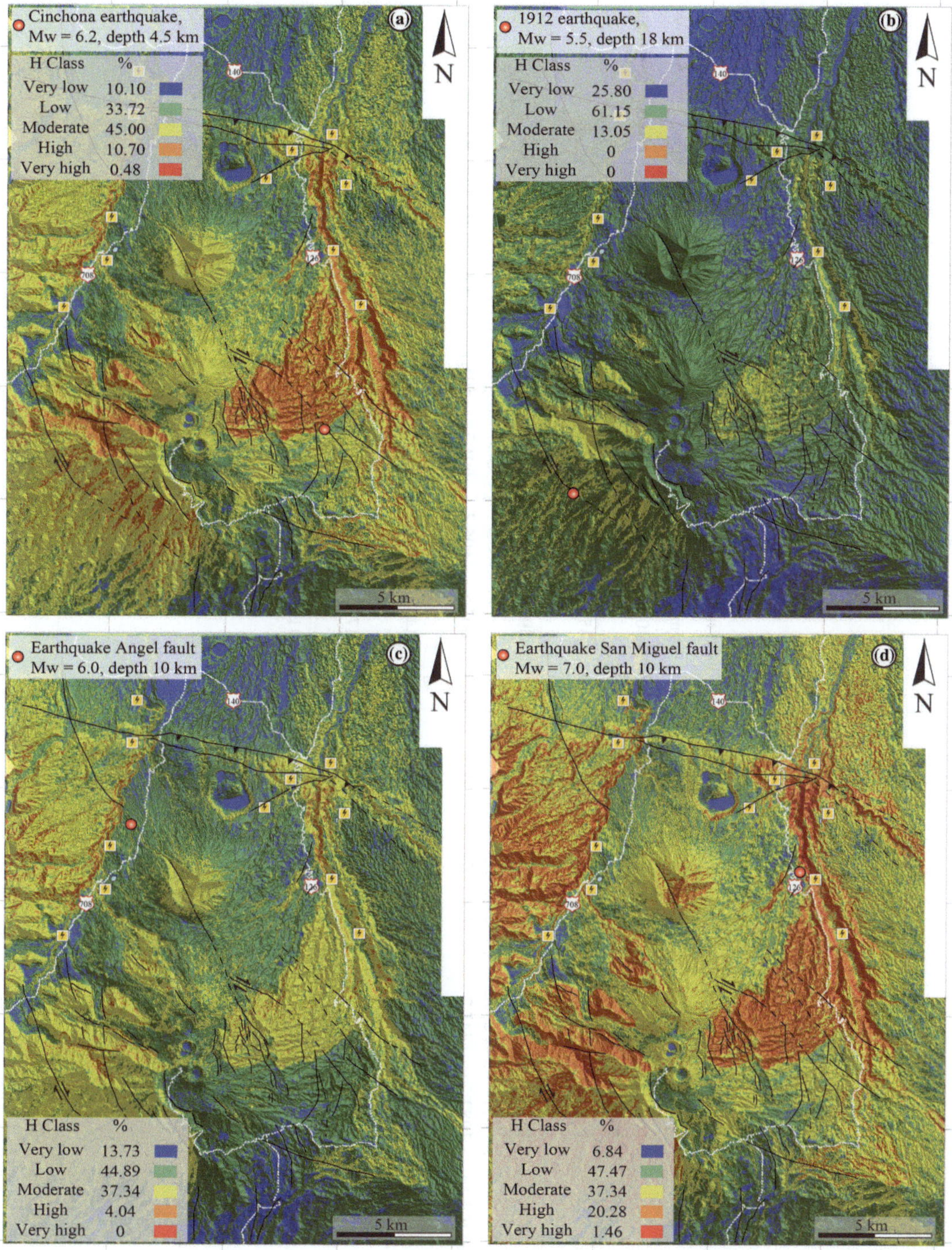

Fig. 10 **a** Susceptibility map for Cinchona earthquake. **b** Susceptibility map for the historical case of the 1912 earthquake. **c** Susceptibility map for a hypothetical earthquake produced by an extension of the Angel fault with its epicenter located on the northwestern flank of Congo volcano. **d** Susceptibility map for a hypothetical earthquake produced by the San Miguel reverse fault with its epicenter located on the east of Congo volcano

Table 11 Monthly rainfall data from seven meteorological stations within the study area used in to obtain the ground moisture susceptibility factor (S_h)

Station (code)	Period	Jan	Feb	Mar	Apr	May	Jun	Jul	Aug	Sep	Oct	Nov	Dec	Annual
Fraijanes (FRA)	1976–2008	11.7	67.8	60.6	124.4	401.3	408.7	336	380.6	506.6	510.1	323.4	170.6	3301.8
Vara Blanca (VBL)	1959–2006	252.7	135	101.2	137.8	324.9	356.9	370	354.8	366.7	408.6	439.8	411	3659.4
Poás (POA)	1972–1979	255	150	100	117	400	390	385	345	365	400	415	398	3720
Colonia Los Angeles (CLA)	1982–2001	448.7	245.2	222.2	198.7	541.5	607.5	606.7	601.7	595.1	628.8	605.5	560.1	5861.7
Quebrada Pilas (QPI)	1986–2001	502.8	277.7	216.6	204.3	392	444	485.5	478.3	470.1	498.5	524.1	615.7	5109.6
Isla Bonita (IBO)	1981–1983	385.2	222.5	180.1	193.4	429.6	473.2	468	469	443.4	515.6	563.3	502.1	4845.4
Los Cartagos (LCA)	1968–1984	250	130	105	140	420	410	379	360	370	410	450	415	3839

location, therefore by using Eq. (10) to calculate the PGAa for each pixel location (latitude-longitude) in our map grid, we obtained different values for the PGAa that decrease as the pixels are located away from the epicenter and in this way we were able to model the attenuation of the PGA in the study area. The lowest value for PGAa obtained using Eq. (10) within the study area was 92.52 cm/s^2. Thus the PGAa values obtained for the Cinchona event gave us a range from 392.92 to 92.52 cm/s^2. Using the ranges for the relationship between peak ground acceleration and the Modified Mercalli Intensity scale from Linkimer (2008) for MMI < VII and Wald et al. (1999) for MMI > VII we obtained the earthquake trigger values that depend on the values of MMI following the classification from Mora et al. (1992). Based on the range of values of PGAa for the Cinchona event, we computed the intensities from VIII to VI and trigger weight values that attenuated from 6 to 4. Using Eq. (11), we obtained the relationship between the peak ground acceleration attenuation ranges and these trigger values. Therefore, we were also able to obtain a specific value for the earthquake trigger factor for each pixel in the study area, modeling in this way the attenuation of earthquake trigger factor and making our susceptibility model more realistic (Fig. 9d).

3.2.5 Slope Susceptibility to Slide (H)

Following Eq. (3), we obtained the slope susceptibility to slide values by multiplying the intrinsic susceptibility factors by the earthquake trigger values of Cinchona event. The values, obtained for each of the three factors of the intrinsic susceptibility to slide, could be considered fixed for the study area, unless they are obtained using a different methodology. The only factor that could change in the area is the trigger event as it depends on the earthquake location and its magnitude. By using the maximum possible value for each of the four factors used to calculated H we obtained the upper limit for the range values of H, the lower limit is = 0 and this only happens when the slope angle of a certain site is zero. Therefore, the range limits for H and their classification in the study area using our susceptibility to slide model with its corresponding classification following the same five categories proposed by Mora et al. (2002), are: 0–93.24 = very low, 93.24–186.48 = low, 186.48–279.72 = moderated, 279.72–372.96 = high and 372.96–466.2 = very high. Using these categories we created a landslides hazard map for the Cinchona event (Fig. 10a), which shows the relative hazard without quantifying the absolute hazard.

In the landslide hazard map for the Cinchona event (Fig. 10a), the areas classified as very low correspond to 10.10% of the total area and are mostly located in the north sector of the study area where the Rio Cuarto Lavas Unit is situated. Factors like distance from the epicenter, low slope angles values and unweathered rocks determine the low susceptibility to slide values for this region. The areas classified as low susceptibility correspond to 33.72% of the study area and are located in the lower flanks of the Congo volcano and Von Frantzius Units and southern sector of the Botos cones. Most of the study area corresponded to the classification moderate (45%), nearly all the area covered by the Paleo-phase units and the steep zones of the Neo phase Units have this zonation or higher. The distribution of the high and very high zones are 10.70 and 0.48% of the total area, and are almost exclusively located in the sector where the La Paz Andesites Unit (Paleo phase) is located. This includes the epicentral area and the river canyons of the Sarapiquí and Toro rivers.

3.3 Methodology Evaluation

Considering that coseismic landslides normally occur in zones classified as moderate, high and very high susceptibility to slide, we performed a statistical analysis to evaluate our methodology. The area from our model where we expect slides comprises 291.57 km^2, i.e. ~56% of the total study area. From the Cinchona landslide catalog, we found that 97.10% of coseismic landslides studied were located within the moderate to very high susceptibility to slide zones. If the same analysis is performed for those areas classified as high or very high susceptibility we obtained that from N_{Lt}, 48.95% are located within these higher susceptibility zones, which only include about 11% of the total study area. The ratio between the area stripped by the Cinchona landslides (21.98 km^2) and the area that resulted from our model as moderate high and very high susceptibility to slide ($\sim$293 km^2) is 0.075. This means that our susceptibility model can determinate satisfactorily the zones where the landslides may occur but there is not a good correlation between sizes of the area stripped by landslides and areas of high susceptibility zones.

Some of the Cinchona coseismic landslides located in the very low and low susceptibility to slide zones are located on road cuts and/or crops and farming areas. This analysis suggests that human activities that generated changes in the original topography might have influenced the occurrence of these landslides.

There is a discrepancy between the PGA obtained from accelerometers (658.0 cm/s^2) within the study area for the Cinchona earthquake and the maximum value for PGAa obtained using Eq. (10) with the data from the same earthquake (392.92 cm/s^2). This difference is reflected in an incongruity of one degree in the MMI values between those reported for the study area after the earthquake (RSN: ICE-UCR 2009) and those associated with our model. Therefore, the trigger values that we used could also be considered one degree lower than what they are supposed to be, if we based the earthquake trigger on the PGA measured and not the calculated with Eq. (10).

These differences between the values for the PGA measured within the study area and those obtained with Eq. (10) could be explained by the fact that this is a general equation used to calculate the peak ground attenuation for crustal earthquakes in Costa Rica and does not include the type of faulting originating the earthquake. Therefore, directivity and site effects could affect the PGA values measured, making them higher than the calculated values.

We decided to use the PGA values obtained from Eq. (10) and not the relatively limited measured data, to be consistent and use only values from our calculations, since we also propose to use this new methodology to recreate historical earthquakes and estimate future earthquakes that lack measured PGA data.

3.4 Use of the Coseismic Landslide Susceptibility Model in Other Events Near Poás Volcano

Here, we present the results obtained using our coseismic landslide susceptibility model in three other events near Poás. In the first case we recreate an historical event that generated landslides by using the earthquake data from Montero et al. (2010) for the 1912 earthquake (Table 3), which was located in the southwest sector of the study area (Fig. 10b). In the second case we create a hypothetical earthquake located in the northern flank of Congo volcano (Fig. 10c), between the Ángel and the Venecia faults (Fig. 3). Finally, in the third case we created an earthquake located on northeast flank of Poás volcano (Fig. 10d) that could be associated with an earthquake generated by the San Miguel fault.

3.4.1 The Sarchí Earthquake (June 12-1912, 5.5M_w, Depth 18 Km)

This earthquake was relocated by Montero et al. (2010) in the southern flank of Poás volcano and associated with the Carbonera fault (Fig. 10b). Most of the infrastructure damages and coseismic landslides reported (Peraldo and Montero 1994) were located in the southwestern flank of Poás volcano outside of our study area. However, our model is in agreement with the historical reports (Peraldo and Montero 1994) that described how the region, located close to the springs of the Sarchí and Anonos rivers and their valleys, was affected by coseismic landslides. Also, our model shows that in the northern sectors from the epicenter there are zones that could be affected by landslides. However, because these zones were isolated at that time, there are no specific reports of the occurrence coseismic landslides. The distribution of the susceptibility to slide zones for this case was (Fig. 10b): the areas classified as very low correspond to 25.80%, the zones classified as low susceptibility correspond to 61.15% and only 13.05% of the study area corresponds to a moderate susceptibility to slide. There are no zones classified as high and very high zones. The areas classified as moderate susceptibility to slide are located close to the epicenter, specifically the zone of the non-vegetated steep corridor located west from the Poás main crater, also some patches in the Platanar northeastern flank and the canyons of the Sarapiquí and Toro rivers (Fig. 10b). Due to the location of this earthquake in relation to our study area (almost at the southwestern edge) we missed the data that this event could generate outside our limits especially on the Platanar volcano and southwestern flank of Poás volcano.

3.4.2 Hypothetical Earthquake Case: Extension of Ángel Fault (6.0 M_w, Depth 10 km)

Based on the historical ruptures for the local faults and the seismic hazard proposed by Montero et al. (2010) for the study area, the Ángel fault could have an extension to the northwest that could rip and produce a shallow earthquake with magnitude between 5.5 and 6.0 M_W. For this hypothetical case we ran our model for an earthquake located on the northwestern flank of Congo volcano, between Ángel and Venecia faults (Fig. 10c), with a focal depth of 10 km and $M_W = 6.0$. The distribution of the susceptibility to slide zones for this case was (Fig. 10c): the areas classified as very low correspond to 13.73%, low = 44.89%, moderate = 37.34% and only 4.04% corresponded to high susceptibility. No areas were classified as very high susceptibility. The high susceptibility zones are mostly located in the Toro and lower part of the Sarapiquí river canyons. Additionally the northeast flank of Platanar could severely be affected by landslides. Since the Congo volcano and von Frantzius summit cones were classified as moderate susceptibility, and based on what we observed from the Cinchona event, the landslides that could occur in these cones are debris flows, which could severely affect the lower areas and river valleys.

3.4.3 Hypothetical Earthquake Case: San Miguel Fault (7.0 M_w, Depth 10 km)

There are no historical records of activity for the San Miguel fault. However, it could have the potential to produce one of the worst possible scenarios in our study area due to its length and fault type. We modeled an earthquake produced by this fault with a depth = 10 km, $M_W = 7.0$ and an epicenter located east of Congo volcano (Fig. 10d). The results of our susceptibility to slide model are: only 6.84% of the study area was cataloged with very low susceptibility, 23.95% low, 47.47% moderate, 20.28% high and 1.46% very high (Fig. 10d). This means that about 70% of the study could be affected by landslides, and only the low angle slopes located north of the study area can be regarded as safe. If an earthquake like the one modeled here would occur, a very dangerous situation could be expected since landslides can be triggered even for areas located over 20 km from the epicenter (Fig. 10d). The Toro and Sarapiquí canyons, the summit of Congo volcano, and the zones where La Paz Andesites unit are located, could be the zones most affected by coseismic landslides from this hypothetical earthquake.

4 Discussion

The combination of LiDAR data and orthophotos to create the Cinchona earthquake coseismic landslides catalog, the analyses of intrinsic susceptibility factors and the new methodology proposed in this work to model the earthquake trigger event provided a broad framework to create the first coseismic susceptibility to slide model for the Poás volcano. The fact that our susceptibility to slide model was able to recreate satisfactorily the distribution of the Cinchona coseismic landslides allow us to venture to use it for modeling other events. Since in the study area we might reasonably expect future earthquakes to produce landslides with characteristics similar to those triggered by the Cinchona earthquake, we tested our model for two hypothetical events and one historical event.

The morphometric data from the Cinchona coseismic landslides catalog provided useful insights on how frequency, sizes and types of coseismic landslides are related to the differences in weathering level and age of the geologic units. The geological units grouped as the Paleo-temporal phase are predisposed to present higher frequency of coseismic landslides than those of the Neo-temporal phase. However, we discovered that since the most common type of landslide occurring on the Neo-phase units is debris flow, size and area affected by these types of coseismic landslides are considerable, especially on the Congo volcano and Von Frantzius units.

Data from the Cinchona coseismic landslides catalog also provided different causes to explain the distribution and style of the coseismic landslides. It is interesting to note how the differences in rainfall rates between the Caribbean and Pacific slopes prior to the event might have affected these parameters. The Caribbean slope had ~408 mm of rain more than that in the Pacific slope one month before the earthquake, in addition to a previously intense rainy season (Table 4). Therefore, the surface soils of the Caribbean side were saturated at the time of the earthquake due to the combination of rain surplus and volcanic soil features. Meanwhile, high moisture in soils on the Pacific slope of Poás volcano was not present, explaining the low frequency of landslides on this side. The saturated soils in the Caribbean side might also explain why some of the earthquake-induced slope failures on this side were very fluid. From our observations we determined that slope failures commonly started in the head streams of the different river basins, and as thin slips, which rapidly turned into very fluid debris flows (lahars) that traveled through the river valleys, using the smooth surface of the fresh lava flows as a slide plane and affecting the river valley walls all the way until lower altitudes, where the flows lost their energy and were able to spread out.

The morphometric data also indicate the range of slope angles on which the coseismic landslides occurred. The majority of the triggered landslides occurred on slopes ranging from 30 to 60° (Fig. 6). Although some coseismic landslides occurred on much more gentle slopes, we observed that some of them where located in places affected by human activities like road cuts and/or farming zones. We proposed that a site effect and a reverse pendulum effect took place in the vicinity of the epicenter, explaining the high frequency of landslides in this area. Most of the coseismic landslides occurred in areas where the land use was primary tropical rain forest and secondary gallery tropical rain forest (Fig. 4). The epicenter of the earthquake was located near the limits of the Poás Volcano National Park where the vegetation cannot be modified by human activities and the gallery forests as they are located in zones with steep slopes.

In Costa Rica, landslide triggered by earthquakes with MMI from IX to VII may cause at least one landslide per km^2 in an area between 1000 and 90 km^2 (Mora and Mora 1994). However, we observed for the Cinchona event that in the study area only 348 km^2 had at least one landslide per km^2. This number may be higher if regions outside the limits of the study area are included and where isolated landslides occurred. Nevertheless, it is still lower than the 1000 km^2 proposed by Mora and Mora (1994). The same intensities may cause an area between 6 and 90 km^2 to present a destruction rate of 60% (Mora and Mora 1994). However, this was not the case for the Cinchona event for which we found only 1 km^2 with that rate of destruction, 18 km^2 presented rates between 30 and 59% of destruction, 27 km^2 had rates between 15 and 29% and 308 km^2 with <15% destroyed. The zones with destruction >15% were located close to the epicenter and/or in zones with steep slopes, like Congo volcano. The possible explanation of this difference is that since the Cinchona earthquake was shallow (4.5 km depth) and the energy was dissipated very quickly, creating a much smaller impacted area.

The comparison of the Cinchona coseismic landslide catalog with another catalog from a similar magnitude, depth and focal mechanism earthquake but located in a dryer location (Fig. 8), showed similarities in the number of landslides generated by the shaking, but significant differences in the affected areas and volumes. Different amount of rainfall and occurrence of debris flows at Cinchona can likely explain the recorded differences. Such factors may also reasonably explain why coseismic landslides in Central America affect more area (at least an order of magnitude) when compared to other world regions (Keefer 1984; Rodríguez et al. 1999; Bommer and Rodríguez 2002).

The use of our new methodology to model the attenuation of the trigger earthquake events allowed us to include data in the susceptibility to slide model what before was left aside, though strongly related to occurrence of landslides. The use of the epicenter location, moment magnitude, focal depth and the application of an equation that follows the same graph as the attenuation of the peak ground acceleration of the earthquake provided a better and more realistic landslide distribution. Furthermore, the possibility of changing the parameters for different earthquakes opens the possibility to test multiple scenarios. However, our model can further be improved by adding fault type and energy distribution.

5 Summary

The use of LiDAR data and orthophotos to create the Cinchona coseismic landslide catalog allowed us to obtain detailed measurements and valuable insights into the characteristics and causes of the coseismic landslides that occurred at Poás volcano.

The distribution of the Cinchona coseismic landslides was the result of a combination of key factors that coincided with the earthquake's epicenter location, these factors are: (a) the age (>0.5 Ma) and therefore the erosion and highly weathering level of the rocks and soils from the paleo-phase units, (b) the steep slopes of the la Paz Andesites unit, and (c) the high moisture level of the soils in the Caribbean slope of the Poás volcano.

The Cinchona earthquake caused about 4,946 ± 100 landslides that striped an area of 22.14 ± 0.27 km^2 and removed a total volume between 0.24 ± 0.0013 and 0.39 ± 0.10 km^3. The occurrence of debris flows was relevant, because this type of landslides contributed for about 37% of the total area striped by landslides and the largest debris flows were mostly originated from the Neo-phase units. We demonstrated that an earthquake in a tropical setting can approximately generate the same number of landslides than an equivalent earthquake located in a subtropical dryer region but the landslides from the first case will strip larger areas due to the differences in the ground moisture that allows the generation of debris flows in the tropical and more humid zones.

The used of the new DEM based on LiDAR data and our landslide catalog allowed us to determinate in great detail the values for the intrinsic susceptibility factors (lithology, slope and ground moisture) of the study area. The new methodology proposed here to obtain the earthquake trigger event values and then its application to obtain the susceptibility to slide zones from an earthquake resulted in high congruence (97%) for the Cinchona event. Because of this high congruence and since the intrinsic susceptibility factors are fixed for our study area, we felt confidence enough to change the epicenter location and other earthquake characteristics (M_w and depth) to model three additional events.

From the four run models we found that the Toro and Sarapiquí river canyons are zones that may suffer high susceptibility to slide values, because of their high slope angles and the lithology present on their walls (Paleo-phase units). Other sites where high susceptibility values computed were the non-vegetated corridor located west from the main crater of Poás volcano and the areas where the La Paz Andesites unit is exposed.

The northern part of the study area, where the Río Cuarto Lavas unit outcrops always showed the lowest values of susceptibility due to the low slope angles and low weathering level of its rocks. The landslides located on the very low and low susceptibility zones could be attributed to changes in the original topography by recent human activities. The effective use of the information generated in this chapter by planners and developers could reduce the impact of future coseismic landslides on the population and on the important civil infrastructures located in the study area.

Acknowledgements Field work logistics and acquisition of the LiDAR data were possible only by the efforts and support offered by the Costa Rican Institute for Electricity (ICE), which is highly acknowledged. The rainfall data of the study area were obtained thanks to the Costa Rican Meteorological Institute (IMN). Reviews and suggestions by Sergio Mora and Scott Burns and an anonymous reviewer are warmly thanked.

References

Alvarado GE, Morales LD, Montero W, Climent A, Rojas W (1988) Aspectos sismológicos y morfo tectónicos del extremo occidental de la Cordillera Volcánica Central, Costa Rica. Rev Geol Amér Central 9:75–98

Alvarado GE, Carr MJ (1993) The Platanar-Aguas Zarcas volcanic centers, Costa Rica: spatial-temporal association of Quaternary calc-alkaline and alkaline volcanism. Bull Volcanol 55:443–453

Alvarado GE (2009) Los volcanes de Costa Rica: Geología, historia, riqueza natural y su gente. 3rd edn, 32, Editorial Universidad Estatal a Distancia, Costa Rica, 330 pp (In Spanish)

Alvarado GE (2010) Aspectos Geohidrológicos y sedimentológicos de los flujos de lodo asociados al terremoto de Cinchona (M_w 6,2) Del 8 de enero del 2009, Costa Rica. Rev Geol Amér Central 43:67–96 (In Spanish with English abstract)

Arredondo SG, Soto GJ (2006) Edad de las lavas del Miembro Los Bambinos y sumario cronoestratigráfico de la Formación Barva, Costa Rica. Rev Geol Amér Central 34–35:59–71 (In Spanish with English abstract)

Barquero R, Peraldo G (1993) El temblor de Pejibaye de Turrialba del 10 de Julio de 1993: aspectos sismológicos, neotectónicos y geotécnicos. Informe interno del Instituto Costarricense de Electricidad 1–32 (In Spanish)

Bommer JJ, Rodríguez CE (2002) Earthquake-induced landslides in Central America. Eng Geol 63:189–220

Bonell M, Gilmour DA, Sinclair SF (1981) Soil hydraulic properties and their effect on surface water transfer in tropical rainforest catchment. Hydrol Sci Bull 26:1–18

Borgia A, Burr J, Montero W, Morales LD, Alvarado GE (1990) Fault propagation folds induced by gravitational failure and slumping of the Central America Costa Rica volcanic range: implications for large terrestrial and Martian volcanic edifices. J Geophys Res 95(B9):14357–14382

Boschini I, Alvarado G, Rojas W (1988) El terremoto de Buena Vista de Pérez Zeledón (Julio 3, 1983): Evidencia de una fuente sismológica intraplaca desconocida en Costa Rica. Rev Geol Am Central 8:111–120 (In Spanish with English abstract)

Carr MJ, Saginor I, Alvarado GE, Bolge L, Lindsay F, Milidakis K, Turrin B, Feigenson MD, Swisher III CC (2007) Element fluxes from the volcanic front of Nicaragua and Costa Rica. Geochem Geophys Geosyst 8(6):1–22 Q06001. https://doi.org/10.1029/2006gc001396

CIESIN, Center for International Earth Science Information Network, Columbia University; FAO, United Nations Food and Agriculture Programme, CIAT, Centro Internacional de Agricultura Tropical (2005) Gridded Population of the World, Version 3 (GPWv3): Population Density Grid, Future Estimates. Palisades, NY: Socioeconomic Data and Applications Center (SEDAC), Columbia University, Nov 2011. Available at http://sedac.ciesin.columbia.edu/gpw

Clift P, Vannucchi P (2004) Controls on tectonics accretion versus erosion in subduction zones: implications for the origin and recycling of the continental crust. Rev Geophys 42, RG2001. https://doi.org/10.1029/2003rg00127

Climent A, Moya A (2009) Registros acelerográficos obtenidos durante el terremoto de Cinchona del 8 de enero de 2009, Costa Rica. Memoria electrónica, 10th national congress of geotechnics and 5th Central-American congress of geotechnicians, Aug 2009, San José Costa Rica, pp 19–21

Climent A, Rojas W, Alvarado GE, Benito B (2009) Costa Rica. In: Benito B, Torres Y (eds) Amenaza sísmica en América Central. Entinema, Madrid, pp 229–251

Evans SG, Bent AL (2004) The Las Colinas landslide, Santa Tecla: a highly destructive flowslide triggered by the January 13, 2001, El Salvador earthquake. In: Rose WI, Bommer, JJ Lopez DL, Carr MJ, Major JJ (eds), Natural hazards in El Salvador, Geol Soc Am, Special Paper 375:25–38

DeMets C, Gordon RG, Argus DF (2010) Geologically current plate motions. Geophys J Inter 181:1–80. https://doi.org/10.1111/j.1365-246X.2009.04491.x

Fernández-Arce M, Mora-Amador R (Chapter 5) Seismicity of Poás volcano, Costa Rica. In: Tassi F, Mora-Amador R, Vaselli O, (eds) Poás volcano (Costa Rica): The pulsing heart of Central America Volcanic Zone. Springer, Heidelberg (Germany)

Garwood NC, Janos DP, Brokaw N (1979) Earthquake-caused landslides: A major disturbance to tropical forests. Science 205:997–999

Gazel E, Ruiz P (2005) Los Conos piroclasticos de Sabana Redonda: Componente magmático enriquecido del volcán Poas, Costa rica. Rev Geol Am Central 33:45–60 (In Spanish with English abstract)

GVN (2009) Fatalities from a large earthquake; slide and minor eruption in crater. Smithsonian Institution, BGVN 34:01

Holbrook WS, Lizarralde D, McGeaary S, Bangs N, Diebold J (1999) Structure and composition of the Aleutian island arc and implications for continental crustal growth. Geology 27:31–34

Hovius N, Stark CP, Allen PA (1997) Sediment flux from mountain belt derived by landslide mapping. Geology 25:231–234

ICE (Instituto Costarricense de Electricidad), Sector de energía UEN proyectos y servicios asociados (2008) Manejo de los sedimentos del embalse del P.H. Cariblanco. Internal Report ICE, San José, pp 25–50 (In Spanish)

IMN (Instituto Meteorológico Nacional-Costa Rica (2008a) Boletín Meteorológico Año XXXIII November 2008 ISSN-1659-0465:1-29 (In Spanish)

IMN (Instituto Meteorológico Nacional-Costa Rica (2008b) Boletín Meteorológico Año XXXIII December 2008 ISSN-1659-0465:1-30 (In Spanish)

Julian BR, Sipkin SA (1985) Earthquake Processes in the Long Valley Caldera Area, California. J Geophys Res 90:11155–11169

Keefer DK (1984) Landslides caused by earthquakes. Geol Soc Am Bull 95:406–421

Kerle N, van Wyk de Vries (2001) The 1998 debris avalanche at Casitas volcano, Nicaragua investigation of structural deformation as the cause of slope instability using remote sensing. J Volcanol Geotherm Res 1105:49–63

Laporte G (2009a) Taludes y sismos. Comportamiento dinámico de taludes durante el sismo de Cinchona y sus implicaciones al diseño geotécnico. Boll Geotécn Notisuelos 10:6–7 (In Spanish)

Laporte G (2009b) Efectos de los sismos en el comportamiento de las laderas naturales, cortes y rellenos: caso del sismo de Cinchona. 10th National Congress of Geotechnics, 5th Central-America Meeting of Geotechnicians, San José, Costa Rica, August 2009, Memoria digital: 10 (In Spanish)

Larsen IJ, Montgomery DR, Korup O (2010) Landslide erosion caused by hillslope material. Nature Geosci 3:247–251

Linkimer L (2008) Relationship between peak ground acceleration and modified Mercalli Intensity in Costa Rica. Rev Geol Am Central 38:81–94 (In Spanish with English abstract)

Malamud BD, Turcotte DL, Guzzetti F, Reichenbach P (2004) Landslides, earthquake, and erosion. Earth Planet Sci Lett 229:45–59

Marshall AJ (1937) Northern New Guinea, 1936. Geograph J 89:489–506

Marshall JS, Fisher DM, Gardner TW (2000) Central Costa Rica deformed belt: kinematics of diffuse faulting across the Western Panama block. Tectonics 19:468–492

Méndez J, Soto GJ, Zamora N, Vargas A, Sjöbohm L, Bonilla E, Barahona D, Solís L, Kycl P, Baroň I (2009) Geología de los deslizamientos provocados por el Terremoto de Cinchona, Costa Rica (M_w 6,2; 8 de enero del 2009) en la Ruta 126 (Varablanca-San Miguel). 10th National Congress of Geotechnics, 5th Central-America Meeting of Geotechnicians, San José, Costa Rica, August 2009, Memoria digital: 22 (In Spanish)

Montero W (1999) El terremoto de 1924 (Ms 7, 0): Un gran temblor intraplaca relacionado al límite incipiente entre la placa Caribe y la microplaca de Panamá? Rev Geol Am Central 22:25–62 (In Spanish with English abstract)

Montero W (2001) Neotectonica de la región central de Costa Rica: frontera oeste de la microplaca de Panamá. Rev Geol Am Central 24:29–56 (In Spanish with English abstract)

Montero W, Rojas W, Boschini I, Barquero R, Flores H (1991) Neotectonica de la región de Puriscal. Origen de la sismicidad de mayo-diciembre de 1990. Memorias 51 Seminario Nacional de Geotecnia—Ier Encuentro Centroamericano de Geotecnistas, pp 4.38–4.51

Montero W, Alvarado GE (1995) El terremoto de Patillos del 30 de diciembre de 1952 (Ms = 5.9) y el contexto geotectónico de la región del volcán Irazú, Costa Rica. Rev Geol Am Central 18:25–40 (In Spanish with English abstract)

Montero W, Soto GJ, Alvarado GE, Rojas W (2010) Division del deslizamiento tectónico y transtensión en el macizo del volcán Poas (Costa Rica), Basado en Estudios Neotectónicos y de sismicidad histórica. Rev Geol Am Central 43:13–36 (In Spanish with English abstract)

Mora R, Mora S, Vahrson W (1992) Microzonificación de la amenaza de deslizamientos y resultados obtenidos en el área del valle central de Costa Rica. Escala 1:286 000 Cepredenac, San José Costa Rica (In Spanish)

Mora R, Chávez J, Vásquez M (2002) Zonificación de la susceptibilidad al deslizamiento: Resultados obtenidos para la Península de Papagayo mediante la modificación del método Mora & Vahrson (Mora et al., 1992). Memoria del tercer curso internacional sobre microzonificación y su aplicación en la mitigación de desastres. Lima, Perú, pp 38–46

Mora S (1985) Las laderas inestables de Costa Rica. Rev Geol Am Central 3:131–161

Mora S (1989) Extend and social-economic significance of slope instability in Costa Rica. In: Brabb E, Harrod L (eds) Landslides: extend and economic significance. Balkema, Rotterdam, pp 93–99

Mora S, Vahrson G (1993) Macro-zoning landslide hazards. Manual for zonation on seismic geotechnical hazards. Jap Geotechn Soc pp 58–61, 128–136

Mora C, Vahrson WG (1994) Macrozonation methodology for landslide hazard determination. Bull Ass Eng Geol 31:49–58

Mora S, Mora R (1994) Los Deslizamientos por el Terremoto de Limón: Factores de Control y comparación con otros eventos en Costa Rica, Rev Geol Am Central, Special Issue, Terremoto de Limón, pp 139–152 (In Spanish with English abstract)

Pain CF (1972) Characteristics and geomorphic effects of earthquake-initiated landslides in the Adelbert Range, Papua New Guinea. Eng Geol 6:261–274

Paniagua S, Soto GE (1986) Reconocimiento de los riesgos volcánicos potenciales de la cordillera central de Costa Rica. Rev Cienc Tecn 10:49–72 (In Spanish with English abstract)

Parker RN, Densmore AL, Rosser NJ, de Michele M, Yong L, Runqiu H, Whadcoat S (2011) Mass wasting triggered by the 2008 Wenchuan earthquake greater than orogenic growth. Nat Geosci 4:449–452

Peel MC, Finlayson BL, McMahon TA (2007) Updated world map of Köppen-Geiger climate classification. Hydrol Earth Syst Sci 11:1633–1644

Peraldo G, Montero W (1994) Los temblores del periodo colonial de Costa Rica. Tecnológica de Costa Rica, Cartago, 162 p (In Spanish)

Peraldo G, Montero W (1999) Sismología histórica de América Central. Inst. Panamericano de Geogra. Historia, México, 347 p (In Spanish)

Peraldo G, Rojas E (2000) Catálogo de deslizamientos históricos de Costa Rica, periodo 1772–1960. IGN, Semestral Report II (In Spanish)

PNUD (Programa de las Naciones Unidas para el Desarrollo), IMN (Instituto Meteorológico Nacional), MINAET (Ministerio de Ambiente y Telecomunicaciones) (2009) Diagnostico Biofísico para Costa Rica, 478 pp (In Spanish)

Prosser JT, Carr MJ (1987) Poás Volcano, Costa Rica: geology of the summit region and spatial and temporal variations among the most recent lavas. J Volcanol Geotherm Res 33:131–146

Rodríguez CE, Bommer JJ, Chandler RJ (1999) Earthquake-induced landslides: 1980–1997. Soil Dynam Earthq Eng 18:325–346

Rojas W, Montero W, López A, Alvarado G, Vargas A, Taylor W (2009) In: RSN: ICE-UCR: Informe del terremoto de Cinchona del jueves 8 de enero de 2009. Univ de Costa Rica, San José, pp 26–44 (In Spanish)

RSN (Red Sismológica Nacional) (UCR-ICE) (2009) El Terremoto de Cinchona del jueves 8 de enero de 2009. Rev Geol de Am Central 40:91–95 (In Spanish with English Abstract)

Ruiz P, Gazel E, Alvarado GE, Carr MJ, Soto GJ (2010) Caracterización geoquímica y petrográfica de las unidades geológicas del macizo del volcán Poás, Costa Rica. Rev Geol Am Central 43:37–66 (In Spanish with English Abstract)

Ruiz P, Mana S, Gazel E, Soto GJ, Carr M, Alvarado GE (Chapter 2) Geochemical and geochronological characterisation of the Poas stratovolcano stratigraphy. In: Tassi F, Mora-Amador R, Vaselli O (eds) Poás volcano (Costa Rica): the pulsing heart of Central America Volcanic Zone. Springer, Heidelberg (Germany)

Schmidt V (2010) Avances para estudios del riesgo a escala regional y local: Aplicación a América Central y a la bahía de Cádiz (Sur de España). Ph.D. thesis, Univ Politécnica de Catalunya (In Spanish with English Abstract)

Schuster RL, Nieto AS, O'ouke TD, Crespo E, Plaza-Nieto G (1996) Mass wasting triggered by the 5 March 1987 Ecuador earthquakes. Eng Geol 42:1–23

Skinner DJ, Porter SC (1992) The dynamic earth: an introduction to physical geology, 2nd edn. Wiley, New York, p 570

Small C, Naumann T (2001) The global distribution of human population and recent volcanism. Environ Haz 3:93–109

Soto GJ (1994) Volcanología Física. In: Denyer P, Kussmaul S (eds) Atlas Geológico Gran Área Metropolitana. Editorial Tecnológica de Costa Rica, pp 131–146

Soto GJ (1999) Geología Regional de la Hoja Poás (1:50 000). In: Alvarado GE, Madrigal LA (eds) Estudio Geológico-Geotécnico de Avance a la Factibilidad del P.H. Laguna Hule. Internal Report ICE, San José, pp 15–45

Van Westen CJ, Soeters R (2000) Remote sensing and geographic information systems for natural disaster management. In: Roy PS, Van Westen J, VK, Lakhera RC, Champati RPK (eds) Natural disasters and their mitigation. A remote sensing and GIS perspective. Indian Inst Rem Sens, Nat Rem Sens Ag, India, pp 31–76

Van Zuidam RA (1986) Aerial photointerpretation in terrain analysis and geomorphologic mapping. Smits Publishers, The Hague, 442 p

Vannucchi P, Mason JP (Chapter 1) Overview of the tectonics and geodynamics of Costa Rica. In: Tassi F, Mora-Amador R, Vaselli O (eds) Poás volcano (Costa Rica): the pulsing heart of Central America Volcanic Zone. Springer, Heidelberg (Germany)

Varnes DJ (1978) Slope movement types and processes. In: Schuster RL, Krizek RJ (eds) Landslides: analysis and control-special report, 176, Transport Research Board, Nat Acad Sci, Washington, DC (USA), pp 11–33

Wald DJ, Quitoriona V, Heaton TH, Kanamori H (1999) Relationship between peak ground acceleration, peak ground velocity, and modified Mercalli Intensity in California. Earthq Spectra 15:557–564

Seismicity of Poás Volcano, Costa Rica

Mario Fernández-Arce and Raúl Alberto Mora Amador

Abstract

This chapter summarizes the results of previous studies on the seismicity of Poás volcano and shows its behavior from 1980 to 2006, the last year of analogical recording at the seismic station of the Red Sismológica Nacional (RSN: ICE-UCR) located on the summit of the volcano. A new estimate of the signal frequency for those events occurring between 2007 and 2015 is also shown. The overview is based on the need to understand and classify the seismic signals of the Poás volcano according to the available records, modern instrumentations, and current knowledge about volcanic seismology. This will help to improve the management of volcanic emergencies and interpret changes in seismic activity in the volcano. To carry out the work, we first revised previous publications related to the seismicity of Poás volcano. Then, we analyzed seismic records of the seismic station VPS2 of the Red Sismológica Nacional (RSN: ICE-UCR) and other data generated by temporary seismic networks that were installed in the volcano. Three seismic signals have been clearly identified at Poás volcano: low-frequency events, tremors, and tectonic earthquakes. The most typical ones are low-frequency signals. The low-frequency events occur inside and around the main crater, at shallow depths, in response to vapor pressure. Tremors sporadically occur at Poás before and during the period of perturbation inside the volcano as a consequence of degassing. Swarms of small tectonic earthquakes (frequency = 1–2.5 Hz) also occur within the volcanic area in response to movements in local faults. The seismicity of Poás volcano, especially the low-frequency ones, increases before each eruptive cycle.

Keywords

Volcanic seismic signals · Poás · Tremors · Low-frequency events

1 Introduction

The Poás is an active stratovolcano and one of largest and most active volcanoes in Costa Rica. Three large structures developed in the volcanic edifice: Von Frantzius cone, Main Crater and Botos cone (Fig. 1). Von Frantzius is a highly eroded composite cone with a south-facing breached crater (Prosser and Carr 1987). Next

M. Fernández-Arce (✉)
Escuela de Geografía, Universidad de Costa Rica (UCR), Preventec - UCR, Apdo. 35-2060, San José, Costa Rica
e-mail: mario.fernandezarce@ucr.ac.cr

R. A. Mora Amador
Escuela Centroamericana de Geología, Universidad de Costa Rica, Red Sismológica Nacional (ICE-UCR), San José, Costa Rica
e-mail: raulvolcanes@yahoo.com.mx

F. Tassi et al. (eds.), *Poás Volcano*, Active Volcanoes of the World,
https://doi.org/10.1007/978-3-319-02156-0_5

Fig. 1 Aereal photo of Poas volcano. From top to bottom: Cerro Congo, Von Frantzius cone, the Main Crater and Botos cone (see Fig. 2). The cones and craters form a roughly NS alignment Courtesy of Gino González

to Von Frantzius, to the south, is the Main Crater, a structure of approximately 1 km in diameter with two smaller craters inside. One of these craters is the site of a hot acid lake (Rouwet et al. Chapter "39 Years of Geochemical Monitoring of Laguna Caliente Crater Lake, Poás: Patterns from the Past as Keys for the Future" and references therein). Two km south of the Main Crater is the Botos cone, which hosts a cool lake. In Fig. 2 the location of the Poás volcano is reported.

This chapter is a summary of the seismic activity at Poás based on analog and digital records of the VPS2 seismic station of the Red Sismológica Nacional (RSN: ICE-UCR) and data from the seismic networks installed in the summit area of the volcano. Using the seismic records, we characterized the seismic signals. The seismic networks data supply information about the source of B-type volcanic signal and tectonic earthquakes in volcanic areas. This work reviewed different studies on seismicity, i.e. Casertano et al. (1987); Morales et al. (1988); Alvarado et al. (1988); Fernández (1990, 1999); Rowe et al. (1992); Montero et al. (2000). Using digital records, we performed spectral analysis to some seismic signals.

Seismological investigations at Poás volcano started in 1965 when one station was installed in San Pedro de Poás, 13 km SW from the active crater (Taylor and Rojas 1991). Since then, few studies have been carried out at Poás. Casertano et al. (1987) classified Poás seismicity in local or A-type events, impulsive-explosive events, dispersive tremors, and harmonic tremors. Morales et al. (1988) classified Poás low frequency seismicity, as follows: VT events ($f > 10$ Hz with clear P and S phases), B-type events ($1 < f$ 5 Hz), harmonic tremor ($1 < f \leq 5$ Hz), impulsive-explosive events ($f > 10$ Hz) and high frequency tremor ($f > 5$ Hz). Fernández (1990) located a set of low frequency events of Poás

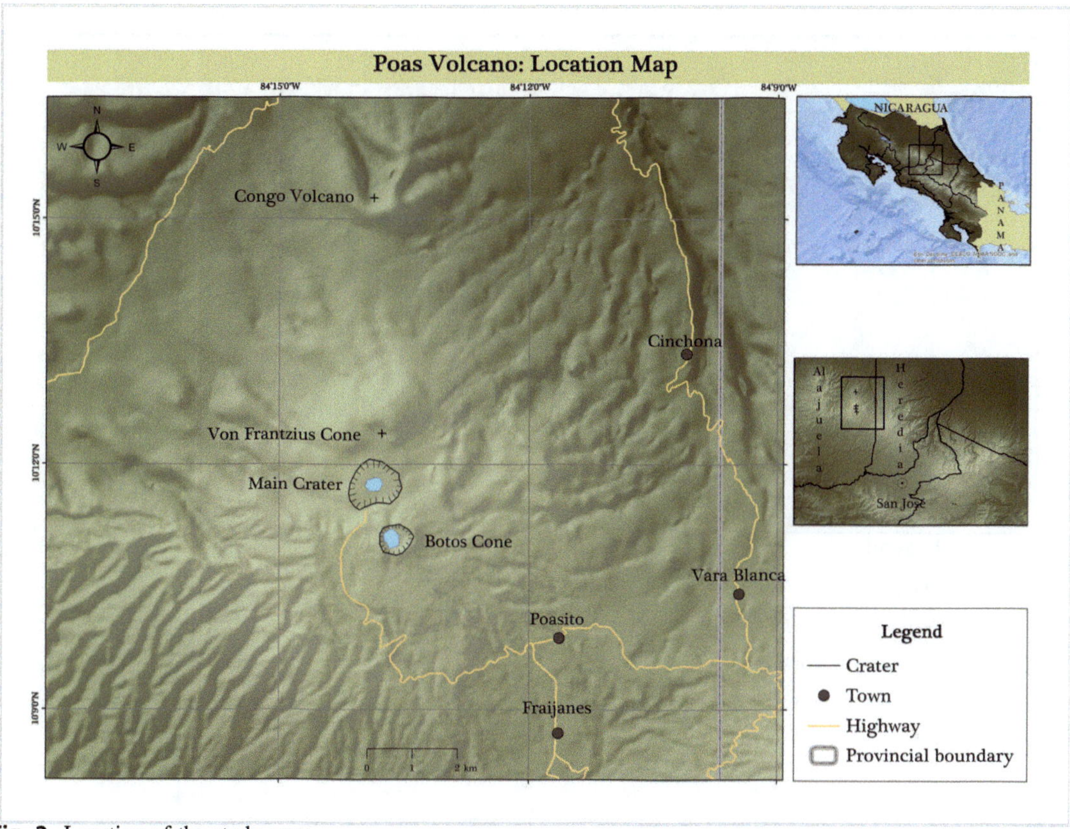

Fig. 2 Location of the study area

volcano while Rowe et al. (1992) used seismic data to determine hydro-fracturing/magma ascent at Poás.

New investigations generated more data about the activity of Poás in the last 15 years, which included seismic swarms and short eruptive cycles. Two seismic networks installed at Poás in 1999 and 2004 provided useful data. More recent studies (Montero et al. 2000) yielded new data on the faulting of the area. Since 1994, digital records of the Poás seismic station are available that allowed a more detailed characterization of the seismic signals.

This chapter includes complete and detailed information about seismicity at Poás according to data and observations collected during the last 35 years. Our goal is to promote and contribute to further seismological investigation at Poás.

2 Data and Processing

In this chapter, we used analog and digital seismic data gathered by three seismic networks of 5 or more stations and the permanent station VPS2 located at 0.5 km from the Main Crater (Fig. 3). The recording system of VPS2 was analog until 2006, and the resulting seismograms were used to make daily counts of seismic signals. In 2006 the analog system changed into a digital recording, and the daily count of volcanic seismic signal stopped. Consequently, since then spectral analysis of seismic signals of Poás became easier and faster.

The first network, operated from 1989 through 1990, consisted of 5 portable MEQ-800 seismic stations. Seismic sensors in operation during this

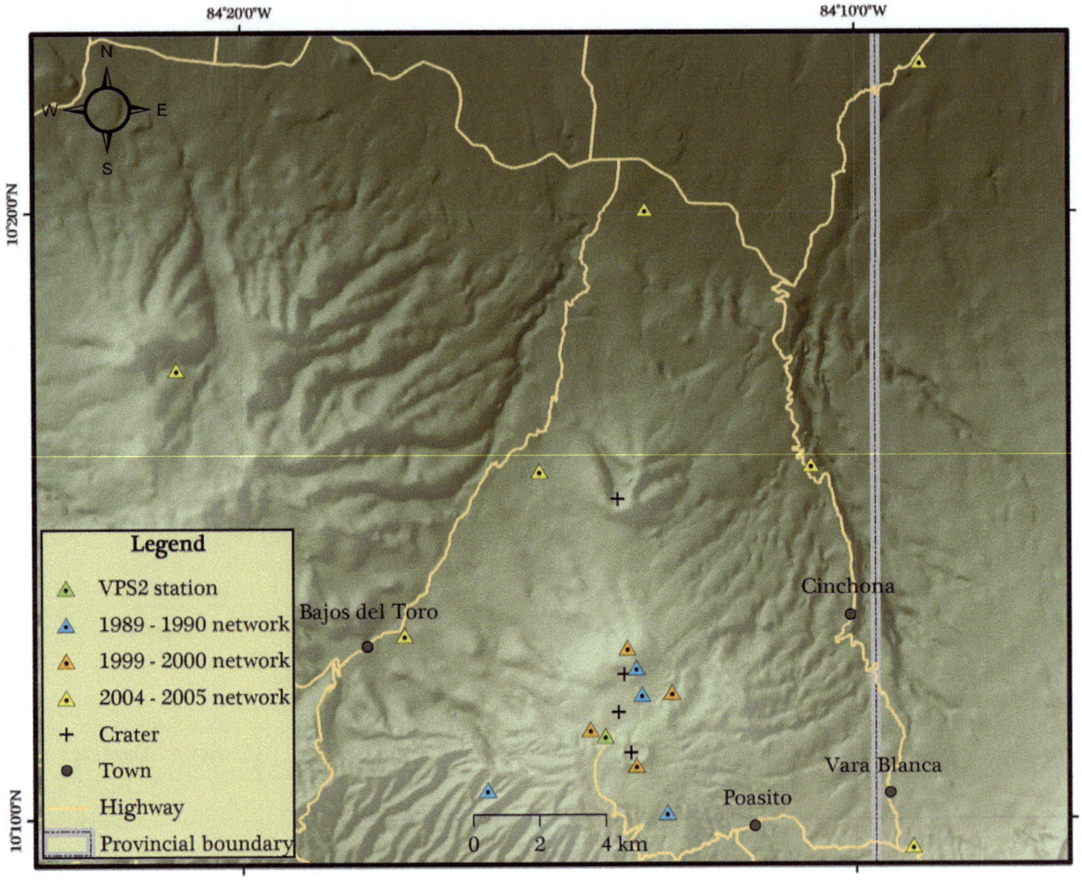

Fig. 3 Seismic stations and networks installed at Poás volcano

period were short-period vertical component seismometers (Mark Products L4C) with a one-second natural period. The nearest station was 0.5 km from the crater and the farthest was placed 3.4 km away from the Main Crater (Fernández 1990).

A second network of 5 stations was deployed at the summit area in 1998. The seismic sensors were short-period vertical component seismometers (Mark Products L4C) with an one-second natural period. The seismic signals were transmitted via VFH radio to communication hubs located at the University of Costa Rica. The data were telemetered to the Red Sismological Nacional (RSN: ICE-UCR) recording center and were then processed along with data from the rest of the RSN Seismic Network. Due to the extreme weather conditions in the area, only 4 stations of a total of 5 were operating. The network operated until 2000 since technical, administrative and financial difficulties to maintain radio-telemetry transmissions made it impossible to keep the network working at Poás.

In July 2004 the last network was installed by the Costa Rican Electricity Company to monitor the seismicity near the volcano where there is a hydroelectrical project. This network consisted of 10 seismic stations, 3 of them set in the summit area and used Lennartz 3-components, short-period seismometers, and REFTEK data loggers. The network operated until September 13, 2005. Presently, only one station of this network is still operating at Poás summit.

Earthquakes hypocenters were calculated using the program HYPO71 (Lee and Lahr 1975) and the SEISAN system, which includes a modified version of Hypocenter (Havskov and Ottomoller 2003).

3 Seismicity

Seismicity at Poás Volcano can be classified in low-frequency events, tremors, and tectonic earthquakes. Here below a description of each seismic signal is reported.

Low-Frequency Seismicity (B-Type)

The generation of low-frequency volcanic seismicity is attributed to: conversion of liquid water into vapor (Latter 1979; Chouet 1985), collapse of bubbles (Kieffer 1984), excitation of the volcanic conduit (Nabyl et al. 1997), expansion of gas pockets in the conduit filled with fluid (Uchida and Sakai 2002), simple shear-induced rock fracture (Sherbun and Scott 1993) and magma (Soosalu et al. 2006; Neuberg et al. 2006; Lin et al. 2007). Mcnutt (2005) pointed out that this kind of volcanic seismic signal is thought to be caused by fluid processes that are not well understood yet.

Low frequency seismicity at Poás is characterized by a large variety of events that can be sub-classified into 3 groups (Fernández 1990). The signals of the first group start with an emergent low amplitude phase with a characteristic time of about 9 s. The second phase has larger amplitude with a variable characteristic time. S-waves cannot be recognized in these signals. Their emergent character may correspond to slow steam release in cavities (Fernández 1990). Araña and Ortiz (1984) suggested that these features could be due to water releasing in fractures. Events belonging to the second group have impulsive P-wave arrivals. Some of them have energy at higher frequencies. The signals of the third group have a higher amplitude, which reaches a maximum and then, begins to gradually decrease. The form of these signals resembles a fish skeleton.

This seismicity of low frequency is extremely abundant at Poás and is one indicator of its activity. Low-frequency events occur up to hundreds per day (Fig. 4), and the number can vary depending on the state of the hydrothermal system. A total of over 950,000 low-frequency events have been recorded at VPS2 seismic station since 1980.

This is a seismic activity that permanent and began to be detected since VPS2 station was installed at the top of the volcano in the 1980s. Its origin could be linked to the magmatic activity of 1953, when the magma ascended and the emplacement of a lava dome of approximately 300 m in diameter in the center of the active crater took place. Before that event, a lagoon completely covered the semicircular crater floor, which had 1,320 m in diameter. The dome divided the lake in 2 parts and reached 40 m above the level of the water (Alvarado 2008). From that year on, the southern lagoon dried up and only the northern sector of the lake remained. It is likely that the hot root of the dome and the magma under it started to heat up a large water reservoir. The lake at that time had a volume of $\sim 1{,}000{,}000\ m^3$, although a groundwater system developed under the surface, which contributes to the production of steam and consequently to the high seismicity.

Fernández (1990) located low-frequency events (Fig. 5) of magnitudes between 0.6 and 2.6 (Ml). The calculated hypocenter for more than 31 reliable locations was within the perimeter of the network, roughly beneath the Main Crater at a depth ranging from 0.09 to 1.0 km. It is not possible to appreciate the 31 aforementioned epicenters because they were basically grouped into 4 sites. For example, at the point tagged with 1 in Fig. 5, 20 epicenters with the same coordinates were located.

There is a clear and direct relationship between the number of low-frequency events and the level of the water in the hot lake. This became evident in 1989, when the number of seismic events declined from 10,000 in February

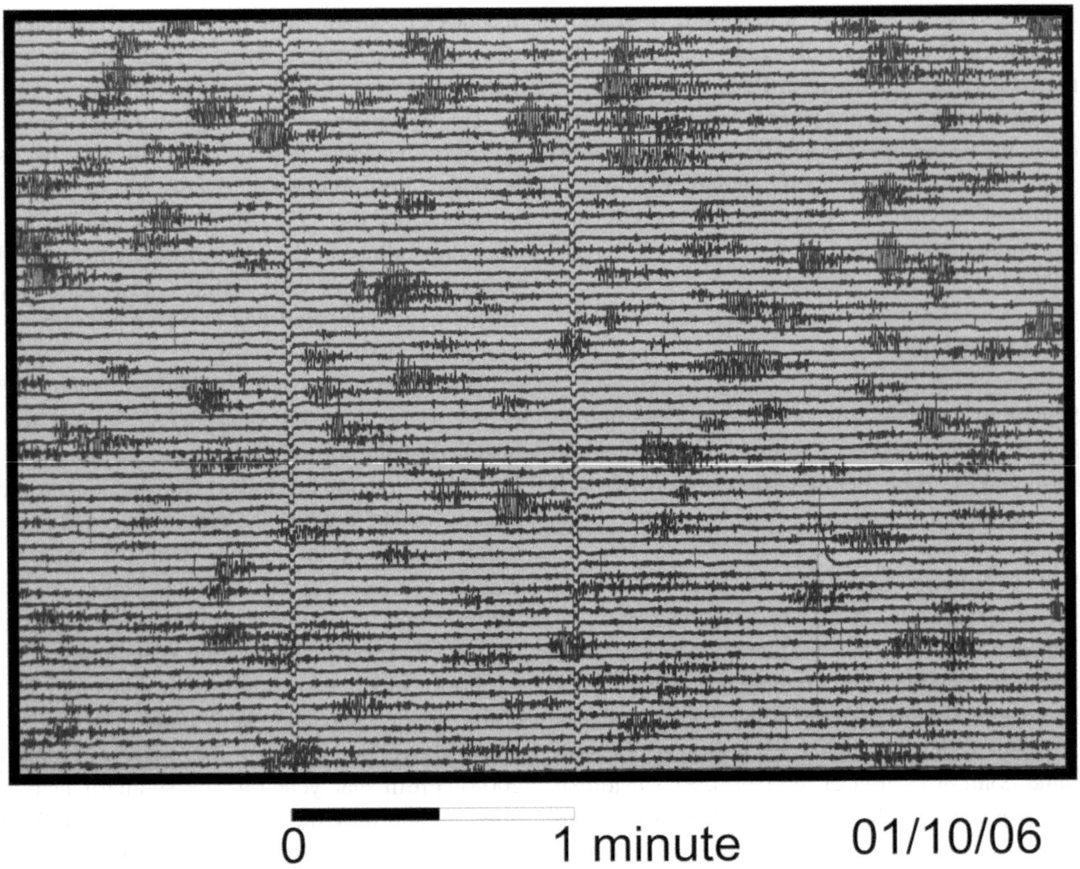

Fig. 4 Seismograms of the VPS2 seismic station showing low-frequency events recorded on January 10, 2006. Thousands of these events were recorded monthly at VPS2

to 2,000 in June and simultaneously a rapid decrease of the water level of the hot lake occurred. The water level dropped down to 40 m and eventually the lake dried out in April 1989. Seismicity reached a very low level (20 events per day), lower than the daily average of low-frequency events (55) in 1985 when the activity at Poás was considered normal. In June, when rainfall started and the hot lake recharged, low-frequency seismicity increased. This behavior repeated in 1990, 1991, and 1994.

Such a behavior suggests that low-frequency seismicity at Poás needs water to occur and, therefore, the main mechanism responsible for this seismicity is the conversion of water into vapor. Rainfall supplies the water necessary for the generation of vapor, which in its turn it is able to influence the formation of low-frequency events. Summarizing when the lake dries out and low-frequency seismicity events are low. Once precipitations allow the lake to reform, evaporation and steam, as well as the gas pressure in cavities below the surface, increase and seismicity also increases.

Although the magmatic activity is responsible for the seismicity of the low frequency in other volcanoes (Soosalu et al. 2006; Neuberg et al. 2006; Lin et al. 2007), this factor does not seem to have a direct effect on the permanent low-frequency seismicity of Poás. This is based on the fact that this type of volcanic signals continuously occurs at the volcano and during long periods of complete calm and normalcy. The function of the magma in the conduit seems

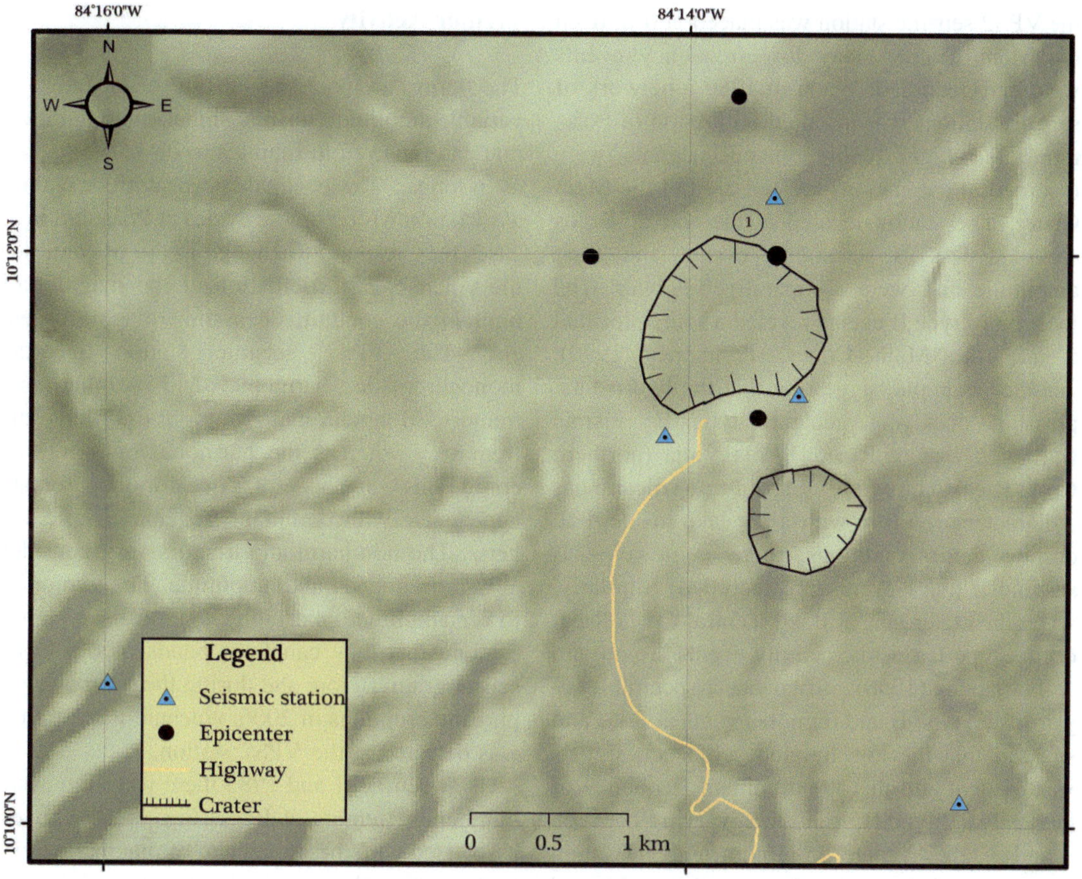

Fig. 5 High-quality epicenters of low-frequency volcanic signal located by Fernández (1990). Such events were recorded in five stations

to be limited to heating the ground and the bottom of the lake water to generate steam. Although the magma remains in the duct, low frequency signals disappear if there is no water in the lake, suggesting that the hydrothermal system of Poás water has a greater influence in the production of low-frequency events than that of the magma. When the heat increases, i.e. ascent of magma or magmatic fluids, as happened in 1989, the heat flow allows the lake to dry out completely. Contrary, when the hydrothermal system is not able to facilitate the cooling mechanism, the heat source is not capable to generate the usual seismicity of low frequency at Poás. In other volcanoes magma likely generates low-frequency earthquakes and perhaps they are also generated at Poás when there is a significant rise of magma in the duct that may reach the surface, but the permanent seismicity of low frequency of Poás does not seem to be related to magma.

A higher water level is also responsible for a higher hydrostatic pressure and pore pressure into the hosting rocks and, therefore, the high B-type seismicity could be generated by the high hydrostatic pressure caused by the lake. If this were the cause of type B seismicity, then below the crater Botos should also constitute a high seismicity of the same type. This crater is located 1 km south of the Main Crater and has 750 m in diameter while the lake water cover of surface that has 400 m in diameter. Comparing the diameters of the two lakes, that of Botos is 100 m wider than that of the Main Crater. Since the VPS2 seismic station was placed 500 m from the Main Crater, many low frequency events were here recorded. When in 1989 a network of seismic stations was installed on the top of Poás, more and high quality seismic signals were recorded at the stations located around the Main Crater. The station located at the rim of Botos recorded less B seismicity events and their amplitude and were comparatively lower. The location of type B events reveled a source located around the Main Crater (Fernández 1990). Another example suggesting that the hydrostatic pressure does not generate type B seismic activity is that of Rincón de la Vieja (northern Costa Rica). The active crater of this volcano has a diameter of 700 m with a hot lake inside, but, despite these conditions, there is no type B seismic activity events are recorded. Similarly, when the crater lake at Irazu (central Costa Rica) occurs type B seismic activity events are absent.

Fernández (1990) carried out a spectral analysis of the type B and tremors events at Poás and stressed that the low frequency of such signals was evident and that the dominant frequency of the analyzed type B events was varying between 1 and 3 Hz. Average, minimum and maximum frequency of these signals were 1.9, 0.35 and 3.6 Hz respectively. Morales et al. (1988) found an average frequency of 1.7 Hz, a minimum frequency of 0.35 Hz and a maximum frequency of 2.25 Hz, which agrees with the results obtained by Fernández (1990). Accordingly, the strong correspondence of spectral peaks with low frequencies suggests a unique source process for the type B events at Poás.

Tremor Activity

The term "tremor" denotes a wide range of variable amplitude seismic signals with durations that can range from minutes to days, weeks, and even longer. These signals occur with or without association with eruptive cycles at Poás, but in all cases they respond to change in the dynamics of the volcano, e.g. oscillation of the magma column in the conduit, degassing, magma ascent, etc. The VPS2 seismic station recorded monochromatic, harmonic, and non-harmonic tremors. Monochromatic tremors only have one frequency whereas the harmonic ones refer to continuous signals of quasi-monochromatic appearance with dominant frequencies of a few hertz. The non-harmonic tremors vary considerably in frequency and amplitude. The tremors of lower amplitude and high duration have signals so long that they can form bands of 6 h in the seismogram. Before and during the short cycle of phreatic eruptions in 2006, volcanic-noise tremor was detected at the VPS2 station.

Between 1981 and 1983 there was an unusual number of tremors at Poás (Fig. 6). They were associated with the movement of magmatic fluids to the surface through the fractured dome located inside the Main Crater (Fernández 1990). In December 1981, the fumaroles at the dome reached a temperature of 1020 °C (Global Volcanism Program 1981) and incandescence was observed in the northern wall. However, the volcano did not erupt. Morales et al. (1988) recorded 2 events whose frequencies were 1.3 and 2.8 Hz, respectively. Volcanic tremors were

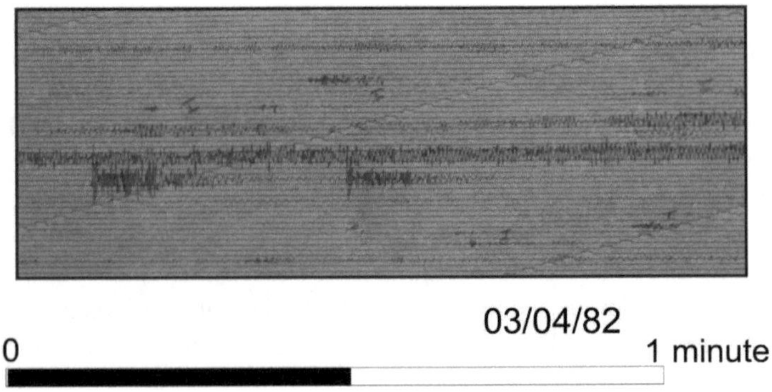

Fig. 6 Tremor of Poás recorded at the VPS2 seismic station. The duration of these signals vary between 5 and 10 min. They were recorded in March 4, 1982

also recorded at Poás during periods of high phreatic activity, as observed since 1994.

We postulate that the causes of the known tremor episodes at Poás are movement of volatiles or small magma oscillations in the conduit. This argument arises from the fact that during those episodes there were no lava emissions. Opening of cracks, conversion of thermal to mechanical energy and gas content are the proposed mechanisms for the generation of volcanic tremors (McNutt 2005).

Tectonic Earthquakes

Although the focus of this work is the volcanic seismicity of Poás volcano, brief information about seismic sources of historical tectonic earthquakes from the Poás area is provided because these data could be useful to the scientific community. After all, Poás seems to have grown over a NS fracturing (e.g. Fernández 1990). In addition, such information has relevance because volcanic activity can be influenced by changes in the pattern of strains related to tectonic earthquakes (Walter et al. 2007; Watt et al. 2009; Brenguier et al. 2014).

In the Poás volcanic area there are 4 major faults (Fig. 7) named Viejo-Aguas Zarcas, Carbonera, Ángel, and San Miguel (Montero 2001; Denyer et al. 2009; Fernández 2013). The first one is a major, young, north-striking, normal fault with a deep and linear valley that extends for about 25 km from the west of Poás to the alluvial plain. Carbonera is a 6.5 km long NNW trending fault with prominent scarps: a deep linear valley slightly displaced dextrally. The Angel fault consists of several segments that run through the eastern slopes of the Poás volcano from the Von Frantzius cone to the vicinity of Vara Blanca for a length of ≈15 km; scarps, linear valleys and rivers adapted or displaced can be considered evidence of the Angel fault (Barquero 2009). San Miguel is a 15 km long fault with a linear scarp relatively little dissected whose maximum height is 160 m in its central part. Along the scarp and the nearby rivers running along deep canyons hot springs seep out.

Such faults have been mapped as active faults. According to Denyer et al. (2009), Viejo-Aguas Zarcas is normal, Carbonera is a strike-slip and San Miguel is a reverse fault. Angel fault is a normal fault with a dextral component (Barquero 2009). Five historical earthquakes (Table 1; Fig. 7) are possibly linked to those faults.

Inspections carried out in the Main Crater of Poás after the 1888 earthquake showed repeated streams of intermittent muddy geyser-like water springing up with violence in the midst of the boiling lagoon, where the temperature was higher than in other times (González 1910). The authors have no knowledge of the effect of other earthquakes, neither if there are records of inspections made by geologist or local authorities in the crater after those earthquakes.

Swarms and Sequences

Several swarms and sequences of tectonic earthquakes occurred in the volcanic area of Poás, especially in the western and eastern flanks. Most earthquakes of the swarms occurring near the summit had a signature typical of tectonic earthquakes with impulsive, high frequency P and S waves. Due to the relationship between tectonic earthquakes and volcanic eruptions, it is important to analyze the tectonic swarms in this area.

The 1982 Swarm

Barquero and Alvarado (1989) reported a swarm of earthquakes at Poás in February 1982. Some events were located near the Angel fault at a depth of approximately 5 km.

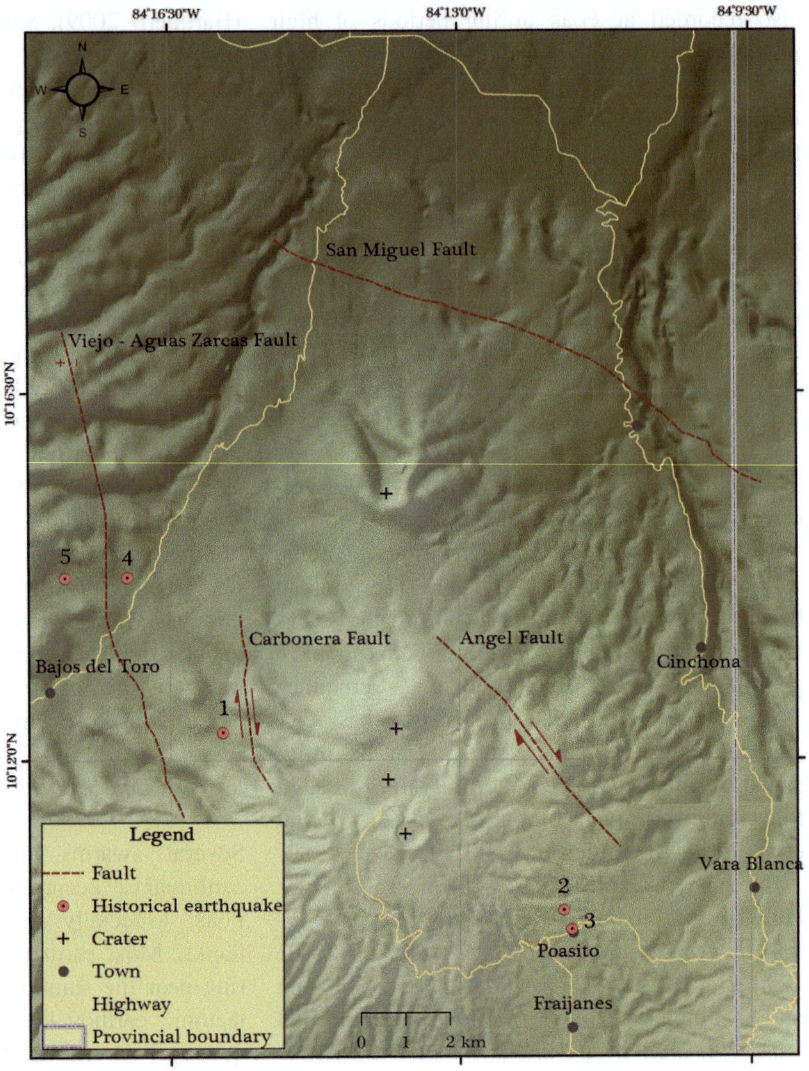

Fig. 7 Faults and historical earthquakes at Poás volcano

The 1990 Swarm

A swarm of earthquakes took place in the Poás area on July 19, 1990 (Rojas and Barquero 1990). Most events occurred within a period of three hours and were located SW of the Main Crater. The magnitude of the largest event was 3.6 (Mc). According to Rojas and Barquero (1990), the source of this activity was the faulting on the SW flank of the volcano.

The 1999 Sequence

At the beginning of July 1999, a new seismic activity began on the western and eastern flanks of the volcano. This new activity was similar to the one that had been previously observed since the VPS2 seismic station began operating. In July 10 1999, an increase in tectonic seismicity was observed at Poás. After July 10, the rate of events was 1/day until July 18, when a magni-

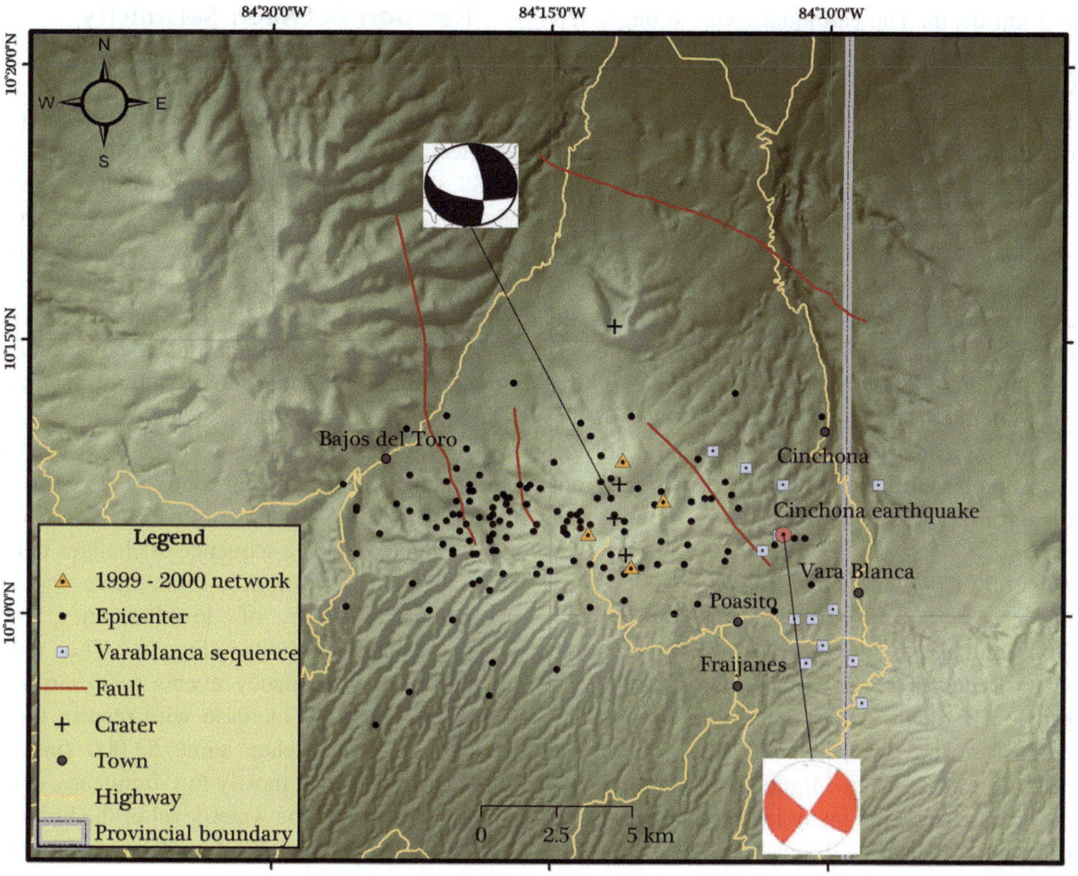

Fig. 8 Epicenters of earthquakes of the 1999–2000 sequence, Vara Blanca sequence and Cinchona earthquake. Fault plane solutions are shown. The focal mechanism of the Cinchona earthquake is from Harvard and was taken from Barquero (2009)

tude 3.4 earthquake occurred 2 km north of the Main Crater and at 4.8 km depth (Fig. 8); it marked the beginning of an earthquake sequence that lasted five months, during which 5 earthquakes reached magnitude 2.5 and above.

Many earthquakes of this sequence were recorded and located by the seismic stations of the 1999–2000 network. Because the earthquake sources were located out the network covering the hypocenter, determinations reported high location errors. The main event location was fairly good and the fault-plane solution showed strike-slip motion. One plane stroke NS. This result supports the existence of a NS fracturing in the summit.

The seismic sequence of 1999 was accompanied by an intense fracturing inside the Main Crater and in particular in its northeastern side. This fracturing was coincident with a significant increase of fumaroles activity in that part of the crater. Based on the earthquake clusters, Fernández (1999) pointed out that the Carbonera, Viejo, and Angel faults were associated with this unusual seismic activity. A local stress pattern instead of volcanic processes would have been responsible for the activation of the mentioned faults.

The 2005 Vara Blanca Activity

On Thursday, June 16, 2005 at 11:30 pm local time, an earthquake of magnitude 4.0 occurred in the Poás volcanic area (Vara Blanca earthquake). The epicenter was located 1 km East of Poasito at 5 km depth. The earthquake was strongly felt in Fraijanes and Poasito. It was also felt in central Costa Rica, though no damages were reported. Ten aftershocks of magnitude 2 or higher were recorded, most within the first 8 h after the earthquake. The largest aftershock was a magnitude 3.7 earthquake three weeks later. The location of the epicenter set the Vara Blanca earthquake on the Angel fault. Slip occurred along a nearly vertical fault. The fault-plane solution and the location of aftershocks indicated that the fault motion was NW-SE oriented with a right-lateral strike-slip motion. The focal mechanism was consistent with a right-lateral slip on the Angel fault. This seismic activity did not trigger any volcanic activity.

The 2009 Cinchona Earthquake

On January 8, 2009, a 6.2 magnitude (Mw) struck the Poás volcanic area, killing 25 people (Barquero 2009) and destroying many houses, several bridges, and a section of the national route to San Miguel de Sarapiquí. The earthquake triggered many landslides (Ruiz et al. Chapter "Coseismic Landslide Susceptibility Analysis Using LiDAR Data PGA Attenuation and GIS: The Case of Poás Volcano, Costa Rica, Central America"), damming the flow of some rivers that raised the level of the water at Sarapiquí River, when the dams broke up some areas along the river were flooded. The community of Cinchona was severely damaged and was relocated. The economic losses were estimated in $ 492,000,000 (Laurent 2009). This moderate-magnitude earthquake was located in the eastern flank of the volcano (Fig. 8) at 4 km depth and its source was the Angel fault (Barquero 2009). The focal mechanism suggested an oblique fault with a high component of normal slip. Although the shock was strong and the epicenter was located near the Main Crater, the Cinchona earthquake did not change the behavior of the volcano, although gas chemistry was modified (Vaselli et al. Chapter "The Last Eighteen Years (1998–2014) of Fumarolic Degassing at the Poás Volcano (Costa Rica) and Renewal Activity").

4 Relation Between Seismicity and the Volcanic Activity

In Fig. 9 earthquakes of low frequency (B type) and tectonic earthquakes (type A) detected by VPS2 station during the period January 1980–March 2006 are reported. The events were counted in analog seismograms. The reported values could be greater because there were moments in which the seismic station was not operating for technical reasons. The counting stopped in 2006 because the recording system of VPS2 switched to digital.

The low-frequency seismicity at Poás varied with changes in the volcanic activity as observed during the 1986–1990 eruptive cycle. Since 1986, there has been a sequence of changes in the behavior of Poás volcano. First, there was an anomalous increase of low-frequency and volcano-tectonic activity (Fig. 9). The daily average of low-frequency events for 1986 was 122. This is a 52% increase with respect those occurred in 1985, when some 53,000 (in one year) seismic events, mostly low-frequency ones, were recorded, and the water level of the hot lake started to slowly but steadily decrease, mostly due to intense evaporation. Intense evaporation caused increasing seismic activity and the decrease in the volume of water in the hot lake.

Such a water loss allowed acid gases to pour out into the atmosphere, thus increasing acidic rains. The temperature of the hot lake also increased. After being absent since 1980, in 1987 geyser-like eruptions reappeared, and the water loss in the hot lake increased, while the low-frequency seismicity remained a high level. The loss of water from the hot lake, as well as the increase of acidity surrounding the volcano, continued during 1988. In April 1989, the hot lake dried out completely, and a 22-day cycle of gas and ash eruptions commenced. With no water in the hydrothermal system, there was no steam production and therefore, the B-type seismicity disappeared.

In 1991, low-frequency events dropped to 40,674 per year, and the low-frequency seismic activity at Poás returned almost to background levels in 1994 (26,862). The subsequent return to

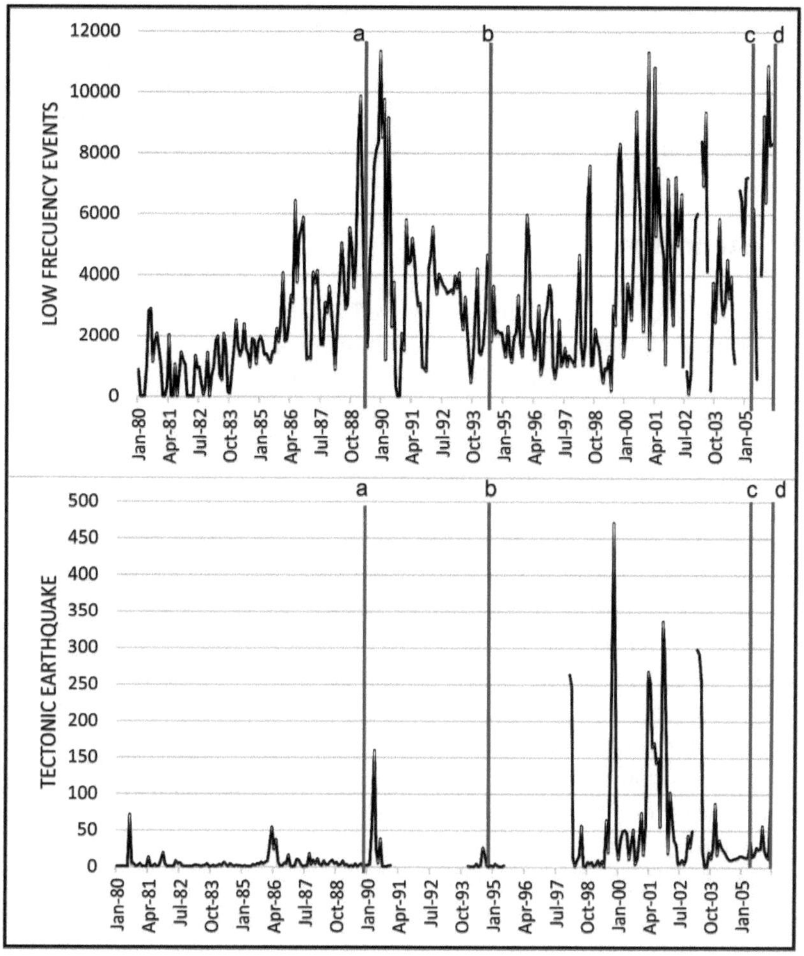

Fig. 9 Earthquakes histograms. Time histograms showing number of events per month. The main peak of low-frequency activity was centered in 1989 and 2001, both coinciding with eruptive activity. Upper figure: low-frequency events and lower figure: tectonic earthquakes. The red vertical lines are: **a** drying of the hot lake and beginning of a 22-day long cycle of eruptions in April 1989; **b** drying of the hot lake in April 1994; **c** Sulfur flows in May 2005 and **d** end of the timeline and start of a new cycle of phreatic eruptions

a condition of dynamic equilibrium between heat and water was characterized by a monthly average of 2,400 earthquakes. During the 1994–1997 period, the average of low-frequency earthquakes was 18,300/year.

There was another increase in the activity of low-frequency earthquakes that began slowly in 1998 and increased markedly in 1999. The shape of the distribution for low-frequency earthquakes (Fig. 10) between 1998 and 2005 showed two peaks, one in 2001 and the other one in 2005. The 2001 peak consisted of 61,000 events, and the one of 2005 reached 60,000 events. However, the total number of low frequency events in 2005 could be higher because the seismic station at Poás failed to partially operate in 2005. In December 2005, the total number of low-frequency earthquakes was 10,888. This amount was only surpassed in December 1989, when 11,344 low-frequency events occurred at Poás. The increase in the occurrence of low-frequency earthquakes since 1998 could be related to the small cycle of phreatic eruptions in March 25, 2006. It is possible that an increase in the heat flow favored the degassing and the low-frequency events, and consequently led to those sporadic eruptions. Because of this eruptive activity, the national park was closed to the public for two weeks.

The tectonic seismicity within the volcanic area increased fracturing and fumarole activity inside the crater, but no eruptions were triggered.

We realized that there was no link between the moderate-magnitude earthquakes at Poás and the eruptive cycles. The 2009 event did not yet trigger any large eruption, but it slightly increased the degassing after the shock.

5 Summary

There are three seismic signals very well identified at Poás: low frequency events, tremors, and tectonic earthquakes. Low-frequency signals represent the typical seismicity. The frequency of these signals varies between 0.8 and 2.5 Hz; they are generated in and around the active crater at shallow depths.

That seismicity probably began when a magmatic intrusion moved up and was emplaced at 500 m depth. Since then, the shallow water was heated up by a heat source. When the cold water sunk down, part of the water heated up and converted into vapor, while the rest remained liquid within fractures and pores. In this manner, both phases could coexist. The gas phase gave rise to steam-rich bubbles containing other gases generated by the magma; in loosely consolidated and fractured pyroclastic materials, a high probability of the presence of small cavities exists, although they can also possibly be created by pressure as vapor expansion occurred. Once a critical pressure was reached, the surrounding media failed, the cavity collapsed because of either fracturing or gas escaping, and the energy was released in the form of small low-frequency waves.

We propose that the causes of the tremor episodes at Poás are related to the movement of magma volatiles or small oscillations of magma in the conduit.

The results of this investigation suggest unequivocally that seismic observations represent a reliable method of volcanic monitoring at Poás volcano.

Acknowledgements We would like to thank the Red Sismológica Nacional (RSN: ICE-UCR) for the seismograms used in this investigation. Thanks are also due to the personnel of the Instituto Costarricense de Electricidad (ICE) and Parque Nacional Volcán Poás for their help during this study. Our gratitude to Vilma Barboza from the Observatorio Vulcanológico y Sismológico de Costa Rica (OVSICORI) for providing seismic phases of earthquakes. We are graceful to Dennis Lindwall and an anonymous reviewer for their comments and suggestions.

References

Alvarado G (2008) Los Volcanes de Costa Rica: geología, historia, riqueza natural y su gente, 3rd edn. San José, Costa Rica, 386 p (in Spanish)

Alvarado G, Morales G, Montero W, Climent A, Rojas W (1988) Aspectos sismológicos y morfotectónicos en el extremo occidental de la Cordillera Volcánica Central de Costa Rica. Rev Geol Am Central 9:75–98 (in Spanish with English abstract)

Araña V, Ortiz R (1984) Vulcanología. Consejo Superior de Investigaciones Científicas. Rueda Madrid, Spain, 150 p (in Spanish)

Barquero R (ed) (2009) El terremoto de Cinchona, 8 de enero de 2009. Report RSN (National Seismological Network), 138 p (in Spanish)

Barquero J, Alvarado G (1989) Los enjambres de temblores en el arco volcánico de Costa Rica. Bol Obs Sism Vulcanol Arenal 2:18–39 (in Spanish)

Brenguier F, Campillo M, Takeda T, Aoki Y, Shapiro N, Briand X, Emoto K, Miyake H (2014) Earthquake dinamics. Mapping pressurized volcanic fluids from induced crustal seismic velocity drops. Science 345:80–82

Casertano L, Borgia A, Cigolini C, Morales L, Montero W, Gómez Fernández J (1985) Investigaciones geofisicas y caracteristicas de las aguas hidrotermales: Volcan Poas, Costa Rica. Geofis Int 24(2):315–332

Chouet B (1985) Excitation of a buried magmatic pipe: a seismic model for volcanic tremor. J Geophys Res 90 (B2):1881–1893

Denyer P, Montero W, Alvarado G (2009) Atlas Tectónico de Costa Rica, 2nd edn. University of Costa Rica, San José, Costa Rica, p 55

Fernández M (1990) La actividad del Volcán Poás (Costa Rica): Analisis sísmico durante el period 1980-1989. Msc Thesis, Escuela Centroamericana de Geología, Universidad de Costa Rica (in Spanish)

Fernández M (1999) Determinación de fallas activas y localización de sismos de baja frecuencia en el volcán Poás. Final Report, Vicerrectoría de Investigación, Universidad de Costa Rica, 18 pp (in Spanish)

Fernández M (2013) Seismotectnic and the hypothetical Tectonic Boundary of Central Costa Rica. In: D'Amico S (ed) Earthquake research and analysis/Book 2 ISBN 980-953-307-437-0, InTech—open science | open minds Rijeka, Croatia

González C (1910) Temblores, terremotos, inundaciones y erupciones volcánicas en Costa Rica 1608–1910, 1st edn. Cartago, Costa Rica, Tecnológica de Costa Rica, 240 p (in Spanish)

Global Volcanism Program (GVP) (1981) Report on Poas (Costa Rica). In: McClelland L (ed) SEAM (Scientific Event Alert Network Bulletin) vol 6, p 8. http://dx.doi.org/10.5479/si.GVP.SEAN198108-345040

Havskov J, Otomoller L (2003) SEISAN: the earthquake analysis software. University of Bergen, Norway, p 250

Kieffer SW (1984) Seismicity at old faithful geyser. An isolated source of geothermal noise and possible analogue of volcanic seismicity. J Volcanol Geotherm Res 22:59–95 (Osharsson NG (ed) Volcano monitoring)

Latter J (1979) Volcanological observations at Tongarino National Park. Report No. 150. Geophysics Division, Department of Scientific and Industrial Research, New Zealand

Laurent J (2009) Evaluación económica de pérdidas y daños. In: Barquero J (ed) El terremoto de Cinchona, 8 de enero. RSN (National Seismological Network), pp 101–127 (in Spanish)

Lee W, Lahr J (1975) HYP07: a computer program for determining hypocenter, magnitude, and first motion pattern of local earthquakes. U.S. Geol Survey, Open File Report 75–311, 113 p

Lin C, Hsu L, Ho M, Shin T, Chen K, Yeh Y (2007) Low-frequency submarine volcanic swarm at the southwestern end of the Okinawa Trough. Geophys Res Lett 34:L06310. https://doi.org/10.1029/2006GL029207

McNutt S (2005) Volcanic seismology. Ann Rev Earth Planet 32:461–491

Montero W (2001) Neotectónica de la región central de costa rica: frontera oeste de la microplaca de Panamá. Rev Geol Am Central 24:29–56 (in Spanish with English abstract)

Montero W, Rojas W, Schimidt V (2000) Estudio de amenaza sísmica y peligro volcánico para el proyecto eólico Vara Blanca en la provincia de Heredia. Final Report, Empresa de Servicios Públicos de Heredia (ESPH), Costa Rica, 66 p (in Spanish)

Morales LD, Soley JF, Alvarado GE, Borgia A, Soto G (1988) Análisis espectral de algunas señales sísmicas de los volcanes Arenal y Poás (Costa Rica), y su relación con la actividad eruptiva. Bol Obs Vulcanol Arenal 1:1–25 (in Spanish)

Nabyl A, Dorel J, Lardy M (1997) A comparative study of low-frequency seismic signals recorded at Stromboli volcano, Italy, and at Yasur volcano, Vanuatu. New Zealand J Geol Geophys 40:549–558

Neuberg J, Tuffen H, Collier L, Green D, Powell T, Dingwell D (2006) The trigger mechanism of low-frequency earthquakes on Montserrat. J Volcanol Geotherm Res 153:37–50

Prosser J, Carr M (1987) Poas volcano, Costa Rica: geology of the summit region and spatial and temporal variations among the most recent lavas. J Volcanol Geotherm Res 33:131–146

Rojas W (1993) Catálogo de sismicidad histórica y reciente en América Central: Desarrollo y Análisis. Tésis de Licenciatura en Geología, Universidad de Costa Rica, 91 p (in Spanish)

Rojas W, Barquero R (1990) Sismos sentidos y actividad volcánica en Costa Rica. Bol Mens Red Sismol Nac, 10 p (in Spanish)

Rowe G, Brantley S, Fernández F, Borgia A, Barquero J (1992) Fluid-volcano interaction in an active stratovolcano: the cráter lake system of Poás volcano, Costa Rica. J Volcanol Geotherm Res 49:23–51

Rouwet D, Mora-Amador R, Sandri L, Ramírez-Umaña C, González G, Pecoraino G, Capaccioni B (Chapter 9). 39 years of geochemical monitoring of Laguna Caliente crater lake, Poás: Patterns from the past as keys for the future. In: Tassi F, Mora-Amador R, Vaselli O (eds) Poás volcano (Costa Rica): the pulsing heart of Central America Volcanic Zone. Springer, Heidelberg (Germany)

Ruiz P, Carr MJ, Alvarado GE, Soto GJ, Mana S, Feigenson MD, Sáenz LF (Chapter 4). Coseismic landslide susceptibility analysis using LiDAR data PGA attenuation and GIS: the case of Poás volcano, Costa Rica, Central America. In: Tassi F, Mora-Amador R, Vaselli O (eds) Poás volcano (Costa Rica): the pulsing heart of Central America Volcanic Zone. Springer, Heidelberg (Germany)

Sherburn S, Scott B (1993) B-type volcanic earthquakes at White Island volcano, New Zealand. J Volcanol Geotherm Res 56:351–355

Soosalu H, Lippitsch R, Einarsson P (2006) Low-frequency earthquakes at the Torfajökull volcano, south Iceland. J Volcanol Geotherm Res 153:187–199

Taylor M, Rojas W (1991) Información general sobre la Sección de Sismología. Vulcanología y Exploración Geofísica. Escuela Centroamericana de Geología, San José (in Spanish)

Uchida N, Sakai T (2002) Analysis of peculiar volcanic earthquakes at Satsuma-Iojima volcano. Earth Plan Space 54:197–209

Vaselli O, Tassi F, Fischer TP, Tardani D, Fernandez Soto E, Duarte E, Martinez M, De Moor MJ, Bini G (Chapter 10). The last eighteen years (1998-2015) of

fumarolic activity at the Poás volcano (Costa Rica) and the renewed activity. In: Tassi F, Mora-Amador R, Vaselli O (eds) Poás volcano (Costa Rica): the pulsing heart of Central America Volcanic Zone. Springer, Heidelberg (Germany)
Walter TR, Wang R, Zimmer M, Grosser H, Lühr B, Ratdomopurbo A (2007) Volcanic activity influenced by tectonic earthquakes: static and dynamic stress triggering at Mt Merapi. Geophys Res Lett 34: L05304. https://doi.org/10.1029/2006GL028710
Watt S, Pyle D, Mather T (2009) The influence of great earthquakes on volcanic eruption rate along the Chilean subduction zone. Earth Planet Sci Lett 277:340–399

Diffuse CO_2 Degassing and Thermal Energy Release from Poás Volcano, Costa Rica

Gladys V. Melián, Nemesio M. Pérez, Raúl Alberto Mora Amador, Pedro A. Hernández, Carlos Ramírez, Hirochicka Sumino, Guillermo E. Alvarado and Mario Fernández

Abstract

During the period 2000–2003 four soil CO_2 efflux surveys were carried out at Poás volcano (Costa Rica) to investigate the spatial distribution and evaluate the diffuse CO_2 emission as well as its associated thermal energy. Inspection of soil CO_2 efflux maps showed that the highest values were always identified inside the Main Crater of Poás, being the 2002 survey the one with the highest number of anomalous observed values. The spatial distribution of soil CO_2 efflux and the $\delta^{13}C$–CO_2 values in soil gas samples showed a positive correlation with the main volcanic-structural features of the area. Main soil CO_2 efflux anomalies were identified close to fumaroles, where several acidic hot springs and soils with high permeability were recognized. Temporal evolution of diffuse CO_2 emissions showed the lowest emission rate in 2000 (164 ± 15 t d^{-1}), followed by a significant increase inside the active crater during 2001 and 2002 (423 ± 54 and 537 ± 69 t d^{-1}, respectively) and with a relatively constant value in 2003 (542 ± 63 t d^{-1}). These data correlated with the observed changes in the $\delta^{13}C$–CO_2 mean value of collected soil gases. To estimate the thermal energy release associated with the diffuse CO_2 degassing, we considered the diffuse CO_2 emission released from the active crater as the most representative of a deep-seated source. Calculated thermal energy released through soil was estimated in 255, 548 and 831 MW for 2000, 2001 and 2003 surveys, respectively. Temporal variations of the diffuse CO_2 degassing and thermal energy release also showed a good correlation with the $\delta^{13}C$–CO_2 values and $^3He/^4He$ ratios measured in the fumarolic discharges of Poás during the same

G. V. Melián (✉) · N. M. Pérez · P. A. Hernández
Environmental Research Division, Instituto Tecnológico y de Energías Renovables (ITER), 38611 Granadilla de Abona, S/C de Tenerife, Spain
e-mail: gladys@iter.es

G. V. Melián · N. M. Pérez · P. A. Hernández
Instituto Volcanológico de Canarias (INVOLCAN), INtech La Laguna, 38320 San Cristóbal de La Laguna, Tenerife, Canary Island, Spain

G. V. Melián · N. M. Pérez · P. A. Hernández
Agencia Insular de Energía de Tenerife (AIET), 38611 Granadilla de Abona, S/C de Tenerife, Canary Islands, Spain

R. A. Mora Amador
Escuela Centroamericana de Geología, University of Costa Rica (UCR), 1000 San José, Costa Rica

C. Ramírez · M. Fernández
Centro de Investigaciones en Ciencias Geológicas (CICG), University of Costa Rica (UCR), 1000 San José, Costa Rica

H. Sumino
Department of Basic Science, Graduate School of Arts and Sciences, University of Tokyo, 3-8-1 Komaba, Meguro-ku, Tokyo 153-8902, Japan

G. E. Alvarado
Área de Amenazas y Auscultación Sismológica y Volcánica, Instituto Costarricense de Electricidad (ICE), Apdo, 10032-1000 San Jose, Costa Rica

F. Tassi et al. (eds.), *Poás Volcano*, Active Volcanoes of the World,
https://doi.org/10.1007/978-3-319-02156-0_6

period, with a significant mantle-derived contribution. These observations evidenced the occurrence of changes in the shallow magmatic-hydrothermal system of Poás that were likely related to a potential magmatic intrusion during the period 2000–2003.

Keywords
Soil CO_2 efflux · Carbon isotopes · Thermal energy · Poás volcano · Costa rica

1 Introduction

Diffuse CO_2 emission studies have become an important geochemical tool to investigate the chemical/physical processes that occur at depth in volcanoes and to evaluate the level of volcanic activity (Carbonelle et al. 1985; Allard et al. 1987, 1991; Baubron et al. 1990; Chiodini et al. 1996; Hernández et al. 2001a, 2012a; Carapezza et al. 2004; Melián et al. 2004, 2012, 2014; Pérez et al. 2004, 2006, 2013; Lecinsky et al. 2007; Padrón et al. 2008a, b, 2012, 2015; Rizzo et al. 2009; Arpa et al. 2013). Among the diffuse degassing phenomena, CO_2 efflux studies at volcanoes have played an important role due to its special characteristics: it is the major gas species after water vapor in both volcanic fluids and magmas and it is an effective tracer of sub-surface magma degassing due to its low solubility in silicate melts at low-to-moderate pressure (Gerlach and Graeber 1985).

Energy balances in volcanic areas have mainly derived from mass balance calculations at crater lakes (Sheperd and Sigurdsson 1978; Hust and Dibble 1981; Brown et al. 1989; Rowe et al. 1992a; Pasternack and Varekamp 1997; Hernández et al. 2007; Taran and Rouwet 2008). However, during the last decade, several studies have demonstrated that diffuse CO_2 phenomena are associated with the release of huge energy fluxes (Brombach et al. 2001; Caliro et al. 2004; Chiodini et al. 1996, 2001, 2004, 2005, 2007; Fridriksson et al. 2006, Frondini et al. 2009; Hernández et al. 2012b; Granieri et al. 2014; Dionis et al. 2015). Calculations of the thermal energy released by active volcanoes in quiescent periods can constitute a powerful monitoring tool for volcanic activity. This is based on the relatively high thermal energy outputs estimated for many volcanic-geothermal systems (Chiodini et al. 2001, 2007; Fridriksson et al. 2006; Frondini et al. 2009; Hernández et al. 2012b; Granieri et al. 2014; Dionis et al. 2015) and their potentially high variability as a result of magmatic and tectonic processes.

In this chapter, we present the first detailed study of soil CO_2 efflux and thermal energy release carried out at the surface of Poás volcano (Costa Rica) during the period 2000–2003. Spatial and temporal distribution of diffuse degassing of CO_2 efflux, the isotopic composition of carbon in the soil CO_2 ($\delta^{13}C\text{–}CO_2$), as well as the total output of CO_2 emitted to the atmosphere from the studied area and the thermal energy release associated with the diffuse degassing process were evaluated during this period in order to relate these parameters to the level of volcanic activity.

2 Geological Setting and Volcanic Activity at Poás Volcano

Poás is a large basaltic-andesitic stratovolcano with an elevation of 2,708 m above sea level, located within the Cordillera Central of Costa Rica, which belongs to the Central America Volcanic chain (Fig. 1). The volcanic activity in this region is related to the tectonic activity produced by the subduction of the Cocos plate beneath the Caribbean plate (Vannucchi and Mason Chapter "Overview of the Tectonics and Geodynamics of Costa Rica"). The main geological feature of Poás is the existence of three different eruptive centers aligned in the N–S direction. All surface

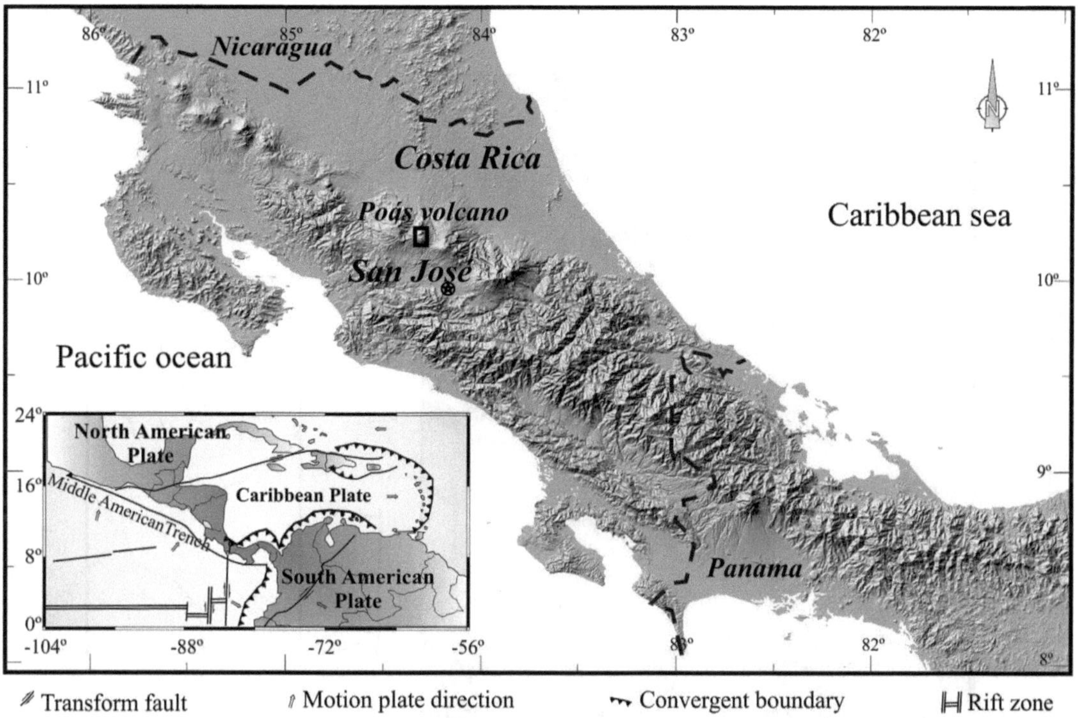

Fig. 1 Tectonics of Central America and the Caribbean area and location of Poás volcano, Costa Rica. Modified from Alvarado 1984

geothermal activity at Poás occurs in and around the main crater, located between the Von Frantzius and Botos craters (Fig. 2).

The main crater is the main surface geothermal expression of an active hydrothermal system (Casertano et al. 1987) supplied with heat and volatiles from a small magma body located at a relatively shallow depth (about 500 m) at Poás (Rymer and Brown 1989). The summit crater of Poás hosts a hyper-acidic (pH < 1) sulfate-chloride lake, with waters with a temperature between 30 and 40 °C (Brown et al. 1989, 1991; Rowe et al. 1992a, b; Rouwet et al. Chapter "39 Years of Geochemical Monitoring of Laguna Caliente Crater Lake, Poás: Patterns from the Past as Keys for the Future"). The lake water chemistry is strongly affected by the fumarolic activity and evaporation processes of the lake. During the last decades, Poás has shown different levels of volcanic activity. Since 1994, the system has been characterized by a constant fumarolic activity that has migrated from one place to another inside the main crater (Martinez et al. 2000; Vaselli et al. 2003; Hilton et al. 2010; Fischer et al. 2015). During the period 2005–2015, strong fumarolic degassing has been observed around the dome (T > 900 °C) together with the occurrence of intermittent phreatic activity (4–9 phreatic eruptions per month; OVSICORI 2013; Global Volcanism Network 2006–2014; see: www.volcano.si.edu).

Several variables can affect the chemical composition of volcanic gas discharges at summit crater of Poás, e.g. rock permeability, lake depth, gas-gas, rock-fluid and water-gas interactions among others (Casertano et al. 1987; Rowe et al. 1992a, b, 1995). The chemical composition of fumarolic gases at Poás is mainly

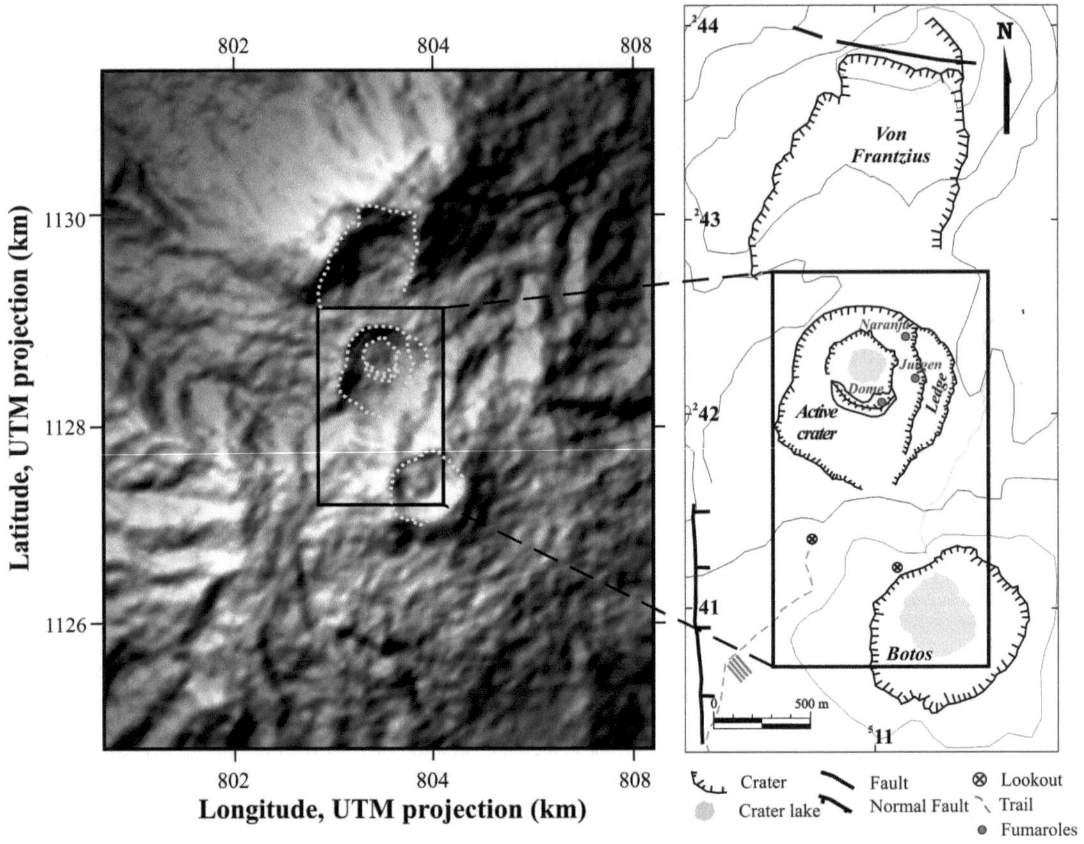

Fig. 2 Map of summit region of Poás volcano. The black square represents the area under study. Gray circles showed the approximate locations of fumarolic fields

characterized by relatively high contents of SO_2, HF, HCl, H_2 and CO, which are typically magmatic species (Vaselli et al. 2003; Vaselli et al. Chapter "The Last Eighteen Years (1998–2014) of Fumarolic Degassing at the Poás Volcano (Costa Rica) and Renewal Activity"). The reactions SO_2–H_2S and SO–SO_2 gas buffers control the content of the C–S–H–O gas species in the fumarolic emission at Poás volcano, at magmatic temperatures > 800 °C (Fischer et al. 2015). The presence of a crater lake plays a critical role in determining type and intensity of the gas emissions into the atmosphere and, therefore, the acidity of rainfall. In the period in which this study was conducted (between 2000 and 2003), the summit crater of Poás was characterized by significant variations in terms of fumarolic activity. Significant changes were also observed in the intensity of the gas discharges and small differences in their temperature, in both the pyroclastic dome and the different fumarolic fields (Hilton et al. 2010; Fischer et al. 2015; Vaselli et al. Chapter "The Last Eighteen Years (1998–2014) of Fumarolic Degassing at the Poás Volcano (Costa Rica) and Renewal Activity"), well as changes in the $CO_2/^3He$, C/S, H_2/Ar ratios are reported by Hilton et al. (2010) and Fischer et al. (2015).

Finally, Poás volcano is characterized by frequent changes in the crater lake temperature, seismicity and chemical composition of volcanic fluids (Casertano et al. 1987; Rowe et al. 1992a; Vaselli et al. 2003, Hilton et al. 2010; Fischer et al. 2015). The observed variations in microgravity and seismicity, together with the variations of the chemical composition of the fumarolic discharges and diffuse H_2 emissions detected during the 2000–2003 period, have suggested that pressure and temperature changes in the volcanic-hydrothermal system of Poás at depth are responsible of the observed increases in volatile fluxes at the main crater during this period, as described in the literature (e.g. Mora et al. 2004; Vaselli et al. 2003; Rymer et al. 2005; Melián et al. 2007; Hilton et al. 2010; Fischer et al. 2015).

3 Procedures and Methods

3.1 Soil CO_2 Efflux and Soil Temperature

During 2000, 2001, 2002 and 2003, four soil gas surveys were carried out at Poás volcano with the aim of measuring diffuse CO_2 efflux and sampling of soil gases at the surface environment of Poás main crater for chemical and isotopic analysis. The selection of the studied area was done according to geological and structural features of the study area as well as to accessibility, in order to obtain the best homogeneous spatial distribution. In total, an area of about 3.2 km^2 was covered (Fig. 2). Measurements were always performed during the dry season and under similar meteorological conditions to avoid unwanted influences of external factors. The number of sampling sites was 135, 182, 244 and 238 for the 2000, 2001, 2002 and 2003 surveys, respectively. Measurements at the crater lake surface were not performed due to logistical problems.

Soil CO_2 efflux measurements were performed in situ according to the accumulation chamber method (Parkinson 1981) by means of a portable CO_2 efflux instrument provided with a non-dispersive infrared (NDIR) CO_2 analyzer RIKEN KEIKI system in the 2000 survey and a LICOR-800 system in the 2001, 2002 and 2003 surveys. The measurement procedure consisted in placing the chamber on the ground, forcing the gas from the soil to circulate in a closed loop between the chamber and the CO_2 analyzer, which was connected to a hand size computer. The increase of CO_2 concentration in the õchamber was recorded as a function of time, allowing the operator to easily calculate the CO_2 efflux at each measuring site. The accuracy of the measures of soil CO_2 efflux in the range 10–35,000 g m^{-2} d^{-1} was estimated to be 10%. Soil temperature was determined by inserting a thermocouple in each sampling site at a depth of 15 cm.

3.2 Soil Gas Concentration and $\delta^{13}C$–CO_2 Isotopic Composition

Soil gas samples were collected at 40 cm depth using a metallic probe following the method described by Hinkle and Kilburn (1979) to have a representative chemical composition of the soil gas. The samples were stored in 10 mL glass Vacutainer^{TM} vials with rubber stoppers. A selection of samples collected in homogenously distributed sampling sites along the studied area was analyzed with a Finnigan MAT Delta-S mass spectrometer at the University of Tokyo (Japan) for the determination of the $\delta^{13}C$–CO_2 isotopic signature. The $^{13}C/^{12}C$ ratios are reported as $\delta^{13}C$ values (±0.1‰) with respect to V-PBD standard.

The soil CO_2 efflux and temperature data were used to construct spatial distribution maps using sequential Gaussian simulation (sGs) provided by the sgsim program from GSLIB (Deutsch and Journel 1998). This approach is commonly used in soil diffuse degassing studies from volcanic zones (Cardellini et al. 2003; Chiodini et al. 2007; Frondini et al. 2009; Fridriksson et al. 2006; Hernández et al. 2012a, b, Melián et al. 2012, 2014; Padrón et al. 2008a, b; Pérez et al. 2013). The spatial distribution maps of CO_2 efflux were used to quantify the uncertainty of the total CO_2 efflux, which is an important task for a correct interpretation of the temporal variations of this parameter. Based on the variogram model, 100 simulations were performed over a grid of 42,752 cells (7 m × 10 m). The average values obtained in each cell by the simulation were used to construct the spatial distribution maps of the studied variable.

3.3 Heat Flux Involved in the Diffuse Degassing Process

Heat flux involved in the diffuse degassing process was estimated following the method proposed by Chiodini et al. (2001) and Frondini et al. (2004) based on the use of CO_2 as a tracer of hydrothermal fluids, and assuming that the H_2O/CO_2 ratio of hydrothermal fluids, before steam condensation, is recorded by fumarolic discharges. This method calculates the heat flux assuming the following contributions: (a) heat released by H_2O vapor moving from the hydrothermal reservoir to the steam condensation zone; (b) heat given off by CO_2 passing from the hydrothermal reservoir to atmospheric conditions; (c) enthalpy of steam condensation at 100 °C and (d) heat lost by liquid water on cooling from 100 °C to the average seasonal temperature value.

With the aim of identifying thermal anomalies at the surface environment of Poás volcano, a handheld FLIR (Forward Looking Infrared Radiometer) ThermaCAM P65 thermal camera (Ashtead Technology) was used to collect thermal images of the crater lake and the fumarolic areas. The FLIR system consists of an uncooled microbolometer detector with a thermal sensitivity of 0.05 °C (50/60 Hz 50 mK at 30 °C), able to show small temperature differences. Internal calibration and atmospheric correction based on user input for reflected ambient temperature, distance, relative humidity, atmospheric transmission, and external optics, allow the FLIR built-in software to calculate realistic source temperatures. The accuracy of the instrument was estimated in ±2 °C or ±2%. FLIR data were collected on a daily basis between Mach 20 and 30, 2002.

3.4 Carbon and Helium Isotopic Composition of Fumarolic Discharges

In this chapter, the isotopic composition of C (in CO_2) and He was determined in the fumarolic discharges from the Jurgen fumarole field situated in the E of the Poás active crater (Fig. 2). For this purpose, a glass funnel was introduced into the fumarolic vent and the fumarolic gas was dragged into a high-vacuum lead-glass container with a syringe for further analysis in the lab. The $^{13}C/^{12}C$ ratios in CO_2 (expressed as $\delta^{13}C$–CO_2 ‰V-PDB) were determined at the University of Tokyo (Japan), following a procedure similar to that previously described for the soil gas with a Finnigan MAT Delta S mass spectrometer. The analytical error for $\delta^{13}C$–CO_2 in the fumarolic gases was ±0.1‰. Helium concentration and $^3He/^4He$ ratios (expressed as $R_A = R/R_{air}$, where R is the measured $^3He/^4He$ ratios and R_{air} is that of air, 1.39×10^{-6}; Mamyrin and Tolstikhin 1984) were analyzed with a high-precision noble gas mass spectrometer (VG5400, VG Isotopes) following the procedure of Sumino et al. (2001). The correction factor for the He isotope ratio was determined by measurements of inter-laboratory He standard named HESJ, with a recommended $^3He/^4He$ value of 20.63 ± 0.10 R_A (Matsuda et al. 2002). The analytical error for R/R_{air} determination was <2%. The measured $^3He/^4He$ ratios were corrected for the addition of air on the basis of the $^4He/^{20}Ne$ ratios measured with the mass spectrometer, assuming that Ne has an atmospheric origin (Craig and Lupton 1976; Sano and Wakita 1985). Chemical composition of Poás fumarolic discharges during the period of study was taken from different literature sources (Vaselli et al. 2003; Zimmer et al. 2004; Hilton et al. 2010; Fischer et al. 2015).

4 Results and Discussion

4.1 Soil Diffuse CO_2 Degassing

Soil CO_2 efflux values ranged from instrumental detection limit (<0.5 g m^{-2} d^{-1}) to 4,712 g m^{-2} d^{-1} with average values of 56, 88, 185 and 76 g m^{-2} d^{-1} for 2000, 2001, 2002 and 2003 surveys, respectively (Table 1). In order to distinguish the presence of different geochemical populations in the soil CO_2 efflux data sets, statistical probability plots (Sinclair 1974) were applied to each of them (Fig. 3). Soil CO_2 efflux

Table 1 Statistics of soil gas data at Poás volcano, Costa Rica, for the 2000, 2001, 2002 and 2003 surveys

	2000			2001			2002			2003		
	Min.	Max.	Med.	Min.	Max.	Med.	Min.	Max.	Med.	Min.	Max.	Med.
CO_2 (g m^{-2} d^{-1})	<0.5	2361	56	<0.5	2129	88	<0.5	2145	185	<0.5	4712	76
$\delta^{13}C$–CO_2 (‰)	−36	−3.97	−28.4	−38	0.92	−26.2	−36	+0.74	−21.9	n.m.	n.m.	n.m.
T soil (ºC)	7.2	92.7	17.9	8.3	92	18.1	10	94.3	18.1	10.4	93.3	19.4

n.m. not measured

data from each survey were log-normally separated in two normal modes (normal I and normal II), implying the existence of three different geochemical populations (Tennant and White 1959; Sinclair 1974): background, intermediate and peak. Background values represent pure biogenic contribution to the CO_2 efflux, while both the intermediate and peak populations can equally be considered as contributions to the deep CO_2 diffuse emission. The statistical parameters of the individual partitions, estimated by the Sichel t estimator (David 1977) are reported in Table 2. Due to the high variability found in the CO_2 efflux peak population for 2002 survey, Sichel t estimator could not be applied (David 1977).

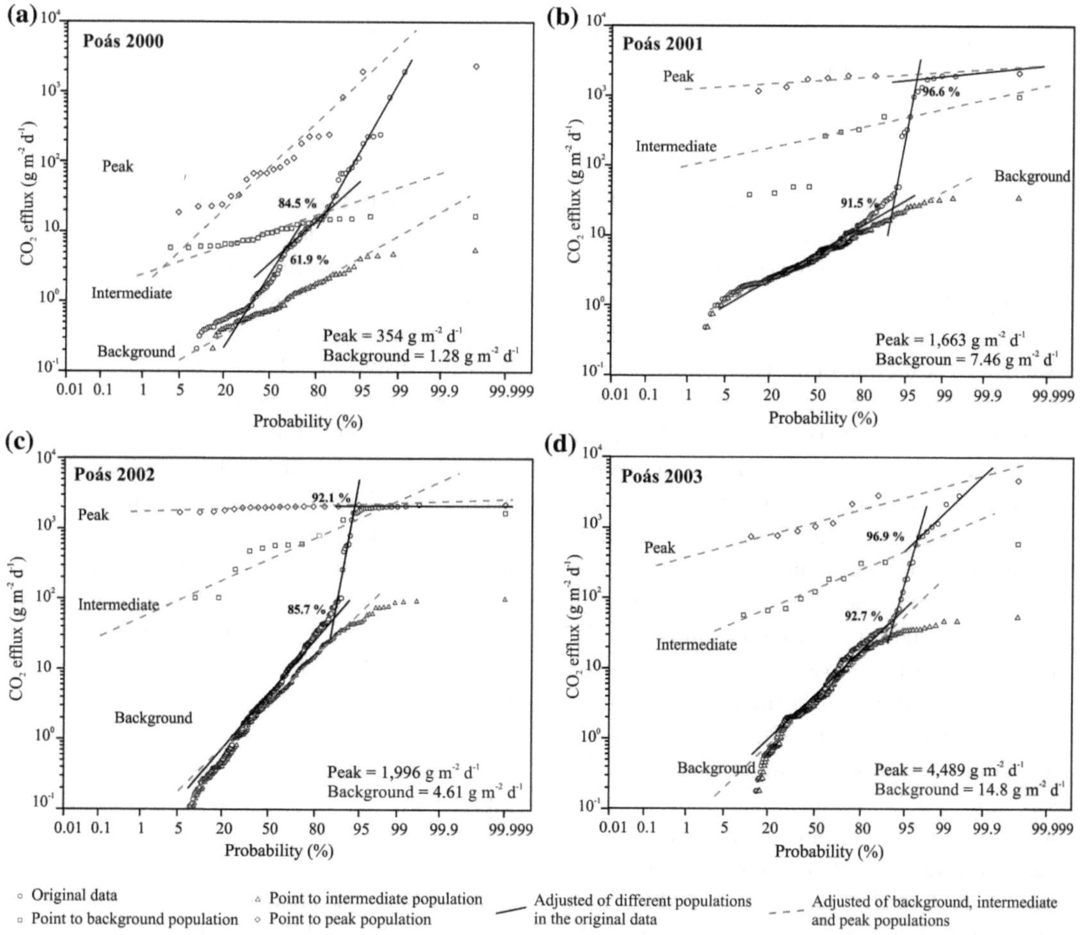

Fig. 3 Probability-log plot of soil CO_2 efflux values observed at Poás volcano in **a** 2000, **b** 2001, **c** 2002 and **d** 2003. Three different geochemical populations are present in the data set: background, peak and intermediate group

Table 2 Estimated parameters of partitioned populations for CO_2 efflux values at Poás volcano, Costa Rica

Survey	Population	No. of points	Proportion (%)	Average CO_2 efflux ($g\ m^{-2}\ d^{-1}$)	Standard deviation 95% confidence interval 95% ($g\ m^{-2}\ d^{-1}$)
2000	Background	79	61.9	1.3	1.1 – 1.6
	Intermediate	28	22.6	12.4	11.2 – 14.1
	peak	22	15.5	354	242 – 591
	Total	129	100		
2001	Background	165	91.5	7.5	6.4 – 9.1
	Intermediate	9	5.1	275	258 – 296
	Peak	7	3.4	1663	1651 – 1674
	Total	181	100		
2002	Background	211	85.7	4.6	3.6 – 6.2
	Intermediate	16	6.4	250	208 – 314
	Peak	18	7.9	1996	n.a.
	Total	245	100		
2003	Background	223	92.7	14.8	10.1 – 25.2
	Intermediate	10	4.2	181	158 – 213
	Peak	8	3.1	4498	3134 – 7512
	Total	241	100		

n.a. not available

Background soil CO_2 efflux values ranged from 1.3 to 14.8 $g\ m^{-2}\ d^{-1}$ (average ~10 $g\ m^{-2}\ d^{-1}$), suggesting a biogenic source (i.e., root respiration, microbial decomposition of soil organic matter, etc.), as reported for other worldwide volcanic systems (e.g. Chiodini et al. 1996, 2007; Hernández et al. 2001a, b, 2012a; Cardellini et al. 2003; Notsu et al. 2005; Padrón et al. 2008a; Melián et al. 2014). The average values calculated for the peak populations were two orders of magnitude higher than the background values (Table 2), indicating the presence of an additional deep contribution for the CO_2. The existence of a peak population suggests the contribution of different sources for CO_2 and the influence of Poás volcanic-hydrothermal system in the diffuse degassing phenomena.

Inspection of soil CO_2 efflux distribution maps (Fig. 4) shows that background CO_2 efflux values (~10 $g\ m^{-2}\ d^{-1}$) are identified at most of the studied area, whereas efflux anomalies are mainly observed inside the active summit crater, over an area of 0.86 km^2. During the 2000 survey, anomalous CO_2 efflux values (> 1,500 $g\ m^{-2}\ d^{-1}$) were measured at the southwest part of the crater dome, which is located inside the summit crater. In fact, this area showed the strongest fumarolic activity during the 2000 survey with soil CO_2 efflux values higher than 800 $g\ m^{-2}\ d^{-1}$. Other areas also displayed relatively high values, such as south of the crater dome (> 400 $g\ m^{-2}\ d^{-1}$) and northern side out of the active crater (> 600 $g\ m^{-2}\ d^{-1}$). Regarding the 2001 survey, a significant change in the location of the CO_2 efflux anomalies inside the active crater was observed, with the highest values measured along the east inner side of the crater (~1,500 $g\ m^{-2}\ d^{-1}$). Few measuring sites showed soil CO_2 efflux values close to 2,100 $g\ m^{-2}\ d^{-1}$. During the 2002 survey, an increase on the CO_2 efflux was observed inside the active crater, showing four areas with particularly high values: (1) along the east rim of the crater, with values > 1,500 $g\ m^{-2}\ d^{-1}$, where in 2001 an increasing of CO_2 efflux was observed with recorded values > 800 $g\ m^{-2}\ d^{-1}$; (2) at the crater dome, where an increase of CO_2 efflux (from 800 $g\ m^{-2}\ d^{-1}$ in 2000 to 500 and $g\ m^{-2}\ d^{-1}$ in

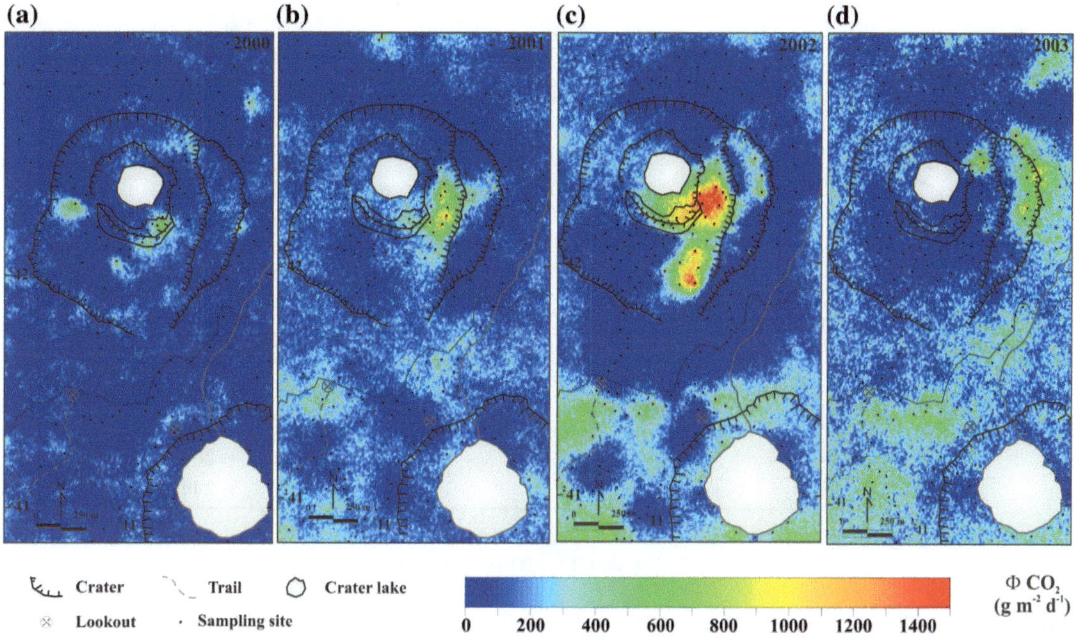

Fig. 4 Contour maps of the soil CO_2 efflux values (g m^{-2} d^{-1}) measured at Poás volcano, Costa Rica **a** April 2000, **b** March 2001, **c** March 2002 and **d** April 2003. Black dots indicate sampling sites; white solid areas shown crater lakes

2001 and then to values > 1,300 g m^{-2} d^{-1} in 2002) was observed; (3) in the southern part of the crater dome, with values close to 1,400 g m^{-2} d^{-1} and (4) at the east ledge inside the active crater, with values close to 1,200 g m^{-2} d^{-1}. With respect to the 2003 survey, a change in the location of CO_2 efflux anomalies was again observed, with a decrease on the extension and magnitude of the values. However, several sites showed high CO_2 efflux values up to 4,710 g m^{-2} d^{-1}, the highest value recorded throughout all this study. Inside the active crater, two sectors with relatively high CO_2 efflux values were identified: (1) northeast and east sides of the active crater with values close to 1,500 g m^{-2} d^{-1} and (2) east ledge inside the active crater (> 1,000 g m^{-2} d^{-1}).

To evaluate the temporal evolution of the diffuse soil CO_2 emission at Poás and its relationship with the volcanic activity, the value of the diffuse CO_2 emission rate was considered for each survey. The obtained average values of total diffuse CO_2 released by thee Poás volcanic system and the standard deviation for the 2000, 2001, 2002 and 2003 surveys were 164 ± 15, 423 ± 54, 537 ± 69 and 542 ± 63 t d^{-1}, respectively. Assuming an area of 3.2 km^2, total diffuse CO_2 normalized values ranged from 51 to 169 t km^{-2} d^{-1}. These values are in the same order of magnitude than those computed for other similar volcanic systems, such as Vesuvio (83 t km^{-2} d^{-1}; Frondini et al. 2004) and Vulcano (537 t km^{-2} d^{-1}; Granieri et al. 2006) in Italy, Miyakejima (167 t km^{-2} d^{-1}; Hernández et al. 2001b) in Japan, and Teide (360 t km^{-2} d^{-1}; Pérez et al. 2013) in Spain.

4.2 Isotopic Signature of the Diffuse CO_2 Emissions

Carbon isotopic data of CO_2 in soil gas from the three first surveys carried out in the present study (2000, 2001 and 2002) presented a large range of values, from −37.5 to +0.92‰ versus VPDB (Table 1). Figure 5 shows the isotopic composition of CO_2 in the soil gas at Poás volcano obtained for each survey versus their respective CO_2 efflux values. Carbon isotopic composition of the hydrothermal CO_2 at Poás volcanic system was well defined by the carbon isotopic data

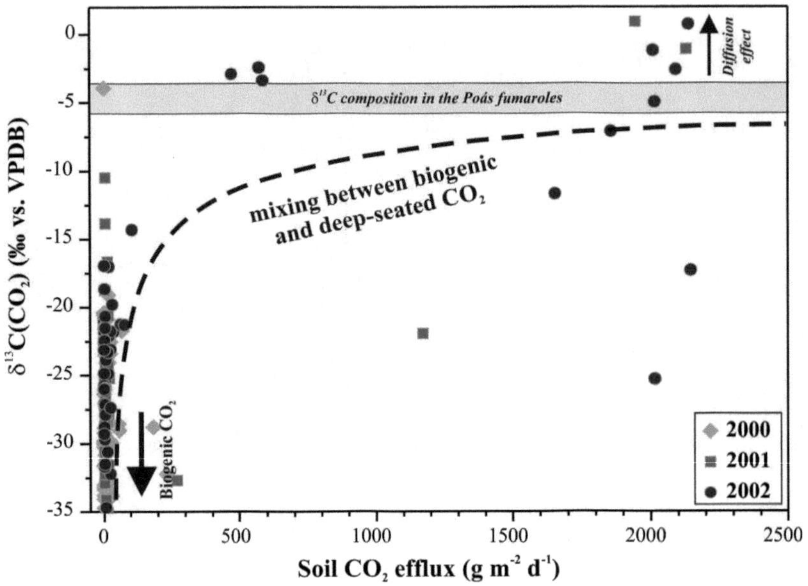

Fig. 5 Bivariate plot of $^{13}\delta C\text{-}CO_2$ (‰) versus soil CO_2 efflux (g m^{-2} d^{-1}). The carbon isotopic of CO_2 at Poás fumaroles "deep CO_2" are also shown (horizontal gray band)

obtained from the samples collected in the fumaroles during this geochemical study (horizontal gray band in Fig. 5), which is also in accordance with the work of other authors (Hilton et al. 2010). The $\delta^{13}C\text{–}CO_2$ values of the fumaroles from 2000 to 2003 had a range from −3.62 to −5.75‰ with an average of −4.42‰ (Table 3), which is marked in Fig. 5 as a horizontal gray band defined as "deep CO_2".

Most high soil CO_2 efflux values at Poás volcano showed a positive agreement with the heaviest carbon isotope composition, representative of an important endogenous CO_2 contribution. The summit crater of Poás is characterized by highly eroded soils due to acidic gases released from existing fumaroles inside the crater, consequently most of the area under study lacks of vegetation. Nevertheless, the vegetated areas are characterized by C3 and C4 plants which have a wide range of isotopic composition ($\delta^{13}C$ from −35 to −20‰ for C3 plant and from −15 to −7‰ for C4 plant; Cheng, 1996). This fact makes

Table 3 Helium and carbon isotopic composition of fumarolic gas at Poás volcano

	Fecha	$\delta^{13}C\text{-}CO_2$ (‰ vs. PDV)	$^4He/^{20}Ne$	$^3He/^4He$ (R/R_{air})	$^3He/^4He_{corr.}$ (R/R_{air})	Atm. He (%)	Crust He (%)	MORB He (%)
Fumarole[a]	01/03/1999	+4.17	23.38	7.08	7.08	1.4	10.34	88.3
Fumarole Jugen	29/04/2000	−4.9	16.94	6.28	6.37 ± 0.12	1.9	19.9	78.2
Fumarole of Domo	29/04/2000	−5.75	5.533	6.4	6.68 ± 0.15	5.7	15	79.3
Fumarole Jugen	27/02/2001	−4.46	–		–	–	–	–
Fumarole Jugen	20/03/2002	−3.62	99	7.08	7.10 ± 0.06	0.3	11.2	88.5
Fumarole Jugen	20/03/2002	–	93.7	7.03	7.05 ± 0.06	0.3	11.9	87.8
Fumarole Jugen[b]	31/03/2003	−3.5	–	7.07	7.14	–	–	–

[a]Snyder et al. (2001), [b]Fischer et al. (2015)

difficult to assess the biogenic origin of the emitted CO_2. Regarding the samples collected in the study areas with lower CO_2 efflux values, their isotopic carbon signature ranged from −35 to −10‰ (Fig. 5). The relatively wide dispersion of the carbon isotopic values can be attributed to the simultaneous occurrence of different processes such as: (i) natural isotopic variability of the biogenic CO_2 produced in the soils, (ii) mixture between biogenic CO_2 and variable amounts of hydrothermal gases, and (iii) uncertainty of the method which was higher at low CO_2 efflux values (Chiodini et al. 2008). Figure 5 shows that samples from the 2002 survey (solid dark gray circles in Fig. 5) presented a major endogen contribution with respect to those collected in 2001 (solid medium gray cube in Fig. 5) and at the same time, those from 2001 highlighted a major contribution with respect to those of 2000 (solid light gray rhombus in Fig. 5). The differences in the diffusion coefficients between $^{12}CO_2$ and $^{13}CO_2$ may explain the heaviest $\delta^{13}C$ values (~ -1‰) observed. The isotopic composition of soil CO_2 is indeed enriched in $\delta^{13}C$ relative to soil respiration CO_2 by at least 4.4‰ (Cerling et al. 1991), due to the aforementioned difference in the diffusion coefficients. This could be a possible explanation for the soil CO_2 isotopic composition values, which were heavier than the fumarolic end-member.

The classed map of $\delta^{13}C$–CO_2 isotopic values for the 2000, 2001 and 2002 surveys is shown in Fig. 6. Most of the studied area had $\delta^{13}C$–CO_2 values < −25‰, whereas heaviest values were measured always inside the active crater ($\delta^{13}C$–CO_2 > −4‰). It is important to point out that an increase in the number of samples with heavier $\delta^{13}C$–CO_2 values was observed from 2000 to 2002, suggesting an intensification of the magmatic contribution to the soil degassing.

The dynamics of the diffuse CO_2 emission at Poás volcanic system during this study showed a clear pattern related to the fumarolic activity. The Pearson correlation factors between soil CO_2 efflux and $\delta^{13}C$–CO_2 increased from 2000 to 2002 (0.013, 0.55 and 0.62 for 2000, 2001 and 2002 surveys, respectively), indicating a contribution of deep-seated gases to the diffuse emissions. This positive correlation was also observed in the measured concentrations of soil gases (H_2, CO_2, Hg^0 and H_2S) and soil temperature (Nolasco et al. 2003; Melián et al. 2007) during the same period of study. As a matter of fact, the spatial distribution of CO_2 efflux, soil CO_2, H_2, $\delta^{13}C$–CO_2 and soil temperature showed a similar spatial and temporal behavior that suggests an increase in permeability through vertical structures, thus favoring the uprising of deep fluids to the surface (Nolasco et al. 2003; Melián et al. 2007; Melián 2008). On the other hand, the observed temporal variation of the diffuse CO_2 emission at this volcanic system seems to be related to changes in the pressure and temperature of the volcanic-hydrothermal system at depth (Vaselli et al. 2003; Melián et al. 2007; Hilton et al. 2010; Fischer et al. 2015), as suggested by Rymer et al. (2005), who detected significant microgravimetry changes and postulated a potential intrusion at Poás between 2000–2004.

4.3 Dynamic of Diffuse CO_2 Emission

On the basis of the SO_2 emission and the gas concentrations measured in the fumarolic vents, Zimmer et al. (2004) estimated the total fumarolic CO_2 flux at Poás volcano in 2001 and 2003, being 34 and 33 t d^{-1}, respectively. The values estimated in this work of the diffuse CO_2 emission (423 ± 54 and 542 ± 63 t d^{-1} for 2001 and 2003, respectively) are one order of magnitude higher than those calculated by Zimmer et al. (2004) for the total visible CO_2 emissions, with diffuse/plume CO_2 mass ratios of 3.7 and 2.7 for 2001 and 2003, respectively. These mass ratios are similar to those calculated for volcanoes with lower eruptive rates: ~1.5 for Sierra Negra in Galapagos Islands, Ecuador (Padrón et al. 2012) or ~2.8 for Vulcano in Italy

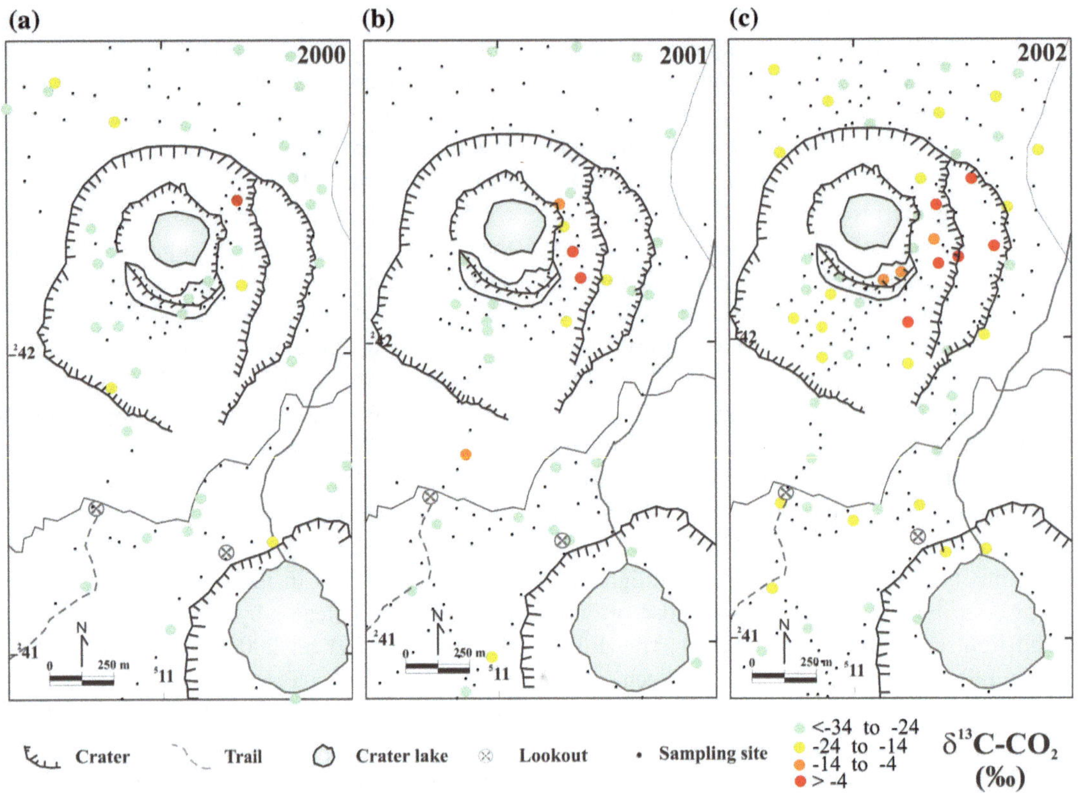

Fig. 6 Classed map of CO_2 isotopic composition (‰) measured at Poás volcano, Costa Rica **a** April 2000, **b** March 2001 and **c** March 2002. Black dots indicate sampling sites

(Carapezza et al. 2011), but significantly higher than the values obtained for other volcanoes with strong and recent magmatic activity, such as Etna (~0.57; Hernández et al. 2015) or Stromboli (~0.01; Hernández et al. 2015), both in Italy. This parameter has been considered as a plausible indicator of the current state of volcanic activity (Hernández et al. 2015). As previously mentioned, CO_2 emissions from the crater lake were not considered since no survey was carried out due to the vigorous discharges of acidic gases from the lake itself.

A parameter that could help to further understand the status of volcanic activity in a certain edifice is the so-called "acceleration of diffuse CO_2 degassing", which gives information about the dynamics and/or the rate of changes in the diffuse degassing process. The acceleration (in t d^{-2}) was calculated as the change in total CO_2 emission rate (t d^{-1}) divided by the elapsed time period (d). The temporal variation of the total CO_2 diffuse emission at Poás volcano and the acceleration of the diffuse CO_2 emission during the same period are reported in Fig. 7. The highest acceleration value, 0.82 t d^{-2}, corresponded to the period 2000–2001, while in the period 2001–2002 the acceleration was 0.29 t d^{-2}. From 2002 to 2003, a notable decrease in the acceleration was observed, with a value of 0.013 t d^{-2}. The observed increase during the period 2001–2002 can be explained by a sharp increase on the diffuse CO_2 emission rate and accompanied by both a shift of the fumarolic discharges and soil gas $\delta^{13}C–CO_2$ towards heavier values (Fig. 7) and a sharp increase of the $^3He/^4He$ ratio and changes in the $CO_2/^3He$ ratio as reported by Hilton et al. (2010), as well as variations in the C/S and H_2/Ar ratios (Vaselli et al. 2003; Fisher et al. 2015), reflecting an increase in the mantle wedge contribution to

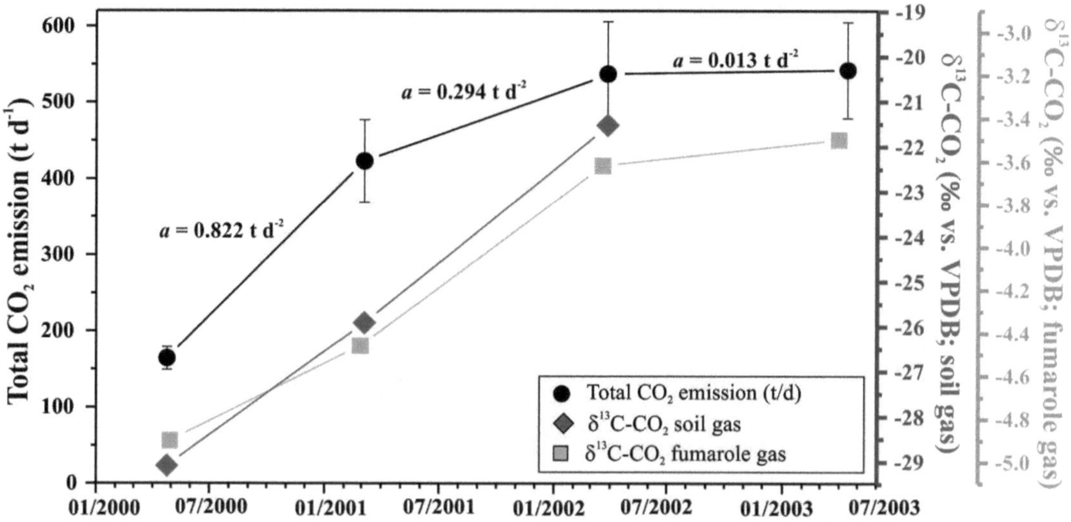

Fig. 7 Temporal evolution of total diffuse CO_2 emission and the acceleration of diffuse CO_2 degassing processes at Poás volcano (black circles). Temporal evolution of average values of $\delta^{13}C$-CO_2 in the soil gas for each survey (dark gray diamond) and for gas fumarolic discharges (light gray squares) at Poás volcano are also shown

the CO_2 emission. This observed increase on the diffuse CO_2 emission during the period of study does not seem to be related to an enhanced biogenic CO_2 production in the soils of Poás, because the average of isotopic composition of the CO_2 from soil gases during each survey did not follow the expected trend for a biogenic CO_2 addition (Fig. 7).

4.4 Thermal Energy Release

The highest soil temperatures (> 60 °C) were measured inside the active summit crater for all surveys (Melián et al. 2007), with a significant increase in the intensity and magnitude of soil temperature anomalies between 2000 and 2002 with a relative decrease in 2003 (Melián et al. 2007). Based on the soil temperature data, IR thermal images of the main crater were taken in 2002. Figure 8 shows a thermal image of the studied area for four different portions of the Poás crater and surroundings characterized by high surface temperatures (10–20 °C higher than ambient temperature): (1) the crater lake, with a mean temperature (T_{mean}) of 38.8 °C, (2) the dome, with a T_{mean} of 14.8 °C, (3) the area along the entire wall of the northeast crater, with a T_{mean} of 30.5 °C, where Jugen and Naranja fumarole fields are located (Fig. 2) and (4) the terrace of the main crater with a T_{mean} of 50.4 °C. The main fumarolic areas are located around the zones that showed the maximum apparent soil temperatures obtained from the IR images. In fact, the most evident surface geothermal manifestations occur inside the summit crater, coinciding with the areas where soil gas anomalies of H_2, CO_2 and Hg^0 were detected (Nolasco et al. 2003; Melián et al. 2007).

Several studies have been carried out in the past to estimate the heat output from Poás volcano (Brown et al. 1989; Rowe et al. 1992b). These authors calculated the total heat flux based on the existence of different types of contributions: radiation, conduction and evaporation heat loss from the lake surface, run-off rainwater heating, summit crater heat flow, energy loss through fumarole emissions and phreatic activity (Brown et al. 1989). The temporal variation of the heat output from Poás was studied from 1978 to 1989 by Brown et al. (1989), who estimated a background value of the total energy dissipated at the surface of Poás volcano (crater lake and

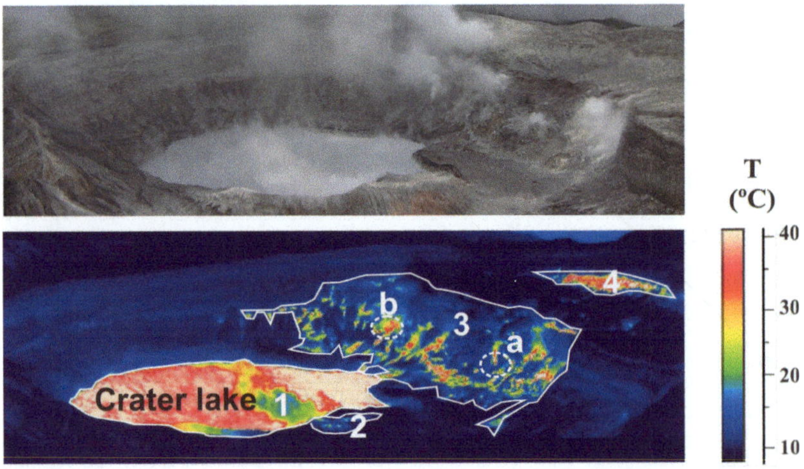

Fig. 8 Thermal infrared image of main crater at Poás volcano taken in March 2002. The numbers are referred to the different areas characterized by high surface temperatures: (1) the crater lake, (2) the dome, (3) the area along the entire wall of the northeast crater, where Jugen and Naranja fumarole fields are located and (4) the terrace of the main crater

active crater) in about 200 MW. During high-temperature fumarole events or lake phreatic activity, peaks of the total power output were observed (~600 MW; Rowe et al. 1992b), while in their absence Rowe et al. (1992b) considered that the heat loss from the lake surface was approximately equal to the total heat output from the Poás volcano. As a result of the destruction of the meteorological station during the volcanic activity of 1989, data are not available for the estimation of heat output from Poás volcano since that time (Rymer et al. 2009). It is worth mentioning that the studies of Rowe et al. (1992b) and Brown et al. (1989) did not consider the energy associated with the diffuse degassing processes. However, they observed that changes in the energy balance might be related to changes in the volcanic activity. In this way, the estimation of the thermal energy released from volcanic systems could be considered a useful tool for the evaluation of the volcanic activity.

In this chapter, in order to estimate the heat flow associated to the diffuse degassing process at Poás volcanic system, the method proposed by Chiodini et al. (2001) and Frondini et al. (2004) was applied. This method uses CO_2 as a tracer of hydrothermal fluids, assuming that the main contribution in the H_2O/CO_2 ratio of hydrothermal fluids, before steam condensation, is recorded by the fumarolic discharges. Most of the condensed steam might flow towards the most permeable structures of Poás contributing to the thermal water outflows of the area. The values of energy release due to the diffuse degassing process were obtained by combining the following two main contributions: (1) the enthalpy of steam condensation at 100 °C, given by the product of the total amount of steam condensed in one day times the enthalpy of evaporation at 100 °C (Keenan et al. 1969), and (2) the heat lost by liquid water on cooling from 100 °C to the average seasonal value of 10 °C, which is given by the product of the enthalpy lost by 1 g of water times the mass of water. The results of the energy release associated with the CO_2 degassing processes from Poás volcano system between 2000 and 2003 are listed in Table 4. Thermodynamic calculation of these heat fluxes indicates that 255, 548 and 831 MW were released through the surface environment of the study area at Poás system during the 2000, 2001 and 2003 surveys, respectively (data not available for 2002). The summit crater of Poás (0.86 km^2) represents between 30 and 40% of the energy associated with the diffuse degassing process (Table 4). These herein estimated values are of

Table 4 Total CO_2 output and Energy release from Poás volcano systems

Survey	H_2O/CO_2 fumaroles	Poás volcano (3.2 km^2)			Poás summit crater (0.86 km^2)		
		Total CO_2 diffuse emission (t d^{-1})	Steam (t d^{-1})	Energy release (MW)	Total CO_2 diffuse emission (t d^{-1})	Steam (t d^{-1})	Energy release (MW)
2000	51.7[a]	164 ± 15	8,474 ± 775	255 ± 23	61 ± 16	3,119 ± 255	95 ± 25
2001	42.9[b]	423 ± 54	18,129 ± 2,314	548 ± 70	123 ± 22	5,254 ± 298	159 ± 29
2002	–	537 ± 69	–	–	192 ± 31	–	–
2003	50.8[c]	542 ± 63	27,520 ± 3,199	831 ± 97	89 ± 15	5898 ± 230	136 ± 23

[a]Vaselli et al. (2003), [b]Zimmer et al. (2004), [c]Fischer et al. (2015); Data not available for 2002

the same order of magnitude than those calculated for other volcanic systems from Italy: Campi Flegrei, Solfatara (100.8 MW; Chiodini et al. 2005), Donna Rachele, Ischia (40.1 MW; Chiodini et al. 2005) and Vulcano (17 MW; Frondini et al. 2004); Greece: Nisyros in Greece, (42.6 MW; Chiodini et al. 2005); Iceland: Reykjanes (17 MW; Fridriksson et al. 2006) and Hengil (1,237 MW; Hernández et al. 2012b); Yellowstone mud volcanoes (30.4 MW; Chiodini et al. 2005) and Volcano do Fogo crater in Cabo Verde (10.3 MW; Dionis et al. 2015). Chiodini et al. (2005) estimated that the total thermal energy released by steam condensation normally varies between 1 and 100 MW when the CO_2 efflux is used as a tracer of deep fluids, these results also agreeing with those obtained in this work. It is also interesting to emphasize that the thermal energy values estimated in the present study are of the same order of magnitude than the background value estimated by Rowe et al. (1992b) and Brown et al. (1989) without taking into account the energy released by the degassing process ($\sim$200 MW).

Finally, the energy released through the fumaroles at Poás volcano was studied by combining the fumarole chemical composition and the total visible SO_2 emission data, similarly to what has been described for the calculation of the energy release related to CO_2 diffuse emission. In this manner, taking the fumarole composition data from Zimmer et al. (2004) and Fischer et al. (2015) and the SO_2 flux calculated by Zimmer et al. (2004) in 2001 and 2003, it was possible to estimate the water flux through the Jugen fumarole in 1,457 t d^{-1} for the 2001 survey and in 1,676 t d^{-1} for that of 2003, at which thermal energies of 44 and 51 MW for 2001 and 2003, respectively, can be associated. These values are in the same order of magnitude as those estimated by Brown et al. (1989) and Rowe et al. (1992b) for the 1968–1985 period, but significantly lower than those estimated for the heat released by the diffuse degassing fluxes through the surface environment during the herein described 2000–2003 surveys (255, 548 and 831 MW).

4.5 Temporal Variation of Carbon and Helium Isotopic Composition of Fumarolic Discharges

Table 3 shows the $\delta^{13}C$–CO_2 and $^3He/^4He$ data measured during 1999–2003 period including previously published data. The $\delta^{13}C$–CO_2 values from fumarole gas showed a range from −3.62 to −4.42‰ (Table 3). The temporal evolution of $\delta^{13}C$–CO_2 showed a continuous increase in $\delta^{13}C$–CO_2 towards heavier values during the 2000–2003 period and a similar behavior of the $\delta^{13}C$–CO_2 mean values estimated for the collected soil gases at each survey (Fig. 7) was observed, supporting the hypothesis of a input of deep-seated gases from the magmatic-hydrothermal system of Poás. These data are consistent with those reported by Hilton et al. (2010), who highlighted a steady increase of the

$\delta^{13}C\text{–}CO_2$ values between 2001 and 2005, with the maximum value ($\delta^{13}C\text{–}CO_2 \sim -1.5‰$) occurring 9 months before the event of the first phreatic eruption in 2006.

The observed $^3He/^4He$ isotopic ratios showed a range from 6.28 to 7.08 R_{air}, with $^4He/^{20}Ne$ ratios from 5.5 to 99.0 (Table 3). These values are in agreement with those previously reported for the same fumarole by Hilton et al. (2010). The samples collected from the Poás Volcano fumaroles showed a significant mantle-derived helium component and fall within the range reported for Hilton et al. (2010) from 2001 to 2009. The observed $^3He/^4He$ ratios in the discharging fluids from Poás volcano are explained with a three-component mixing model: atmospheric-type, MORB-type and crust-type helium. This model proposed by Sano and Wakita (1985) can estimate the fraction of helium from each geochemical reservoir using the following equations:

$$\left({}^3He/{}^4He\right)_i = \left({}^3He/{}^4He\right)_a \cdot A + \left({}^3He/{}^4He\right)_c \cdot C + \left({}^3He/{}^4He\right)_m \cdot M$$

$$1/\left({}^4He/{}^{20}Ne\right)_i = A/\left({}^4He/{}^{20}Ne\right)_a + C/\left({}^4He/{}^{20}Ne\right)_c + M/\left({}^4He/{}^{20}Ne\right)_m$$

$$A + C + M = 1$$

where subscripts *i*, *a*, *m*, and *c* indicates the sample, atmospheric, MORB-type, and crust-type, respectively, and A, M, and C are the fraction of helium from atmospheric, MORB, and crust, respectively. By using the values of $^3He/^4He$ = 1 R_{air} and $^4He/^{20}Ne$ = 0.318 for atmospheric-type reservoir, of $^3He/^4He$ = 8 R_{air} and $^4He/^{20}Ne$ = 1000 for MORB-type reservoir and of $^3He/^4He$ = 0.02 R_{air} and $^4He/^{20}Ne$ = 1000 for crust-type reservoir (Sano and Wakita, 1985), the proportion of the helium from each geochemical reservoir in terrestrial fluids from Poás

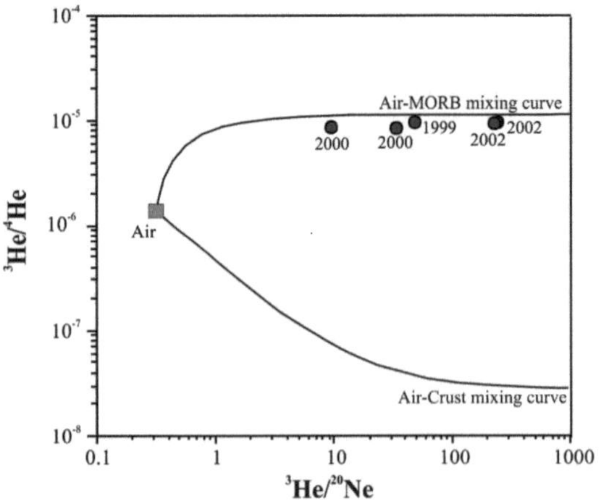

Fig. 9 $^3He/^4He$ versus $^4He/^{20}Ne$ diagram for gas from the Poás volcano fumaroles. The curves represent the theoretical mixing lines between air and two end-members: 7.15 Ra and MORB

Volcano can be estimated (Table 3). In Fig. 9, the corrected $^3He/^4He$ ratios are plotted versus the $^4He/^{20}Ne$ ratios for the Poás fumarole gas together with data from Snyder et al. (2001) and Hilton et al. (2010). We assumed the He isotopic ratio ($^3He/^4He$ = 7.15 R_{air}) as representative of the helium isotope composition of the Poás magmatic end-member. The curves represent the theoretical mixing lines between air and two end-members: 7.15 Ra and MORB. MORB-type is the predominant fraction of helium in most of the terrestrial fluid discharges from the Poás fumaroles, with an estimated range between 78.2 to 88.5% (Fig. 9). Atmospheric contribution is negligible (lower than 2%) in all collected samples, excepting the fumarole domo sample collected in 2000 (Table 3). Contribution of crust-type helium is observed at all the samples (Table 3), and seems to be responsible of the variations observed on the $^3He/^4He$ ratios.

Rymer et al. (2009) published a time series of microgravity data collected from Poás between 1985 and 2008. They reported the occurrence of a new cycle of activity started in 2002 with increments of gravity in the southern crater followed by changes in lake level, temperature and sulfate concentration in 2006. They concluded that the gravity changes were caused by sub-surface mass changes of magmatic and/or hydrothermal (vapor/fluid interface) origin, indicating that a further intrusive event was occurring. These data are in agreement with our observations on the diffuse CO_2 emission rates as well as on the carbon isotopic composition of soil and fumarole gases and heat flux from 2000 to 2003. Therefore, volatile rich magma pulses escaping periodically through fractures in the magma carapace and form upwelling dendritic intrusions (finger-like injections of gas-rich magmatic fluids) seem to precede increases on the eruptive activity, as observed from April 2006.

5 Summary

Volatile rich magma pulses were observed before the series of phreatic eruptions occurred at Poás Volcano from April 2006 suggesting the injection of magmatic volatiles as trigger mechanism for this activity volcano. The increase on the volcanic activity at Poás was preceded by changes in the diffuse CO_2 emission, isotopic composition of fumarolic gases and soil gas as well as in the thermal energy released and other geophysical parameters, e.g. microgravimetry (Rymer et al., 2009). The diffuse CO_2 discharged through the surface environment of Poás showed an increase between 2000 and 2002, and a relatively constant value between 2002 and 2003. However the acceleration of diffuse CO_2 degassing presented its maximum value in the 2001–2002 interval, indicating changes in volcanic activity produced by the injection of new rich magma volatiles in these years.

This is the first time that the release of energy related to the diffuse degassing processes was estimated for the Poás volcanic system, resulting in a very significant contribution that was not considered in previously published works. In fact, the energy associated with the diffuse degassing process is comparable to the total energy dissipated at the surface of Poás volcano (crater lake and active crater) during a previous period of increasing volcanic activity (1985–1990) when a magmatic intrusion occurred beneath the crater lake.

In summary, the results obtained in the present work demonstrate that the diffuse CO_2 emission rate and the energy released by diffuse degassing process constitute a powerful tool to monitor the activity of Poás volcano.

Acknowledgements We are indebted to Juan Dobles and all the staff of the Poás National Park for their assistance during the fieldworks. We would also like to thank COVIRENAS, students of the Escuela Centroamericana de Geología, Universidad de Costa Rica, and Juan Carlos Mesa for their help during this study. This research was supported by the Cabildo Insular de Tenerife, CajaCanarias (Canary Islands, Spain) and the CGL2005-07509/CLI project of the MICINN, Spain.

References

Allard P, Le Bronec J, Morel P, Vavasseur C, Faivre-Ferret R, Robe MC, Roussel C, Zettowog P (1987) Geochemistry of soil gas emanations from Mt. Etna, Sicily. Terra Cognita 7(407):G17–52

Allard P, Carbonelle J, Dajlevic D, Le Bronec J, Morel P, Robe MC, Maurenas JM Faivre-Ferret R, Martin D, Sabroux JC, Zettowog P (1991) Eruptive and diffuse emissions of CO_2 from Mount Etna. Nature 351:387–391

Arpa MC, Hernández PA, Padrón E, Reniva P, Padilla GD, Bariso E, Melián GV, Barrancos J, Nolasco D, Calvo D, Pérez NM, Solidum RU Jr (2013) Geochemical evidence of magma intrusion at Taal volcano, Philippines, in 2010–2011 from diffuse carbon dioxide emissions. Bull Volcanol 75:747. https://doi.org/10.1007/s00445-013-0747-9

Baubron J, Allard P, Toutain J (1990) Diffuse volcanic emissions of carbon dioxide from Vulcano Island, Italy. Nature 344:51–53

Brombach T, Hunziker J, Chiodini G, Cardellini C, Marini L (2001) Soil diffuse degassing and thermal energy flux from the southern Lakki plain, Nisyros (Greece). Geophys Res Lett 28:67–72

Brown G, Rymer H, Dowden J, Kapadia P, Stevenson D, Barquero J, Morale LD (1989) Energy budget analysis for Poás Crater Lake: implications for predicting volcanic activity. Nature 339:370–373

Brown G, Rymer H, Stevenson D (1991) Volcano monitoring by microgravity and energy budget analysis. J Geol Soc London 148:585–593

Caliro S, Chiodini G, Galluzzo G, Granieri D, La Rocca M, Saccorotti G, Ventura G (2004) Recent activity of Nisyros volcano (Greece) inferred from structural, geochemical and seismological data. Bull Volcanol 67:358–369

Carapezza M, Inguaggiato S, Brusca L, Longo M (2004) Geochemical precursors of the activity of an open-conduit volcano: the Stromboli 2002–2003 eruptive events. Geophys Res Lett 31:L07620

Carapezza M, Barberi F, Ranaldi M, Ricci T, Tarchini L, Barrancos J, Fischer C, Perez N, Weber K, Di Piazza A, Gattuso A (2011) Diffuse CO_2 soil degassing and CO_2 and H_2S concentrations in air and related hazards at Vulcano Island (Aeolian Arc, Italy). J Volcan Geotherm Res 207:130–144

Carbonelle J, Dajlevic D, LeBronce J, Morel P, Obert JC, Zettwoog P (1985) Etna: Composantes sommitales et pariétales des émission de gaz carbonique. Bull Programme Interdisciplinaire de Recherche sur la Prévision et la Surveillance des Eruptions Volcaniques—Centre National de la Recherche Scientifique 108:64

Cardellini C, Chiodini G, Frondini F (2003) Application of stochastic simulation to CO_2 flux from soil: mapping and quantification of gas release. J Geophys Res 108(B9):2425

Casertano L, Borgia A, Cigolini C, Morales LD, Montero W, Gomez M, Fernandez JF (1987) A integrated dynamic model for the volcanic activity at Poás Volcano, Costa Rica. Bull Volcanol 49:588–598

Cerling TE, Solomon DK, Quade J, Bowman JR et al (1991) On the isotopic composition of carbon in soil carbon dioxide. Geochem Cosmochim Acta 55:3403–3405

Cheng WX (1996) Measurement of rhizosphere respiration and organic matter decomposition using natural ^{13}C. Plant Soil 183:263–268

Chiodini G, Frondini F, Raco B (1996) Diffuse emission of CO_2 from the Fossa crater, Vulcano Island (Italy). Bull Vulcanol 48:41–50

Chiodini G, Frondini F, Cardellini C, Grenieri D, Marini L, Ventura G (2001) CO_2 degassing and energy release at Solfatara volcano, Campi Flegrei, Italia. J Geophys Res 106(B8):16.213–16.221

Chiodini G, Avino R, Brombach T, Caliro S, Cardellini C, De Vita S, Frondini F, Granieri D, Marotta E, Ventura G (2004) Fumarolic and diffuse soil degassing west of Mont Epomeo, Ischia, Italy. J Volcanol Geotherm Res 133:291–309

Chiodini G, Granieri D, Avino R, Caliro S, Costa A, Werner C (2005) Carbon dioxide degassing and estimation of heat release from volcanic and hydrothermal systems. J Geophys Res 110:B08204

Chiodini G, Baldini A, Barberi F, Carapezza ML, Cardellini C, Frondini F, Granieri D, Ranaldi M (2007) Carbon dioxide degassing at Latera caldera (Italy): evidence of geothermal reservoir and evaluation of its potential energy. J Geophys Res 112:B12204

Chiodini G, Caliro S, Cardellini C, Avino R, Granieri D, Schmidt A (2008) Carbon isotopic composition of soil CO_2 efflux, a powerful method to discriminate different sources feeding soil CO_2 degassing in volcanic-hydrothermal areas. Earth Planet Sci Lett 274:372–379

Craig H, Lupton JE (1976) Primordial neon, helium and hydrogen in oceanic basalts. Earth Planet Sci Lett 31:369–385

David M (1977) Geoestatistical ore reserve estimation, development in geomathematics 2. Elsevier Sci New York, 364 pp

Deutsch C, Journel A (1998) GSLIB: geostatistical software library and users guide, 2nd edn. Oxford University Press, New York, p 369

Dionis S, Melián G, Rodríguez F, Hernández PA, Padrón E, Pérez NM, Barrancos J, Padilla G, Sumino H, Fernandes P, Bandomo Z, Silva S, Pereira JM, Semedo H (2015) Diffuse volcanic gas emission and the termal energy release from the Summit crate of Pico do Fogo, Cape Verde. Bull Volcanol 77:1–13

Fischer TP, Ramírez C, Mora-Amador RA, Hilton DR, Barnes JD, Sharp ZD, Le Brun M, de Moor JM, Barry PH, Füri E, Shaw M (2015) Temporal variations in fumarole gas chemistry at Poás volcano, Costa Rica. J Volcan Geotherm Res 294:56–70

Fridriksson T, Kristjánsson BR, Ármannsson H, Margretadottir E, Olafsdottir S, Chiodini G (2006) CO_2 emissions and heat flow trough soil, fumaroles, and steam heated mud poor at the Reykjanes geothermal area, SW Iceland. Appl Geochem 21:1551–1569

Frondini F, Chiodini G, Caliro S, Cardellini C, Granieri D, Ventura G (2004) Diffuse CO_2 degassing at Vesuvio, Italia. Bull Volcanol 66:642–651

Frondini F, Caliro S, Cardellini C, Chiodini G, Morgantini N (2009) Carbon dioxide degassing and thermal energy release in the Monte Amiata volcanic-geothermal area (Italy). Appl Geochem 24:860–875

Gerlach T, Graeber E (1985) Volatile budget of Kilauea Volcano. Nature 313:273–277

Granieri D, Carapezza M, Chiodini G, Avino R, Caliro S, Ranaldi M, Ricci T, Tarchini L (2006) Correlated increase in CO_2 fumarolic content and diffuse emission from La Fossa crater (Vulcano, Italy): evidence of volcanic unrest or increasing gas release from a stationary deep magma body? Geophys Res Lett 33: L13316.1–L13316.4. https://doi.org/10.1029/2006gl026460

Granieri D, Chiodini G, Avino R, Caliro S (2014) Carbon dioxide emission and heat release estimation for Pantelleria Island (Sicily, Italy). J Volcan Geotherm Res 275:22–33

Hernández PA, Notsu K, Salazar J, Mori T, Natale G, Okada H, Virgili G, Shimoike Y, Sato M, Pérez NM (2001a) Carbon dioxide degassing by advective flow from Usu Volcano, Japan. Science 292:83–86

Hernández PA, Salazar J, Shimoike Y, Mori T, Notsu K, Pérez NM (2001b) Diffuse emission of CO_2 from Miyakejima Volcano, Japan. Chem Geol 177:175–185

Hernández PA, Perez N, Varekamp J (2007) Crater Lake temperature change of the 2005 Eruption of Santa Ana

volcano, El Salvaor, Central America. Pure appl Geophys 164:2507–2522
Hernández PA, Padilla G, Padrón E, Pérez NM, David C, Dácil N, Melián G, Barrancos J, Dionis S, Fátima R, Hirochika S (2012a) Analysis of long- and short-term temporal variations of the diffuse CO_2 emission from Timanfaya volcano, Lanzarote, Canary Islands. Appl Geochem 27:2486–2499
Hernández PA, Pérez N, Fridriksson T, Jolie E, Ilyinskaya E, Thárhallsson A, Ívarsson G, Gíslason G, Gunnarsson I, Jónsson B, Padrón E, Melián G, Mori T, Notsu K (2012b) Diffuse volcanic degassing and thermal energy release from Hengill volcanic system, Iceland. Bull Volcanol 74:2435–2448
Hernández PA, Melián G, Giammanco S, Sortino F, Barrancos J, Pérez NM, Padrón E, López M, Donovan A, Mori T, Notsu K (2015) Contribution of CO_2 and H_2S emitted to the atmosphere by plume and diffuse degassing from volcanoes: the Etna volcano case study. Suv Geophys 36:327–349
Hilton DR, Ramirez C, Mora Amador RA, Fischer TP, Füri E, Barry OH, Shaw AM (2010) Monitoring of temporal and spatial variations in fumarole helium and carbon dioxide characteristics at Poás and Turrialba volcanoes, Costa Rica (2001–2009). Geochem J 44:431–440
Hinkle M, Kilburn J (1979) The use of vacutainer tube for collection of soil sample for helium analysis. U. S. Geol Surv Open File Rep 79:1441
Hust A, Dibble RR (1981) Bathymetry, heat output, and convection in Ruapehu Crater Lake, New Zealeand. J Volcanol Geotherm Res 9:215–236
Keenan JH, Keyes FG, Hill PG, Keyes FG, Moore JG (1969) Steam tables—thermodynamic properties of water including vapor, liquid and solid phases. International Edition Metric Units. Wiley, New York, 162 pp
Lecinsky J, Hilley G, Tosha T, Aoyagi R, Yamamoto K, Benson SM (2007) Dynamic coupling of volcanic CO_2 flow and wind at the Horseshoe Lake tree kill, Mammoth Mountain, California. Geophys Res Lett 34:L03401. https://doi.org/10.1029/2006gl028848
Mamyrin BA, Tolstikhin IN (1984) Helium isotopes in nature: developments in geochemistry 3. Elsevier, Amsterdam, p 273
Martinez M, Fernandez E, Valdés J, Barboza V, Van der Laat R, Duarte E, Malavassi E, Sandoval L, Barquero J, Marino T (2000) Chemical evolution and volcanic activity of the active crater lake of Poás volcano, Costa Rica, 1993–1997. J Volcanol Geotherm Res 97:127–141
Matsuda J, Matsumoto T, Sumino H, Nagao K, Yamamoto J, Miura Y, Kaneoka I, Takahata N, Sano Y (2002) The $^3He/^4He$ ratio of the new internal He standard of Japan (HESJ). Geochem J 36:191–195
Melián G (2008) Emisión difusa de dióxido de carbono y otros volátiles en el volcán Poás, Costa Rica. Thesis Univ La Laguna Canary Islands, América Central (In Spanish)
Melián G, Pérez NM, Hernández PA, Salazar JML, Yock A, Sánchez E, Alvarado GE, Sumino H, Notsu K (2004) Emisión Difusa de Dióxido de Carbono y Vapor de Mercurio en el volcán Miravalles, Costa Rica, Eds. G. Soto and G. Alvarado, La Vulcanología y su entorno Geoambiental. Rev Geol Am Central 30:179–188. (In Spanish with English abstract)
Melián G, Galindo I, Pérez NM, Hernández P, Fernández M, Ramírez C, Mora-Amador R, Alvarado GE (2007) Diffuse emission of hydrogen from Poás volcano, Costa Rica, America Central. PAGEOPH Special Issue “Terrestrial Fluids, Earthquakes and Volcanoes: The Hiroshi Wakita vol. II” 164(12):2465–2487. https://doi.org/10.1007/s00024-007-0282-8
Melián G, Tassi F, Pérez N, Hernández P, Sortino F, Vaselli O, Padrón E, Nolasco D, Barrancos J, Padilla G, Rodríguez F, Dionis S, Calvo D, Notsu K, Sumino H (2012) A magmatic source for fumaroles and diffuse degassing from the summit crater of Teide volcano (Tenerife, Canary Islands): geochemical evidence for the 2004–05 seismic-volcanic crisis. Bull Volcanol 74:1465–1483
Melián G, Hernández PA, Padrón E, Pérez NM, Barrancos J, Padilla G, Dionis S, Rodríguez F, Calvo D, Nolasco D (2014) Spatial and temporal variations of diffuse CO_2 degassing at El Hierro volcanic system: relation to the 2011–2012 submarine eruption. J Geophys Res Solid Earth 119:6976–6991
Mora R., Ramírez C, Fernández M (2004) La actividad de los volcanes de la Cordillera Central, 1998–2002, Costa Rica. In: Soto G, Alvarado GE (eds) La Vulcanología y su entorno Geoambiental. Rev Geol Am Central 30:189–197
Nolasco D, Melián G, Galindo I, Hernández PA, Salazar JML, Pérez NM, Fernández M, Ramírez C, Alvarado GE (2003) Diffuse mercury degassing at Poás Volcano, Costa Rica, Central America. EOS, Trans Am Geophy Union 84, Fall Meet Suppl Abstract F11513
Notsu K, Sugiyama K, Hosoe M, Uemura A, Shimoike Y, Tsunomori F, Sumino H, Yamamoto J, Mori T, Hernández PA (2005) Diffuse CO_2 efflux from Iwojima volcano, Izu-Ogasawara arc, Japan. J Volcanol Geother Res 139:147–161
Padrón E, Melián G, Marrero R, Nolasco D, Barrancos J, Padilla G, Hernández PA, Pérez NM (2008a) Changes in the diffuse CO_2 emission and relation to seismic activity in and around El Hierro, Canary Islands. Pure appl Geophys 165:95–114
Padrón E, Hernández PA, Toulkeridis T, Pérez NM, Marrero R, Melián G, Virgili G, Notsu K (2008b) Diffuse CO_2 emission rate from Pululahua and the lake-filled Cuicocha calderas, Ecuador. J Volcanol Geotherm Res 176:163–169
Padrón E, Hernández PA, Pérez NM, Toulkeridis T, Melián G, Barrancos J, Virgili G, Sumino H, Notsu K (2012) Fumarole/plume and diffuse CO_2 emission from Sierra Negra caldera, Galapagos archipelago. Bull Volcanol 74:1509–1519

Padrón E, Pérez N, Rodríguez F, Melián G, Hernández PA, Sumino H, Padilla G, Barrancos J, Dionis S, Notsu K, Calvo D (2015) Dynamics of diffuse carbon dioxide emissions from Cumbre Vieja volcano, La Palma, Canary Islands. Bull Volcanol 77:1–15. https://doi.org/10.1007/s00445-015-0914-2

Parkinson K (1981) An improved method for measuring soil respiration in the field. J Appl Ecol 18:221–228

Pasternack G, Varekamp J (1997) Volcanic lake systematics: I. Physical constraints. Bull Volcanol 58:528–538

Pérez N, Salazar J, Hernández P, Soriano T, López DL, Notsu K (2004) Diffuse CO_2 and ^{222}Rn degassing from San Salvador volcano, El Salvador, Central America. Bull Geol Soc Am Special Paper 375:227–236

Pérez N, Hernández P, Padrón E, Cartagena R, Olmos R, Barahona F, Melián G, Salazar P, López DL (2006) Anomalous diffuse CO_2 emission prior to the January 2002 short-term Unrest at San Miguel Volcano, El Salvador, Central America. Pure Appl Geophys 163 (4):883–896

Pérez N, Hernández P, Padrón E, Melián G, Nolasco D, Barrancos J, Padilla G, Calvo D, Rodríguez F, Dionis S, Chiodini G (2013) An increasing trend of diffuse CO_2 emission from Teide volcano (Tenerife, Canary Islands): geochemical evidence of magma degassing episodes. J Geol Soc London. https://doi.org/10.1144/jgs2012-125

Rizzo A, Grassa F, Inguaggiato S, Liotta M, Longo M, Madonia P, Brusca L, Capasso G, Morici S, Rouwet D, Vita F (2009) Geochemical evaluation of observed changes in volcanic activity during the 2007 eruption at Stromboli (Italy). J Volcanol Geotherm Res 182:246–254

Rowe G, Brantley S, Fernández M, Fernández JF, Barquero J, Borgia A (1992a) Fluid-volcano interacting in an active stratovolcano: the Crater Lake system of Poás volcano, Costa Rica. J Volcanol Geotherm Res 49:23–51

Rowe G, Ohsawa S, Takano B, Brantley SL, Fernández JF, Barquero J (1992b) Using Crater Lake chemistry to predict volcanic activity at Poás volcano, Costa Rica. Bull Volcanol 54:494–503

Rowe G, Brantley S, Fernández J, Borgia A (1995) The chemical and hydrologic structure of Poás volcano, Costa Rica. J Volcanol Geotherm Res 64:233–267

Rouwet D, Mora-Amador R, Sandri L, Ramírez-Umaña C, González G, Pecoraino G, Capaccioni B (Chapter 9) 39 years of geochemical monitoring of Laguna Caliente crater lake, Poás: Patterns from the past as keys for the future. In: Tassi F, Mora-Amador R, Vaselli O (eds) Poás volcano (Costa Rica): the pulsing heart of Central America Volcanic Zone. Springer, Heidelberg (Germany)

Rymer H, Brown G (1989) Gravity changes as a precursor to volcanic eruption at Poás volcano, Costa Rica. Nature 342:902–905

Rymer H, Locke C, Brenes J, Williams-Jones G (2005) Magma plumbing processes for persistent activity at Poás volcano, Costa Rica. Geophys Res Lett 32: L08307. https://doi.org/10.1029/2004GL022284

Rymer H, Locke CA, Borgia A, Martinez M, Brenes J, Van der Laat R, Williams-Jones G (2009) Long-term fluctuations in volcanic activity: implications for future environmental impact. Terra Nova 21:304–309

Sano Y, Wakita H (1985) Geographical distribution of $^3He/^4He$ ratios in Japan: Implication for arc tectonics and incipient magmatism. J Geophy Res 90:8719–8741

Sheperd J, Sigurdsson H (1978) The Soufriere Crater Lake as a calorimeter. Nature 271:344–345

Sinclair A (1974) Selection of thresholds in geochemical data using probability graphs. J Geochem Expl 3:129–149

Snyder G, Poreda R, Hunt A, Fehn U (2001) Regional variations in volatile composition: Isotopic evidence for carbonate recycling in the Central American volcanic arc. Geochem Geophys Geosys 2:U1–U32

Sumino H, Nagao K, Notsu K (2001) Highly sensitive and precise measurement of helium isotopes using a mass spectrometer with double collector system. J Mass Spectrom Soc Jap 49:61–68

Taran Y, Rouwet D (2008) Estimating thermal inflow to El Chichón crater lake using the energy-budget, chemical and isotope balance approaches. J Volcanol Geother Res 175:472–481

Tennant C, White M (1959) Study of the distribution of some geochemical data. Econ Geol 54:1281–1290

Vaselli O, Tassi F, Fischer TP, Tardani D, Fernandez Soto E, Duarte E, Martinez M, De Moor MJ, Bini G (Chapter 10). The last eighteen years (1998–2015) of fumarolic activity at the Poás volcano (Costa Rica) and the renewed activity. In: Tassi F, Mora-Amador R, Vaselli O, (eds) Poás volcano (Costa Rica): The pulsing heart of Central America Volcanic Zone. Springer, Heidelberg (Germany)

Vaselli O, Tassi F, Montegrossi G, Duarte E, Fernández E, Bergamaschi F (2003) Fumarole migration and fluid chemistry at Poás volcano (Costa Rica) from 1998 to 2001. In: Oppenheimer C, Pyle DM, Barkley J (eds) Volcanic degassing. Geol Soc London, Special Publications 213:247–262. https://doi.org/10.1144/gsl.sp.2003.213.01.15

Vannucchi P, Mason JP (Chapter 1) Overview of the tectonics and geodynamics of Costa Rica Eds. In: Tassi F, Mora-Amador R, Vaselli O (eds) Poás volcano (Costa Rica): the pulsing heart of Central America Volcanic Zone. Springer, Heidelberg (Germany)

Zimmer MM, Fischer TP, Hilton DR, Alvarado GE, Sharp ZD, Walker JA (2004) Nitrogen systematics and gas fluxes of subduction zones: insights from Costa Rica arc

Behaviour of Polythionates in the Acid Lake of Poás Volcano: Insights into Changes in the Magmatic-Hydrothermal Regime and Subaqueous Input of Volatiles

María Martínez-Cruz, Manfred J. van Bergen, Bokuichiro Takano, Erick Fernández-Soto and Jorge Barquero-Hernández

Abstract

In this chapter, we document an extensive record of concentrations and speciation of polythionates (PTs: $S_4O_6^{2-}$, $S_5O_6^{2-}$, and $S_6O_6^{2-}$), which form in the warm (21–60 °C) and hyper-acidic (pH < 1.8) waters of the crater lake of Poás volcano (Costa Rica) through interaction with gaseous SO_2 and H_2S of magmatic origin. Our data set, together with earlier published results, covers the period 1980–2006 during which lake properties and behavior were marked by significant variations. Distinct stages of activity can be defined when combining PT distributions with geochemical, geophysical and field observations. Between 1985 and mid-1987, when fumarolic outgassing was centered on-shore, the total concentration of PTs in the lake was consistently high (up to 4,200 mg/kg). Mid-1987 was the start of a 7-year period of vigorous fumarolic activity with intermittent phreatic eruptions from the lake, which then dried out. Concentrations of PTs remained below or close to detection limits throughout this period. After mid-1994, when a new lake formed and fumarolic outgassing shifted to the dome, the total PT concentrations returned to relatively stable intermediate levels (up to 2,800 mg/kg) marking more quiescent conditions. Since early 1995, numerous weak fumarole vents started, opening up at several other locations in the crater area. During short intervals (November 2001–May 2002 and October 2003–March 2005), PTs virtually disappeared. After April 2005, PTs re-appeared in large amounts (up to more than 3,000 mg/kg) until February 2006, one month before the onset of the March 2006–2017 cycle of phreatic eruptions, when concentrations dropped and remained below 100 mg/kg. The observed behavior of PTs records changes in the input and SO_2/H_2S ratios of subaqueous fumaroles. The prevailing distribution of PTs is $S_4O_6^{2-} > S_5O_6^{2-} > S_6O_6^{2-}$, which is common for periods when total PT concentrations and SO_2/H_2S

M. Martínez-Cruz (✉) · E. Fernández-Soto · J. Barquero-Hernández
Observatorio Vulcanológico y Sismológico de Costa Rica Universidad Nacional OVSICORI-UNA, Heredia, Costa Rica
e-mail: maria.martinez.cruz@una.cr

M. Martínez-Cruz · M. J. van Bergen
Faculty of Geosciences Universiteit Utrecht, Utrecht, The Netherlands

B. Takano
Graduate School of Arts and Sciences, The University of Tokyo, Komaba 3-8-1, Meguro, Tokyo 153-8902, Japan

F. Tassi et al. (eds.), *Poás Volcano*, Active Volcanoes of the World,
https://doi.org/10.1007/978-3-319-02156-0_7

ratios of the gas influx into the lake are relatively high. PTs are virtually absent as a consequence of thermal or sulphitolytic break-down during periods of strong fumarolic outgassing in response to shallow intrusion of fresh magma or fracturing of the solid envelope around a pre-existing body of cooling magma. They are also low in abundance or undetected during quiescent periods when subaqueous fumarolic output is weak and has low SO_2/H_2S ratios, resulting in a concentration sequence $S_5O_6^{2-} > S_4O_6^{2-} > S_6O_6^{2-}$. The onset of phreatic eruptions are preceded by an increase in PT concentrations, accompanied by a change in the dominance from penta- to tetrathionate, and followed by a sharp drop in total PT content, up to several months before. Periods of phreatic eruptive activity that started in 1987 and 2006 followed these PT signals of increased input of sulfur-rich gas, in both cases possibly in response to shallow emplacement of fresh magma or hydrofracturing.

Keywords
Polythionates · Sulphur chemistry · Acid volcanic lakes · Magmatic volatiles · Magmatic-hydrothermal system · Volcano monitoring · Volcanic unrest

1 Introductory Remarks

1.1 Polythionates: Research and Discovery in Waters of Volcanic Origin

A variety of chemical constituents of volcanic lakes and associated fluids is classically employed to monitor volcanic activity, assess the flow rate and nature of volcanic gases expelled from cooling magmas, and study water-gas-rock interactions in shallow magmatic-hydrothermal systems (e.g. Giggenbach 1974; Sigurdsson 1977; Casadevall et al. 1984a; Rowe et al. 1992a, b; Christenson and Wood 1993; Takano et al. 1994a, b; Delmelle and Bernard 1994; Delmelle 1995; Pasternack and Varekamp 1994, 1997; Armienta et al. 2000; Christenson 2000; Delmelle and Bernard 2000; Delmelle et al. 2000; Martínez et al. 2000; Xu et al. 2000; Sriwana et al. 2000; Varekamp et al. 2001; Tassi et al. 2009; Stimac et al. 2003; Bernard et al. 2004; Takano et al. 2004, 2008; Ohba et al. 2008; Rouwet et al. 2014). Among the more exotic geochemical parameters with monitoring potential are the natural oligomeric sulfur compounds, known as polythionates (PTs), which have the general formula $(O_3SS_nSO_3)^{2-}$, where "n = 2" in tetrathionate (Steudel and Holdt 1986; Shriver and Atkins 1999). They are named according to the total number of sulfur atoms: trithionate ($S_3O_6^{2-}$), tetrathionate ($S_4O_6^{2-}$), pentathionate ($S_5O_6^{2-}$), etc. Trithionate, the shortest PT, will decompose if exposed to acid but tetrathionate, and higher polythionates, are stabilized by acid conditions. Trithionate in low pH < 1 solutions is formed by rearrangement reactions of tetrathionate but it is rapidly oxidized converting tetrathionate to sulfate. Thus, trithionate is not considered in this chapter because it decomposes in acid conditions whereas tetrathionate and higher polythionates are favored by chemical conditions occurring at Laguna Caliente. Oxidation kinetics of PTs studied through Fe^{3+}/O_2 catalysis experiments, indicate that under low pH conditions and excess of Fe^{3+} and O_2, the rate of polythionate oxidation is fast and Fe^{3+} and O_2 somehow control some of the reaction pathways of the oxidation kinetics of the PTs (Druschel et al. 2003).

Polythionates were found in waters of volcanic origin for the first time in New Zealand when MacLaurin (1911) discovered pentathionate in the crater lake of White Island Volcano. Day and Allen (1925) reported the presence of pentathionate and possibly that of tetrathionate, and trithionate in the hot springs of Lassen Peak whilst Wilson (1941, 1953, 1959) recognized various PTs in waters from the Black Pool near Rotorua and the Black Geyser at Ketetahi thermal springs. The boiling springs of Yellowstone National Park have also been studied for total PTs and other sulfur species (Xu et al. 2000). However, the research on the complex chemistry of the polythionates started earlier in the 19th century. Dalton in 1808 in his "A New System of Chemical Philosophy" commented on the chemical nature of the constituents of the liquid, which later came to bear the name Wackenroder (Foss 1960). The three lowest acids were discovered between 1840 and 1846 by Langlois, Fordos and Gelis (Janitzki 1969), and later Debus (1888) from Wackenroder's liquids isolated a salt that analyzed for potassium hexathionate. The Wackenroder's solution is a milky liquid produced by the reaction of SO_2 and H_2S gases in an aqueous solution originating a mixture of colloidal sulfur, precipitated sulfur, sulfur in solution, traces of sulfate, and at least three polythionates (tri-, tetra- and pentathionate) (Koh 1990). Chemists discovered that tetrathionate was the predominant product and that the pentathionic acids in the Wackenroder's solution decompose into sulfur and tetrathionates under alkaline conditions but are more stable in presence of other acids such as H_2SO_4 and HCl (Debus 1888).

Polythionates as well as a variety of sulfur species, including sulfate (mainly as bi-sulfate), bi-sulfite, are transient sulfur compounds formed in acid aqueous environments such as acid volcanic lake systems by reaction between SO_2 and H_2S gases (Takano and Watanuki 1990; Takano et al. 1994b) from subaqueous fumaroles entering the lake at relatively high SO_2/H_2S ratio (Goehring 1952). Polythionates reach maximum concentrations when this ratio is around 0.13, but they are decomposed via sulfitolysis at higher ratios, yielding elemental sulfur and sulfate as main products (Takano et al. 2001) according to:

$$SO_{2\,(aq)} + H_2O_{(l)} \rightleftharpoons HSO_3^{-}{}_{(aq)} + H^{+}_{(aq)} \quad (x = 4 \text{ to } 6) \qquad \text{(a)}$$

$$3\,S_xO_6^{2-}{}_{(aq)} + 3\,HSO_3^{-}{}_{(aq)} \rightleftharpoons 3\,S_{(x-1)}O_6^{2-}{}_{(aq)} + 2\,SO_4^{2-}{}_{(aq)} + 4\,S^{\circ}{}_{(s)} + H^{+}{}_{(aq)} + H_2O_{(l)} \quad (x = 4 \text{ to } 6) \qquad \text{(b)}$$

Decomposition will also occur as a result of increasing temperature:

$$3\,S_xO_6^{2-}{}_{(aq)} + 2\,H_2O_{(l)} \rightleftharpoons 5\,SO_4^{2-}{}_{(aq)} + (3x-5)\,S^{\circ}{}_{(s)} + 4\,H^{+}{}_{(aq)} \quad (x = 4 \text{ to } 6) \qquad \text{(c)}$$

Hence, SO_2-induced or thermal decomposition of PTs will contribute to an (extra) increase in the total sulfate concentration of an acidic crater lake. For example, significant increases in the concentration of sulfate in the Yugama crater lake were accompanied by breakdown of PTs during periods of phreatic activity (Takano and Watanuki 1990). Polythionates as well as other sulfur species can also be produced by the hydrolysis of elemental sulfur, which abounds in the form of molten sulfur at the interface between the bottom of the hyper acid lakes and their underlying hydrothermal systems (Delmelle and Bernard 2015).

Polythionates in acid volcanic lakes may range in abundance from trace or minor amounts to several hundreds or thousands of mg/l. They have been detected in acid volcanic lakes with pH < 3 such as those at Kusatsu-Shirane (Takano 1987; Takano and Watanuki 1990; Takano et al. 2008), Ruapehu (Takano et al. 1994b), Poás (Rowe et al. 1992b), Keli Mutu Tiwu Nua Muri Koohi Fah (Pasternack and Varekamp 1994), Patuha (Sriwana et al. 2000), Kawah Ijen (Delmelle and Bernard 1994; Takano et al. 2004), Santa Ana (Bernard et al. 2004), Maly Semiachik (Takano et al. 1995), El Chichón (Casas et al. 2015), and Rincón de la Vieja (B. Takano and M. Martínez, unpublished data). Despite the reported presence of PTs in a number of volcanic acid lakes and their monitoring potential (see Delmelle and Bernard 2015 for a recent review), a long-term behavior of PT species was only investigated at Kusatsu-Shirane, Ruapehu and Poás acid lakes. The main reason for that is that polythionates are labile sulfur species at pH > 3, sensitive to Eh-pH, ionic strength, temperature, dissolved SO_2 and H_2S and microbial metabolism in the waters of the acid lakes. Nowadays, chemists continue investigating polythionates as testified by a recent method to produce stable non toxic salts of polythionic acids based on the Smolyaninov reaction for its potential application in safe gold leaching instead of cyanation or mercury amalgamation (Smolyaninov et al. 2014). This is a promising approach and we hope these new stable polythionates will soon be available. Thus, separation and quantification of polythionate species could be continued for the benefit of volcano surveillance and research on magmatic fluids worldwide.

1.2 The *Laguna Caliente* at Poás

1.2.1 Summary of Historical Activity

Poás (10° 11′ 26″N–84° 13′ 56″W), one of the most active volcanoes in central Costa Rica where about 60% of the inhabitants of the country live (Fig. 1), has shown persistent fumarolic degassing with large fluctuations, periods of phreatic explosions in its hyperacid lake and rare strombolian to vulcanian eruptions during at least the last two and a half centuries. The largest historic eruption (vulcanian type) occurred in 1910 when a major steam and ash cloud rose up to ca. 8 km above the summit (Calvert and Calvert 1917). A period of moderate phreatic and phreatomagmatic eruptions occurred between 1952 and 1955, drying out the crater lake and forming a dome, a small lava flow, and a pit crater that hosted an acid lake in the 60s, (Krushensky and Escalante 1967; Bennett and

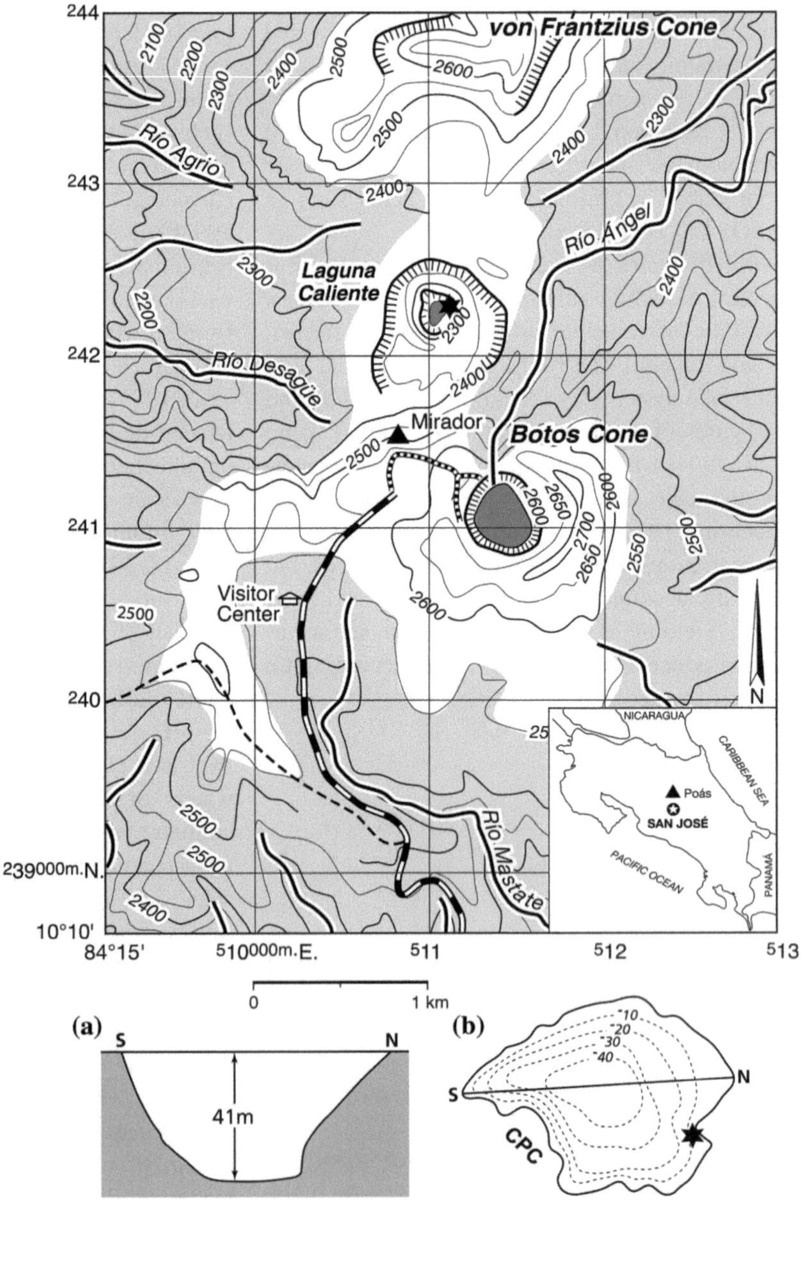

Fig. 1 Overview of the summit area of Poás and sampling sites. Cross section (**a**) and bathymetric contours (**b**) are from a survey carried out in early 2001 (adapted from Tassi et al. 2009). Black star in diagram B is the location where lake waters were regularly collected

Raccichini 1978a, b). Periods with phreatic explosions from the lake ranging from gas bursts to ejection of gas and vapor columns mixed with sediments, ballistics, and lake water up to several dozens and hundreds of meters high have occurred intermittently since then. The most recent eruptive period with phreatic, strombolian and vulcanian phases with erupted products ranging between basalt-andesitic and andesitic composition (Martínez et al. 2017), occurred between 23 April and late June 2017, preceded by conspicuous geophysical and geochemical changes ca. 1 month in advance: rapid increase in seismic activity, ground deformation (inflation), sharp rise in the magmatic volatile and thermal power output, enrichment of the hyperacid lake in halogens of volatile origin, and remarkable and rapid increase in the lake level and the water table at the bottom of the active crater, despite of absence of rainfall. This eruption triggered small lahars in several of the rivers draining toward the Atlantic slope, destroyed the dome formed in the previous magmatic eruptive period of 1952–1955, and dried out the hyperacid lake (Laguna Caliente) allowing fumaroles enriched in magmatic volatiles and particles to discharge directly into the atmosphere. The overall activity ceased gradually throughout the second half of 2017 and the first half of 2018, and only low temperature fumaroles (92 °C) and boiling mud pools remained a the dried bottom of the crater.

1.2.2 Laguna Caliente and First Polythionate Data

The warm and hyperacid lake of Laguna Caliente, that was stable between mid 1994 and early April 2017, registered temperature, pH and TDS values ranging between 19 °C and 65 °C, below zero and 1.8, and 10^4–10^3 mg/kg, respectively. The lake reached about 300–350 m in diameter and up to 40–50 m deep when at its high level. Sub-aerial fumarolic activity was mainly centered at the dome that was at the southern edge of the lake (Fig. 1) and recently destroyed by the eruptions of April 2017. The dome usually presented low temperature fumaroles (at about boiling point of water at the altitude of the summit: 92 °C) but occasionally it reached up to ca. 900 °C (in 2011) and around 1000 °C (in 1981 and 2017). Other low temperature fumarole fields have appeared elsewhere within the crater as well (e.g. Vaselli et al. 2003, Chap. 10; Martínez 2008). The intermittent, though frequent, eruptive behavior, and strong temporal variability in lake properties and fumarolic manifestations testify that the Poás volcanic-hydrothermal system is highly dynamic and sensitive to input from magmatic and meteoric sources.

Rowe et al. (1992b) reported the first PT data for Poás's Laguna Caliente, covering the period 1980-early 1990 when the lake evolved from relative quiescence to an interval of high activity (see Martínez 2008 for a summary of the historic activity of Poás). In the early 1980s the authors noticed relatively moderate concentrations of total polythionates with the pentathionate as the predominant species as well as a sharp decrease in PT concentrations three months prior to the recurrence of phreatic eruptions in 1987. They suggested that rapidly changing PT stability and speciation could be a precursor signal of changes in the magmatic-hydrothermal system that affect subaqueous fumarolic input and trigger phreatic activity. Since then, several other periods of eruptions have occurred (phreatic in 2006–2016 and phreatomagmatic to strombolian in the first half of 2017), as well as frequent, sometimes drastic, fluctuations in the physicochemical properties of the crater lake (Martínez et al. 2000; Venzke et al. 2002; OVSICORI open reports; Martínez 2008; Rouwet et al. Chapter "39 Years of Geochemical Monitoring of Laguna Caliente Crater Lake, Poás: Patterns from the Past as Keys for the Future"). This unstable behavior of the volcano provides a unique opportunity to evaluate the response of PTs to marked changes in the input of magmatic volatiles and heat, and to test their value in forecasting upcoming phreatic events. In this paper, we document PT results for the period 1990–2006 that, together with those of Rowe et al. (1992b), constitute a monitoring record of about 25 years. It will be shown that the PT budget in Laguna Caliente is characterized by a remarkable variability, which is directly linked to changes in the flow rate, composition and location of fumarole emissions within the crater that hosts the lake. In conjunction with routinely monitored geochemical signatures, geophysical parameters and field observations, our findings provide new insights into the conditions of the subsurface magmatic-hydrothermal system, and the mechanisms that control eruptive events.

2 Sampling Strategy and Analytical Procedures

2.1 Sampling

Selected samples for PT analysis were routinely collected on the NE side of the acidic crater lake (Fig. 1). Most samples were stored in dark high-density polyethylene bottles at ambient temperature without filtration, dilution or addition of preservatives before carrying out the PT analyses. The available PT record covers the period 1980–2006 and represents an approximate sampling frequency of once every two months.

2.2 Polythionate Analysis

About 190 lake water samples were analyzed for tetra-, penta-, and hexathionate ions at the Department of Chemistry of Tokyo University (Japan) and at the Laboratory of Atmospheric Chemistry (LAQAT) of the Universidad Nacional (Costa Rica) using similar ion-pair chromatographic techniques described in the supplementary data to this article and described at http://www.ovsicori.una.ac.cr/index.php/vulcanologia/geoquimica/investigacion-geoquimica.

Overall deviations from the mean values were better than 20% RSD (3–20% for tetra-; 2–15% for penta-, and 1–20% for hexathionate. As most analyses in Tokyo and Costa Rica were carried out with time differences of about 3 years, it is possible to state that the PT time series presented in this chapter are largely unaffected by potential inter-laboratory differences or chemical instability. Repetition of sulfate and chloride analyses suggested that some of the oldest samples might have experienced a certain degree of evaporation or mineral precipitation during storage. On the basis of the recorded increasing concentrations of SO_4 and Cl, those of PTs may have raised in these cases by 15–20% at most.

2.3 Determination of Major Anions, Dissolved Gases, PH and Other Solutes

All of the samples collected were analyzed for sulfate, chloride and fluoride by suppressed ion chromatography, using a fully automated microbore ion suppressed Dionex chromatographic sytem at the Laboratory of Volcano Geochemistry "Dr. Eduardo Malavassi Rojas" of OVSICORI Universidad Nacional (Costa Rica) (Martínez 2008). Based on the analysis of a synthetic solution, precision was about 0.1, 0.3 and 4% for sulfate, chloride, fluoride, respectively. Detection limits were 0.3, 0.1 and 0.05 mg/kg, respectively.

Dissolved unreacted SO_2 and H_2S gases in the lake water were measured in situ on an irregular basis during 1999–2006, using a gas detection tube method (Takano et al. 2008). Detection limits for dissolved SO_2 and H_2S were 1 and 0.2 ppm, respectively.

The pH measurements were performed on untreated samples at ambient temperature (24 ± 2 °C) in the OVSICORI geochemistry laboratory using a WTW Multi 340i potentiometer. Combination of the new results with previously available data (e.g. Casertano et al. 1985; Rowe et al. 1992a, b; Nicholson et al. 1992, 1993; Martínez et al. 2000) constitutes a record for major anion concentrations, pH and temperature that covers the period 1980–2006.

All field-related data (lake volume, temperature, color, seismic records, etc.) are from the database of OVSICORI (Venzke et al. 2002; Martínez et al. 2000; Martínez 2008). Most monthly rainfall data are from the Poás volcano summit rain gauge of the Centro de Servicios y Estudios Básicos de Ingeniería of Instituto Costarrincense de Electricidad ICE located at 2,564 m a.s.l.

Table 1 Polythionate and major anion concentrations (mg/kg) of Poás crater lake for the period 1980–2006

Date	Temp. (°C)	pH_{lab} 24 ± 2 °C	$S_4O_6^{2-}$	$S_5O_6^{2-}$	$S_6O_6^{2-}$	Total $S_xO_6^{2-}$	$S_4O_6^{2-}/S_5O_6^{2-}$	$S_4O_6^{2-}/S_6O_6^{2-}$	S_4O^{2-}	Cl^-	F^-	PT lab
Stage II												
31-0ct-80	45	0.11	190	270	130	590	0.7	1.5	65,500	26,900	1,320	(1)
28-Nov-84	48	0.07	83	410	100	593	0.2	0.9	57,700	23,700	1,500	(2)[a]
24-Jan-85	44	0.14	1,190	2,120	770	4,080	0.6	1.5	49,500	25,400	1,660	(2)[a]
20-Mar-85	44	0.13	1,120	1,760	510	3,390	0.6	2.2	48,200	24,700	15,60	(2)[a]
06-May-85	44	0.14	450	1,180	360	1,990	0.4	1.3	55,000	22,000	1,400	(2)[a]
22-Aug-85	45	0.17	970	1,740	520	3,230	0.6	1.8	54,200	21,100	1,280	(2)[a]
09-Oct-85	45	0.18	600	1,270	460	2,330	0.5	1.3	52,900	20,700	1,260	(2)[a]
04-Feb-86	39	0.26	1,160	1,420	520	3,100	0.8	2.2	36,900	16,500	1,010	(2)[a]
Stage III												
02-May-86	38	0.20	1,730	1,810	630	4,170	1.0	2.7	40,100	16,500	970	(2)[a]
23-Aug-86	52	0.20	1,700	1,700	520	3,920	1.0	3.3	41,200	18,900	1,020	(2)[a]
31-Oct-86	54	0.14	1,040	1,310	420	2,770	0.8	2.5	52,000	23,200	1,240	(2)[a]
10-Jan-87	58	−0.01	1,560	1,370	620	3,550	1.1	2.5	64,400	30,400	1,590	(2)[a]
27-Feb-87	62	−0.03	170	630	120	920	0.3	1.4	78,600	33,700	1,750	(2)[a]
19-Mar-87	62	0.00	53	70	b.d.l.	123	0.8	n.a.	82,900	35,900	1,820	(2)[a]
27-July-87	68	−0.27	2	6	b.d.l.	8	0.3	n.a.	103,000	44,900	2,110	(2)[a]
10-Sep-87	65	−0.23	b.d.l.	b.d.l.	b.d.l.	n.a.	n.a.	n.a.	108,000	43,400	2,060	(2)[a]
27-Nov-87	60	−0.24	b.d.l.	b.d.l.	b.d.l.	n.a.	n.a.	n.a.	101,000	48,000	1,910	(2)[a]
16-Jan-88	64	−0.37	b.d.l.	b.d.l.	b.d.l.	n.a.	n.a.	n.a.	127,000	59,300	2,250	(2)[a]
02-Mar-88	60	−0.31	b.d.l.	b.d.l.	b.d.l.	n.a.	n.a.	n.a.	118,000	51,300	2,100	(2)[a]
24-June-88	65	−0.61	b.d.l.	b.d.l.	b.d.l.	n.a.	n.a.	n.a.	175,000	73,100	2,970	(2)[a]
02-Sep-88	65	−0.53	b.d.l.	b.d.l.	b.d.l.	n.a.	n.a.	n.a.	196,000	44,200	3,810	(2)[a]
26-Oct-88	65	−0.43	b.d.l.	b.d.l.	b.d.l.	n.a.	n.a.	n.a.	192,000	47,400	4,220	(2)[a]
07-Feb-89	75	−0.87	b.d.l.	b.d.l.	b.d.l.	n.a.	n.a.	n.a.	286,000	28,000	10,400	(2)[a]
03-Mar-89	82	−0.55	b.d.l.	b.d.l.	b.d.l.	n.a.	n.a.	n.a.	233,000	27,200	6,310	(2)[a]

(continued)

Table 1 (continued)

Date	Temp. (°C)	pH_{lab} 24 ± 2 °C	$S_4O_6^{2-}$	$S_5O_6^{2-}$	$S_6O_6^{2-}$	Total $S_xO_6^{2-}$	$S_4O_6^{2-}/S_5O_6^{2-}$	$S_4O_6^{2-}/S_6O_6^{2-}$	S_4O^{2-}	Cl^-	F^-	PT lab
31-May-89	85	−0.09	110	6	b.d.l.	116	18	n.a.	124,000	25,100	6,150	(2)[a]
22-June-89	80	−0.44	b.d.l.	b.d.l.	b.d.l.	n.a.	n.a.	n.a.	175,000	40,600	8,420	(2)[a]
26-July-89	82	−0.55	250	90	25	365	2.9	10	175,000	46,500	8,600	(2)[a]
17-Aug-89	87	−0.36	b.d.l.	b.d.l.	b.d.l.	n.a.	n.a.	n.a.	99,300	39,200	7,230	(2)[a]
24-Oct-89	85	−0.56	b.d.l.	b.d.l.	b.d.l.	n.a.	n.a.	n.a.	154,000	64,100	7,910	(2)[a]
26-Jan-90	78	−0.29	22	20	b.d.l.	42	1.2	n.a.	109,000	52,000	6,400	(2)[a]
21-Feb-90	85	−0.56	16	10	b.d.l.	26	1.5	n.a.	103,000	95,200	8,460	(2)[a]
17-Mar-90	75	−0.60	b.d.l.	b.d.l.	b.d.l.	n.a.	n.a.	n.a.	165,000	63,100	8,730	(2)[a]
01-June-90	83	−0.46	b.d.l.	b.d.l.	b.d.l.	n.a.	n.a.	n.a.	67,800	36,600	9,440	(1)
10-July-90	94	−0.44	b.d.l.	b.d.l.	b.d.l.	n.a.	n.a.	n.a.	85,600	79,600	15,400	(1)
13-Sep-90	88	−0.70	b.d.l.	b.d.l.	b.d.l.	n.a.	n.a.	n.a.	90,300	114,000	25,400	(1)
08-Dec-90	73	n.d.	b.d.l.	b.d.l.	b.d.l.	n.a.	n.a.	n.a.	34,800	40,800	4,860	(1)
08-Feb-91	65	−0.40	b.d.l.	b.d.l.	b.d.l.	n.a.	n.a.	n.a.	32,000	52,400	5,930	(1)
19-Apr-91	78	−0.42	b.d.l.	b.d.l.	b.d.l.	n.a.	n.a.	n.a.	43,600	50,700	5,610	(1)
26-July-91	63	−0.38	90	110	20	220	0.8	4.5	51,300	47,900	4,310	(1)
24-Sep-91	71	−0.44	b.d.l.	b.d.l.	b.d.l.	n.a.	n.a.	n.a.	67,800	55,800	6,320	(1)
21-Oct-91	74	n.d.	8	16	9	33	0.5	0.9	70,000	58,500	6,560	(1)
01-Nov-91	70	−0.50	9	b.d.l.	b.d.l.	9	n.a.	n.a.	72,100	61,100	6,790	(1)
12-Dec-91	64	−0.30	105	130	73	308	0.8	1.4	40,700	35,800	4,690	(1)
17-Jan-92	67	−0.50	b.d.l.	b.d.l.	b.d.l.	n.a.	n.a.	n.a.	75,000	58,700	5,750	(1)
07-Feb-92	67	−0.60	75	100	37	212	0.8	2.1	41,600	32,900	3,700	(1)
13-Mar-92	65	−0.70	b.d.l.	b.d.l.	b.d.l.	n.a.	n.a.	n.a.	119,000	87,400	8,800	(1)
03-Apr-92	75	−0.80	b.d.l.	b.d.l.	b.d.l.	n.a.	n.a.	n.a.	147,000	102,000	11,000	(1)
24-July-92	70	−0.60	12	b.d.l.	b.d.l.	12	n.a.	n.a.	94,200	67,800	8,260	(1)
21-Aug-92	75	−0.60	b.d.l.	b.d.l.	b.d.l.	n.a.	n.a.	n.a.	78,900	63,300	6,870	(1)
18-Sep-92	70	−0.60	b.d.l.	b.d.l.	b.d.l.	n.a.	n.a.	n.a.	93,100	74,400	8,390	(1)

(continued)

Table 1 (continued)

Date	Temp. (°C)	pH_{lab} 24 ± 2 °C	$S_4O_6^{2-}$	$S_5O_6^{2-}$	$S_6O_6^{2-}$	Total $S_xO_6^{2-}$	$S_4O_6^{2-}/S_5O_6^{2-}$	$S_4O_6^{2-}/S_6O_6^{2-}$	S_4O^{2-}	Cl^-	F^-	PT lab
07-Oct-92	75	−0.50	30	24	20	74	1.1	1.2	52,400	71,600	5,060	(1)
19-Nov-92	80	−0.50	b.d.l.	b.d.l.	b.d.l.	n.a.	n.a.	n.a.	38,400	73,000	5,720	(1)
23-Jan-93	65	0.00	b.d.l.	b.d.l.	b.d.l.	n.a.	n.a.	n.a.	56,100	54,000	4,510	(2)
25-Feb-93	65	0.00	160	196	110	466	0.8	1.4	58,000	56,200	n.d.	(2)
02-Apr-93	60	0.00	b.d.l.	b.d.l.	b.d.l.	n.a.	n.a.	n.a.	65,000	75,500	7,550	(2)
09-Sep-93	64	0.35	b.d.l.	b.d.l.	b.d.l.	n.a.	n.a.	n.a.	50,400	29,600	5,840	(1)
22-Oct-93	60	0.00	b.d.l.	b.d.l.	b.d.l.	n.a.	n.a.	n.a.	46,000	27,600	5,330	(2)
12-Mar-94	55	0.00	b.d.l.	b.d.l.	b.d.l.	n.a.	n.a.	n.a.	83,100	56,500	7,780	(2)
23-Sep-94	65	−0.18	1,020	530	850	2,400	1.9	1.2	27,000	24,000	2,660	(1)
21-Oct-94	60	0.14	3,700	2,660	990	7,350	1.4	3.7	17,000	8,470	1,220	(2)
15-Nov-94	55	0.38	b.d.l.	b.d.l.	b.d.l.	n.a.	n.a.	n.a.	11,000	7,640	n.d.	(1)
06-Jan-95	50	0.22	2,700	1,530	900	5,130	1.7	3.0	16,000	10,200	1,060	(1)
03-Feb-95	47	0.18	2,340	1,420	880	4,640	1.6	2.7	16,600	10,100	990	(1)
10-Mar-95	42	0.68	1,800	1,230	500	3,530	1.5	3.6	20,500	7,780	950	(2)
Stage IV												
08-Sep-95	34	1.10	294	290	140	724	1.0	2.2	8,710	5,440	530	(2)
20-Oct-95	30	1.21	b.d.l.	b.d.l.	b.d.l.	n.a.	n.a.	n.a.	6,230	3,800	190	(2)
05-Jan-96	29	1.22	40	80	36	156	0.5	1.0	9,060	3,590	200	(2)
23-Feb-96	26	1.55	244	75	40	359	3.3	6.4	4,480	2,630	140	(2)
22-Mar-96	30	1.50	290	130	66	486	2.3	4.3	6,270	3,370	200	(2)
26-Apr-96	36	1.50	315	110	80	505	2.9	3.9	5,290	3,910	200	(2)
10-May-96	42	1.45	450	170	84	704	2.7	5.4	5,180	4,020	330	(2)
14-June-96	45	1.35	690	330	170	1,190	2.1	4.0	7,000	4,390	330	(2)
24-July-96	36	1.75	520	290	110	920	1.8	4.6	7,520	4,460	270	(2)
08-Aug-96	35	1.45	240	180	90	510	1.3	2.8	6,970	4,510	360	(2)
27-Sep-96	40	1.65	580	270	120	970	2.2	4.7	6,780	4,320	220	(2)

(continued)

Table 1 (continued)

Date	Temp. (°C)	pH_{lab} 24 ± 2 °C	$S_4O_6^{2-}$	$S_5O_6^{2-}$	$S_6O_6^{2-}$	Total $S_xO_6^{2-}$	$S_4O_6^{2-}/S_5O_6^{2-}$	$S_4O_6^{2-}/S_6O_6^{2-}$	S_4O^{2-}	Cl^-	F^-	PT lab
28-Nov-96	31	1.55	580	310	96	986	1.9	6.0	6,790	5,110	300	(2)
18-Dec-96	29	1.13	310	170	80	560	1.8	3.8	4,870	3,650	260	(2)
07-Jan-97	32	1.40	490	240	100	830	2.0	4.8	6,880	5,750	290	(2)
04-Feb-97	31	1.40	360	170	100	630	2.1	3.5	5,110	3,850	310	(2)
03-Mar-97	29	1.18	230	130	90	450	1.8	2.6	4,410	3,260	170	(2)
17-Apr-97	27	1.40	270	150	80	500	1.7	3.3	4,640	4,230	230	(2)
14-May-97	29	1.45	510	200	70	780	2.5	7.2	4,640	4,420	310	(2)
04-June-97	32	1.50	610	250	60	920	2.5	10	4,750	4,280	230	(2)
02-July-97	32	1.50	500	200	75	775	2.5	6.7	4,870	4,620	210	(2)
22-Aug-97	31	1.45	700	270	60	1,030	2.6	12	5,770	6,690	200	(2)
05-Sep-97	35	1.30	590	260	50	900	2.3	12	4,060	4,330	170	(2)
17-Oct-97	34	1.06	790	280	56	1,126	2.9	14	7,840	6,520	360	(2)
04-Nov-97	35	1.40	810	280	64	1,154	2.9	13	6,460	7,800	280	(2)
06-Feb-98	38	0.90	840	420	115	1,375	2.0	7.3	7,910	7,920	420	(2)
25-Mar-98	37	0.79	990	430	110	1,530	2.3	9.1	9,120	9,440	490	(2)
17-Apr-98	37	0.68	b.d.l.	b.d.l.	b.d.l.	n.a.	n.a.	n.a.	10,900	8,870	460	(2)
15-May-98	36	0.64	810	370	120	1,300	2.2	6.8	11,500	12,400	620	(2)
25-June-98	36	0.99	260	170	35	465	1.6	7.4	7,500	6,490	500	(2)
30-July-98	35	0.73	890	490	160	1,540	1.8	5.7	10,800	11,800	600	(2)
26-Aug-98	35	0.73	890	420	120	1,430	2.1	7.3	7,970	8,410	390	(2)
21-Sep-98	35	0.74	850	400	110	1,360	2.1	7.6	9,850	10,800	550	(2)
04-Nov-98	35	0.85	350	190	110	650	1.9	3.3	8,140	8,740	480	(2)
23-Dec-98	27	0.90	330	170	80	580	1.9	4.0	7,650	7,840	600	(2)
13-Jan-99	32	0.77	430	210	130	770	2.1	3.4	9,110	9,890	640	(2)
25-Feb-99	33	0.68	470	220	110	800	2.1	4.2	10,600	11,700	660	(2)
17-Mar-99	32	0.81	460	180	120	760	2.5	3.9	9,760	10,900	720	(2)

(continued)

Table 1 (continued)

Date	Temp. (°C)	pH_{lab} 24 ± 2 °C	$S_4O_6^{2-}$	$S_5O_6^{2-}$	$S_6O_6^{2-}$	Total $S_xO_6^{2-}$	$S_4O_6^{2-}/S_5O_6^{2-}$	$S_4O_6^{2-}/S_6O_6^{2-}$	S_4O^{2-}	Cl^-	F^-	PT lab
21-Apr-99	32	0.72	380	200	150	730	1.9	2.6	10,400	11,300	800	(1)
07-May-99	34	0.71	360	200	150	710	1.8	2.4	10,700	11,600	810	(1)
18-June-99	34	0.68	330	130	90	550	2.6	3.7	8,550	7,510	480	(1)
07-July-99	32	0.68	300	170	140	610	1.7	2.1	11,100	12,200	720	(1)
19-Aug-99	35	0.81	290	150	110	550	2.0	2.6	10,200	11,200	720	(1)
08-Sep-99	39	0.62	290	130	90	510	2.2	3.2	8,220	8,120	660	(1)
22-Oct-99	38	0.61	290	190	140	620	1.5	2.1	12,100	13,300	920	(1)
05-Nov-99	39	0.80	380	140	100	620	2.7	3.7	9,250	9,130	730	(1)
27-Dec-99	34	0.72	300	200	140	640	1.5	2.1	10,400	11,400	790	(1)
21-Jan-00	38	0.76	240	100	62	402	2.3	3.9	9,270	9,850	740	(1)
10-Feb-00	44	0.69	210	100	60	370	2.0	3.5	10,200	10,800	830	(1)
13-Mar-00	36	0.71	120	65	50	235	1.9	2.5	9,270	9,620	700	(1)
11-Apr-00	32	0.74	230	110	66	406	2.0	3.4	9,780	10,400	830	(1)
05-May-00	31	0.67	200	100	53	353	2.0	3.8	8,330	8,280	810	(1)
09-June-00	31	0.77	210	97	87	394	2.1	2.4	9,260	11,900	710	(1)
17-July-00	33	0.74	110	52	30	192	2.1	3.7	7,230	9,050	510	(1)
24-Aug-00	31	0.60	150	90	57	297	1.6	2.6	9,240	13,300	700	(1)
12-Sep-00	35	0.60	190	87	62	339	2.2	3.1	9,410	12,700	620	(1)
10-Oct-00	35	0.70	180	84	52	316	2.2	3.5	8,900	11,700	580	(1)
21-Nov-00	34	0.71	160	76	40	276	2.0	4.0	8,930	11,900	670	(1)
01-Dec-00	34	0.90	160	72	50	282	2.2	3.3	8,320	10,900	610	(1)
12-Jan-01	31	0.86	120	80	78	278	1.6	1.6	8,700	11,400	680	(1)
23-Feb-01	31	0.83	120	80	97	297	1.6	1.3	7,870	9,670	540	(1)
08-Mar-01	32	0.65	130	79	38	247	1.7	3.5	9,080	11,600	710	(1)
26-Apr-01	35	0.75	120	66	25	211	1.8	4.7	9,400	11,700	660	(1)
08-May-01	34	0.70	160	80	80	320	2.0	2.0	9,080	12,200	690	(1)

(continued)

Table 1 (continued)

Date	Temp. (°C)	pH_{lab} 24 ± 2 °C	$S_4O_6^{2-}$	$S_5O_6^{2-}$	$S_6O_6^{2-}$	Total $S_xO_6^{2-}$	$S_4O_6^{2-}/S_5O_6^{2-}$	$S_4O_6^{2-}/S_6O_6^{2-}$	S_4O^{2-}	Cl^-	F^-	PT lab
13-June-01	35	0.67	150	74	84	308	2.0	1.8	9,100	11,700	560	(1)
19-July-01	28	0.70	190	70	34	294	2.7	5.5	7,270	8,400	460	(1)
03-Aug-01	24	0.85	160	74	72	306	2.2	2.2	8,090	9,760	510	(1)
07-Sep-01	27	0.39	360	110	80	550	3.3	4.6	8,510	9,590	540	(1)
18-Oct-01	28	0.85	360	160	105	625	2.2	3.4	7,800	8,280	380	(1)
27-Nov-01	30	1.07	b.d.l.	b.d.l.	b.d.l.	n.a.	n.a.	n.a.	3,450	2,720	82	(1)
07-Dec-01	30	1.43	b.d.l.	b.d.l.	b.d.l.	n.a.	n.a.	n.a.	3,310	2,830	67	(1)
31-Jan-02	30	1.51	6	10	9	25	n.a.	n.a.	3,590	2,540	52	(1)
27-Feb-02	23	1.17	9	14	12	35	0.6	0.7	6,050	4,670	160	(1)
13-Mar-02	26	1.06	33	50	7	90	0.6	4.7	6,390	5,760	190	(1)
03-Apr-02	25	1.04	53	80	14	147	0.7	3.8	6,830	6,420	380	(1)
16-May-02	29	1.08	b.d.l.	b.d.l.	b.d.l.	n.a.	n.a.	n.a.	6,060	6,260	230	(1)
11-June-02	29	0.90	67	50	37	154	1.3	1.8	5,060	5,780	130	(1)
30-July-02	32	0.99	140	52	30	222	2.7	4.8	6,800	7,620	320	(1)
04-Sep-02	38	0.90	580	220	70	870	2.7	8.4	7,660	8,720	300	(1)
11-Dec-02	39	0.84	620	280	63	963	2.2	9.9	8,510	10,700	430	(2)
17-Jan-03	34	0.83	460	300	138	898	1.5	3.3	8,780	10,500	480	(2)
28-Feb-03	36	0.77	600	380	143	1,123	1.6	4.2	8,870	10,900	520	(2)
31-Mar-03	39	0.70	940	460	143	1,543	2.1	6.6	9,360	15,000	580	(2)
30-Apr-03	41	0.67	910	560	170	1,640	1.6	5.3	10,500	16,100	620	(2)
20-May-03	41	0.70	810	470	180	1,460	1.7	4.5	11,000	16,200	620	(2)
17-July-03	35	0.64	400	380	184	964	1.1	2.2	11,800	17,100	650	(2)
12-Aug-03	33	0.61	720	450	148	1,318	1.6	4.9	8,860	18,200	720	(2)
30-Sep-03	33	0.62	280	240	112	632	1.2	2.5	13,800	15,300	560	(2)
28-Oct-03	28	0.80	6	4	b.d.l.	10	1.5	n.a.	9,290	12,700	320	(2)
05-Dec-03	24	0.93	b.d.l.	b.d.l.	b.d.l.	n.a.	n.a.	n.a.	6,100	8,310	240	(2)

(continued)

Table 1 (continued)

Date	Temp. (°C)	pH_{lab} 24 ± 2 °C	$S_4O_6^{2-}$	$S_5O_6^{2-}$	$S_6O_6^{2-}$	Total $S_xO_6^{2-}$	$S_4O_6^{2-}/S_5O_6^{2-}$	$S_4O_6^{2-}/S_6O_6^{2-}$	S_4O^{2-}	Cl^-	F^-	PT lab
20-Jan-04	24	1.17	b.d.l.	b.d.l.	b.d.l.	n.a.	n.a.	n.a.	5,260	6,670	180	(2)
03-Feb-04	29	1.11	b.d.l.	b.d.l.	b.d.l.	n.a.	n.a.	n.a.	4,550	5,880	190	(2)
23-Mar-04	25	1.05	b.d.l.	b.d.l.	b.d.l.	n.a.	n.a.	n.a.	5,290	6,530	180	(2)
27-Apr-04	28	1.12	b.d.l.	b.d.l.	b.d.l.	n.a.	n.a.	n.a.	5,300	6,330	160	(2)
25-May-04	28	1.31	b.d.l.	b.d.l.	b.d.l.	n.a.	n.a.	n.a.	3,940	5,380	26	(2)
11-June-04	29	1.20	b.d.l.	b.d.l.	b.d.l.	n.a.	n.a.	n.a.	4,540	6,480	65	(2)
01-July-04	26	1.24	b.d.l.	b.d.l.	b.d.l.	n.a.	n.a.	n.a.	5,680	5,320	n.d.	(2)
12-Aug-04	25	1.10	b.d.l.	b.d.l.	b.d.l.	n.a.	n.a.	n.a.	7,690	6,320	170	(2)
22-Sep-04	29	1.16	b.d.l.	b.d.l.	b.d.l.	n.a.	n.a.	n.a.	7,170	5,860	140	(2)
22-Oct-04	27	1.08	b.d.l.	b.d.l.	b.d.l.	n.a.	n.a.	n.a.	7,940	6,920	160	(2)
2-Dec-04	23	1.17	b.d.l.	b.d.l.	b.d.l.	n.a.	n.a.	n.a.	7,750	5,930	130	(2)
22-Feb-05	22	1.25	b.d.l.	b.d.l.	b.d.l.	n.a.	n.a.	n.a.	3,780	3,680	92	(2)
Stage V												
21-Mar-05	32	1.22	b.d.l.	b.d.l.	b.d.l.	n.a.	n.a.	n.a.	4,670	5,330	130	(2)
12-Apr-05	34	0.97	190	110	10	310	1.6	21	7,340	6,570	220	(2)
4-May-05	41	1.02	470	120	30	620	3.9	15	8,120	6,900	260	(2)
24-June-05	50	0.92	1,160	360	70	1,590	3.3	16	9,310	8,380	470	(2)
14-July-05	50	0.79	540	250	110	900	2.2	4.8	9,900	8,800	530	(2)
24-Aug-05	51	0.75	860	470	172	1,502	1.9	5.0	11,400	10,200	690	(2)
20-Sep-05	52	0.70	1,100	820	270	2,190	1.3	4.1	12,800	11,100	750	(2)
30-Nov-05	54	0.64	1,850	1,390	410	3,650	1.3	4.5	13,600	11,600	920	(2)
31-Jan-06	51	0.65	1,620	1,310	500	3,430	1.2	3.2	15,700	12,000	870	(2)
28-Feb-06	51	0.59	760	670	280	1,710	1.1	2.8	17,700	13,400	990	(2)
1-Apr-06	54	0.63	710	500	275	1,485	1.4	2.6	17,200	15,700	1,050	(2)
25-May-06	47	0.72	90	100	48	238	0.9	1.8	19,500	15,900	1,130	(2)
22-June-06	46	0.59	110	100	50	260	1.1	2.3	23,100	16,300	1,090	(2)

(continued)

Table 1 (continued)

Date	Temp. (°C)	pH_{lab} 24 ± 2 °C	$S_4O_6^{2-}$	$S_5O_6^{2-}$	$S_6O_6^{2-}$	Total $S_xO_6^{2-}$	$S_4O_6^{2-}/S_5O_6^{2-}$	$S_4O_6^{2-}/S_6O_6^{2-}$	S_4O^{2-}	Cl^-	F^-	PT lab
20-July-06	43	0.62	70	70	30	170	0.9	2.2	21,300	13,400	890	(2)
11-Aug-06	43	0.52	120	95	58	273	1.2	2.0	25,400	16,100	1,100	(2)
05-Sep-06	41	0.47	280	240	100	620	1.2	2.8	27,500	17,100	1,100	(2)
28-Oct-06	53	0.50	100	50	22	172	1.9	4.4	29,000	16,300	1,170	(2)
10-Nov-06	57	0.43	85	40	25	150	2.1	3.4	29,100	13,200	950	(2)

PT lab refers to laboratory where PTs were separated and analysed by chromatography: (1) Laboratory of Atmospheric Chemistry School of Chemistry Universidad Nacional, LAQAT-UNA Heredia; (2) Tokyo University

[a]Data from Rowe et al. (1992b); b.d.l.: below detection limit; n.d.: not determined; n.a.: not applicable

3 Results and Discussion

3.1 Polythionate Results

Analytical results of tetra-, penta-, and hexathionate are given in Table 1, together with the concentrations of sulfate, chloride and fluoride and other relevant parameters. For completeness, published data by Rowe et al. (1992b) are also included. As a consequence, Table 1 lists a PT record for the period October 1980–November 2006.

Many chromatograms showed two or three peaks that were eluted after the hexathionate ion, corresponding to higher homologues of PTs such as heptathionate ($S_7O_6^{2-}$), octathionate ($S_8O_6^{2-}$), and nonathionate ($S_9O_6^{2-}$) (Fig. 2) (cf. Steudel and Holdt 1986). Qualitative inspection of all chromatograms suggested that these long-chain PTs were present in the lake water during periods when the shorter homologues were abundant. They were undetectable during periods when the latter were absent or present in low amounts. Higher PTs, including nona- and decathionate, were also detected in the acid crater lakes of Ruapehu (Takano et al. 1994b) and Kusatsu-Shirane (Takano 1987). Because long-chain PTs are more susceptible to thermal and sulfitolytic breakdown than the shorter PTs (Fujiwara et al. 1988; Takano et al. 2001, 2004), and because they are much less abundant and more difficult to be quantified, only the concentrations of the tetra-, penta-, and hexathionate are reported here. Thus, throughout this paper "total PTs" refers to the sum of concentrations of these three species.

Martínez (2008) distinguished five main stages of volcanic activity at Poás since the late 1970s, based on time-series trends for a set of geochemical and geophysical monitoring parameters (see also Rymer et al. 2009). Figures 3, 4 and 5 portray time series of PT concentrations, together

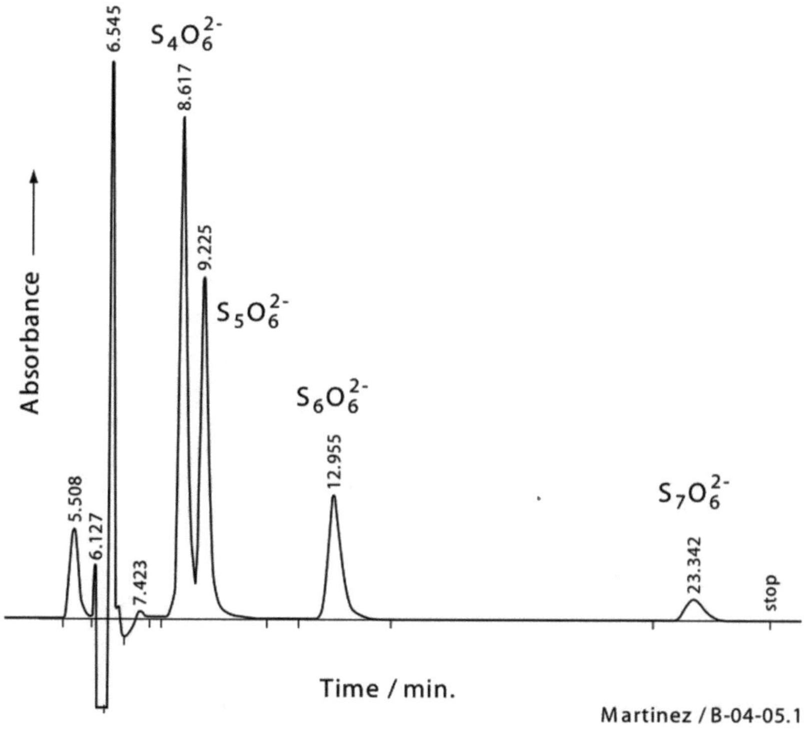

Fig. 2 Example of a chromatogram for a lake-water sample (collected on 21 August 1999) obtained at LAQAT, where separation and quantification of the thionates were performed following Miura and Kawaoi (2000). The sample was diluted 25 times by weighing. Peaks of tetrathionate, pentathionate, hexathionate and heptathionate are labeled. The peaks that eluted before tetrathionate are due to unknown impurities in the sample

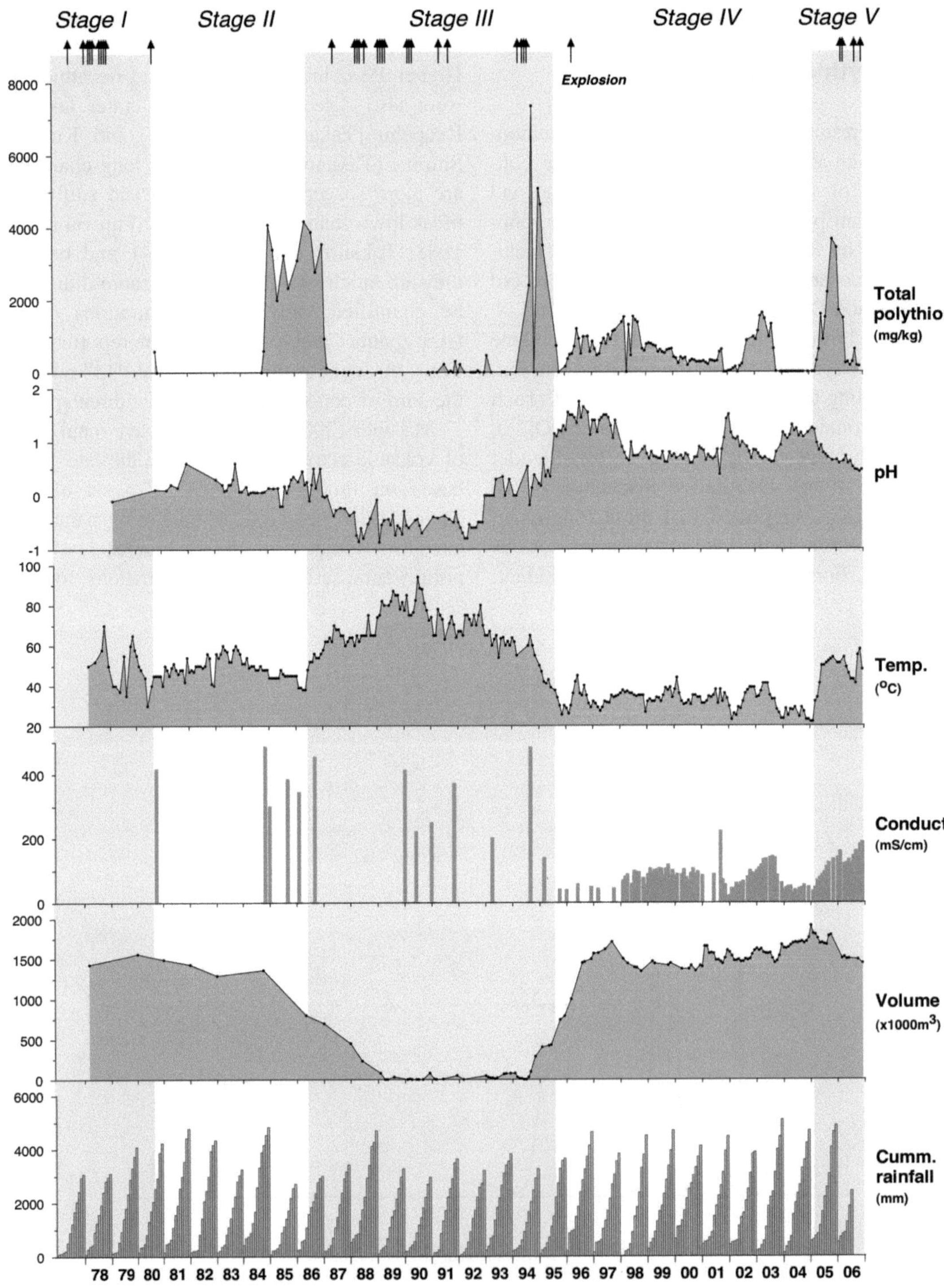

Fig. 3 Time series of total PT concentrations (sum of tetra-, hexa- and pentathionate), pH, temperature, conductivity and volume of the crater lake between 1978 and 2006. Rainfall registered at the summit of Poás (ICE's pluviometer) is expressed as cumulative monthly totals in cycles of calendar years. Arrows on top of the graph indicate phreatic eruptions and a minor hydrothermal explosion at the northern side of the dome that occurred on April 8, 1996, presumably as a result of unclogging of fumarolic vents

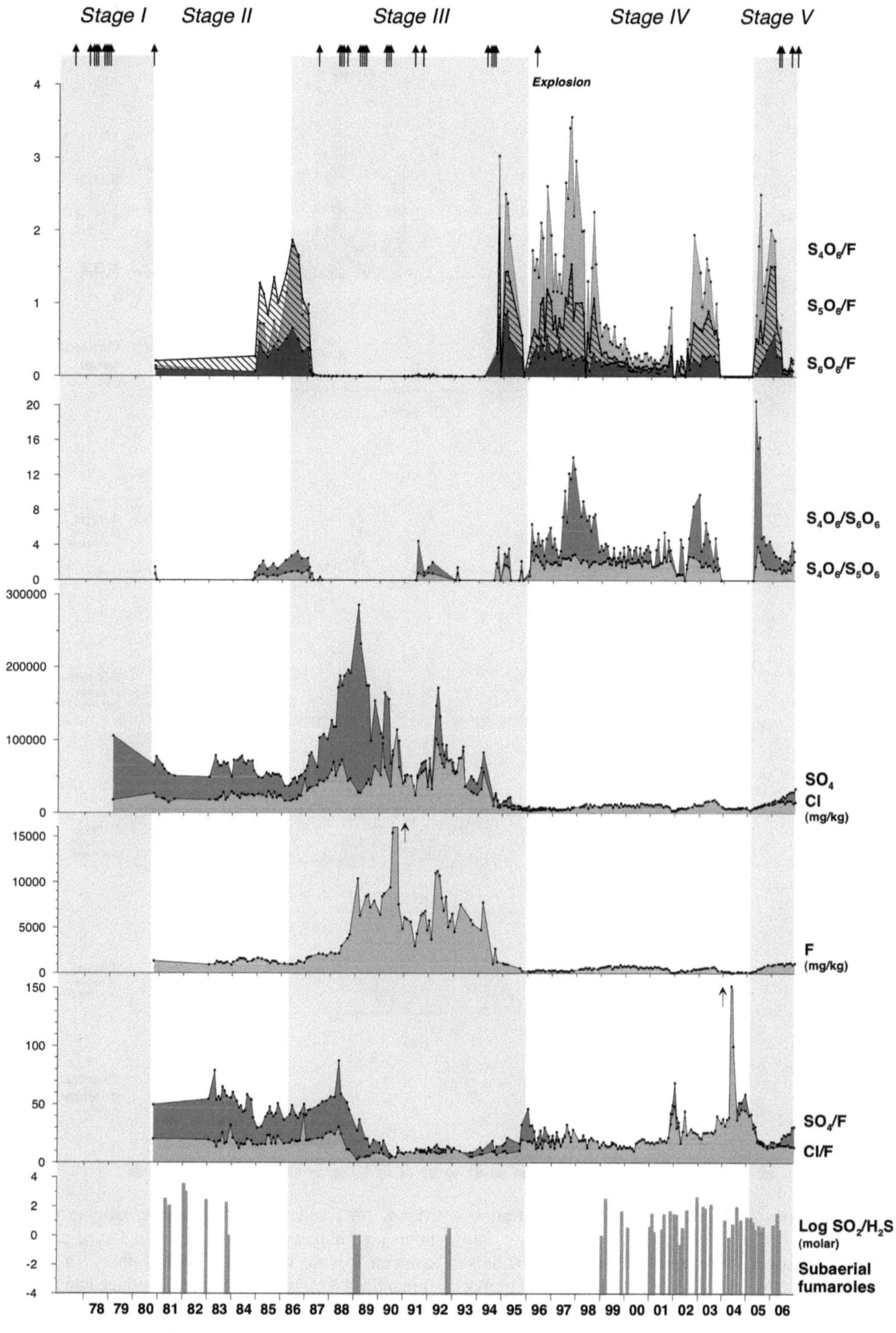

Fig. 4 Time series of PT concentrations normalized to fluoride, S_4O_6/S_5O_6 and S_4O_6/S_6O_6 ratios, concentrations (mg/kg) and ratios of major anions, as well as molar SO_2/H_2S ratios of sub-aerial fumaroles. Except for the latter, all ratios are wt/wt. Arrows on top of the graph indicate phreatic eruptions. Arrows in the F and Cl/F panels indicate values that are out of scale

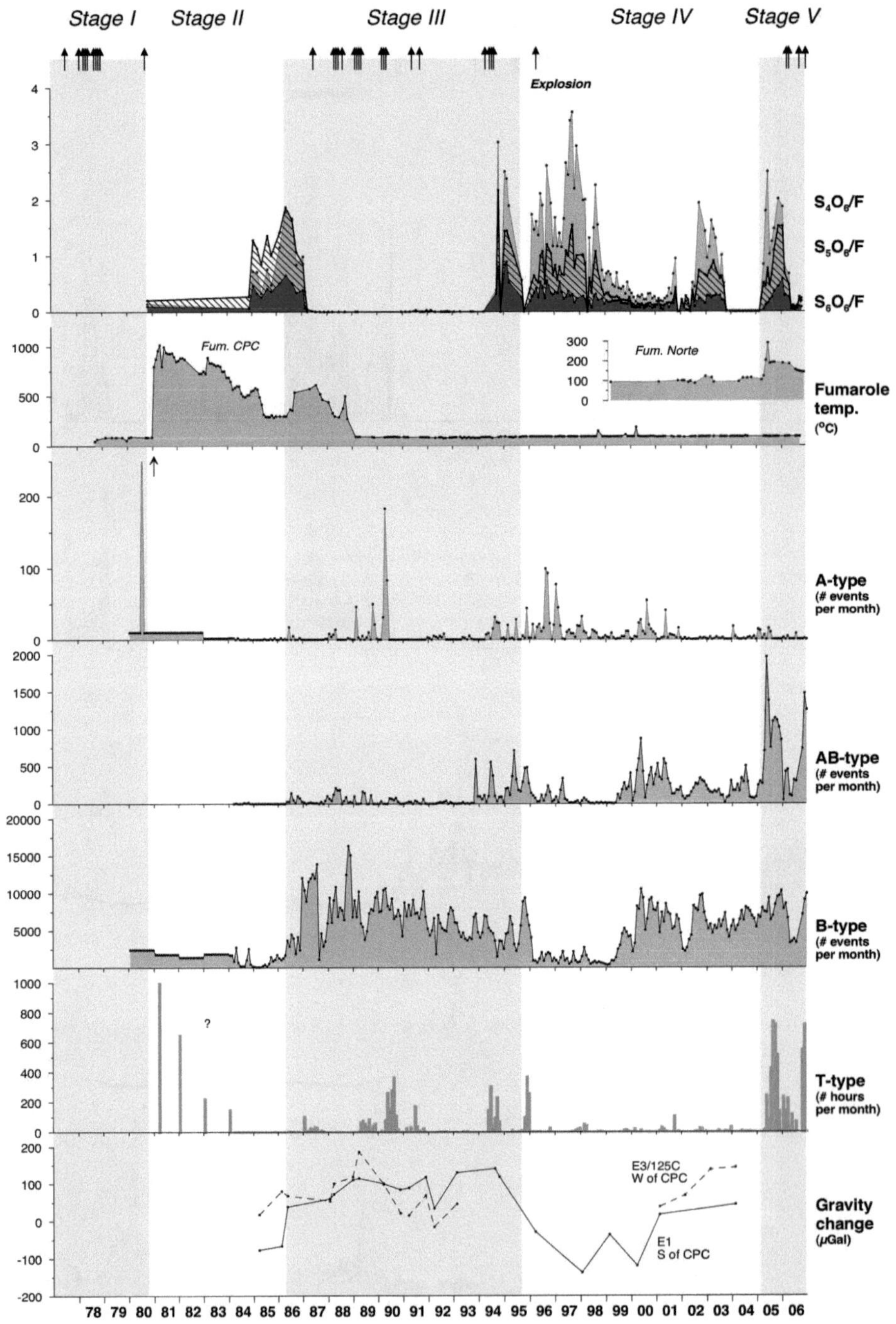

Fig. 5 Time series of F-normalized PT concentrations compared with fumarole temperatures, seismicity and microgravity changes monitored at Poás volcano. Seismicity is distinguished according to A, AB, B and tremor (T)-type characteristics, and is expressed as numbers of monthly events, except for T-type seismicity, which is given in hours per month. Arrow in the A-type panel refers to July 1980 when recorded quakes numbered in the thousands within a two-week period (Casertano et al. 1985). Peaks in volcanic tremor were registered between March 1981 and 1983 but the exact number of monthly hours are unknown (question mark in T-type panel). They coincided with very high temperatures of the dome fumaroles (700–1020 °C), which produced an SO_2 flux of about 800 tons per day (Casadevall et al. 1984b; Fernández 1990). Microgravity data (Rymer et al. 2005) are shown for two stations near the dome and reflect changes relative to a base station 7 km from the crater, normalized to the difference observed in 1987. Arrows on top of the graph indicate months with phreatic eruptions

with key chemical and physical properties of the lake, fumarole temperatures and seismicity observed during these stages. To eliminate potential dilutive or evaporative effects, the PT concentrations are normalized against fluoride rather than chloride or sulfate, because the former behaves conservatively, whereas HCl exerts significant vapor pressure and is easily distilled from acidic lake brines under conditions of high acidity and relatively low temperature, such it was observed in mid 1988-early 1989 (Rowe et al. 1992b) and 1998-late 2003 (Martínez 2008). On the other hand, sulfate tends to precipitate as a solid assemblages (e.g. gypsum/anhydrite, hydrous aluminium sulphates) in saturated to oversaturated lake brines.

The PTs formed a significant part of the anion budget during the transitional period between Stages II and III, and from the end of Stage III on through most of Stages IV and V (Figs. 3, 4, and 5), when concentrations reached several hundreds or thousands of milligrams per kilogram. Commonly, the analyzed species showed $S_4O_6^{2-} > S_5O_6^{2-} > S_6O_6^{2-}$ as concentration sequence (Table 1; Figs. 4, 5 and 6). However, during Stage II and the first fourteen months of Stage III the sequence was $S_5O_6^{2-} > S_4O_6^{2-} > S_6O_6^{2-}$ (Rowe et al. 1992b). This distribution was also observed during brief intervals of relative quiescence in the lake: mid 1991-early 1992, mid 1995–January 1996, the first four months of 2002 and May–July 2006, when overall PT concentrations tended to be rather low (Table 1).

3.2 Polythionates: Relationships with Lake Properties, Fumarolic Activity and Eruptive Phreatic Events

In the following sections, variations in terms of PT distributions are described in relation to the composition and behavior of the lake, on-shore fumarolic activity, eruptive phreatic events and monitored seismicity for the lake activity stages as recognized by Martínez (2008). Since no polythionate data are available for Stage I, the overview will commence with Stage II.

3.2.1 Stage II (September 1980–April 1986): High PT Concentrations, Moderate Convective Activity in the Lake Without Phreatic Eruptions, High-Temperature Degassing Through the Dome

During the last part of Stage II the PT concentrations were distributed in the sequence $S_5O_6^{2-} > S_4O_6^{2-} > S_6O_6^{2-}$ (Rowe et al. 1992b). Single data points for October 1980 and November 1984 suggested that PTs were initially present in relatively low amounts (about 600 mg/kg). From early 1985 on, an irregularly increasing trend culminated in a maximum of total polythionates of about 4,170 mg/kg in May 1986 (Table 1; Fig. 3) was observed. This strong increase coincided with swarms of A-type seismicity (volcano-tectonic quakes) (peaking in June–August), and also marked the start of a period of elevated B-type seismicity (long-period quakes) in May (Fernández 1990; Rowe et al. 1992a), which continued throughout Stage III (Fig. 5). The observed lowering of the lake volume after 1985 was probably partly due to relatively low rainfall in 1985, 1986, and 1987 (Fig. 3), but increased heat input must have significantly contributed to a rapid decline in early-mid 1986 (Rowe et al. 1992a). A marked increase in lake temperature was not observed until June 1986 when 48 °C were measured, i.e. 10 °C higher than in the previous month (Fig. 3).

Trends of individual PT species were not entirely parallel. A conspicuous change in relative abundances was found in February and May 1986 when, compared to the other species, tetrathionate apparently had formed at much higher rates than what previously recorded. Pentathionate remained predominant throughout, but tetrathionate was present in almost equal amount at the peak levels of May 1986 (Fig. 4). As discussed in Rowe et al. (1992b), fluctuations in PT speciation and their stability in the lake reflected changes in the SO_2/H_2S ratio of subaqueous fumaroles. The observed distribution of

$S_5O_6^{2-} > S_4O_6^{2-} > S_6O_6^{2-}$ points to the predominance of H_2S-enriched subaqueous fumaroles with a molar SO_2/H_2S ratio of <0.07, which would favor the formation of pentathionate over tetrathionate (Fig. 7) (Takano et al. 1994b).

Despite the termination of phreatic eruptions by the end of 1980, several thousands of A-type seismic events were recorded between April and August 1980, which preceded an interval of strongly increased well-defined volcanic tremors between September 1980 and 1984 (Casertano et al. 1985; Fernández 1990). This enhanced seismicity coincided with a very high flux of high-temperatures gases (800–1,020 °C, Fig. 5) from the dome in January 1981–September 1983, producing a strong sulfurous odour (Malavassi and Barquero 1982; Barquero and Malavassi 1983). COSPEC measurements recorded SO_2 fluxes of about 800 tons/day in February 1982 (Barquero and Fernández 1983; Casadevall et al. 1984b). This vigorous release of hot SO_2-rich gas from the dome was possibly associated with the emplacement of a small magma body below it (Brown et al. 1991). If so, both the intrusion and the channeling of expelled volatiles must have been confined to the small area below the dome, because properties of the lake were not dramatically affected, as illustrated by the distribution of PT species pointing to an input of H_2S rather than SO_2-rich gas into the lake.

Casertano et al. (1987) suggested that, during maximum heating of the dome in 1981–1983, the lake was still in continuous hydraulic connection with the rising hot fluids, consistent with the steady increase in lake temperature in these years (see Fig. 3). The lake had temperatures around 50 °C, a grayish-turquoise color, and showed strong evaporation and streaks of floating sulfur globules. By the end of October 1983 the water level descended about 10 m (Fig. 3). The detected increase in the $S_4O_6^{2-}/S_5O_6^{2-}$ and $S_4O_6^{2-}/S_6O_6^{2-}$ ratios in 1985-mid 1986 can thus be explained by an increase in the SO_2/H_2S ratio of subaqueous fumaroles after October 1985 (Figs. 4 and 5). This change probably occurred in early 1986 when swarms of A-type earthquakes and elevated B-type seismicity were recorded, possibly induced by hydrofracturing of the magmatic carapace and subsequent release of magmatic volatiles (Rowe et al. 1992a, b). Episodes of harmonic tremor of short duration (<1 min) and low amplitude (<1 mm), which are rare at Poás and not yet well understood, were also recorded in early 1986 (Fernández 1990). Shallow intrusion of fresh magma batches as source of the enhanced SO_2 output was also envisaged from micro-gravity increases in the southern part of the active crater observed between 1985 and 1989 (Fig. 5) (Rymer et al. 2000, 2004, 2005).

3.2.2 Stage III (May 1986–August 1995): PTs in Trace Amounts or Absent, Frequent Phreatic Eruptions, Intense Fumarolic Degassing

After the peak levels of May 1986 PT concentrations showed a strong steady decrease concomitant with an increase in lake temperature, a decrease in pH and a reduction in lake volume (Fig. 3). By mid 1987 the lake temperature rose above 60 °C, the pH dropped to negative values, and the PT concentrations were below the detection limits. These effects roughly coincided with the first phreatic eruptions from the lake in June 1987 (Venzke et al. 2002; Rowe et al. 1992a). Since then, PTs remained undetected for most of Stage III; only occasionally minor amounts were measured (Fig. 3). By the end of Stage III, after cessation of phreatic eruptions in early August 1994, a new lake started to rise, the temperature dropped, the pH increased, and a significant amount of PTs were formed, as observed for the first time between October 1994 and February 1995. The initial recovery of the lake was characterized by high PT concentrations ($\Sigma S_xO_6^{2-}$ up to 7,350 mg/kg in October 1994), similar to those observed in May 1986. Irregular activity variations in the lake occurred in view of their temporary disappearance in November 1994 (Figs. 3 and 4).

If the occasional presence of PTs in the course of Stage III is taken into account, a generalized pattern for the PT distribution can be discerned. Available data between 1989 and early 1990

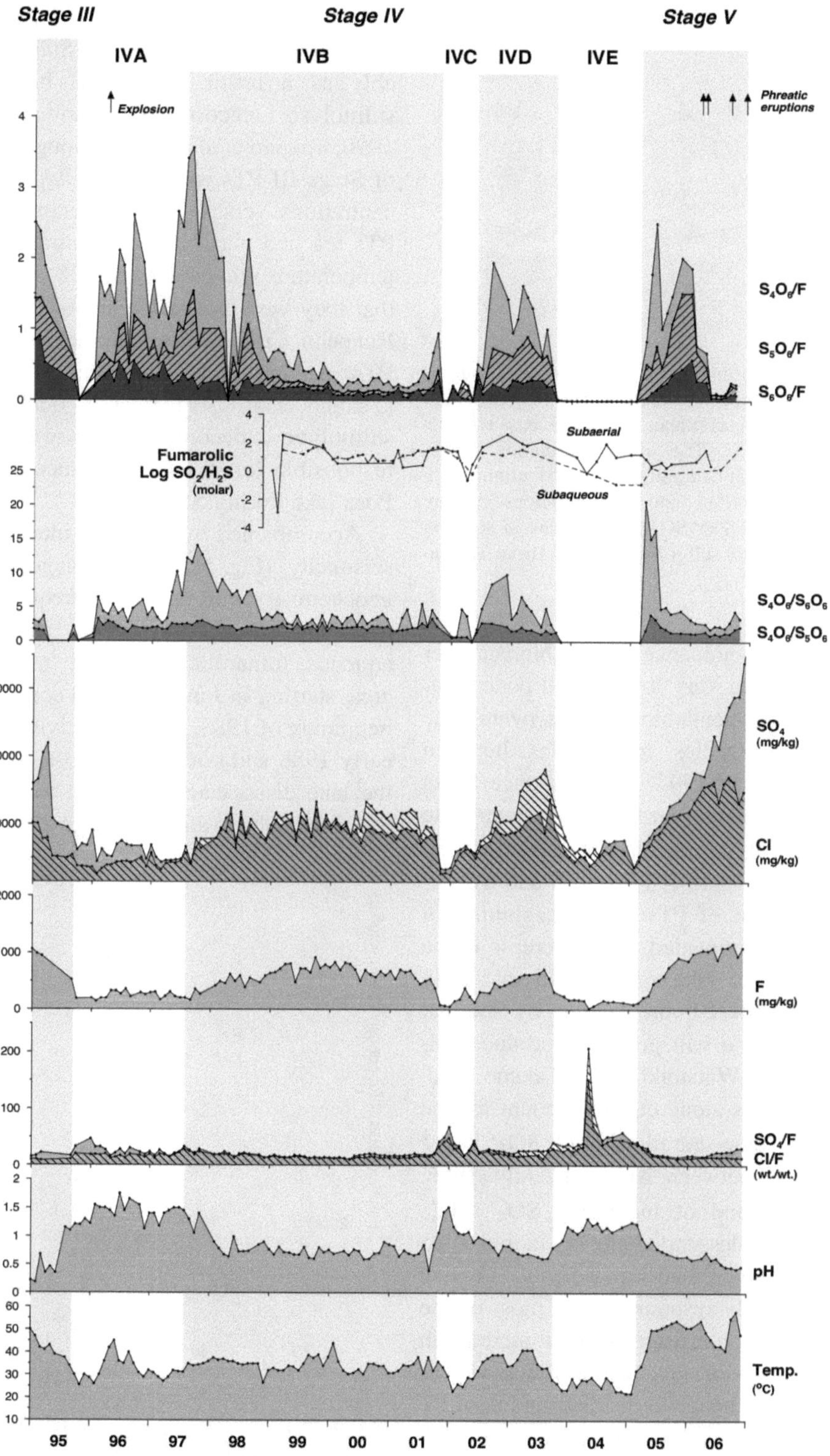

Fig. 6 Detailed time series of PT concentrations (normalized to fluoride) and ratios of other major anions (wt/wt) for part of Stage III, the Stage IV and the first years of Stage V, together with other crater-lake parameters and molar SO_2/H_2S ratios of sub-aerial (measured) and subaqueous (inferred, see text) fumaroles. Note the (near) absence of PTs at the beginning of sub-stage IVA and in sub-stages IVC and IVE, periods of relative quiescence in the lake, whereas the low concentrations of PTs after 2006 is rather related to enhanced fumarolic outgassing through the lake. Concentrations of anions are given in mg/kg

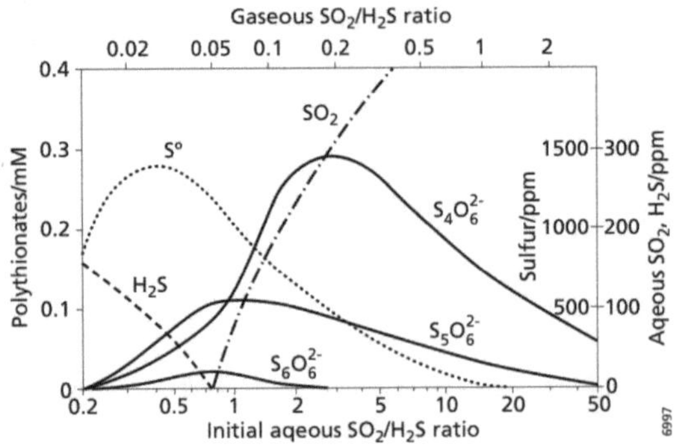

Fig. 7 Predicted distributions of PTs, elemental sulfur, SO_2 and H_2S in an aqueous solution. The SO_2/H_2S ratio of entering fumarolic gas (upper horizontal axis) controls the initial aqueous SO_2/H_2S ratio in solution (lower horizontal axis), which determines the distribution of PT species, elemental sulfur, and concentrations of any excess (unreacted) H_2S or SO_2 that remains in solution after PTs and elemental sulfur have formed (from Takano et al. 1994b)

showed a strong dominance of tetrathionate over the other species at very low overall concentrations. In contrast, pentathionate was overall the most abundant species in samples between mid-1991 and early 1993, whereas once again the distribution changed in favor of tetrathionate at the high $\Sigma S_xO_6^{2-}$ concentrations towards the end of Stage III (Table 1; Figs. 4, 5 and 6).

The breakdown of PTs at the beginning of Stage III was accompanied by an increase in the SO_4^{2-}/F^- ratios of lake samples till mid-1988 (Fig. 4). Although PT breakdown by thermal effects or otherwise will generally produce sulfate (Takano and Watanuki 1990; Takano et al. 2001), this process alone cannot account for the rise of SO_4^{2-} concentrations and SO_4^{2-}/F^{-1} ratios observed between May and July 1986. Moreover, the trend of increasing SO_4^{2-}/F^{-1} ratios in lake samples started already in mid-1985 (Fig. 4). Hence, increased input of hot SO_2-rich volatiles was likely responsible for most of the sulfate production, consistent with an increase in the SO_2/H_2S ratio of subaqueous fumaroles, as inferred from the change in the distribution of PT species since August 1986 (Fig. 4), albeit with temporary reversals in some of the late 1986 and early 1987 samples. The enhanced SO_2 input in August 1986 produced maximum $S_4O_6^{2-}/S_6O_6^{2-}$ ratios at low $\Sigma S_xO_6^{2-}$ concentrations in the first half of Stage III, presumably as a result of thermal breakdown and sulfitolysis reactions (Takano and Watanuki 1990; Takano et al. 2001), although for most part of Stage III PTs were absent. A plot of PT concentrations versus lake temperature shows that PTs reached maximum concentrations when the temperature ranged between 38 and 65 °C, and that they were not stable at >65 °C (Fig. 8). At Ruapehu Crater Lake, complete degradation of PTs occurred at temperatures above 47 °C (Takano et al. 1994b). Thus, both thermal and sulfitolytic decomposition were probably responsible for the disappearance of PTs in the Poás lake during Stage III.

Accompanied by a notable increase in B-type seismicity (Fig. 5), these changes in the lake geochemistry culminated in frequent, moderate to strong phreatic explosions, and strong subaqueous fumarolic degassing within the lake area, starting in June 1987 and continuing till the beginning of 1988. Phreatic activity intensified in early 1988 and continued until early 1989 when the lake disappeared, enabling for the first time the direct observation of large pools of molten

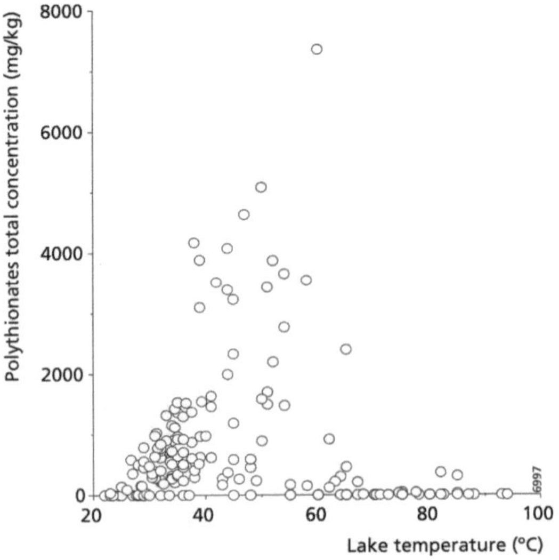

Fig. 8 Total PT concentrations versus lake temperature for the period October 1980–November 2006. Maximum concentrations were observed at temperatures between 38 and 65 °C, whereas higher temperatures result in the breakdown of PTs

sulfur and "sulfur volcanoes" on the dried lake bottom (Oppenheimer and Stevenson 1989). Subsequently, a series of strong phreatic explosions that ejected mud, sulfur, and lake sediments (some up to heights of ~2 km) occurred throughout 1989 and early 1991 (Venzke et al. 2002; Rowe et al. 1989), coinciding with periods of significant T-type seismic activity and A-type seismic swarms (A-type seismic signals were recorded together with T-type events in 1989 and 1990 but not in 1991) (Fig. 5). The strong phreatic activity and intense release of gases and steam reduced the lake to scattered mud pools during the dry seasons (April 1989, April 1990 and March 1991), and prevented it from reaching normal levels during rainy seasons. In addition, the yearly rainfall between 1989 and early 1991 (2,950–3,600 mm) was somewhat less than the average of ~4 m per year (Fig. 3).

The subsequent shift towards a PT distribution of $S_5O_6^{2-} > S_4O_6^{2-} > S_6O_6^{2-}$ (when they were detectable) and the inferred concomitant decrease in the SO_2/H_2S ratio of subaqueous fumaroles marked a period of noticeable lower activity in the lake area. Conditions of only moderate fumarolic degassing and absence of phreatic eruptions (Venzke et al. 2002; Barquero 1998) prevailed from mid-1991 throughout 1992 and 1993. The lake level rose to a maximum of several meters, especially during the rainy seasons of 1991 and 1993 when rainfall was relatively heavy (Fig. 3). Although the lake remained shallow, recorded water levels were higher than in the previous years (depths were ~5 m and 4 m in December 1991 and December 1993, respectively) (Barquero 1998; Venzke et al. 2002). Between mid-1991 and 1993 seismic activity was relatively low, with A-type and T-type (volcanic tremor) seismicity being virtually absent, and a slight decrease in the predominant B-type seismic events (Fig. 5). COSPEC measurements carried out in February 1991 showed an average flux of 91 tons of SO_2 per day (Andres et al. 1992).

As a prelude to renewed activity, the shallow lake evaporated at a rapid rate in early 1994, and was reduced once again to mud pools by March. This desiccation was preceded by an increase in AB-type seismicity (long period medium-frequency quakes, $f = 2.1$–3.0 Hz) in November 1993, and was accompanied by increases in A-type, AB-type, and B-type seismicity. Between June and early August 1994 numerous strong phreatic explosions and strong release of gas and steam occurred within the lake area. Occasionally, a large water-rich plume rose to heights of more than 2,000 m above the crater floor. The largest phreatic explosions were registered in the second half of July (Venzke et al. 2002; Martínez et al. 2000). Swarms of A-type, AB-type, and numerous T-type seismic events preceded or accompanied this explosive activity (Fig. 5), which was further marked by sulfur combustion (suggesting temperatures ≥248 °C, the auto-ignition temperature of elemental sulfur) and high-temperature fumaroles (up to about 700 °C) at the dried lake bottom (Barquero 1998; Venzke et al. 2002).

The increase in seismicity, heat and volatile fluxes that triggered the mid-1994 phreatic activity was possibly related to (hydro)fracturing rather than to a magmatic intrusion. If a fresh magmatic intrusion was emplaced, it must have occurred within a restricted zone underneath the lake area (Martínez et al. 2000), since no drastic increases in fumarole temperatures around the dome were observed since late 1988 (Fig. 5).

A new lake started to form in mid August-September 1994 when the peak of the rainy season was near, but it remained shallow and grey due to the vigor of subaqueous fumarolic degassing (Fig. 3) despite the relatively heavy rainfall and the lowering of fumarolic output and temperature (92 °C; Barquero 1998; Venzke et al. 2002). Between October 1994 and January 1995 the lake depth approached ~15–20 m, and in April 1995 it appeared quiet with its typical milky turquoise color (Barquero 1998; Martínez et al. 2000; Venzke et al. 2002).

The limited data available for 1994–1995 showed that PTs returned in significant amounts, apparently due to considerable input of volatiles. Large fluctuations in their concentrations reflect strong variations in gas fluxes. The high PT concentrations, particularly between September 1994 and March 1995 (except for November 1994 when they were not detected), were marked by a dominance of the tetrathionate ion: $S_4O_6^{2-} > S_5O_6^{2-} > S_6O_6^{2-}$ (Table 1; Fig. 4). These observations suggest that the SO_2/H_2S ratio of the subaqueous fumaroles still remained relatively high. After March 1995 and around the transition between Stage III and Stage IV, input of volatiles and heat seemed to have reduced substantially, as indicated by the sharp decrease in PTs, which coincided with decreasing temperature, acidity, and solute concentrations.

3.2.3 Stage IV (September 1995–February 2005: Relative Quiescence in the Lake, Significant Fluctuations in PT Concentrations and Subaqueous Input of Volatiles and Heat

The three main PTs were present in the lake throughout most of Stage IV, though with variable concentrations (Fig. 6; Table 1). Totals ranged between below the detection limit and 1,640 mg/kg, suggesting drastic changes in the subaqueous fumarolic input. Despite this variability, overall trends shown by the individual species were largely parallel. The tetrathionate ion dominated their distribution ($S_4O_6^{2-} > S_5O_6^{2-} > S_6O_6^{2-}$), except for some brief intervals when pentathionate prevailed ($S_5O_6^{2-} > S_4O_6^{2-} > S_6O_6^{2-}$) and total PT concentrations dropped, as happened in January 1996 and January–April 2002. The unusual distribution of penta-and hexa-thionate dominance over tetrathionate dominance ($S_5O_6^{2-} > S_6O_6^{2-} > S_4O_6^{2-}$) in a few samples collected in November 1984 and January–February 2002 might be due to analytical error or to degradation, since these samples originally contained low amounts of PTs (cf. Takano and Watanuki 1988). The general predominance of the tetrathionate ion throughout Stage IV suggests that the subaqueous fumaroles were relatively rich in SO_2, with a SO_2/H_2S ratio of >0.07 during most the time (cf., Fig. 7).

During Stage IV the total influx of heat and volatiles into the lake was lower than in Stages I–III and V, although in early 2005 strong subaqueous fumarolic activity within the lake turned it into a convective grey and hotter lake. Sites of fumarolic degassing included more sub-aerial vents outside the lake area, extending from the dome to the crater floor and inner crater walls. These conditions may explain the general relative quiescence of the lake during most of this stage, but field observations and the lake chemical and physical properties suggested that it continued to receive a significant input of heat and volatiles during most of Stage IV. The marked fluctuations in PT patterns apparently reflect dynamic changes in activity of subaqueous vents throughout this period. Based on the mode of these variations in PTs and in other monitored parameters, five sub-stages can be distinguished (Fig. 6). Their most salient features are described in the following sections.

Sub-stage IVA (September 1995–July 1997): Relative Quiescence, Large Production of PTs with Predominance of Tetrathionate Due to Renewed Input of Gases, Enhanced Fumarolic Output Through the Dome and Other on-Shore Locations

Following the last months of Stage III when PTs concentrations were high, the transition towards this sub-stage showed first a rapid decline to values near or below detection limits. A concomitant decrease in temperature from 42 to 26 °C, increase in pH from ~0.7 to 1.6, and shifts towards lower $S_4O_6^{2-}/S_5O_6^{2-}$ and $S_4O_6^{2-}/S_6O_6^{2-}$ ratios (Figs. 3, 5 and 6) argue against thermal breakdown or sulfitolysis of PTs in response to increased subaqueous input of heat and volatiles (cf. Takano et al. 1994b). These observations are instead consistent with a weakening of the injection rate and lower SO_2/H_2S ratios of fumarolic gases, which is further supported by decreasing sulfate, chloride and fluoride concentrations (Fig. 6). Field observations supported this interpretation. After the new lake started to form in mid-August–September 1994, the filling rate was initially slow and the color grey, despite heavy seasonal rainfall (Fig. 3), due to still persistent vigorous fumarolic degassing in the lake area (Barquero 1998; Venzke et al. 2002). However, after the lake approached ~20 m depth between October 1994 and January 1995, subaqueous fumarolic activity was weak around April 1995 when the lake was fairly quiet with a typical milky turquoise color (Barquero 1998). This quiescence resulted in a rapid and steady increase of the lake volume from May 1995 until 1996, which continued to grow until reaching a record level in November 1997 (Fig. 3). All of these features strongly suggest that the decline in PT levels was caused by a substantial decrease in the input of heat and volatiles that might be caused by temporal partial blockage of subaqueous vents.

Polythionates reappeared in January 1996, after which significant amounts were produced in the course of this sub-stage. Tetrathionate was the predominant species except during the initial build-up when pentathionate briefly prevailed, possibly due to relatively low SO_2/H_2S ratios of subaqueous fumaroles. The rise of concentrations in early 1996 points to renewal of fumarolic input and the large production of PTs towards the end of substage IVA suggests a sustained influx of gases with moderately high SO_2/H_2S ratios (Fig. 7) (cf. Takano 1987; Takano and Watanuki 1990). Although concentrations of the major anions, pH and lake temperature responded to the initial increase in gas flux, a clear correlation was obscured by dilution effects from the steady increase of the lake volume throughout 1995–1997, partly due to above-average rainfall in 1995–1997 (Fig. 3). Hence, in September–December 1997 the lake reached a record level of more than 1.7×10^6 m^3 and a depth of about 45 m (Martínez et al. 2000).

The renewed PT production coincided with enhanced fumarolic activity at the dome and other on-shore locations, and with visible upwelling in the lake. More vigorous fumaroles around the dome started in September 1995, marking the beginning of Stage IV after a 7-year period of only very weak fumarolic activity around the dome (OVSICORI 1995). Fumaroles with temperatures of ~93 °C escaped from old and new fractures located mostly on its northern face above the lake's surface, increasing the plume height from about 100 m to an average of ~500 m until May 1996. Other features observed at the dome were small flows of molten sulfur, indicating minimum temperatures of 113–119 °C (Shriver and Atkins 1999), minor landslides towards the lake and a constant bubbling below the lake surface near the dome, which persisted until the end of 2001. A small phreatic explosion at the northern face of the dome on 8 April 1996 turned the color of the lake from turquoise to grayish for almost six weeks (OVSICORI 1996). It possibly marked the clearance of fumarolic vents, which were clogged by sediments and solidified sulfur, gypsum and other hydrothermal minerals that accumulated since the previous phreatic periods between 1987 and 1994. Moreover, between 1995–1996, an increase in the amount of soil diffuse CO_2 gas from 0.01 to 2.1% and a shift in the $\delta^{13}C$ values from −21.1 to −6.2‰ measured at the dome (Williams-Jones et al. 2000) indicated a larger permeability, consistent with the more vigorous fumarolic degassing and the occurrence of minor landslides around the dome. The dome remained one of the main sites of fumarolic degassing since then, although degassing rates were variable throughout Stages IV and V.

Elsewhere near the lake shores, initially in the southern sector, later also in the western and north-western sectors, new sub-aerial fumaroles giving off minor columns of vapor and gases (plume heights <100 m, temperatures <97 °C), started manifesting as well (the first sub-aerial fumaroles opened by April 1995). The opening of sub-aerial vents coincided with enhanced A-, AB- and B-type seismicity in 1995 (Fig. 6). They remained active for about 5 years until they ceased completely in June 2000 (Venzke et al. 2002; Mora and Ramírez 2004; Martínez 2008).

During the months immediately before the reappearance of the PTs a remarkable increase in volcanic tremor was registered (on the order of several hundreds of hours per month in October–December 1995), as well as a significant increase in AB and B-type seismicity (Fig. 5). This upsurge in seismic activity was possibly related to build up of pressure in the magmatic-hydrothermal system, opening of cracks or conduits and movement of fluids (McNutt 2000). Another notable observation is the sharp ~75% drop in B-type seismic events to a level similar to that observed before 1985–1986 (Fernández 1990), after the re-appearance of strong fumarolic activity on the dome in late 1995—early 1996. Thus, sub-stage IVA was marked by a low number of B-type events, which were thought to reflect a lower degree of interaction between the subsurface heat source and the shallow aquifer beneath the crater (Barboza V. pers. comm. 2005). This agrees with the decrease of the lake temperature and anion concentrations, the rise in pH and lake volume, and with enhanced sub-aerial fumarolic manifestations around the dome and onshore locations around the acid lake observed between late 1995 and early 1996.

Sub-stage IVB (August 1997–October 2001): Relative Quiescence But Largest PT Concentrations of the 1980–2006 Period Due to Influx of S-Rich Fumarolic Gas, Predominance of Tetrathionate

From mid-1997 on, PT concentrations first increased until reaching peak levels in early 1998, and then showed sharp drops in April (below detection limits) and June 1998 (Table 1; Fig. 6). This brief interval of low concentrations ended with a rise to values similar to those observed before April. It occurred right after maxima in the $S_4O_6^{2-}/S_5O_6^{2-}$ and $S_4O_6^{2-}/S_6O_6^{2-}$ ratios (Fig. 6), which can be attributed to influx of volatiles with increased SO_2/H_2S ratios or higher flux rates (Takano and Watanuki 1990). The combined trends can be interpreted to result from the continuous injection of gas carrying increasing amounts of SO_2. First a SO_2/H_2S ratio was reached that favored an optimum production of total PTs, then further increased ratios resulted in a more efficient production of tetrathionate compared to the longer PTs (Takano et al. 1994b, 2001), while an excess of unreacted SO_2 finally caused a temporary breakdown of PTs due to sulfitolysis. A subsequent shift back to lower SO_2/H_2S ratios then allowed their formation (Figs. 6 and 7). The inferred increasing volatile input at the beginning of sub-stage IVB is supported by the increasing trend of sulfate, chloride, and fluoride concentrations in the lake (Fig. 6), accompanied by a drop in the pH, while relatively high levels of these anions confirmed that it received a significant influx throughout this sub-stage. Hence, the brief disappearance of PTs in April 1998 was probably the result of their breakdown by sulfitolysis. It is important to note that this absence was not followed by phreatic activity, unlike what happened in March–June 1987 and most the Stage III (Fig. 5).

From November 1998 to October 2001 the PT concentrations showed a steady decline, coinciding with a period when many new fumarolic vents opened, especially along the southern, western and eastern inner crater walls. A brief sharp increase in the concentrations of PTs, $S_4O_6^{2-}/S_5O_6^{2-}$ and $S_4O_6^{2-}/S_6O_6^{2-}$ ratios, major anions, and temperature in September–October 2001, just before the PTs dropped down to below detection limits in November, pointed to a sudden short-lived influx of SO_2-rich fumarolic gases into the lake. This is consistent with peaks in unreacted dissolved SO_2 in the lake, which showed up to 300 ppm in 1999 and up to 255 ppm in the second half of 2001 (Fig. 6).

From August 1997 on, when the lake composition started showing a remarkable change, the concentration of chloride was practically always higher than that of sulfate (except for the interval November 2001–April 2002), pointing to the input of chlorine-rich magmatic gas into the lake, whereas from July 2004 the concentrations of sulfate resulted to be higher than those of chloride (Table 1; Fig. 6). The lake level dropped by about 8 m during the first half of 1998, and remained more or less constant throughout the next two and a half years, despite the significant amount of rainfall in 1999 and 2000, which ultimately led to a lake-level rise in the first months of 2001 (Fig. 3). In Fig. 9 an aerial view of Laguna Caliente with a vapor-rich plume rising from the dome is shown.

Apart from the apparent relationships between PT behavior, lake chemistry and supply of volatiles during sub-stage IVB, there were also a number of noticeable results from various geophysical signals and surface manifestations monitored during this interval as pointed in the Supplementary data found at http://www.ovsicori.una.ac.cr/index.php/vulcanologia/geoquimica/investigacion-geoquimica.

Sub-stage IVC (November 2001–May 2002): Relative Quiescence, Sharp Drop in PT Concentrations with Predominance of Pentathionate, Fairly Low Influx of Volatiles

A sharp decline in PTs concentrations to below detection limits characterized this short interval during which the lake activity was relatively calm and visible upwelling ceased. Only very low amounts of PTs were sporadically detected in November 2001–May 2002, with a distribution that was dominated by pentathionate. The $S_4O_6^{2-}/S_5O_6^{2-}$ and $S_4O_6^{2-}/S_6O_6^{2-}$ ratios were occasionally lower than in the previous years (Fig. 6), suggesting that during this sub-stage the subaqueous fumaroles were significantly depleted in SO_2. Thus, the SO_2/H_2S ratios of the subaqueous fumarolic input shifted from SO_2-excess (unreacted) (>0.07) to H_2S-excess (Takano et al. 1994b) (Fig. 7).

The drop in PT concentrations coincided with a sharp decrease in the concentrations of other major anions, a significant increase in pH and a gradual cooling of the lake water (Fig. 6). Free dissolved SO_2 or H_2S gases were not detected. In addition, the lake color changed from milky turquoise to unusual milky dark bluish, suggesting the presence of less suspended particulate

Fig. 9 Aerial view of the active crater of Poás volcano (~1.3 km diameter along its E–W axis). The vapor-rich plume rising from the dome, located on the southern shore of the lake (with its typical blue turquoise color), is related to fumaroles with temperatures around 90–100 °C. Photograph taken from an aircraft by Federico Chavarría Kopper in mid-1999

matter in the water column and much less subaqueous fumarolic activity.

Because the lake level volume rather decreased by about 8% between January–June 2002 (Fig. 3), it is clear that these drastic changes can be attributed to dilution of the lake waters, signaling a substantial decrease in heat and volatile supply, presumably because of temporary blockage of the subaqueous fumarolic vents. Hence, the gas output through the lake area was too weak to produce appreciable amounts of PTs.

At the same time, fumarolic activity around the dome abruptly decreased and only a very weak plume of ~50 m or less was observed, although vigorous activity continued in the fumarolic field on the eastern side of the crater (Fig. 10). Noticeable decreases in B and AB-type seismic events and absence of high-frequency earthquakes and volcanic tremor were recorded as well (Fig. 5). All these observations point to a significant decrease in the influx of volatiles and heat into the lake.

Sub-stage IVD (June 2002–September 2003): Relative Quiescence, Enhanced Input of Volatiles and Heat, Reappearance of Significant Amounts of PTs, Dominance of Tetrathionate

This sub-stage was marked by renewed heating of the lake and the reappearance of PTs in such amounts that concentrations similar to those observed in 1997–1998 were reached. Overall increasing trends of PT concentrations as well as $S_4O_6^{2-}/S_5O_6^{2-}$ and $S_4O_6^{2-}/S_6O_6^{2-}$ ratios

Fig. 10 View of the active crater from the lookout point (El Mirador) at the southern rim, just a few hours after renewal of phreatic activity some time between the evening of 23 March and the morning of 24 March 2006. Note the gray color of the lake and the dense whitish columns of vapor and gases issued vigorously and simultaneously from the lake surface and from the Fumarole Norte near the NE edge of the lake. Rock fragments and sediment lumps on the flat crater floor were ejected from the lake. Photograph taken by Dr. Geerke Floor on 24 March 2006 around 3:30 p.m.

coincided with substantial increases in the lake-water temperatures (from 29 to 41 °C) and SO_4, Cl and F concentrations, and with a concomitant decrease of pH (Fig. 6). It is noteworthy that temperature and PT concentrations showed parallel trends with peaks around September 2002 and June 2003, whereas those of SO_4^{2-}, Cl^- and F^- continued to accumulate until July–September 2003. This period of high PT concentrations terminated by a sudden decline to below detection limits in October 2003.

Dominance of tetrathionate throughout the sub-stage and presence of unreacted dissolved SO_2 in the lake water point to relatively high SO_2/H_2S ratios of subaqueous fumaroles. The patterns of the individual PTs fluctuated and were not always parallel, especially during the second half of this interval (Fig. 6), suggesting variations in the injection rates and proportions of SO_2 and H_2S.

It is of interest to note that during this sub-stage the PT concentrations did not follow the steadily increasing trends of SO_4^{2-}, Cl^- and F^-, but showed peaks that rather tended to correspond with peaks in the lake temperature (Fig. 6). The decrease of PT concentrations when major anions still continued to rise may be the result of sulfitolytic breakdown or reflect an influx of (recycled?) brine water into the lake. The latter alternative would be supported by the corresponding early decrease of the temperature and by a gradual increase in the Cl^-/F^- and Cl^-/SO_4^{2-} ratios, which were observed in the second half of this sub-stage. This might also fit with an overall pattern of increasing importance of Cl^- relative to the other anions observed throughout Stage IV (Fig. 6).

Simultaneous with the return of PTs, the appearance of the lake changed significantly. In late May-early June 2002, it recovered its typical milky sky-blue color after the previous months when it was milky dark blue, apparently due to the small load of suspended particles. In late July 2002 the lake turned dark greenish due to convective circulation that transported abundant sulfur globules and bottom sediment to the surface in the central part of Laguna Caliente. Convection and floating sulfur globules were observed practically throughout the entire sub-stage, but the activity was particularly strong in September–October 2003 when some small unusual blackish globules appeared together with the commonly observed yellow sulfur globules and fragments of molten sulfur with a 'soft elastomeric' texture emerged from the lake bottom, suggesting an intensification of subaqueous fumarolic output (cf. Takano et al. 1994a). At the same time, especially between September 2002 and December 2003, a persistent emanation of acidic fumes (HCl evaporation?) formed a white acidic fog over the lake, which covered at times the entire surface. Constant bubbling was observed at the interface between the lake and the north face of the dome, where water "spurts" of about 1–5 m length were sporadically ejected between August 2002 and September 2003. In spite of the large and relatively constant volume ($\sim 1.5 \times 10^6$ m^3), the lake level showed some fluctuations, possibly as a result of lower yearly rainfall in 2002 (Fig. 3) in combination with stronger evaporation in 2002–2003 due to the increased input of heat.

Throughout this sub-stage, the fumarolic emissions from the dome remained modest and fairly constant, producing a less than 400 m high plume, except for a short-lived increase up to 600 m in February 2003. New thermal springs, fumaroles and cracks continued to appear in the eastern sector of the crater, and their plumes reached heights between 50 and 300 m similar to the previous sub-stage. In December 2002, the temperature of the Fumarola Norte rose from boiling point to 122 °C (Fig. 5). Bright reddish-orange and white hydro-sulfosalts were observed along the Norte-Este and Este and acid sulfate-rich springs in the eastern side of the crater. In early January 2003, an OVSICORI field survey provided indications about a minor hydrothermal explosion, which occurred within the area where the Este springs seeped out in the previous month. The concentric cracks that appeared along the eastern inner terrace in July 1999 became wider, and crater wall collapses were observed. All these manifestations thus indicate that, during this period, the release of fumarolic gas was concentrated in an area comprising the dome, the lake and the eastern inner crater wall.

Following a substantial drop in the number of events recorded by the end of the previous interval, B-type and AB-type seismic activity increased during this sub-stage, reaching maximum intensity between September–October 2002. The increase accompanied the appearance of new fumaroles on the dome and in the eastern sector of the crater. At the end of the sub-stage, AB-type seismicity decreased substantially, whereas B-type earthquakes continued to be recorded in large quantities.

Rare types of short-duration monochromatic tremors, which are distinct from the continuous long-duration tremors typically recorded at Poás, as well as "banded" tremors (a rarity for this volcano) and "intrusive" signals were registered by the OVSICORI POA2 seismic station between mid-2002 and the end of 2003. Short-duration (<1 min) monochromatic tremors and "intrusive" signals were also recorded in 1986 before renewal of geyser-like phreatic activity in the crater lake, and related to the ascent of minor magmatic intrusions (Fernández 1990; Rowe et al. 1992a).

Microgravity surveys showed a rise in sub-surface mass at stations to the west and south of the dome between 2000 and 2009, which was attributed to a highly localized dendritic intrusion beneath the crater (Rymer et al. 2005, 2009), similar to other inferred events associated with microgravity rises over the last 20 years (Fig. 5). The enhanced input of heat and SO_2-rich volatiles into the lake system, in combination with the accompanying manifestations, might be consistent with an ongoing minor intrusion of magma during sub-stage IVD, but without triggering phreatic explosions.

Sub-stage IVE (October 2003–February 2005: Relative Quiescence, Sharp Drop in PT Concentrations, Strong Weakening of Fumarolic Outgassing, Record Lake Level

A sharp decline in PT concentrations to below detection limits in October–November 2003, comparable to that observed in sub-stage IVC, was the start of this period of quiescence, which was marked by much less input of heat and volatiles into the lake. Polythionates were monitored and remained undetected until April 2005. In the few determinations conducted during this period, only traces of dissolved SO_2 and no H_2S were detected between June 2004 and March 2005. The lake temperature, which had decreased gradually from a maximum of 41 °C in April–June 2003, ranged between 23 and 35 °C during the second half of 2003 and throughout 2004. Anion concentrations decreased substantially, with fluoride showing the lowest values ever measured at Poás, which were related to the very high SO_4^{2-}/F^- and Cl^-/F^- ratios observed in May–June 2004. The pH increased by about 0.7 units meaning that the lake was about 5 times less acidic (Fig. 6).

From late 2003 on, the lake volume increased significantly and continued growing until at least 1.9 million m^3 between late 2004 and early 2005, reaching the highest volume recorded since the late 70s and culminating in the flooding of part of the eastern crater floor. Sub-aerial fumaroles in this sector remained submerged for several weeks (OVSICORI 2005). This expansion of the lake was caused by extremely heavy rainfall in the wet seasons of 2003, 2004 and 2005 (Fig. 3), which were among the rainiest years recorded at the summit of Poás volcano in the last 15 years (ICE rainfall data; Aguilar et al. 2005). The volume increase was accompanied first by a subtle color change towards the end of 2003, but at the end of April 2004 it turned from bluish green to emerald green (see photos at the Supplementary data to this article available online at http://www.ovsicori.una.ac.cr/index.php/vulcanologia/geoquimica/investigacion-geoquimica). After decades of monitoring, this was the only time that such a color was seen (Martínez 2008). A similar color was also observed in other acid crater lakes during periods of reduced gas emission (e.g. White Island, Venzke et al. 2002).

A conspicuous reduction in the vigor of fumarolic outgassing was observed around the dome, which was one of the main sites with persistent strong gas and vapor emission since late 1995. The plume that typically reached a height of 500 m or more remained particularly low (<200 m) throughout most of 2004, occasionally allowing to observe some of the vents. Outlet temperatures of fumaroles on the dome remained rather constant, i.e. around boiling point at the altitude of the crater (93 °C). In contrast, vigorous sub-aerial fumaroles (up to 120 °C) continued in the eastern fumarolic field. The Fumarole Norte (Figs. 5 and 10) showed a temperature rise up to 116 °C, and some new fumaroles opened, which is consistent with the recorded high number of B and AB-type earthquakes during this sub-stage (Fig. 5). During the peak of the dry season of 2005 (March–April), deposition of bright yellow-orange insoluble or readily soluble acid alteration minerals such as hydroxysulfates (e.g. alunite, alunogen, jarosite), and some water-soluble hydroxysulfates (mohrite, pickeringite, dietrichite, kaolinite, halotrichite, apjohnite, voltaite, bilinite, römerite), were once again observed along the Norte-Este and Este acid sulfate-rich springs (Martínez 2008; Rodríguez and van Bergen 2015, 2017). Landslides continued to accompany the vigorous fumarolic activity in this sector.

The large decrease in the flux of fumarolic gases and heat within the area between the lake and the dome might be the consequence of a structural change which led to a localized blockage of the gas conduits beneath the lake, without affecting the fumarolic field in the eastern region that started forming in late 1998 and 1999. Another cause of the weakening fumarolic input into the lake to be considered is a stronger gas scrubbing (Symonds et al. 2001) due to an increased amount of groundwater table, in agreement with a significant rise of the lake water level between mid-2003 and early 2005.

3.2.4 Stage V (From March 2005 to February 2019): Reappearance of Large Quantities of PTs, Dominance of Tetrathionate, Subsequent PT Breakdown at Onset of the Current Cycle of Phreatomagmatic to Magmatic Activity

This period is marked by peak activity, culminating in phreatic explosions in March, April, September and December 2006, and includes the most recent PT data available (Table 1). After their absence for more than a year, the PTs reappeared in April–May 2005, after which their abundance increased sharply until reaching maximum total concentrations of about 3,500 mg/kg between November 2005 and January 2006 (Figs. 3 and 6). Subsequently, between April–May 2006 shortly after the onset of phreatic activity in March, a steep decrease was observed down to values of 150–620 mg/kg. Excess (un-reacted) dissolved SO_2 was detected in the lake water between May 2005 and late 2006, and the order of species abundance was virtually always $S_4O_6^{2-} > S_5O_6^{2-} > S_6O_6^{2-}$, indicating that the injected gas was again characterized by a relatively high SO_2/H_2S ratio. An exception was observed in May–July 2006 when pentathionate was dominant, coinciding with the presence of ca. 2 mg/L of dissolved H_2S in July, the first time that this species was detected since 1999. The $S_4O_6^{2-}/S_6O_6^{2-}$ ratios at the onset of Stage V were the highest recorded in 25 years, while the $S_4O_6^{2-}/S_5O_6^{2-}$ ratios were close to stage IV values (Figs. 4 and 6). Tetra- and pentathionate showed fluctuating but parallel trends, whereas hexathionate was initially produced in very small amounts and gradually increased in the course of 2005-early 2006. The strong decrease of PTs that started in late 2005-early 2006 and their low concentrations in 2006 are apparently due to sulfitolytic (and thermal?) breakdown caused by a sustained influx of SO_2-rich gases, since the decline occurred during a period of steadily increasing SO_4–Cl–F concentrations and temperature as well as decreasing pH (Fig. 6). Starting from 22 °C in February 2005, the lake temperature rose sharply in the next months until reaching peaks of 50–54 °C in the second half of 2005 (Fig. 6), similar to temperatures registered during the latest period of phreatic eruptions in 1994.

The intensification of activity was accompanied by marked changes in the appearance of the lake. It turned from a calm green reservoir in 2004 to a grey fuming lake in early 2005 (Fig. 10), showing several convective cells that brought up abundant dark gray and large (~0.5 cm diameter) sulfur globules. The lake level dropped several meters throughout 2005 despite of heavy rainfall at the summit of the volcano in 2004 and 2005 (Fig. 3). Strong evaporation of acidic fumes affected the sparse vegetation within the crater.

All these observations pointed to a significant new influx of condensing magmatic gases and heat that affected the entire lake system. From early 2005 on, the combined changes in geochemical and seismic behavior could be interpreted as precursory signals to pending phreatic outbreaks, presumably associated with new gas releases provoked by either fracturing of the brittle envelope of a cooling magma body or the intrusion of a fresh magma batch. The latter alternative is supported by increases in micro-gravity in the south-western part of the crater floor since 2001 (Rymer et al. 2004, 2005, 2009), which would imply that magma emplacement at shallow levels started already about halfway Stage IV (Fig. 5).

In early 2006, further changes in monitored signals, pointing to a continued increase in the input of hot gases into the lake and conduits of sub-aerial fumaroles, were the prelude of phreatic eruptions that took place in March, after a period of almost 12 years of relative quiescence. Several mild phreatic eruptions from the lake were observed in March–April 2006. Later that year, at least two small phreatic eruptions occurred on 25–26 September and a visitor of Poás Volcano National Park reported another (about 30–50 m high) on 16 December.

It is conceivable that deep penetration of large amounts of meteoric groundwater into the magmatic/hydrothermal system played a role in triggering the phreatic activity, given the high record of rainfall registered between 2003 and 2005, the flooding of the eastern crater floor and the rate of degassing, fracturing and wall collapses in the eastern sector (cf. Vaselli et al. 2003 and Section on Substage IVB in this paper for a similar mechanism to explain the expanding fumarolic activity in 1998–2000). However, the strong rise in all magmatic volatiles, particularly SO_2 and HF, argues for new releases from a magma source as main controlling factor. A possible connection between subaqueous and sub-aerial fumarolic degassing will be discussed below.

During Stage V, fumarolic degassing was centered within the lake area and the fumarolic field in the eastern sector. The Fumarola Norte at the NE inner crater wall (Fig. 10) showed sudden changes in temperature (from 105 °C in March 2005 to 200 °C in May 2005), as well as changes in gas composition, especially between May 2005 and early 2006. It released large amounts of sulfur crystals that were deposited over a large area of the northeastern inner crater wall, and molten sulfur ponds formed in the vents and flowed towards the vicinity of the acid lake. Emissions from the dome remained moderately low at <95 °C, similar to the previous stage (Fig. 5).

In April 2006, shortly after the climax of renewal of phreatic activity, an average SO_2 flux of approximately 100 tons/day was recorded by mobile Mini-DOAS (OVSICORI 2006; Barrancos et al. 2008). This flux did not reach the Stage-II levels of 500–800 tons/day measured in 1981–1983 (pers. comm. Stoiber, cited in Prosser and Carr 1987; Casadevall et al. 1984b; Stoiber et al. 1986) but was higher than the 29 tons/day in March 2003 and similar to the 90 tons/day measured in early 1991 (Andres et al. 1992).

The onset of increased activity in the lake coincided with a strong increase in seismicity between April and November 2005, particularly AB-type events and polychromatic as well as monochromatic tremors. This was only the second time that such an exceptional amount of volcanic tremor was recorded at Poás since the period of peak heating in 1981–1983 (Stage II), which was presumably due to a magmatic intrusion beneath the dome (Brown et al. 1991) where fumaroles reached temperatures of more than 1,000 °C (Fig. 5). Moreover, it was the first time for AB-type events to reach such a high record, but it is unknown if they were similarly abundant in Stage II because they were not classified as such before 1984. Clear increases in AB-type quakes and polychromatic tremor preceded and accompanied the different intervals of phreatic eruptions in 2006 (Fig. 5). Swarms of A-type quakes were registered in January–

February, April and June–July 2005, but their number remained relatively modest compared to other periods with increased seismicity of this type.

By the end of 2006, dissolved volatiles in the lake reached the highest levels of Stage V. Highest temperatures were recorded in November (58 °C) after a minor dip in the previous months. The lake retained its grey color and continued to show strong convective activity. A few minor phreatic explosions, "mud plumes" and molten sulfur floating on the surface further characterized its behaviour until 2008. Between 2009 and 2017, there were numerous phreatic explosions with jetting columns up to m high and fumaroles which temperature ranged from above 100 °C (in 2008–2009) to around 625 °C (in 2011 and 2015, and ca. 1000 °C (in 2017) (Vaselli et al. Chapter "The Last Eighteen Years (1998–2014) of Fumarolic Degassing at the Poás Volcano (Costa Rica) and Renewal Activity"). Stage V reached a peak activity in April–June 2017, with phreatomagmatic and magmatic eruptions. From then on, the volcano is showing subaerial degassing and frequent weak phreactic eruptions impacting the ambient around, mainly downwind.

3.3 Lake Behavior from Polythionate—Sulfate Relationships

Polythionate concentrations in the Poás lake have rapidly declined several times over the 1980–2006 period. These events are not obviously related to corresponding increases in sulfate concentrations (Figs. 4 and 6) because it can be inferred that different mechanisms, associated with either increased or decreased activity, have played a role.

The PT decrease at the start of the active Stage III was accompanied by an increase in the sulfate concentrations, pointing to sulfitolysis and/or thermal breakdown. This can only account for about 16% of the increase in sulfate observed between May and July 1986 (Fig. 4). Hence, increased input of hot SO_2-rich volatiles was likely responsible for most of the sulfate production, consistent with an increase in the SO_2/H_2S ratio of subaqueous fumaroles, as inferred from the change in the distribution of PT species since August 1986 (Fig. 4) (albeit with temporary reversals in some of the late 1986 and early 1987 samples). On the other hand, PT decay may account for the sulfate increase of March–April 1998 (Sub-stage IVB), and November 2005–May 2006 (Stage V). It is of interest to note that precipitation of gypsum/anhydrite, barite, celestite, aluminium-hydroxysulfate minerals, and native sulfur may obscure the inverse relation between dissolved sulfate ions and PT decomposition, since the lake water is usually saturated or nearly saturated with respect to these sulfate mineral phases (Rodríguez 2016).

The three other intervals in Stage IV with a short-lived decline in PTs (Fig. 6) require a different explanation. Decreased SO_4^{2-}, Cl^-, F^- and dissolved SO_2 concentrations, increased pH and a drop in temperature during these periods point to a reduction or complete cessation of subaqueous fumarolic venting. Marked peaks in SO_4^{2-}/F^- and Cl^-/F^- ratios reflect a stronger decline of F relative to the other anions, which is consistent with less magmatic input, given the preferential reaction of HF with wall-rocks at relative low temperatures of volcanic gases (cf. Symonds et al. 1990; Stimac et al. 2003). The observations thus provide evidence for temporary clogging of subaqueous fumarolic vents and a decrease in the heat flux. A parallel short-lived decrease in SO_2/H_2S ratios of subaqueous and sub-aerial fumaroles during Sub-stage IVC suggests that this blockage affected large parts of the subsurface system feeding the entire crater area. During Sub-stage IVE, however, lower SO_2/H_2S ratios were only observed in the lake water, while sub-aerial fumaroles showed a subtle systematic change.

Furthermore, the cessation of convective activity would promote the development of stratification in the water column, including an aerated shallow layer. According to experiments on Yugama lake water, PT concentrations are relatively insensitive to aeration (Takano and Watanuki 1990), but the combination of oxygen and Fe^{3+} at low pH oxidizes PTs to sulfate to an order of magnitude faster than oxygen alone (Druschel et al. 2003). A representative reaction describing this oxidative decomposition (cf. Takano and Watanuki 1988; Takano et al. 2001; Druschel et al. 2003) is, as follows:

$$S_4O_6{}^{2-}{}_{(aq)} + 3Fe^{3+}{}_{(aq)} + 2.75\,O_{2\,(aq)} + 4.5\,H_2O_{(l)} \rightleftharpoons 4\,SO_4{}^{2-}{}_{(aq)} + 3\,Fe^{2+}{}_{(aq)} + 9\,H^+{}_{(aq)} \quad \text{(d)}$$

Because an influx of oxygen-rich fresh rainwater into the lake will promote the oxidation of Fe^{2+} into Fe^{3+}, and because low Fe^{2+}/Fe_{Total} ratios (~ 0.01) were indeed found by us in samples of surface water during periods with stagnant fumarolic input (unpublished data), it is conceivable that this mechanism has further enhanced the PT breakdown.

Finally, microbial decomposition of PTs may also have contributed to their disappearance at the start of Sub-stage IVA and during Sub-stages IVC and IVE when bacterial activity could have been boosted by relatively high pH conditions, as was also observed in the ultra acid crater lake of Maly Semiachik volcano (Takano et al. 1995, 1997). So far, the presence of *Acidithiobacillus* sp. in the lake water of Poás (Sugimori et al. 1995, 2001) and some rod-shaped and spherical acidiphilic bacteria in the bottom sediments has been reported and are likely to be new species (Sugimori et al. 2002). Recently, Hynek et al. 2018, found in the acid lake of Poás bacteria of the genus acidiphilium such as: acidiphilium angustum, acidiphilium rubrum and A. acidophilum (formerly Thiobacillus acidophilus). However, the potential effect of micro-organisms on the PT budgets remains to be assessed.

3.4 SO_2/H_2S Ratios and Common Feeding Controls of Subaqueous and Sub-aerial Fumarolic Output

Molar SO_2/H_2S ratios in the lake brines were estimated from the concentrations of dissolved SO_2 and H_2S gases that are in excess (unreacted) in the lake waters and that were discontinuously measured in situ since 1999. Since SO_2 is more soluble in water than H_2S, the SO_2/H_2S ratio of gas species dissolved in the lake water will be larger than that of the injected fumarolic gases (e.g. at 20 °C, SO_2 is about 15 times more soluble than H_2S in pure water, Takano et al. 1994b; IUPAC 1983, 1988). Thus, the calculated SO_2/H_2S ratios of subaqueous fumaroles are based on the solubility of both gas species in pure water at 1 atmosphere for the lake temperatures measured (IUPAC 1983, 1988; Xia et al. 1999, 2000), assuming that the solubility values are also valid for the acid and highly saline brines of Poás lake. To estimate the SO_2/H_2S ratios of the subaqueous fumarolic gases a H_2S concentration of 0.2 ppm (detection limit) was adopted, because H_2S was only detected a few times over the period during the surveys of dissolved gases (about 2 ppm in July 2006, May 2009, Sept–Oct 2012, and Oct–Nov 2014). Hence, the calculated ratios reported here (Fig. 6) should be taken as minimum values. Dissolved unreacted SO_2 was always detected in significant amounts, except in November–December 2001 and June 2004–March 2005, when traces were measured at most.

The inferred SO_2/H_2S ratios of subaqueous and sub-aerial fumaroles are similar in magnitude and, with the exception of Sub-stage IVE, show parallel behavior in the time series since 1999 (Fig. 6), which indicates that gas entering at the lake bottom has largely the same source as that emitted from the on-shore vents. The disappearance of PTs coincided with a parallel decrease in the SO_2/H_2S ratios during Sub-stage IVC, whereas in Sub-stage IVE a decrease was only seen in the subaqueous fumaroles, suggesting a decoupling with pathways to the sub-aerial vents during this period. This may imply that in Sub-stage IVE (partial) blockage of rising gases was confined to relatively shallow levels below the lake.

Common feeding controls and gas source for sub-aerial and subaqueous fumaroles are further suggested by field observations. A simultaneous increase in vigor of vapor and gas release from the lake surface, the dome fumaroles and Fumarola Norte was observed in March 2006

(Fig. 10), and both the lake water and the Fumarola Norte showed a simultaneous conspicuous temperature increase in 2005 (Fig. 5). These observations suggest that the fumarolic field on the eastern terrace resulted from fractures going deep below the crater floor (cf. Fournier et al. 2004).

4 Summary

1. According to a 25-years monitoring record, polythionates (tetra-, penta-, and hexathionate), together with other sulfur-bearing species, are commonly present in the acid crater lake of Poás despite wide fluctuations in temperature, chemical composition and volume. Although PTs usually constitute a significant part of the total sulfur budget of the lake, they were only detected as long as the lake temperature did not exceed 60–65 °C and the rate of injection of SO_2 gas were not excessive. In general, large amounts (up to 4,100 mg/kg) were present during periods when strongest fumarolic output occurred sub-aerially and was centered mainly outside the lake area (i.e. Stages II and IV). In contrast, PTs were virtually absent when strong fumarolic activity was predominantly located within the lake area (i.e. Stage III and part of Stage V), thus promoting thermal and sulfitolytic breakdown. Polythionates were also absent during quiescent periods when subaqueous fumarolic release was very weak, presumably due to partial sealing of feeding conduits (i.e., early sub-stage IVA, and sub-stages IVC and IVE).
2. The prevailing distribution of PTs is $S_4O_6^{2-} > S_5O_6^{2-} > S_6O_6^{2-}$. This order was predominantly observed in periods when PT concentrations were particularly high, reflecting substantial influx of gas with relatively high SO_2/H_2S ratios into the lake, as was the case during large parts of Stages III, IV and V. Relatively modest concentrations of PTs were usually associated with the sequence $S_5O_6^{2-} > S_4O_6^{2-} > S_6O_6^{2-}$, which is typical for weakened flux rate and/or decreased SO_2/H_2S ratios of gases entering the lake, as observed during Stage II and some short intervals later.
3. The strong fluctuations in PT concentrations and speciation observed over decades reflect the dynamic character of the magmatic-hydrothermal system of Poás. Processes that control the underlying changes in volatile input include shallow intrusions of fresh magma, fracturing of the brittle envelope of cooling magma, and opening, closure or spatial shifts of conduits and fractures that determine the location of fumarolic activity and thus the degree to which volatiles are intercepted by the lake.
4. Polythionates are sensitive signals of renewed input of sulfur-rich gas into the lake, caused by either an intrusion of fresh magma or fracturing of the brittle carapace around cooling magma that was earlier emplaced. An increase in PT concentrations, a change in their speciation from penta- to tetrathionate dominance, possibly followed by a sharp concentration drop caused by sulfitolytic breakdown, likely precede phreatic eruptions several months before the eruptive activity takes place. Intervals during which PT formation was probably associated with such a sequence are early 1984—early 1986 (Stage II–III transition) and early 2005—early 2006 (beginning of Stage V). In both cases, the build-up was accompanied by an increase in temperature and a decrease in pH and volume of the lake. It is noteworthy, however, that a rise of PTs, followed by a drop is not necessarily a prelude to phreatic eruptions, as was illustrated by the disappearance of PTs during Sub-stage IVE following their rise in Sub-stage IVD (mid 2002–mid 2003). Apparently, this episode was marked first by a reduction of sulfur-rich gas input, and then by a virtual cessation of volatile and heat supply, as a temperature drop and pH increase confirmed.

5. The release of sulfur-bearing gas from the magmatic-hydrothermal system can be monitored reliably through time-series analysis of PT species in the lake. This approach is complementary to traditional gas monitoring at sub-aerial fumarolic vents, and has practical advantages in terms of sampling and analytical techniques, except for the non availability of commercial penta- and hexathionate stable salts for the calibration standards. Interoretatuibs will benefit from simultaneous monitoring of other chemical and physical parameters, since our time series demonstrate that changes in the composition and flux of volatiles into the lake, inferred from PTs, are also reflected by changes in other dissolved components and in the level and/or nature of seismic activity.
Supplementary data to this article can be found online at http://www.ovsicori.una.ac.cr/index.php/vulcanologia/geoquimica/investigacion-geoquimica.

Acknowledgements We gratefully acknowledge the expert assistance of several colleagues with field and laboratory work, in particular Wendy Saénz Vargas (OVSICORI-UNA), Helen de Waard, Ronald Miltenburg, and Erick van Vilsteren (Utrecht University), as well as support from personnel of Poás Volcano National Park. MM thanks co-investigator Dr. Boku Takano for giving to a group of Costa Rica chemists and volcanologists the know-how on the chemistry of polythionates in acid lakes, and also for guiding us during field and laboratory work to learn on sampling and analysis of polythionate species using HPLC-UV techniques. Thanks also to Dr. Yasuyuki Miura (Tokai University) for advice on best practices to separate the thionates and other sulfur species by HPLC-UV. MM is very grateful with Dr. Takeshi Ohba (Tokai University) for training the chemists of OVSICORI-UNA on sampling and analyses of volcanic fluids from Poás and Turrialba volcano. Our sincere thanks also to the Japan International Cooperation Agency (JICA) for bringing to Costa Rica such a fine group of Japanese experts, Dr. Takano, Dr. Miura, and Dr. Ohba, to work with us at field and laboratory level. We are indebted to Dr. Juan Valdés González of the Laboratory of Atmospheric Chemistry of the School of Chemistry of Universidad Nacional (LAQAT-UNA) for providing access to its HPLC facility where part of the separation and analysis of polythionates were performed. We also thank the Instituto Costarricense de Electricidad (ICE) for providing rainfall data used in this chapter. MM acknowledges financial support from Utrecht University, Universidad Nacional, and the Ministry of Science, Technology, and Telecommunications of Costa Rica (MICIT). MM also thanks the financial contribution given by Dr. Karen McNally, Dr. Marino Protti, Dr. Ronnie Quintero, Lic. Jorge Brenes, Mr. Federico Chavarría Kopper, and the Roegiers Family (inSausalito California EEUU) for supporting this novel and exciting study.

Appendix: Sampling Strategy and Analytical Procedures

Sampling

Samples selected for PT analysis were routinely collected at a location on the NE side of the acidic crater lake (Fig. 1). Temperatures of the lake were measured at the same site using thermocouples. Most of the samples were stored in dark high-density polyethylene bottles at ambient temperature without filtration, dilution or addition of preservatives before the PT analyses were conducted. Between March 1990 and 1995 the sampling intervals were irregular in part because of the high activity of the volcano. From 1996 till November 2006, the data represent a monthly sampling frequency. A single available sample from 1980 was analyzed to record the earliest PT signal. Polythionate data documented by Rowe et al. (1992a) cover the period November 1984–March 1990 and represent an approximate sampling frequency of once every two months. Collectively, the currently available PT record thus covers the period 1980–2006. Some stored samples collected in 1988–1994 that contained grey muddy material were taken from mud pools that were formed during episodes when the lake volume was strongly reduced. These pools had a highly variable chemistry since they were largely mixtures of brine left during evaporation of the lake, fumarolic steam condensing at the bottom of the pools, and rainwater (Rowe 1994). Some pools contained unusual high concentrations of PTs (e.g. $\Sigma S_xO_6^{2-}$ 6,740 mg/kg in a sample of 10 June 1994), but their results are not considered representative for the long-term evolution of the lake system and will therefore be ignored in this study.

Polythionate Analysis

About 190 lake water samples were analyzed for tetra-, penta-, and hexathionate ions at the Department of Chemistry of Tokyo University (Japan) and at the Laboratory of Atmospheric Chemistry (LAQAT) of the Universidad Nacional (Costa Rica) using similar ion-pair chromatographic techniques described below.

At Tokyo University tetra-, penta-, and hexathionate ions were determined following a high-performance microbore ion-pair chromatographic separation technique with ultraviolet absorption detection, and an ion-pair chromatographic separation technique with conductivity detection (Takano 1987; Takano and Watanuki 1988, 1990). The first technique allows the determination of tetra-, penta-, and hexathionate in excess of 0.2 ppm, and the second that of tetra-, and pentathionate in excess of 10 ppm. Standard solutions for penta- and hexathionate were prepared from synthesized salts (Takano and Watanuki 1988) following Goehring and Feldmann (1948). A commercially available potassium tetrathionate salt was re-crystallized for the preparation of tetrathionate standard solutions.

At LAQAT, separation and quantification of main PT species were performed using an ion-pair chromatographic technique with UV absorption detection following Miura and Kawaoi (2000) but using a silica ODS-2 chromatographic column as described below. A Shimadzu Model LC-10AS chromatographic system, equipped with a LC-10AS liquid delivery isocratic pump set at a flow rate of 0.6 mL/min, an automatic SIL-10A auto-injector with a 100 μl sampling loop, a SPD-10AV UV-VIS spectrophotometric detector set at 230 nm, and a SCL-10A system controller were employed. Chromatographic signals were obtained with the following settings: an Alltech/Allsphere ODS-2 analytical column 5 μm particle size (150 × 4.6 mm i.d.) coupled to a Brownlee guard column (50 × 3 mm i.d.); an acetonitrile-water (20:80 v/v) mobile phase (6 mM in tetra-propyl-ammonium hydroxide, TPA, and buffered at a pH = 5.0 by dropwise addition of glacial acetic acid 100%), filtered through a 0.45 μm membrane filter and degassed by vacuum. All of the reagents used for the mobile phase were analytical-grade; the acetonitrile was 99.93% HPLC grade, the TPA ion pair reagent used was an aqueous solution 20% (~1 M). Mobile phase, samples and standard solutions were diluted with distilled and deionized water. Samples were filtered and diluted immediately before injection into the chromatographic system. Chromatographic separations were carried out at ambient temperature (23 ± 2 °C).

A potassium tetrathionate calibration solution was injected several times during a 6 h sequence to monitor the response, the reproducibility of peak heights and retention times. Precision of retention times for standards and unknowns was ±0.9% within one day (n = 12) and ±1% within one week (n = 10) for the ion-pair UV chromatographic technique. This enabled unequivocal identification of the PTs by retention time only. Repeated analysis (n = 6) of the 21 June 2002 lake water sample to determine the reproducibility of the procedure yielded a relative standard deviation (RSD) of 5% at a mean concentration of 190 mg/kg for tetrathionate, 4% at 57 mg/kg for pentathionate, and 4% at 25 mg/kg for hexathionate. Reproducibility for the entire set of samples, analyzed in Costa Rica within a three-month period, was tested on randomly selected samples with a range of concentrations. Results showed average RSD values of 10, 7 and 8% for tetra-, penta-, and hexathionate, respectively. All samples were analyzed in duplicate and the results were averaged. Concentration differences between duplicates were usually <5%. Detection limits were 5 mg/kg for tetra-, 1.6 mg/kg for penta-, and 0.5 mg/kg for hexathionate.

One liter of lake water (original T = 32 °C, pH = 1.1, density = 1.02 g/mL), collected on 17 March 1999, was stored at 5 °C and used as a reference standard solution for PT calibration curves at LAQAT. This sample was periodically analyzed at Tokyo University to monitor its quality, in view of the chemical lability of PTs (Stamm et al. 1942). For instance, the analysis carried out at Tokyo University on 28 July 2002, 3 years and 4 months after collection, yielded 462 ± 8 mg/kg $S_4O_6^{2-}$, 182 ± 3 mg/kg $S_5O_6^{2-}$ and 120 ± 2 mg/kg $S_6O_6^{2-}$, concentrations that were in good agreement with those measured in April 1999 (455, 213 and 110 mg/kg, respectively). Polythionate concentrations in this reference sample remained at acceptable stable levels at least until August–October 2002 when analyses were run at LAQAT.

Some samples were analyzed within a few weeks after collection, but the vast majority was stored for several years at ambient temperature without preservation measures. Although it was demonstrated for natural and synthetic highly acid solutions (pH < 2), free from sulfur-oxidizing bacteria, that no significant decomposition of PTs occurs at concentrations of >100 ppm over at least eight years (Takano 1987; Takano and Watanuki 1988, 1990; Takano et al. 1994), the stability of PTs was verified in some lake water samples at LAQAT. As samples from the 1984–1990 period (Rowe et al. 1992a) were not available for re-analysis, a PT-rich mud pool sample and the reference sample (June 10, 1994 and March 17, 1999), were analyzed for tetrathionate on October 16, 2002 using a fresh synthetic aqueous solution of a 98% sodium tetrathionate dihydrate salt, $Na_2S_4O_6.2H_2O$, to prepare a calibration curve. Results were even somewhat higher than those obtained in Tokyo in 1999 and 2002 (RSD = 14 and 12%), indicating that there is no evidence for significant instability of this species over a period of at least three years. On the same date, all major PTs were also re-analyzed in the 10 June 1994 sample, using the 17 March 1999 reference solution. Original and newly measured values were in reasonable agreement with 2,825 and 3,185 mg/kg for tetra-, 2,067 and 2,600 mg/kg for penta-, and 725 and 880 mg/kg for hexathionate, respectively. Repeated analyses of aliquots of the reference solution in Tokyo in 1999 and 2002 confirmed that it had maintained its quality after three years.

Comparison of concentration data obtained on aliquots of 8 samples (June 1994–September 2001) showed random differences between results from the two laboratories. Overall deviations from the mean values were better than 20% RSD (3–20% for tetra-; 2–15% for penta-, and 1–20% for hexathionate. As most analyses in Tokyo and Costa Rica were carried out with time differences of about 3 years, it is concluded that the PT time series presented here are largely unaffected by potential interlaboratory differences or chemical instability. Re-analysis of sulfate and chloride suggested that some of the oldest samples may have experienced a certain degree of evaporation or mineral precipitation during storage. According to enrichments of both anions found, this may have raised PT concentrations in these cases by 15–20% at most.

Determination of Major Anions, Dissolved Gases, pH and Other Data

Most of the samples collected between 1990 and June 2004 were analyzed for sulfate, chloride and fluoride at Utrecht University by suppressed ion chromatography, using a fully automated Dionex Model DX-120 system. Samples were filtered in the laboratory with 0.45 μm polycarbonate membrane filters prior to analysis. Repeated analysis ($n = 14$) of a lake water sample yielded relative standard deviations better than 4% for all of the anions. Precision was about 0.1, 0.3 and 4% for sulfate, chloride, fluoride, respectively, based on the analysis of a synthetic solution. Detection limits were 0.3, 0.1 and 0.05 mg/kg, respectively. Some samples collected in 2000 were filtered with 0.45 μm polycarbonate filters and diluted in the field to prevent precipitation of gypsum. From comparison with results of samples that were untreated in the field, it was inferred that precipitation of gypsum during storage might have lowered the sulfate concentrations by about 7%. Results of four untreated samples were on average 6% lower than those of filtered and diluted equivalents (Vaselli et al. 2003) that were collected on the same dates in 1998–2001.

Similar effects from storage of untreated samples were also attributed to gypsum precipitation by Rowe et al. (1992b).

The samples collected between July 2004 and November 2006 were analyzed for sulfate, chloride and fluoride at OVSICORI-UNA in June–December 2006, using a fully automated microbore ion suppressed chromatographic system (Dionex ICS-3000) (Martínez 2008). The pH measurements were performed on untreated samples at ambient temperature (24 ± 2 °C) using a WTW Multi 340i potentiometer. Combination of the new results with previously available data (e.g. Casertano et al. 1985; Rowe et al. 1992a, b; Nicholson et al. 1992, 1993; Martínez et al. 2000) constitutes a record for major anion concentrations, pH and temperature that covers the period 1980–2006.

Dissolved unreacted SO_2 and H_2S gases in the lake water were measured in situ on an irregular basis during 1999–2006, using a gas detection tube method (Takano et al. 2008). Detection limits for dissolved SO_2 and H_2S were 1 and 0.2 ppm, respectively.

All field-related data (lake volume, temperature, color, seismic records, etc.) are from the database of OVSICORI-UNA (Venzke et al. 2002; Martínez et al. 2000; Martínez 2008). Most of the monthly rainfall data come from the Poás volcano summit rain gauge of the Centro de Servicios y Estudios Básicos de Ingeniería of ICE located at 2,564 m a.s.l.

Polythionates—Relationships with Lake Properties, Fumarolic Activity and Eruptive Phreatic Events

In the following sub-sections, variations in PT distributions and quantities are described in relation to the composition and behavior of the lake, on-shore fumarolic activity, supply of volatiles, eruptive phreatic (phreatomagmatic?) events, monitored seismicity, and microgravity survey for the lake activity stages distinguished by Martínez (2008).

Microgravity and PTs

Microgravity surveys detected an intriguing increase between 1996 and 2001 at stations near the N and NE lakeshores, possibly signaling a magmatic intrusion at depth, which must have been small and fairly localized since most of the other stations showed continuous decreases (Rymer et al. 2005, 2009). Gas release from this putative intrusion may first have induced the sudden re-appearance of large amounts of PTs in the lake in January 1996, followed by their sulfitolysis-induced disappearance due to substantial injection of fresh SO_2 into the lake between early 1998 and late 2002. The occurrence of a new fresh intrusion is also supported by the high C/S ratios and high equilibrium temperatures determined in the volatile phase by Fischer et al. (2015), enhanced CO2 and H2 diffuse emission rates (Melián et al. 2010; Melián et al. Chapter "Diffuse CO_2 Degassing and Thermal Energy Release from Poás Volcano, Costa Rica".), the appearance of low temperature sub-aerial fumaroles onshore the lake, the dome and the eastern sector of the crater, and the sustained increased in microgravity since 2001 until at least 2009 (Fig. 5) (Rymer et al. 2009; Rymer H pers. comm. 2010). Contemporary with these observations is the occurrence of a hydrothermal explosion in April 1996, and a composition of the lake water which was unusually enriched in chloride respect to sulfate between late 1997 and early 2004 (Fig. 6). The enrichment of the lake water in chloride suggests the degassing of a "fresh body of magma" underneath the region that comprises the lake and the dome that started intruding at shallower levels sometime between the mid 1990s (1996–1999) and the early 2000s that might have triggered the 2006–2014 phreatic cycle (Stage V) (Rymer et al. Chapter "Geophysical and Geochemical Precursors to Changes in Activity at Poás Volcano"). A new magmatic intrusion might have started between 2015 and 2016, according to some seismic patterns (recording of discrete short duration harmonic tremor which is unusual at Poás) and chemical changes of the acid lake. We suggest that the phreatomagmatic eruptions of April 2017 is the surface manifestation of the 2015–2016 intrusion event.

Seismicity

Numerous swarms of A-type earthquakes and an intensification of AB-type seismicity between the end of 1996 and the first months of 1997 had preceded the initial increase in volatile concentrations, whereas remarkably sustained volcanic tremors between October 1997 and March 1998 preceded the disappearance of PTs in April 1998. In addition, a short-lived increase in A and B-type seismicity was observed (Fig. 5). Whereas at Poás tremors generally occur in short discontinuous events, a single episode on 21 February 1998 carried on for 2.5 h, and 55 h of tremor were recorded during the whole month. This incidence of sustained tremor possibly reflected a continuous rise of magmatic/ hydrothermal fluids in conduits that on one hand reached the lake bottom, increasing the input of volatiles and heat, and on the other fuelled the new fumaroles onshore around the lake.

The enhancement of hydrothermal activity (sub-aerial fumaroles around the dome, at other locations around the lake and along the lower part of the eastern edge, appearance of thermal springs) within the crater area between 1999 and October 2001 coincided with increased levels of B- and AB-type seismicity after two and a half years of relative quiescence (Fig. 5), suggesting a stronger interaction between the subsurface heat source and the hydrous system beneath the crater. Tremor, which was rare in 1999, took place for less than 30 min per day in September–November 1999 when the plume from the dome reached maximum heights (Fig. 9). Several unusual low-frequency earthquakes with periods of 40–175 s were recorded in this period as well (V Barboza pers comm 2005). An exceptional increase in tremor (a total of 108 h were recorded in September 2001, of which 7% corresponded to monochromatic tremor) coincided with the peak in PT concentrations observed in September–October.

Upwelling

Although the lake had shown upwelling activity during most of sub-stages IVA and IVB, specifically in its central part and near the dome, with sulfur globules emerging from subaqueous vents, convection was more intense and sulfur globules appeared in larger abundances during September 1997–July 1998. In contrast, in September–November 1998 upwelling activity and evaporation were weak, although more vigorous sub-aerial fumaroles continued appearing on the eastern side of the dome. Throughout 2000 and 2001, the lake was entirely covered by an acidic whitish fog (HCl fumes?). In August–September 2001 large bubbles (ca. 3 m in diameter) were observed in the central part of the lake.

Fumarole Activity

From December 1997 and throughout 1998 when the PTs reached maximum concentrations, the gas plume was up to nearly 600 m above the dome. Between February 1999 and June 2000, an unusually strong plume rose from 0.7 to 2 km above the northern fractured side of the dome (the plume was ~2 km height in September 1999) (Fig. 9), and was seen from the capital city of San José at 35 km southeast of the crater.

By late 1998, the first appearance of weak low-temperature fumaroles (around 83–95 °C) was observed in the eastern sector of the crater. In July 1999, long concentric cracks and some ambient and boiling point springs (15–95 °C) appeared here as well, coinciding with a magnitude 3.2 earthquake (Richter scale) that was felt at the summit of the volcano on 18 July. More fumaroles and springs continued appearing in the same area in 2000 and 2001 (Duarte et al. 2003), most of which remained active till 2006–2007. The widening of cracks produced instability and collapses at the eastern terrace. The fumaroles in the SW sector of the crater became more vigorous during the second half of 1999 when outgassing at the dome increased. In early 2000 these fumaroles weakened and eventually disappeared in June 2000 at the same time when degassing at the dome diminished and the gas-vapour plume returned to heights of 100–500 m (Venzke et al. 2002; Mora and Ramírez 2004). Gas condensates from the dome, collected in late 1999-early 2000, showed a significant increase in sulfate, chloride and fluoride concentrations, as the fumarole temperature did (188 °C in March 2000 on the north side of the dome) (Fig. 5) (Vaselli et al. 2003; Fernández et al. 2003).

Gas Fluxes

Infrequent gas remote-sensing measurements carried out during the IVB sub-stage suggest that the SO2 flux was of the same order of magnitude as the average of 90 tons per day recorded in February 1991 (Andres et al. 1992): In February 2001 the minimum average flux of SO_2 was about 40 tons per day (Fournier et al. 2001, 2002), and in March 2002 the flux averaged 61 tons per day (Galle 2002). Between the second half of 1999 and the first part of 2000 an increase in Rn emission through the crater floor was observed (García et al. 2003). Significant increases in soil temperature and diffuse H_2 and CO_2 emissions in the eastern sector of the crater recorded in 2000–2002 were considered as possible precursor of pending volcanic unrest (Melián et al. 2001, 2003, 2004, Chap. 6). Alternatively, the invasion of deep hot permeable zones by meteoric water becoming enriched in volatiles and undergoing redox reactions may explain the strong fumarolic activity around the dome, the opening of the new field in the eastern sector and the H_2 anomalies (Vaselli et al. 2003). Fracturing and subsequent increase in permeability of the volcanic edifice could have been triggered by ground deformation due to the presence of a dense, already crystallized magma body beneath the southern side of the lake (Fournier et al. 2001) or by local tectonic earthquakes (Mora and Ramírez 2004).

Relationship Between PTs and Seismic Activity

Seismic activity at Poás is complex, due to the dynamics of the magmatic and hydrothermal systems, their interactions and the possible combination with tectonic processes (Fernández 1990; Rowe et al. 1992b; Martínez et al. 2000; Rymer et al. 2000). Predominant seismic signals are the B-type or low-frequency earthquakes and the AB-type or intermediate-frequency quakes (Fig. 5). The B-type earthquakes are presumably generated within a few hundred meters from the surface, are concentrated below the crater floor, and may reflect interaction of liquid-gas phases within fracture conduits or bubble formation/collapse in the hydrothermal system. The AB-type quakes were attributed mostly to fumarole-opening events. The A-type or high-frequency earthquakes (volcano-tectonic quakes) and the T-type or volcanic tremor are not as frequently observed as the B- and AB-type quakes, confirming the shallow character of most of Poás' seismic activity. The A-type quakes at Poás were related to tectonic readjustments within the local fault system that may or may not be triggered by regional subduction-related earthquakes (Fernández 1990). They were also associated with fracturing of the brittle carapace of a shallow magma body underlying the hydrothermal system (Casertano et al. 1987), or to mechanical fracturing of country rock, suggesting pressure build up in the system. The T-type seismicity was interpreted as a result of continuous movement of fluids or magma through a rigid medium (Aki et al. 1977; Fernández 1990).

Relationships between seismicity and PTs are difficult to evaluate for Stage II because of the scarcity of data. Despite high seismic activity and a high flux rate of SO_2 from dome fumaroles in the early 1980s, the lake was probably characterized by low PT concentrations with a dominance of pentathionate (Fig. 5), indicative of H_2S-enriched gaseous input and relatively modest activity. This illustrates a particular feature at Poás, namely that the location of major fumarolic outgassing may distort a straightforward relationship between the overall status and nature of degassing of the volcano and the behavior of the lake.

The strong rise in PT concentrations and the shift from pentathionate to tetrathionate dominance in early 1986 were associated with an initial increase in A-, AB- and B-type earthquakes as well as short-duration harmonic tremors after a relatively quiet period. The sharp drop in PT concentrations to below detection limits in early 1987, approximately 3 months before renewal of phreatic activity (Rowe et al. 1992b), coincided with a more conspicuous increase in seismicity, in particular B-type earthquakes. High levels of B-type seismicity remained throughout Stage III when the locus of fumarolic activity was centered within the lake area, and PTs remained virtually below detection limits (Fig. 5). The enhanced heat and volatile output could have been the result of an ascending magma beneath the lake, some time between 1985 and 1986, and/or subsurface hydrofracturing (Fernández 1990; Rowe et al. 1992a, b; Rymer et al. 2005). Peaks in A-type seismicity and tremor in 1990 and 1991 were interpreted in terms of this hydrofracturing, which did promote degassing but without the intrusion of new magma at that time (Fernández 1990; Rymer et al. 2000).

Some of the swarms of A-type quakes in 1994 coincided with large peaks in tremor (Fig. 5), strong degassing and the occurrence of phreatic explosions that may have been associated with renewed magma ascent (Martínez et al. 2000) or fracturing of the brittle envelope of the cooling magma present. Subsequent large variations in PTs contents in the returning lake were accompanied by fluctuations in seismicity.

Changes in the contents and distribution of PTs during Stage IV, reflecting changes in the flow rate and/or in the SO_2/H_2S ratios of subaqueous fumaroles, coincided with periods of either high or low levels of seismicity. Most conspicuous is the fact that the transition from relatively low-to-much higher B- and AB-type activity around 1999 roughly coincided with the sulfitolytic decline of PTs. The enhanced input of SO_2-rich volatiles may thus have been driven by stronger interaction between the magmatic and lake-hydrothermal systems. Following the brief quiet interval of Sub-stage IVC, the sudden rise of PTs, apparently prompted by an increase in volatile input, was accompanied by an increase in AB- and B-type seismicity and by uncommon tremors (Fig. 5). Some of these were also recorded in 1980 when magma was intruding, but in this case, given the lack of A-type earthquakes, the tremors was interpreted as a signal of fluid movement through pre-existing fractures (V Barboza pers. comm. 2005).

The reappearance of PTs in April 2005, their subsequent sulfitolytic or thermal breakdown and the renewal of phreatic explosions all point to the injection of hot SO_2-rich gas promoted by the opening of conduits and the rise of fluids, as signaled by swarms of A-type events and by a dramatic increase in AB-type seismicity that preceded and accompanied the large numbers of volcanic tremors registered throughout Stage V (Fig. 5). The monochromatic character of some of these tremors is usually related to movement of magma or magmatic/hydrothermal fluids at depth.

In summary, changes in the concentrations and speciation of the PTs in response to fluctuations in the rate and composition of gas/volatile input are often accompanied by changes in the activity or type of seismic signals. Increases in seismicity often come together with enhanced input into the lake, leading either to the production of PTs or to their breakdown when the flow rate of volatile and heat release reaches maximum levels. In other cases, the disappearance of PTs can be caused by weakening or shutdown of subaqueous fumarolic gas release, which is then supported by a decrease of seismic activity and changes in the lake properties as clearly observed during sub-stage IVC.

References

Aguilar E, Peterson TC, Ramírez P, Frutos R, Retana JA, Solera M, Soley J, González I, Araujo RM, Santos AR, Valle VE, Brunet M, Aguilar L, Álvarez L, Bautista M, Castañón C, Herrera L, Ruano E, Sinay JJ, Sánchez E, Hernández GI, Obed F, Salgado JE, Vázquez JL, Baca M, Gutiérrez M, Centella C, Espinoza J, Martínez D, Olmedo B, Ojeda CE, Núñez R, Haylock M, Benavides H, Mayorga R (2005) Changes in precipitation and temperature extremes in Central America and northern South America, 1961–2003. J Geophys Res 110(D23):107. https://doi.org/10.1029/2005JD006119

Andres RJ, Barquero JA, Rose WI (1992) New measurements of SO_2 flux at Poás Volcano, Costa Rica. J Volcanol Geoth Res 49:175–177

Armienta MA, De la Cruz-Reyna S, Macías JL (2000) Chemical characteristics of the crater lakes of Popocatepetl, El Chichón, and Nevado de Toluca volcanoes, México. J Volcanol Geoth Res 97:105–125

Barquero JA (1998) Volcán Poás, Costa Rica, 1st edn. San José, Costa Rica. ISBN 9977-12-298-9

Barquero JA, Fernández E (1983) Poás volcanic activity reports: strong continuous gas emission. SEAN (Scientific Event Alert Network) 8(10)

Barquero JA, Malavassi E (1983) Estado de los volcanes, mayo 1982-abril 1983. Bol Vulcanol 13:3 (in Spanish)

Barrancos J, Rosello JI, Calvo D, Padro E, Melián G, Hernández PA, Pérez NM, Millán MM, Galle B (2008) SO_2 emission from active volcanoes measured simultaneously by COSPEC and mini-DOAS. Pure Appl Geophys 165:115–133

Bennett F, Raccichini S (1978a) Nuevos aspectos de las erupciones del volcán Poás. Rev Geogr Am Central 5–6:37–54

Bennett F, Raccichini S (1978b) Subaqueous sulphur lake in Volcán Poás. Nature 271:342–344

Bernard A, Escobar CD, Mazot A, Gutiérrez RE (2004) The acid volcanic lake of Santa Ana volcano, El Salvador. Geol Soc Am Spec Pap 375:121–133

Brown GC, Rymer H, Stevenson DS (1991) Volcano monitoring by microgravity and energy budget analysis. J Geol Soc London 148:585–593

Calvert AS, Calvert PP (1917) A year of Costa Rican natural history. Mac Millan, New York, Chap XVII, p 577

Casadevall T, de la Cruz-Reyna S, Rose WI, Bagley S, Finnegan D, Zoller WH (1984a) Crater lake and post-eruption hydrothermal activity, El Chichón volcano, México. J Volcanol Geoth Res 97:1–30

Casadevall T, Rose WI, Fuller WH, Hart MA, Moyers JL, Woods DC, Chuan RL, Friend JP (1984b) Sulfur dioxide from Poás, Arenal, and Colima volcanoes, Costa Rica and México. J Geophys Res 89:9633–9641

Casas AS, Armienta MA, Ramos SG (2015) Geochemical monitoring of Chichón volcano (México) through sulphur speciation of the crater lake's water. EGU General Assembly 12–17 Apr 2015, Vienna (Austria), p 182

Casertano L, Borgia A, Cigolini C, Morales L, Montero W, Gómez M, Fernández J (1985) Investigaciones geofísicas y características geoquímicas de las aguas hidrotermales: Volcán Poás, Costa Rica. Geofis Int 24:315–332

Casertano L, Borgia A, Cigolini C, Morales LD, Gómez M, Fernández JF (1987) An integrated dynamic model for the volcanic activity at Poás Volcano, Costa Rica. Bull Volcanol 49:588–598

Christenson BW (2000) Geochemistry of fluids associated with the 1995–1996 eruption of Mt. Ruapehu, New Zealand: signatures and processes in the magmatic-hydrothermal reservoir. J Volcanol Geoth Res 97:1–30

Christenson BW, Wood CP (1993) The evolution of a volcano-hosted hydrothermal system beneath Ruapehu Crater Lake, New Zealand. Bull Volcanol 55:547–565

Day AL, Allen ET (1925) The volcanic activity and hot springs of Lassen Peak. Carnegie Institute of Washington Publications, p 360

Debus H (1888) Chemical investigation of Wackenroder's solution, and explanation of the formation of its constituents. J Chem Soc Trans 53:278–357

Delmelle P (1995) Geochemical, isotopic and heat budget study of two volcano hosted hydrothermal systems: the acid crater lakes of Kawah Ijen, Indonesia, and Taal, Philippines, volcanoes. PhD thesis, Université Libre de Bruxelles (Belgium)

Delmelle P, Bernard A (1994) Geochemistry, mineralogy, and chemical modelling of the acid crater lake of Kawah Ijen Volcano, Indonesia. Geochim Cosmochim Acta 58:2445–2460

Delmelle P, Bernard A (2000) Volcanic lakes. In: Sigurdson H, Houghton B, McNutt SR, Rymer H, Stix J (eds) Encyclopedia of volcanoes. Academic Press, USA

Delmelle P, Bernard A (2015) The remarkable chemistry of sulfur in hyper acid crater lakes: a scientific tribute to Bokuichiro Takano and Minoru Kusakabe. In: Rouwet D, Christenson B, Tassi F, Vandemeulebrouck J (eds) Volcanic lakes. Advances in volcanology. https://doi.org/10.1007/978-3-642-36833-2_10

Delmelle P, Bernard A, Kusakabe M, Fischer TP, Takano B (2000) Geochemistry of the magmatic-hydrothermal system of Kawah Ijen volcano, East Java, Indonesia. J Volcanol Geoth Res 97:31–53

Druschel GK, Hamers RJ, Banfield JF (2003) Kinetics and mechanism of polythionate oxidation to sulfate at low pH by O_2 and Fe^{3+}. Geochim Cosmochim Acta 67:4457–4469

Fernández M (1990) Actividad del Volcán Poás, Costa Rica: Análisis sísmico durante el período 1980–1989. Msc thesis, Escuela Centroamericana de Geología, Universidad de Costa Rica, San José, Costa Rica (in Spanish)

Foss O (1960) Structures of compounds containing chains of sulphur atoms: sulfur chains terminated by sulfonate groups, the polythionates. In: Emeléus HJ, Sharpe AG (eds) Advances in inorganic chemistry and radiochemistry, vol 2. Elsevier Academic Press, NY, USA

Fournier N, Rymer H, Williams-Jones G, Brenes J (2004) High-resolution gravity survey: investigation of subsurface structures at Poás Volcano, Costa Rica. Geophys Res Lett 31:L15602. https://doi.org/10.1029/2004GL020563

Fujiwara Y, Ohsawa S, Watanuki K, Takano B (1988) Determination of polythionates in an active crater lake by nitrate ion-selective electrode. Geochem J 22:249–256

Giggenbach WF (1974) The chemistry of Crater Lake, Mt Ruapehu (New Zealand) during and after the 1971 active period. New Zealand J Sci 17:33–45

Goehring M (1952) Die Chemie der Polythionsäuren. Fortschr Chem Forsch Bd. 2 S:444–483

Hynek BM, Rogers KL, Antunovich M, Avard G, Alvarado GE (2018) Lack of microbial diversity in an extreme Mars analog setting: Poás volcano, Costa Rica. Astrobiol 18(7)

IUPAC (1983) Solubility data series. Sulfur dioxide, chlorine, fluorine and chlorine oxides. In: Young CL (ed) vol 12. Pergamon, Oxford, UK

IUPAC (1988) Solubility data series. Hydrogen Sulfide, Deuterium Sulfide and Hydrogen Selenide. In: Fogg PGT, Young CL (eds) vol 32. Pergamon, Oxford, UK

Janitzki J (1969) Chemistry of polythionates and selenopolythionates. Acc Chem Res 2. https://doi.org/10.1021/ar50022a005

Koh T (1990) Analytical chemistry of polythionates and thiosulfate. A review. Anal Sci 6:3–14

Krushensky RD, Escalante G (1967) Activity of Irazú and Poás volcanoes, Costa Rica, November 1964–July 1965. Bull Volcanol 31:75–84

MacLaurin JS (1911) Occurrence of pentathionic acid in natural waters. Proc Chem Soc London 27:10–12

Malavassi E, Barquero J (1982) Excursión al Volcán Poás. Proceed 1st US - Costa Rica joint seminar in volcanology. OVSICORI-UNA: Heredia-Costa Rica. Bol Vulcanol 14:119–131

Martínez M (2008) Geochemical evolution of the acid crater lake of Poás Volcano (Costa Rica): insights into volcanic-hydrothermal processes. PhD thesis, Universiteit Utrecht, The Netherlands, 161 pp

Martínez M, Fernández E, Valdés J, Barboza V, van der Laat R, Duarte E, Malavassi E, Sandoval L, Barquero J, Marino T (2000) Chemical evolution and volcanic activity of the active crater lake of Poás Volcano, Costa Rica, 1993–1997. J Volcanol Geoth Res 97:127–141

Martínez M, Jiménez JD, Pereira R, Trescott SC, Pacheco JF, Porras H, Brenes G, Vega J, Herrera J (2017) Volcanes Poás y Rincón de la Vieja: Geoquímica y textura de piroclastos eruptados en las explosiones freatomagmáticas del 2017. In: Book of abstracts First Central America Congress on Earth Sciences 2017 Universidad Nacional Costa Rica

McNutt S (2000) Seismic monitoring. In: Sigurdsson H (ed) Encyclopaedia of volcanoes. Academic Press, California, USA

Miura Y, Kawaoi A (2000) Determination of thiosulphate, thiocyanate and polythionates in a mixture by ion-pair chromatography with ultraviolet absorbance detection. J Chrom A 884:81–87

Mora R, Ramírez C (2004) Physical changes in the hyperacidic hot lake of Poás volcano (2002–2004, Costa Rica). 6th Meeting IAVCEI committee on volcanic lakes, Caviahue, Argentina, Nov 2004

Nicholson RA, Howells MF, Roberts PD, Baxter PJ (1992) Gas geochemistry studies at Poás Volcano, Costa Rica, Nov 1991. British Geol Surv Overseas Geology Series, Technical report No WC/92/10

Nicholson RA, Howells MF, Baxter PJ, Clegg SL, Barquero J (1993) Gas geochemistry studies at Poás Volcano, Costa Rica, March 1992–January 1993. British Geol Survey, Overseas Geology Series, Technical report No WC/93/21

Ohba T, Hirabayashi J, Nogami K (2008) Temporal changes in the chemistry of lake water within Yugama Crater, Kusatsu-Shirane volcano, Japan: implications for the evolution of the magmatic hydrothermal system. J Volcanol Geoth Res 178:131–144

Oppenheimer C, Stevenson D (1989) Liquid sulfur lakes at Poás Volcano. Nature 342:90–793

OVSICORI (1995) Boletín Estado de los Volcanes 1995. Open report. http://www.ovsicori.una.ac.cr/index.php/vulcanologia/estado-de-los-volcanes/category/44-estado-de-los-volcanes-1995?download=535:estado-volcanes-1995. Accessed in Nov 2015

OVSICORI (1996) Boletín Estado de los Volcanes 1996. Open report. http://www.ovsicori.una.ac.cr/index.php/vulcanologia/estado-de-los-volcanes/category/43-estado-de-los-volcanes-1996?download=534:estado-volcanes-1996. Accessed in Nov 2015

OVSICORI (2005) Boletín Estado de los Volcanes Enero 2005. Open report. http://www.ovsicori.una.ac.cr/index.php/vulcanologia/estado-de-los-volcanes/category/20-estado-de-los-volcanes-2005?download=184:estado-volcanes-enero-2005. Accessed in Nov 2015

OVSICORI (2006) Experto en medición remota de SO_2 en plumas volcánicas de la Universidad de El Salvador comparte experiencia con vulcanólogos del OVSICORI-UNA, Costa Rica. Open report. http://www.ovsicori.una.ac.cr/informes_prensa/2006/Experto_en_medicion_de_SO2_nos_visita25abril2006.pdf

Pasternack GB, Varekamp JC (1994) The geochemistry of the Keli Mutu crater lakes, Flores, Indonesia. Geoch J 28:243–262

Pasternack GB, Varekamp JC (1997) Volcanic lake systematics I. Physical constraints. Bull Volcanol 58:534–535

Prosser JT, Carr MJ (1987) Poás volcano, Costa Rica: geology of the summit region and spatial and temporal variations among the most recent lavas. J Volcanol Geoth Res 33:131–146

Rodríguez A (2016) Volcanic lake systems as terrestrial analogue for sulphate rich terrains on Mars. PhD

thesis, Universiteit Utrecht, The Netherlands. ISBN/EAN:978-90-6266-436-8
Rodríguez A, van Bergen MJ (2015) Volcanic hydrothermal systems as potential analogues of Martian sulphate rich terrains. Ned J Geosci. http://dx.doi.org/10.1017/njg.2015.12
Rodríguez A, van Bergen MJ (2017) Superficial alteration mineralogy in active volcanic systems: an example of Poás volcano, Costa Rica. J Volcanol Geoth Res 346:54–80
Rouwet D, Tassi F, Mora R, Sandri L, Chiarini V (2014) Past, present and future of volcanic lake monitoring. J Volcanol Geoth Res 272:78–97
Rouwet D, Mora-Amador R, Sandri L, Ramírez-Umaña C, González G, Pecoraino G, Capaccioni B (Chapter 9 of this book) 39 years of geochemical monitoring of Laguna Caliente crater lake, Poás: patterns from the past as keys for the future. In: Tassi F, Mora-Amador R, Vaselli O (eds) Poás volcano (Costa Rica): the pulsing heart of Central America Volcanic Zone. Springer, Heidelberg, Germany
Rowe GL, Brantley SL, Fernández JF, Barquero J, Borgia A (1989) Observaciones preliminares del sistema hidrotermal del Volcán Poás, Costa Rica. Bol Vulcanol OVSICORI-UNA 20:23–31 (in Spanish)
Rowe GL, Brantley SL, Fernández M, Fernández JF, Borgia A, Barquero J (1992a) Fluid volcano interaction in an active stratovolcano: the crater lake system of Poás Volcano, Costa Rica. J Volcanol Geoth Res 49:23–51
Rowe GL, Ohsawa S, Takano B, Brantley SL, Fernández JF, Barquero J (1992b) Using crater lake chemistry to predict volcanic activity at Poás Volcano, Costa Rica. Bull Volcanol 54:494–503
Rymer H, Cassidy J, Locke C, Barboza V, Barquero J, Brenes J, van der Laat R (2000) Geophysical studies of the recent 15-year eruptive cycle at Poás Volcano, Costa Rica. J Volcanol Geoth Res 97:425–442
Rymer H, Fournier N, Williams-Jones G (2004) Persistent activity and its effect on microgravity variations at Poás volcano, Costa Rica. 2004 IAVCEI meeting, Pucón Chile, Nov 2004
Rymer H, Locke C, Brenes J, Williams-Jones G (2005) Magma plumbing processes for persistent activity at Poás volcano, Costa Rica. Geophys Res Lett 32, L08307. https://doi.org/10.1029/2004gl022284
Rymer H, Locke CA, Borgia A, Martínez M, van der Laat R, Williams-Jones G (2009) Long term fluctuations in volcanic activity: implications for future environmental impact. Terra Nova 21:304–309
Shriver DF, Atkins PW (1999) The nitrogen and oxygen groups. In: Inorganic chemistry, 3rd edn. Oxford University Press, UK
Sigurdsson H (1977) Chemistry of the crater lake during the 1971–72 Soufriére eruption. J Volcanol Geoth Res 2:165–186
Smolyaninov V, Shekhvatova G, Vainshtein M (2014) Gold leaching by organic base polythionates: new non toxic and secure technology. SpringerPlus 3:180. https://doi.org/10.1186/2193-1801-3-180
Sriwana T, van Bergen MJ, Varekamp JC, Sumarti S, Takano B, van Os BJH, Leng MJ (2000) Geochemistry of the acid Kawah Putih lake, Patuha Volcano, West Java, Indonesia. J Volcanol Geoth Res 97:77–104
Steudel R, Holdt G (1986) Ion-pair chromatographic separation of polythionates $S_nO_6^{2-}$ with up to thirteen sulfur atoms. J Chromatogr 361:379–384
Stimac JA, Goff F, Counce D, Larocque ACL, Hilton DR, Morgenstern U (2003) The crater lake and hydrothermal system of Mount Pinatubo, Philippines: evolution in the decade after eruption. Bull Volcanol 66:149–167
Stoiber RE, Williams SN, Huebert BJ (1986) Annual contribution of sulfur dioxide to the atmosphere by volcanoes. J Volcanol Geoth Res 33:1–8
Sugimori K, Takano B, Matsuo M, Suzuki K, Fazlullin SM (1995) Activity of sulphur-oxidizing bacteria in the acidic crater lake of Maly Semiachik Volcano, Kamchatka. In: Kharaka YK, Chudaev OV (eds) Proceedings of the 8th international symposium water-rock interaction, Vladivostik, Russia. Balkema, Rotterdam, The Netherlands
Sugimori K, Martínez M, Malavassi E, Fernández E, Duarte E, Segura J, Sáenz W, van Bergen MJ, Valdés J (2001) Microorganisms living in the acid lake of Poás volcano. 54th Annual meeting, Balneology Society Japan, Shirahama, Wakayama, 22–25 Aug 2001
Sugimori K, Igarashi H, Martínez M, Duarte E, Fernández E, Malavassi E, van Bergen MJ, Segura J, Valdés J (2002) Microbial life in the acid lake and hot springs of Poás volcano. Colima volcano international meeting 2002, Colima, Mexico
Symonds RB, Rose WI, Gerlach TM, Briggs PH, Harmon RS (1990) Evaluation of gases, condensates, and SO_2 emissions from Augustine volcano, Alaska: the degassing of a Cl-rich volcanic system. Bull Volcanol 52:355–374
Symonds RB, Gerlach TM, Reed MH (2001) Magmatic gas scrubbing: implications for volcano monitoring. J Volcanol Geoth Res 108:303–341
Takano B (1987) Correlation of volcanic activity with sulfur oxyanion speciation in a crater lake. Science 235:1633–1635
Takano B, Watanuki K (1988) Quenching and liquid chromatography determination of polythionates in natural water. Talanta 35:847–854
Takano B, Watanuki K (1990) Monitoring of volcanic eruptions at Yugama Crater Lake by aqueous sulfur oxyanions. J Volcanol Geoth Res 40:71–87
Takano B, Saitoh H, Takano E (1994a) Geochemical implications of subaqueous molten sulfur at Yugama Crater Lake, KusatsuShirane volcano, Japan. Geochem J 28:199–216
Takano B, Ohsawa S, Glover RB (1994b) Surveillance of Ruapehu crater lake, New Zealand, by aqueous polythionates. J Volcanol Geoth Res 60:29–57
Takano B, Matsuo M, Suzuki K (1995) Bathymetry and chemical investigation of crater lake at Maly

Semiachik Volcano, Kamchatka. In: Kharaka YK, Chudaev OV (eds) Proceedings of the 8th international symposium water-rock interaction, Vladivostik, Russia. Balkema, Rotterdam, The Netherlands
Takano B, Koshida M, Fujiwara Y, Sugimori K, Takayanagi S (1997) Influence of sulfur-oxidizing bacteria on the budget of sulfate in Yugama crater lake, Kusatsu-Shirane volcano, Japan. Biogeochemistry 38:227–253
Takano B, Maekawa T, Zheng Q (2001) Kinetic study on aqueous polythionates and its application to active crater lake systems. In: Cidu R (ed) Proceedings of the 10th international symposium water-rock interaction, Balkema, Rotterdam, The Netherlands, pp 927–930
Takano B, Suzuki K, Sugimori K, Ohba T, Fazlullin SM, Bernard A, Sumarti S, Sukhyar R, Hirabayashi M (2004) Bathymetric and geochemical investigation of Kawah Ijen Crater Lake, East Java, Indonesia. J Volcanol Geoth Res 135:299–329
Takano B, Kuno A, Ohsawa S, Kawakami H (2008) Aqueous sulfur speciation possibly linked to sublimnic volcanic gas-water interaction during a quiescent period at Yugama crater lake, Kusatsu–Shirane volcano, Central Japan. J Volcanol Geoth Res 178:145–168
Tassi F, Vaselli O, Fernández E, Duarte E, Martínez M, Delgado Huertas A, Bergamaschi F (2009) Morphological and geochemical features of crater lakes in Costa Rica: an overview. J Limnol 68(2). doi.org/ https://doi.org/10.4081/jlimnol.2009.193
Varekamp JC, Ouimette AP, Herman SW, Bermúdez A, Delpino D (2001) Hydrothermal element fluxes from Copahue, Argentina: a "beehive" volcano in turmoil. Geology 29:1059–1062
Vaselli O, Tassi F, Fischer TP, Tardani D, Fernandez Soto E, Duarte E, Martinez M, De Moor MJ, Bini G (Chapter 10) The last eighteen years (1998–2015) of fumarolic activity at the Poás volcano (Costa Rica) and the renewed activity. In: Tassi F, Mora-Amador R, Vaselli O, (eds) Poás volcano (Costa Rica): the pulsing heart of Central America Volcanic Zone. Springer, Heidelberg, Germany
Vaselli O, Tassi F, Minissale A, Montegrossi G, Duarte E, Fernández E, Bergamaschi F (2003) Fumarole migration and fluid geochemistry at Poás Volcano (Costa Rica) from 1998 to 2001. In: Oppenheimer C, Pyle DM, Barclay J (eds) Volcanic degassing, vol 213. Geological Society of London, Special Publications, pp 247–262
Venzke E, Wunderman RW, McClelland L, Simkin T, Luhr JF, Siebert L, Mayberry G (eds) (2002) Global volcanism, 1968 to the present. Smithsonian Institute, Global Volcanism Program Digital Information Series, GVP-4 (http://www.volcano.si.edu/reports/). Accessed in Feb 2005
Williams-Jones G, Stix J, Heiligmann M, Charland A, Sherwood Lollar B, Garzón V, G, Barquero J, Fernández E (2000) A model of diffuse degassing at three subduction-related volcanoes. Bull Volcanol 62:130–142
Wilson SH (1941) Natural occurrence of polythionic acids. Nature 148:502–503
Wilson SH (1953) The elemental investigation of the hot springs of the New Zealand thermal region, vol 2. South Pacific Sci Congress, New Zealand, pp 449–469
Wilson SH (1959) Physical and chemical investigations (White Island) 1939–1955. New Zealand Dept Sci Ind Res Bull 127:32–50
Xia J, Rumpf B, Maurer G (1999) The solubility of sulfur dioxide in aqueous solutions of sodium chloride and ammonium chloride in the temperature range from 313 to 393 K at pressures up to 3.7 MPa: experimental results and comparisons with correlations. Fluid Phase Equilib 165:99–119
Xia J, Pérez A, Rumpf B, Maurer G (2000) Solubility of hydrogen sulfide in aqueous solutions of single strong electrolytes sodium nitrate, ammonium nitrate, and sodium hydroxide at temperatures from 313 to 393 K and total pressures up to 10 MPa. Fluid Phase Equilib 167:263–284
Xu Y, Schoonen MAA, Nordstrom DK, Cunningham KM, Ball JW (2000) Sulfur geochemistry of hydrothermal waters in Yellowstone National Park, Wyoming, USA. II. Formation and decomposition of thiosulfate and polythionate in Cinder Pool. J Volcanol Geoth Res 178:145–168

Supplementary References

Aki K, Fehler M, Das S (1977) Source mechanism of volcanic tremor: fluid-driven crack models and their application to the 1963 Kilauea eruption. J Volcanol Geoth Res 2:259–287
Andres RJ, Barquero JA, Rose WI (1992) New measurements of SO_2 flux at Poás Volcano, Costa Rica. J Volcanol Geoth Res 49:175–177
Casertano L, Borgia A, Cigolini C, Morales L, Montero W, Gómez M, Fernández J (1985) Investigaciones geofísicas y características geoquímicas de las aguas hidrotermales: Volcán Poás, Costa Rica. Geofis Int 24:315–332 (in Spanish with English abstract)
Casertano L, Borgia A, Cigolini C, Morales LD, Gómez M, Fernández JF (1987) An integrated dynamic model for the volcanic activity at Poás Volcano, Costa Rica. Bull Volcanol 49:588–598
Duarte E, Fernández E, Sáenz W, Malavassi E, Martínez M, Barquero J (2003) Wall instability and fumarole migration from 1999 to 2002, Poás Volcano, Costa Rica. In: Proceedings of the 8th field workshop on volcanic gases Nicaragua and Costa Rica 2003. IAVCEI-CCVG
Fernández M (1990) Actividad del Volcán Poás, Costa Rica: Análisis sísmico durante el período 1980–1989. Msc thesis, Escuela Centroamericana de Geología, Universidad de Costa Rica, San José (Costa Rica) (in Spanish)

Fernández E, Duarte E, Sáenz W, Malavassi E, Barboza V, Martínez M, Valdés J (2003) Volcanic gas condensates from Poás Volcano 1999–2000. Proceedings of the 8th field workshop on volcanic gases Nicaragua and Costa Rica 2003. IAVCEI-CCVG

Fischer TP, Ramírez C, Mora RA, Hilton DR, Barnes JD, Sharp ZD, Le Brun M, de Moor JM, Barry PH, Füri E, Shaw AM (2015) Temporal variations in fumarole gas chemistry at Poás volcano, Costa Rica. J Volcanol Geoth Res 294:56–70

Fournier N, Williams-Jones G, Rymer H (2001) Sulphur budget at Poás Volcano. EOS Trans AGU 82(47), Fall Meet Suppl, Abstract V22E-12

Fournier N, Williams-Jones G, Rymer H (2002) Sulphur budget at Poás Volcano, Costa Rica: a multidisciplinary approach. In: Proceedings of the volcanic and magmatic studies group annual meeting, 3–4 Jan 2002, University College London, England, UK

Galle B (2002) Preliminary report from studies of volcanic SO_2 emission in Nicaragua and Costa Rica, Mar 2002, using DOAS spectroscopy. Department of Radio and Space Science, Chalmers University of Technology, S-412 96 Gothenburg, Sweden

García R, Fernández E, Duarte E, Castro C, Monnin M, Malavassi E, Barboza V (2003) Radon gas concentration monitoring: anomalies correlation found with volcanic activity after year 2000. In: Proceedings of the 8th field workshop on volcanic gases Nicaragua and Costa Rica 2003. IAVCEI-CCVG

Goehring M, Feldmann U (1948) Neue Verfahren zur Darstellung von Kaliumpentathionat und von Kaliumhexathionat. Z Anorg Allg Chem 257:223–226

Martínez M (2008) Geochemical evolution of the acid crater lake of Poás Volcano (Costa Rica): insights into volcanic-hydrothermal processes. PhD thesis, Universiteit Utrecht, The Netherlands, 161 pp

Martínez M, Fernández E, Valdés J, Barboza V, van der Laat R, Duarte E, Malavassi E, Sandoval L, Barquero J, Marino T (2000) Chemical evolution and volcanic activity of the active crater lake of Poás Volcano, Costa Rica, 1993–1997. J Volcanol Geoth Res 97:127–141

Melián G, Galindo I, Salazar J, Hernández P, Pérez N, Ramírez C, Fernández M, Notsu K (2001) Spatial and secular variations of diffuse CO_2 degassing from Poás Volcano, Costa Rica, Central America. EOS Trans AGU 82(47), Fall Meet Suppl Abstract V31A-0951

Melián GV, Galindo I, Salazar J, Pérez N, Hernández P, Fernández M, Ramírez C, Mora R, Alvarado GE (2003) Anomalous changes in diffuse hydrogen emission from Poás Volcano, Costa Rica, Central America: a premonitory geochemical signature of volcanic unrest? In: Proceedings of the 8th field workshop on volcanic gases Nicaragua and Costa Rica 2003. IAVCEI-CCVG

Melián GV, Galindo I, Pérez N, Hernández P, Salazar J, Fernández M, Ramírez C, Mora R, Alvarado GE (2004) Emisión difusa de hidrógeno en el Volcán Poás, Costa Rica, América Central. In: Soto GJ, Alvarado GE (eds) La Vulcanología y su entorno geoambiental. Rev Geol Am Central 30:167–177 (in Spanish with English abstract)

Melián GV, Pérez NM, Mora-Amador R, Hernández PA, Ramírez C, Sumino H, Alvarado GE, Fernández M (Chapter 6) Diffuse CO_2 degassing and thermal energy release from Poás volcano, Costa Rica. In: Tassi F, Mora-Amador R, Vaselli O (eds) Poás volcano (Costa Rica): the pulsing heart of Central America Volcanic Zone. Springer, Heidelberg, Germany

Melián GV, Pérez N, Hernández PA, Nolasco D, Marrero R, Fernández M, Ramírez C, Mora R, Alvarado GE (2010) Emisión difusa de CO_2 y actividad volcánica en el volcán Poás, Costa Rica. Rev Geol Am Central 43:147–169 (in Spanish with English abstract)

Miura Y, Kawaoi A (2000) Determination of thiosulphate, thiocyanate and polythionates in a mixture by ion-pair chromatography with ultraviolet absorbance detection. J Chrom A 884:81–87

Mora R, Ramírez C (2004) Physical changes in the hyperacidic hot lake of Poás volcano (2002–2004, Costa Rica). In: Proceedings of the 6th meeting IAVCEI committee on volcanic lakes, Nov 2004, Caviahue, Argentina

Nicholson RA, Howells MF, Roberts PD, Baxter PJ (1992) Gas geochemistry studies at Poás Volcano, Costa Rica, Nov 1991. British Geol Survey, Overseas Geology Series, Technical report No. WC/92/10

Nicholson RA, Howells MF, Baxter PJ, Clegg SL, Barquero J (1993) Gas geochemistry studies at Poás Volcano, Costa Rica, March 1992–January 1993. British Geol Survey, Overseas Geology Series, Technical report No. WC/93/21

Rowe GL, Ohsawa S, Takano B, Brantley SL, Fernández JF, Barquero J (1992a) Using Crater Lake chemistry to predict volcanic activity at Poás Volcano, Costa Rica. Bull Volcanol 54:494–503

Rowe GL, Brantley SL, Fernández M, Fernández JF, Borgia A, Barquero J (1992b) Fluid volcano interaction in an active stratovolcano: the crater lake system of Poás Volcano, Costa Rica. J Volcanol Geoth Res 49:23–51

Rowe GL (1994) Oxygen, hydrogen, and sulfur isotope systematics of the crater lake system of Poás Volcano, Costa Rica. Geochem J 28:263–287

Rymer H, Cassidy J, Locke C, Barboza V, Barquero J, Brenes J, van der Laat R (2000) Geophysical studies of the recent 15-year eruptive cycle at Poás Volcano, Costa Rica. J Volcanol Geoth Res 97:425–442

Rymer H, Locke C, Brenes J, Williams-Jones G (2005) Magma plumbing processes for persistent activity at Poás volcano, Costa Rica. Geophys Res Lett 32. https://doi.org/10.1029/2004gl022284

Rymer H, Locke C, Borgia A, Martínez M, van der Laat R, Williams-Jones G (2009) Long term fluctuations in volcanic activity: implications for future environmental impact. Terra Nova 2:304–309

Stamm H, Goehring M, Feldmann U (1942) Zur Kenntnis der Polythionsäuren un ihrer Bildung-Neue Verfahren zur Darstellung von Kaliumtrithionat und von Kaliumtetrathionat. Z Anorg Allg Chem 250:226–228

Takano B (1987) Correlation of volcanic activity with sulfur oxyanion speciation in a crater lake. Science 235:1633–1635

Takano B, Watanuki K (1988) Quenching and liquid chromatography determination of polythionates in natural water. Talanta 35:847–854

Takano B, Watanuki K (1990) Monitoring of volcanic eruptions at Yugama Crater Lake by aqueous sulfur oxyanions. J Volcanol Geoth Res 40:71–87

Takano B, Ohsawa S, Glover RB (1994) Surveillance of Ruapehu crater lake, New Zealand, by aqueous polythionates. J Volcanol Geoth Res 60:29–57

Takano B, Kuno A, Ohsawa S, Kawakami H (2008) Aqueous sulfur speciation possibly linked to sublimnic volcanic gas-water interaction during a quiescent period at Yugama crater lake, Kusatsu–Shirane volcano, Central Japan. J Volcanol Geoth Res 178:145–168

Vaselli O, Tassi F, Minissale A, Montegrossi G, Duarte E, Fernández E, Bergamaschi F (2003) Fumarole migration and fluid geochemistry at Poás Volcano (Costa Rica) from 1998 to 2001. In: Oppenheimer C, Pyle DM, Barclay J (eds) Volcanic degassing, vol 213. Geological Society London, Special Publications, pp 247–262

Venzke E, Wunderman RW, McClelland L, Simkin T, Luhr JF, Siebert L, Mayberry G (eds) (2002) Global volcanism, 1968 to the present. Smithsonian Institution, Global Volcanism Program

Geophysical and Geochemical Precursors to Changes in Activity at Poás Volcano

Hazel Rymer, María Martínez, Jorge Brenes, Glyn Williams-Jones and Andrea Borgia

Abstract

Acidic crater lakes at persistently active volcanoes act as both a moderator and an index of volcanic processes. Poás exhibits cyclic behavior characterized by calm periods alternating with periods of phreatic eruptions that last 2–10 years, which roughly correspond to liquid or vapor dominance. However, what causes the system to move on from one phase to the other remains unclear. By integrating the insights gained from different methods of monitoring, a view of the physical and chemical changes occurring before, during and after alterations in activity can be gleaned.

H. Rymer (✉) · A. Borgia
Department of Environmental, Earth and Ecosystems, Faculty of Science, The Open University, Milton Keynes, UK
e-mail: h.rymer@open.ac.uk

M. Martínez · J. Brenes
Volcanological and Seismological Observatory of Costa Rica Universidad Nacional OVSICORI-UNA, Heredia, Costa Rica

G. Williams-Jones
Department of Earth Sciences, Simon Fraser University, Burnaby, Canada

A. Borgia
SAS—Geological Sciences, Rutgers University, Piscataway, NJ 08854, USA

Keywords

Poás volcano · Geophysical and geochemical precursors · Volcanic monitoring · Phreatic activity

1 Introduction

In the early 1990s, the crater lake at Poás disappeared, and acid aerosols were erupted straight into the atmosphere rather than being buffered by the lake. This had a locally significant effect on public health, the environment and economy. The first indications of this event were retrospectively identified from 1985 in a unique 20 years geophysical and geochemical data time series, which provides evidence for: (1) shallow magma intrusion or, (2) fracturing of the brittle chilled envelope around cooling magma (Rymer et al. 2005), or (3) rupture of an impermeable layer in the rock column between the chilled margin and the acid lake. Data (locations of data sampling shown in Fig. 1) subsequently showed similar trends (Rymer et al. 2005) and it was suggested that Poás had entered another pre-eruption period with enhanced volatile supply due to renewed magmatic intrusion or carapace breaking. Severe environmental damage in this region was predicted within the next two years if the current trend were to continue. The trend did indeed continue and the heightened activity at Poás seems likely to have been fed by a magma intrusion which was first identified through the

F. Tassi et al. (eds.), *Poás Volcano*, Active Volcanoes of the World,
https://doi.org/10.1007/978-3-319-02156-0_8

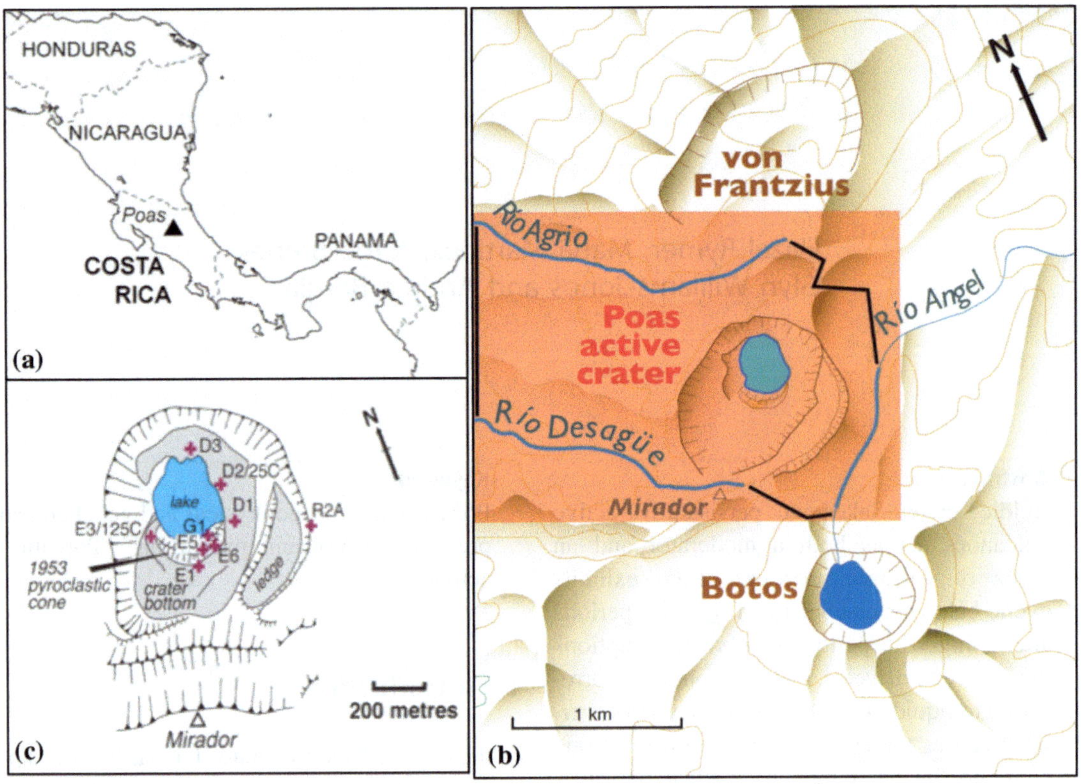

Fig. 1 Map locating Poás volcano in Costa Rica. **a** Map showing location of Poás volcano marked by triangle, in Costa Rica. **b** Topographic map (contour interval 100 m) showing the active crater in relation to surrounding water courses and older craters. The region shown lies within the northern part of a larger summit caldera structure. The groundwater flow model used to correct the gravity data extends between the rivers shown in blue; the area between the rivers at the edge of the model is assumed to be impermeable boundaries (shown in black). **c** Locations of the gravity stations and features of the crater including the pyroclastic cone erupted in 1952. Reference gravity station is located several kilometres south of the active region. From Rymer et al. (2009)

appearance of large amounts of polythionates in the hyper-acid lake (Martínez et al. Chapter "Behaviour of Polythionates in the Acid Lake of Poás Volcano: Insights into Changes in the Magmatic-Hydrothermal Regime and Subaqueous Input of Volatiles" and references therein) and microgravity data.

2 Sub-surface Structure

A high definition Bouguer gravity survey of the shallow sub-surface structure within the crater region at Poás revealed that the magma body feeding the conduit beneath the active crater is slightly offset towards the NNE (Fournier et al. 2004). This is a vestige of the system, which fed earlier activity, now evidenced in the eroded remains of the von Franzius cone. The region of 'cooled magma' identified by Fournier et al. (2004), shown in Fig. 2, is the zone within which density changes have been detected using microgravity measurements over a period in excess of 20 years (Rymer et al. 2005, 2009). The region is thought to comprise a complex zone some 500 m thick (Rymer and Brown 1989) beneath the present crater that gives way to a fractured carapace capping the magma.

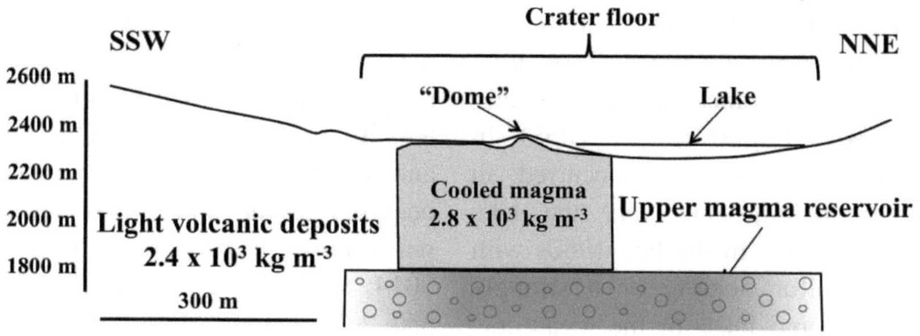

Fig. 2 Schematic cross-section of the active crater at Poás volcano inferred from a high-resolution Bouguer gravity survey. Different material densities are indicated in kg/m^3. The cooled magma body is offset from the actual hyper-acid crater lake. The "dome" is referred here as the CPC. From Fournier et al. (2004)

3 Activity Over the Last 30 Year Period

A cycle of peak activity at Poás began in mid 1980-early 1981 when fumarole temperatures rose to 1020 °C, and a red glow appeared on the northern face of a composite pyroclastic cone (CPC) that is located on the southern edge of the hyper-acid lake. A-type seismicity, high sulfate levels, fumarole temperatures on the CPC and power output are consistent with a gradual intrusion(s?) bringing gas-rich magma close to the surface beneath the CPC between early 1980/early 1981. Estimates of falling power output and sulfate levels from 1983 indicate that the effects of the intrusive episode(s?) were beginning to wane by this time (Rymer et al. 2009). Lake level fell over the next few years until the lake bottom was exposed in 1989. Activity in the late 1980s and early 1990s was consequently characterized by intense degassing and sulfur eruptions (Oppenheimer and Stevenson 1989). Acid gases and particles rained out across the southern and western regions of the volcano, seriously impacting everyday life in nearby residential communities, agriculture and tourism.

By 1995, the lake had re-established to its pre-crisis level. The lake level then remained higher than at any time since the late 1970s until early 2005, when the level again began to fall. Fumarolic activity in the crater walls developed in the mid-late 1990s and intensified over the next 10 years (Martínez 2008). By 2006, energetic phreatic eruptions [i.e. with column heights >150 m above the lake surface and seismically induced displacements higher than 2 cm^2/s (de Moor et al. 2016; McNutt 2002)] began to occur again from the deep lake, ejecting lake sediment and pyrite-rich globular sulfur, plastering the inner walls of the main active crater. During a particularly intense phreatic explosion in March 2006, fumaroles on the eastern inner crater walls and around the CPC, the latter located on the southern edge of the acid lake, simultaneously emitted a distinct pulse of vapor, suggesting a direct, shallow depth coupling of the whole hydrothermal system.

Phreatic activity resumed in March 2006 after a 10 years period of relative quiescence. The peak activity was preceded by enhanced levels of seismicity, lake and fumarole temperature increase, and increases in the concentration of dissolved volatiles and rock forming elements. Phreatic explosions were observed in March,

April, September and December 2006. No phreatic explosions were observed throughout 2007 although degassing and convection within the lake remained strong (Martínez 2008). In 2008, only one phreatic eruption occurred, in January, and throughout that year the lake resembled its appearance in the late 1980s with slicks of molten sulfur floating on a strongly convecting grey-colored lake. In the meantime, the CPC developed some fumaroles with temperatures slightly above boiling (109 °C) signaling a transition of the volcanic system toward its monophasic period dominated by degassing. It is noteworthy that between 1988 and 2007, fumarole temperatures around the CPC remained around the boiling point (95 °C).

In January 2009, two shallow earthquakes occurred along fault systems close to Poás: the M4.5 event on 7 January near Fraijanes Sabanilla, Alajuela, and the M6.2 event on 8 January near Cinchona, Poás Alajuela (Pacheco and Segura 2009). Over the next few months, convection cells within the hyper-acid lake appeared to become more vigorous and the acidity and concentration of dissolved volatiles and rock forming elements increased further. Phreatic activity was scarce during 2009, however, on 12 January, 2009 a small phreatic eruption occurred that may have been triggered by the Cinchona earthquake. Later on that year, a large eruption occurred suddenly on 25 December, 2009 (OVSICORI 2009). In June 2009, the CPC started to show an increase in the heat output and level of degassing, with temperatures as high as 123 °C measured at some of the accessible fumaroles (OVSICORI 2009). By September 2009, the CPC fumaroles had reached ~500 °C with degassing columns rising to ~1 km in height. The lake was characterized by stronger convective activity and a rapid drop in its level and appeared to be repeating part of a cycle previously seen in the late 1980s when the lake disappeared. Although the lake did not disappear, over the next few years, the CPC fumarole temperatures remained high (~800 °C in 2010, 890 °C in 2011, and 600 °C in 2013 and 2014) and occasional energetic phreatic explosions occurred in the lake. After several explosions with column heights exceeding 150 m in October 2014, the level of seismicity and fumarole and lake temperatures decreased and subsequently, Poás returned to a period of 20 months of relative quiescence that has been interrupted by renewed phreatic activity in June 2016.

4 Geophysical and Geochemical Measurements

A range of physical observations was made throughout this period and some continue to the present day. The names of key monumented locations in and around the crater are shown in Fig. 1. Details of the methodologies and rationale for interpretation of the data sets are given in the bibliography in the caption and the changes observed are shown in Fig. 3. The observed gravity changes are those before the correction for water table effects has been made. Although application of this correction, also shown in Fig. 3, does not alter the general trend of the observations, it does in some cases magnify the changes.

There is a clear cyclicity in some of the data (Rymer et al. 2009). The microgravity data collected between 1985 and 2008 and expressed relative to a reference point off the summit area (corrected for Earth tides, deformation and water table fluctuations) shows two characteristic signatures (Fig. 3a, b). Gravity variations at stations in the southern part of the crater (E1, E3/125c, E5, E6 and G1; Fig. 3c) are clearly related as are the variations at the north crater stations (D2/2a/25c and D3/3a; Fig. 3b, d). Gravity variations at station D1 (in the east) correlate with those at the southern stations until 1994, with the northern stations from 2000 to 2004, and finally with the southern stations again until 2008. The volume, temperature, and sulfate concentration of the acid lake over the same period are also shown in Fig. 3 with estimates of the power output of the system from 1978 to 1989. Throughout this period, ground deformation in the summit region, measured by EDM theodolite and dual frequency GPS, was negligible (i.e., <a few cm deflation) (Rymer et al. 2000).

The gravity changes greater than 30 microGal are deemed significant (Rymer et al. 2009) and since there is negligible deformation and the changes in water table have been removed, these residual gravity changes must reflect subsurface mass changes. An increase in gravity would reflect a mass increase while a gravity decrease reflects a mass decrease. Gravity increases seen at Poás in this context could therefore reflect mass accumulation, which might be magmatic intrusion. Conversely, the gravity decreases reflect mass wastage, possibly dissolution by the hydrothermal system or downwards convection of magma.

The lake volume and temperature also showed fluctuations, which demonstrated some cyclicity as well as significant variations in the sulfate concentration in the lake. In 2009, Rymer et al. (2009) predicted that if these trends continued, there would be another period of environmental degradation due to the increased degassing of the volcanic system without the buffering effect of the lake. It was suggested that the situation had been brought about by pulses of magmatic or hydrothermal fluids into the conduit beneath the southern lake area or beneath the CPC (marked as 1953 pyroclastic cone in Fig. 1c). It was also noted that the gravity changes preceded changes in the lake chemistry and increased explosive activity by about 4 years; the concentration of polythionates appear to signal the arrival of fresh sulfur-rich magmatic gas even before detection of the full subsurface mass changes (Martínez et al. Chapter "Behaviour of Polythionates in the Acid Lake of Poás Volcano: Insights into Changes in the Magmatic-Hydrothermal Regime and Subaqueous Input of Volatiles"). A summary of the processes potentially responsible for the observations during the whole period is presented in Fig. 4.

Although there have been no further gravity measurements since 2009, continued observations of the crater lake show that the phreatic cycle persisted. A time series, which includes that shown in Fig. 3 and extends to the present time (Fig. 5), demonstrates that the sulfate concentration continued to increase substantially from 2010 and the pH remains low. The level of hydrothermal activity has been consistently high throughout this recent period and the sulfate to fluoride ratio has risen to levels not seen since the late 1980s when the system was so active that the hydrothermal system became depleted.

Analysis of variations in the chemical and isotopic composition of fumaroles at Poás over the period 2001–2014 have been interpreted in terms of an injection of magma in late 2000 which then led to a hydrofracturing event in 2006 (Fischer et al. 2015). This interpretation is consistent with evidence from the gravity data, which also suggest that the injection (considered to be a dendritic intrusion) started in 2000. However, the behavior of polythionates observed in the acid lake suggests that the injection of a fresh batch of magma at depth might have occurred even earlier, sometime between 1996 and 1999 (Martínez et al. Chapter "Behaviour of Polythionates in the Acid Lake of Poás Volcano: Insights into Changes in the Magmatic-Hydrothermal Regime and

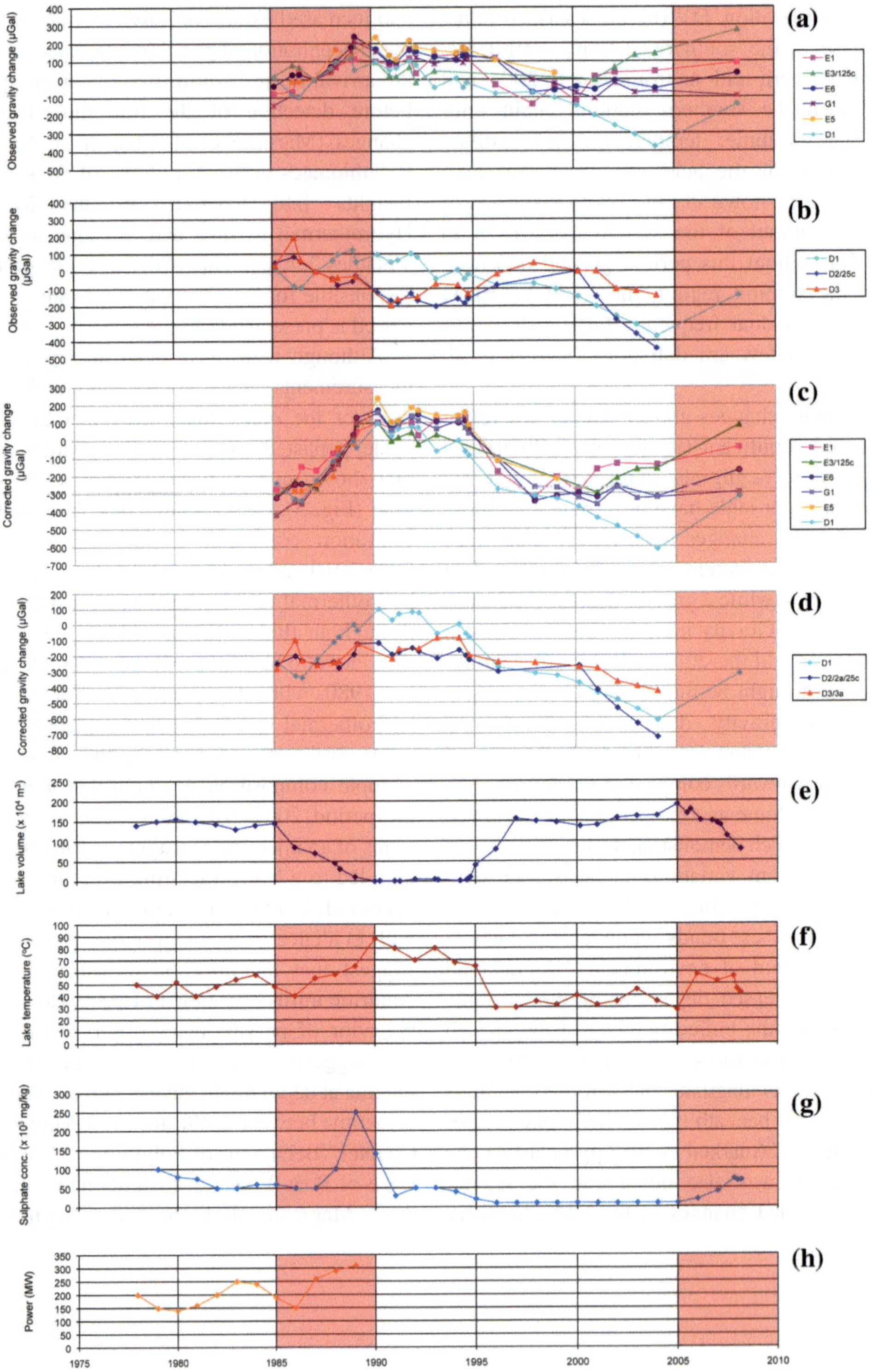
(a)
Observed gravity change (μGal)
E1
E3/125c
E6
G1
E5
D1
(b)
Observed gravity change (μGal)
D1
D2/25c
D3
(c)
Corrected gravity change (μGal)
E1
E3/125c
E6
G1
E5
D1
(d)
Corrected gravity change (μGal)
D1
D2/2a/25c
D3/3a
(e)
Lake volume (x 10⁴ m³)
(f)
Lake temperature (°C)
(g)
Sulphate conc. (x 10³ mg/kg)
(h)
Power (MW)
1975
1980
1985
1990
1995
2000
2005
2010

◄ **Fig. 3** Variations between 1978 and 2008 in **a** observed gravity data from stations in the southern crater corrected for Earth tides and deformation; **b** observed gravity data from stations in the northern crater corrected for Earth tides and deformation; **c** corrected gravity data from stations in the southern crater corrected for Earth tides, deformation and variations in lake level and water table; **d** corrected gravity data from stations in the northern crater corrected for Earth tides, deformation and variations in lake level and water table; **e** lake volume; **f** lake temperature; **g** sulfate concentration; and **h** power output. These graphs include previously published data from Rymer et al. (2009) and Casertano et al. (1987), Brown et al. (1989), Martinez et al. (2000), Rymer et al. (2005), OVSICORI (2008). At times of low lake volume, temperatures of the vestiges of the lake are given. Power output calculations are based on using the lake as a calorimeter (Brown et al. 1989) and using measurements of temperature of lake and air, volume, surface area, wind speed and rainfall. Volcanic activity in 1989 resulted in the closure of the meteorological station on Poás and so data are not available to calculate power output since that time. The repeat gravity survey was carried out using previously published techniques (Rymer et al. 2000). Data are corrected for Earth tide and calibration effects; the resulting uncertainty in the data presented in Fig. 2 is approx. 20 μGal. Lake temperature, sulfate concentrations (mostly determined by ion chromatography) and lake volume data include data from Rymer et al. (2009), Rowe et al. (1992), Martinez et al. (2000)

Date	Observations	Interpretations		
		Magmatic Processes	Hydrothermal Processes	
1948-1952	• Intermittent phreatic explosions and intense fumarolic activity			
1953-1955	• Magmatic eruption, growth of CPC	Intrusion and formation of CPC	Extensive vapour zone	
1968-1980	• Activity focussed on lake • Vigorous geysering (phreatic explosions) • Lake level high • Intense swarm of A-type earthquakes in July 1980	Intrusion cooling.	The aquifer floods the conduit. Vapour zone persists around intrusion.	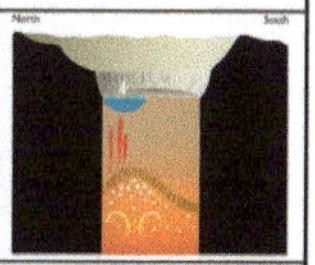
1981-1984	• Focus of activity moved to CPC • Intense fumarole activity and incandescence on CPC • Lake level remained high • No A-type earthquakes • Sharp rise in harmonic tremor and B-type earthquakes in March 1981 which declined through 1985.	*This marks the beginning of a new magmatic cycle* Breakage of the magma carapace and intrusion beneath CPC.	Vapour zone begins to vent substantially to the atmosphere.	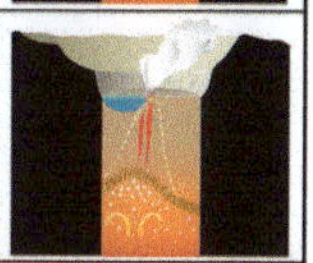
1985-1990	• Focus of activity moves from CPC back to lake • Phreatic activity continued • Lake level fell revealing sulphur pools • Increasing B-type earthquakes; some A-type swarms in 1989-1991 • Deformation in crater <5cm • Gravity increased by up to 600 μGal in south and 100 μGal in north • Sulphate concentration increased increased from 1987 to a maximum of 250 x 10³mg/kg and then declined • Lake temperature increased to 90°C	Ongoing intrusion perhaps now beneath lake.	Vapour zone vents directly to the atmosphere, including through the lake bottom. Volume of vapour zone may shrink due to large loss of thermal energy.	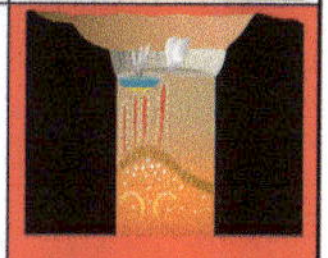
1990-1995	• Lake essentially dry • Intense degassing and acidification of the environment • Negligible deformation • Gravity in south decreased by ~100 μGal and then stabilised; remained stable in north and decreased by 200 μGal in east • Sulphate concentration decreased and then stabilised at 50 x 10³mg/kg • Lake temperature decreased; briefly falling to 26°C in November 1995	Possible drainage of magma beneath CPC and lake.	Vapour zone continues to vent directly to the atmosphere, shrink and flood at depth	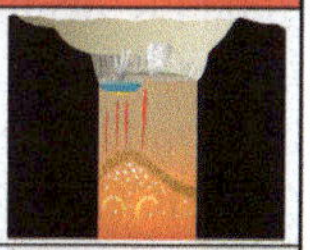
1995-2000	• Activity confined to fumaroles in crater; new fumaroles develop on the CPC, inner crater walls and the crater floor • Lake level rose rapidly • Gravity in south decreased by ~400 μGal and then stabilised; gravity stable at 100 μGal below 1990-1995 level in north and east • Sulphate concentration low and stable at ~10 x 10³mg/kg • Lake temperature decreased to ~40°C	Possible drainage of magma beneath CPC and lake.	Vapour zone reforms at depth	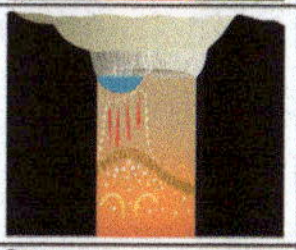
2000-2005	• Fumarolic activity focussed on the innercrater walls, mainly the eastern wall and later beneath the lake • Phreatic eruptions recommenced • Lake level rose rapidly • Gravity in south decreased by ~400 μGal and then stabilised; gravity stable at 100 μGal below 1990-1995 level in north and east • Sulphate concentration low and stable at ~10 x 10³mg/kg • Lake temperature decreased to ~40°C	*2nd cycle begins* Breakage of the magma carapace and intrusion west/south of CPC.	Vapour zone grows toward the north and east, and begins to vent to the atmosphere.	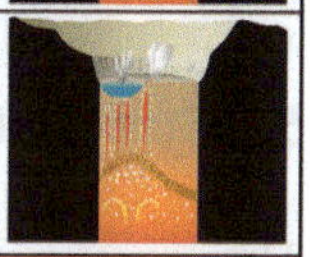
2005-2008	• Intermittent vigorous phreatic eruptions • Fumaroles in the northern and eastern side of the crater weakened after 2006 • Lake level fell rapidly • Deformation negligible - just 3cm subsidence locally on crater floor • Gravity increased by ~250 μGal in south • Sulphate concentration increasing to 98 x 10³mg/kg • Lake temperature increased to ~58°C	Ongoing intrusion west of CPC; intrusion east of CPC.	Vapour zone venting and shrinking	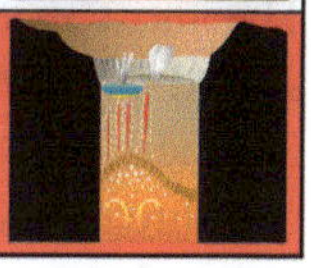

Fig. 4 Summary of observations from Poás volcano from 1953 to 2008 interpreted in terms of cyclical magmatic and hydrothermal processes. Includes data from Martinez et al. (2000), Rymer et al. (2005), OVSICORI (2008). Cartoons (some after Rymer et al. 2005); North–South sections through the active crater area showing the relationship between the lake and the CPC. Crater is approx. 1 km across and 200 m deep. Molten gas-rich magma is capped by a fractured carapace. Upwelling and downwelling magma dendrites transport heat and mass. Broken line indicates schematically the extent of the vapor zone. Variations in lake level and fumarolic activity are also shown. From Rymer et al. (2009)

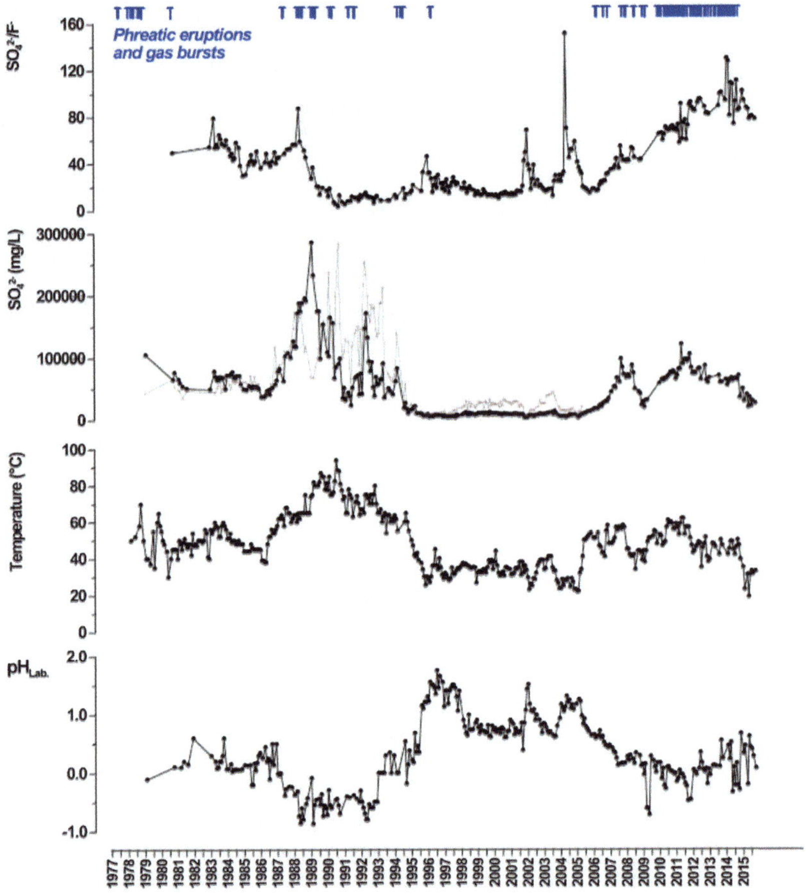

Fig. 5 Lake observations from 1979 to 2015. Blue arrows indicate periods of significant hydrothermal activity (geysering). The ratio of sulfate to fluoride, the sulfate concentration (mg/l), temperature and pH are plotted against date

Subaqueous Input of Volatiles"). For about a year from October 2014, the lake level was relatively high and the level of activity was consistently low (OVSICORI 2015). Presently the cycle is continuing with phreatic eruptions and the periodic disappearance of the lake, with the accompanying environmental degradation.

5 Summary

The most recent cycle of phreatic activity that started in 2006 and lasted ca. 8 years until October 2014 and was characterized by numerous phreatic events, with the highest frequency of events between 2010 and October 2014. Combining the geophysical and geochemical evidence, there appears to have been an intrusive event, which triggered a cycle of activity characterized by increasingly active fumaroles, increased temperatures and phreatic explosions. The evidence is consistent with dendritic (dyke) intrusions of fresh magma to shallow levels beneath the lake and composite pyroclastic cone region sometime in the mid-1990s-early 2000s.

Over the last ca. 40 years, Poás has experienced 4 episodes of phreatic activity that alternate with relatively calm periods in cycles of about 2–10 years. These seem to correspond to marked changes in the regime of the magmatic-hydrothermal system that feeds the lake and the fumaroles of the active crater leading to dominance of either vapor (dendritic intrusions) or liquid zones (when the dendritic intrusions have cooled sufficiently), respectively, within the magmatic-hydrothermal system. A new cycle of phreatic activity started in June 2016, after 20 months of relative quiescence.

References

Brown GC, Rymer H, Dowden J, Kapadia P, Stevenson D, Barquero J, Morales LD (1989) Energy budget analysis for Poás crater lake: implications for predicting volcanic activity. Nature 339:370–373

Casertano L, Borgia A, Cigolini C, Morales LD, Montero W, Gomez M, Fernandez JF (1987) An integrated dynamic model for the volcanic activity at Poás volcano, Costa Rica. Bull Volcanol 49:588–598

de Moor JM, Aiuppa A, Pacheco J, Avard G, Kern C, Liuzzo M, Martínez M, Giudice G, Fischer TP (2016) Short-period volcanic gas precursors to phreatic eruptions: insights from Poás volcano, Costa Rica. Earth Planet Sci Lett 442:218–227

Fischer TP, Ramírez C, Mora-Amador RA, Hilton DR, Barnes JD, Sharp ZD, Le Brun M, de Moor JM, Barry PH, Füri E, Shaw AM (2015) Temporal variations in fumarole gas chemistry at Poás volcano, Costa Rica. J Volcanol Geothermal Res 294:56–70

Fournier N, Rymer H, Williams-Jones G, Brenes J (2004) High-resolution gravity survey: investigation of subsurface structures at Poás Volcano, Costa Rica. Geophys Res Lett 31:L15602. https://doi.org/10.1029/2004gl020563

Martínez M (2008) Geochemical evolution of the acid crater lake of Poás Volcano (Costa Rica): insights into volcanic-hydrothermal processes. Ph.D. thesis, Universiteit Utrecht, The Netherlands. 161 pp

Martinez M, Fernandez E, Valdes J, Barboza V, Van der Laat R, Duarte E, Malavassi E, Sandoval L, Barquero J, Marino T (2000) Chemical evolution and volcanic activity of the active crater lake of Poás volcano, Costa Rica, 1993–1997. J Volcan Geotherm Res 97:127–141

Martínez M, van Bergen MJ, Takano B, Fernández E, Barquero J (Chapter 7) Long term behaviour of polythionate polymers in the hyperacid crater lake of Poás volcano: insights into the subaqueous input of magmatic gases. In: Tassi F, Mora-Amador R, Vaselli O (eds) Poás volcano (Costa Rica): the pulsing heart of Central America Volcanic Zone. Springer, Heidelberg (Germany)

McNutt S (2002) Volcano seismology and monitoring for eruptions. In: Lee W, Kanamori H, Jennings P, Kisslinger C (eds) International handbook of earthquake and engineering seismology, 81A, pp 383–406

Oppenheimer C, Stevenson D (1989) Liquid sulfur lakes at Poás volcano. Nature 342:790–793

OVSICORI (2008) http://www.ovsicori.una.ac.cr/vulcanologia/estado_volcanes.htm

OVSICORI (2009) Open reports: Estado de los Volcanes de Costa Rica. http://www.ovsicori.una.ac.cr/index.php/vulcanologia/estado-de-los-volcanes/category/4-ev2015#

OVSICORI (2015) Open reports: Estado de los Volcanes de Costa Rica. http://www.ovsicori.una.ac.cr/index.php/vulcanologia/estado-de-los-volcanes/category/4-ev2015#

Pacheco J, Segura J (2009) El sismo de Cinchona del 8 de enero del 2009. Open report OVSICORI-UNA. http://www.ovsicori.una.ac.cr/index.php/extension/informes-prensa/category/29-informes-de-prensa-2009?download=346:el-sismo-de-cinchona-del-8-de-enero-del-2009

Rowe GL, Brantley SL, Fernandez JF, Borgia A, Barquero J (1992) Fluid volcano interaction in an active stratovolcano: the crater lake system of Poás volcano, Costa Rica. J Volcan Geotherm Res 49:23–51

Rymer H, Brown GC (1989) Gravity changes as a precursor to volcanic eruption at Poás volcano, Costa Rica. Nature 342:902–905

Rymer H, Cassidy J, Locke C, Barboza V, Barquero J, Brenes J, van der Laat R (2000) Geophysical studies of the recent 15-year eruptive cycle at Poás volcano, Costa Rica. J Volcanol Geotherm Res 97:425–442

Rymer H, Locke CA, Brenes J, Williams-Jones G (2005) Magma plumbing processes for persistent activity at Poás Volcano, Costa Rica. Geophys Res Lett 32: L08307. https://doi.org/10.1029/2004GL022284

Rymer H, Locke CA, Borgia A, Martínez M, Brenes J, van der Laat R, Williams-Jones G (2009) Long-term fluctuations in volcanic activity: implications for future environmental impact. Terra Nova 21:304–309

39 Years of Geochemical Monitoring of Laguna Caliente Crater Lake, Poás: Patterns from the Past as Keys for the Future

Dmitri Rouwet, Raúl Alberto Mora Amador, Laura Sandri, Carlos Ramírez-Umaña, Gino González, Giovannella Pecoraino and Bruno Capaccioni

Abstract

Since 1978 water chemistry of the Laguna Caliente crater lake has been used to monitor volcanic activity at Poás, Costa Rica, making it arguably the best studied hyper-acidic crater lake on Earth. During these 39 years, three phases of unrest occurred, manifested through frequent phreatic eruptions, with each a duration of several years to over a decade (1978–1980, 1986–1996, 2006–2016). We here present a novel technique to deal with the long time series of the chemical composition of water of Laguna Caliente, independent on previous deterministic research and resulting conceptual models. Common patterns of chemical parameters in relation with phreatic eruptive activity for the period 1978–September 2014 are sought, applying the objective statistical method of Pattern Recognition. This resulted in the definition of the strongest precursory signals and their respective thresholds. Numerical outcomes often confirm findings based on geochemical models (e.g. SO_4, SO_4/Cl and pH are strong monitoring parameters). However, some surprising parameters (opposite behavior of Mg/Cl ratios, decreases in Ca and Mg concentrations, increasing Al/Mg ratios) still need a geochemical explanation and should be a focus for future research strategies. The obtained parameters and thresholds were retrospectively applied for the "test period" of the Pattern Recognition method (November 2014–February 2016). This test provided hints that suggested that eruptive activity at Poás was not yet over, despite apparent quiescence in early 2016. Indeed, after new phreatic eruptions since May 2016, the 2006–2016 phreatic eruptive cycle culminated in phreatomagmatic activity in April 2017. We conclude that evaluating time series of chemical composition of crater lakes framed in the Pattern Recognition method can be a useful monitoring approach. Moreover, increased sampling frequency can provide more

The author Bruno Capaccioni deceased.

D. Rouwet (✉) · L. Sandri
Istituto Nazionale di Geofisica e Vulcanologia, Sezione di Bologna, Italy
e-mail: dmitri.rouwet@ingv.it

D. Rouwet · G. Pecoraino
Istituto Nazionale di Geofisica e Vulcanologia, Sezione di Palermo, Italy

R. A. Mora Amador · C. Ramírez-Umaña
Escuela Centroamericana de Geología, Universidad de Costa Rica, San José, Costa Rica

R. A. Mora Amador
PREVENTEC, Universidad de Costa Rica, San José, Costa Rica

C. Ramírez-Umaña
Red Sismológica Nacional, Universidad de Costa Rica, San José, Costa Rica

G. González
Volcanes Sin Fronteras-VSF, San José, Costa Rica

B. Capaccioni
BIGEA, Università degli Studi di Bologna, Bologna, Italy

F. Tassi et al. (eds.), *Poás Volcano*, Active Volcanoes of the World,
https://doi.org/10.1007/978-3-319-02156-0_9

details and more adequate prediction of phreatic activity at Poás. Comparing Laguna Caliente with other two well monitored acidic crater lakes (Ruapehu Crater Lake, New Zealand and Yugama, Kusatsu-Shirane, Japan) Poás results unique in many ways and undoubtedly the most active crater lake of the three during the past four decades.

Keywords
Poás volcano · Laguna Caliente · Lake water chemistry · Volcano monitoring · Phreatic eruptions · Unrest · Pattern recognition

1 Introduction

Tracking variations in the chemical-physical properties of waters from peak-activity crater lakes has been the most obvious way to monitor the underlying actively degassing magmatic-hydrothermal systems (Rowe et al. 1992a, b; Christenson 2000; Ohba et al. 2008; Rouwet et al. 2014a). Crater lakes offer the advantage to preserve changes in the degassing regime by absorption of gases into the lake water (i.e. scrubbing, Symonds et al. 2001), whereas open-conduit volcanoes instantaneously lose these markers to the atmosphere. The acidic gas species are released from magma in a specific order depending on their solubility ($CO_2 < SO_2 < HCl < HF$), and have their own behavior when they hit the water phase (Rouwet et al. 2017), depending on the strength of the acid and thermodynamic and chemical properties in water. They share the capacity to acidify water to more or less extent, and to heat the overlying crater lake as the gas is accompanied by vapor that condenses in the colder lake (Hurst et al. 2015). This is the reason why anion concentrations, water temperature and acidity (pH) are the most indicative parameters to track during monitoring setups (Varekamp et al. 2000), as they directly reflect degassing dynamics.

Volcano monitoring intrinsically implies a time window of observation that should be synchronized with the kinetics of magmatic processes, such as degassing, magma intrusion and (phreatic) eruption. To decipher "how much time ago" a variation in degassing regime actually occurred before eventually being observed in the lake water is key, and depends on the duration the lake water resides in the basin before being flushed out by infiltration, or being lost through evaporation (residence time dependent monitoring time window, Rouwet et al. 2014a). Crater lakes are prototype settings for phreatic eruptive activity, as water is excessively available in these "wet volcanoes" (i.e. pressurized hydrothermal systems; Caudron et al. 2015). Volcanologists agree on the fact that the physical-chemical response of the crater lake upon a dynamically changing degassing regime will reveal the enigmatic precursor for so-far unpredictable phreatic eruptions (Rouwet et al. 2014a, b; de Moor et al. 2016). This, though, is rather another assumption than a fact. Many efforts have been made to better understand the complex dynamics and trigger mechanisms of phreatic eruptions (Browne and Lawless 2001; Rouwet and Morrissey 2015, and references therein). Beyond the usefulness of these models, a general consent does not, and probably will never exist, as phreatic eruptions are probably instigated through several mechanisms that can probably even vary for a single volcano.

The above reasoning assumes that gas is preserved as anions in the lake water (HCO_3^-, CO_3^{2-}, SO_4^{2-}, Cl^-, F^-). In other words, scrubbing of acid gases is complete and irreversible (Symonds et al. 2001). Nevertheless, recent work has confirmed, by direct MultiGAS measurements from evaporative plumes coming off lakes, that even the strongest acids in liquid medium (i.e. HCl, SO_2) degas from hyper-acidic crater lakes (Shinohara et al. 2015; Tamburello et al. 2015; Gunawan et al. 2016; de Moor et al. 2016, for Aso, Copahue, Kawah Ijen and Poás crater lakes, respectively). The stronger acid HCl has long been recognized as being more volatile than hydrophyle in extremely acidic solutions (pH near 0, or lower), in geothermal wells (Truesdell et al. 1989), or through a long-term steady increase in SO_4/Cl ratios in the vigorously evaporating crater lake of Poás volcano, the study object of this volume (Rowe et al. 1992a). We now know that acidic gases flush through hyper-acidic crater lake brines, but we don not know to which extent (completely or partially?), and with which speed (kinetics?). The chemical composition of the lake water hence only reflects a transient phase of the gas flushing through the lake. In terms of volcanic surveillance this brings the advantage that the monitoring time window is definitely shorter than defined by the water mass balance, but yet, we do not know how much shorter. Following a review on HCl degassing from water bodies by Rouwet and Ohba (2015), empirical experiments by Capaccioni et al. (2016) have tried to tackle this "kinetic problem" for HCl degassing from a "lab-lake" on the short-term (2 days). The monitoring rules established since decades, based only on water chemistry, have thus somehow become obsolete, and need revision.

With this state-of-the-art in mind, two new monitoring strategies can be proposed to seek precursory signals of phreatic eruptions from crater lakes:

(1) Tracking variations in gas chemistry from the evaporative plume above lakes can only be insightful if related to the chemistry and thermodynamic conditions of the water, in order to time-frame the process of gas flushing through the lake towards the atmosphere;
(2) The second method adopts a time window ideally synchronized with the speed and dynamics of phreatic eruptions, and sticks to the classical principle in geology of "the past is the key for the future". How did lake chemistry parameters vary during the various stages of unrest and eruption, on a purely mathematical basis? Can we recognize patterns in the numerical values related to the changes in volcanic activity?

The aim of this chapter is to give an overview of the three stages of unrest and phreatic eruptive activity at Poás volcano, Costa Rica (1978–1980, 1986–1995 and 2005–2016), by compiling data available from literature and personal research (Rouwet et al. 2016, and unpublished data for 2011–2016). A fluid geochemical study by Rouwet et al. (2016) on the most recent unrest period has demonstrated the limits of the "classical approach" and the need for the alternative strategies described above, whereas the first MultiGAS analyses on Poás showed some trends with phreatic activity, but remain observations rather than an explicative cause-effect relation (de Moor et al. 2016).

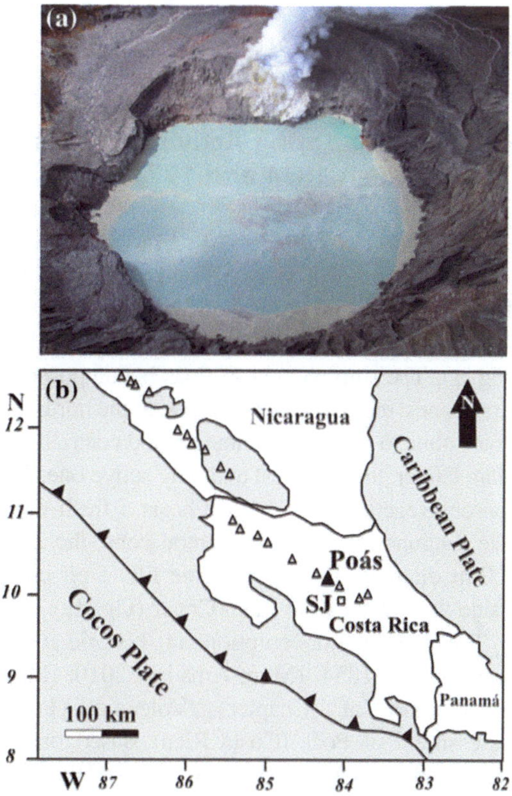

Fig. 1 **a** Aerial view of Poás' Laguna Caliente during the 2006–2016 period of eruptive unrest. Picture by R. M-A. The yellow-grayish patches are floating mats of sulfur spherules. The diameter of Laguna Caliente is approximately 300 m. Strong fumarolic degassing from the 1953–1955 dome is evident; **b** location map of Poás volcano

We here apply a pattern recognition procedure (Mulargia et al. 1991; Sandri et al. 2004) on the Poás data base for Laguna Caliente (Fig. 1a) water chemistry, and, in addition, to a similar data base of two other well-monitored and erupting crater lake hosting volcanoes: Yugama of Kusatsu-Shirane volcano (Japan), and Crater Lake of Ruapehu volcano (New Zealand). Pattern recognition can lead to defining thresholds for signals of unrest, or precursors for phreatic eruptions, that can become useful for geochemical and probabilistic monitoring. The above-mentioned strategy (1) is left open for future research, on Poás and elsewhere, and is suspected to become a major "white paper" topic regarding actively degassing crater lakes.

2 Unrest and Phreatic Eruptions at Poás

2.1 Pre-monitoring Activity of Poás Between 1834 and 1978

Poás (10°12′00″N–84°13′58″W; 2,708 m a.s.l.) is a structurally complex stratovolcano located within the Cordillera Volcánica Central of Costa Rica, ~35 km northwest of the capital San José (Fig. 1). The summit area of Poás is composed of three cones: the Von Frantzius cone to the north, the Botos cone to the south-southeast, and centrally the Main Crater, the youngest and only active one. The densely vegetated Botos cone hosts a fresh-water lake, Laguna Botos. In the central cone, the 250–300 m diameter Laguna Caliente fills a pit crater inside the 800 m wide Main Crater (Fig. 1a).

The first major eruption in historic times occurred in 1834 (Mora-Amador 2010; Mora Amador et al. Chapter "Volcanic Hazard Assessment of Poás (Costa Rica) Based on the 1834, 1910, 1953–1955 and 2017 Historical Eruptions"). All the activity since the beginning of the twentieth century took place at the Main Crater. A major Surtseyan eruption occurred on January 25, 1910, when a dark tephra column rose up ~8 km above the lake (Rudín et al. 1910; Mora-Amador 2010; Mora Amador et al. Chapter "Volcanic Hazard Assessment of Poás (Costa Rica) Based on the 1834, 1910, 1953–1955 and 2017 Historical Eruptions"). Prior to 1951, a hot grey lake (T from 39 to 64 °C) filled the bottom of the crater. Phreatic eruptions at Laguna Caliente took place in 1953–1955 (Raccichini and Bennett 1978; Casertano et al. 1983; Prosser and Carr 1987; Brown et al. 1989; Mora-Amador 2010; Mora Amador et al. Chapter "Volcanic Hazard Assessment of Poás (Costa Rica) Based on the 1834, 1910, 1953–1955 and 2017 Historical Eruptions"). On May 17, 1953, after a period of phreatic eruptions (March–April 1953), a Surtseyan phreatomagmatic eruption expulsed Laguna Caliente. This activity, accompanied by mild Strombolian explosions and small lava flows, led to a dome extrusion within Laguna Caliente (Mora Amador et al. Chapter "Volcanic Hazard Assessment of Poás (Costa Rica) Based on the 1834, 1910, 1953–1955 and 2017 Historical Eruptions", and references therein). By August 1954 the dome had grown 20–30 m in height and remained since then the main locus of fumarolic activity. This eruptive phase lasted until 1955. Afterwards, cycles of major phreatic activity occurred, sometimes accompanied by crater lake dry-out during periods of enhanced heat flow (1953–1965, 1973–1979; Brantley et al. 1987; Brown et al. 1989; Martínez et al. 2000; Mora-Amador et al. 2004; Martínez 2008; Mora-Amador 2010; Mora Amador et al. Chapter "Volcanic Hazard Assessment of Poás (Costa Rica) Based on the 1834, 1910, 1953–1955 and 2017 Historical Eruptions"). Below we describe the three unrest stages since the start of institutionalized monitoring.

2.2 Between 1978 and 1980

Since the start of a monitoring program the first phreatic explosions at Poás occurred in 1978, breaching a grey-colored Laguna Caliente and expulsing molten sulfur droplets and spherules (Mora Amador et al. Chapter "The Extraordinary Sulfur Volcanism of Poás from 1828 to 2018"). Degassing from the lake and 1953–1955 dome continued, but decreased steadily, without phreatic activity. In July 1980 seismic activity indicated hydrofracturing of the brittle margins of a shallow magma body. The last phreatic explosion occurred in September 1980.

2.3 Between 1986 and 1996

A period of high-temperature degassing from the 1953–1955 dome from mid-1981 (1020 °C) to May 1985 (420 °C) probably marked a phase of efficient heat, gas and water mass dispersion from the magmatic-hydrothermal plumbing system throughout Laguna Caliente (Brantley et al. 1987; Brown et al. 1989). As such, during this period no phreatic explosions occurred (Martínez 2008).

A new unrest phase was marked by earthquake swarms (February–May 1986) and heating of dome fumaroles up to 800 °C (early-1987; Martínez 2008). Contemporaneously, the temperature of Laguna Caliente increased from 58 °C (January 1987) to 70 °C (June 1987), coinciding with a lake level drop. Phreatic eruption activity resumed in June 1987, steadily culminating towards strong explosions in April–June 1988 (up to 1–2 km above the lake surface). Lake level drastically declined until complete lake disappearance in 1989. Oppenheimer and Stevenson (1989) and Oppenheimer (1992) described the small sulfur lakes and ponds and spectacularly erupting pyroclastic sulfur cones at the dried up 1989 Laguna Caliente bottom (https://www.youtube.com/watch?v=TUvkqsldfcQ). During the following years (1989–1992), the lake only re-appeared during the rainy season. Phreatic eruptions and "sulfur volcanism" intensified the impact of the Poás eruptions on the surrounding areas (Mora Amador et al. Chapter "The Extraordinary Sulfur Volcanism of Poás from 1828 to 2018"). Strong phreatic eruptions expulsed the at that time shallow Laguna Caliente 6 h after a Mw 7.0 tectonic earthquake (Nicoya, 170 km SW of Poás) in March 1990 (Martínez et al. 2000).

In late-1993 Laguna Caliente somehow restored its volume, with a depth of 6 m, and a temperature of 62 °C (Martínez et al. 2000). Phreatic activity ceased until April 1994, when level dropped again until almost complete lake disappearance. The 1986–1996 unrest climaxed in the spring-summer of 1994 with powerful phreatic eruptions up to 2 km height. Since mid-August 1994 a "new" Laguna Caliente was formed. A last phreatic eruption occurred in April 1996 before returning to quiescent conditions for the next decade.

2.4 Between 2005 and 2016

After a decade of quiescence, characterized by low-T, but "migrating" fumaroles (Vaselli et al. 2003; Mora-Amador et al. 2004; Vaselli et al. Chapter "The Last Eighteen Years (1998–2014) of Fumarolic Degassing at the Poás Volcano (Costa Rica) and Renewal Activity") and a low-T Laguna Caliente, drastic changes in the hydromagmatic dynamics of the crater system started at the beginning of 2005 (Rymer et al. 2009). From January to mid-April that year, Laguna Caliente flooded the eastern terraces of the crater floor. After this maximum, the level of Laguna Caliente decreased about 11 m until March 2006. Fumarolic activity at the northern Naranja fumarole increased in May 2005 and culminated in a spectacular sulfur flow that cascaded toward, and almost flowed, into Laguna Caliente (Mora-Amador and Ramírez-Umaña 2008). After the sulfur flow ceased, the activity of the Naranja fumarole steadily decreased. The color of Laguna Caliente changed from turquoise-blue to opaque grey. In December 2005, tailed, yellow sulfur spherules were observed, floating on the lake surface (Mora Amador et al. Chapter "The Extraordinary Sulfur Volcanism of Poás from 1828 to 2018"). Afterwards, dark mud plumes charged with these sulfur spherules rose continuously from the centre of the lake (Fig. 1a). According to Takano et al. (1994) the presence of sulfur spherules with tails indicates the occurrence of high temperature gas condensation (119–150 °C), and an increase in the SO_2/H_2S ratios of the injected magmatic fluid. Corroboration of an increase in the SO_2 flux was given by mini-DOAS measurements in January 2006 (~500 t/d, C. Oppenheimer personal communication). Floating sulfur spherules, drops in lake level and increases in water temperature are considered precursors for an upcoming cycle of phreatic eruption activity (phase III of Takano et al. 1994).

On March 24, 2006, the first phreatic eruptions since April 1996 occurred at the centre of Laguna Caliente. Between 24 March, 2006 and

June 2010, at least 110 eruptions were reported (Rouwet et al. 2016). Mora-Amador et al. (2010) classified the phreatic eruptions in A-, B- or C-type, depending on the height of the eruption column: (A) from 2 to 50 m, (B) from 51 to 250 m, and (C) >250 m (Fig. 2). The A- and B-type eruptions are generally vertical jets of mud, and can generate seiches on the lake surface. The C-type eruptions expulse a whitish spray that disperse as far as 8 km from the eruptive centre. In 2006, 17 eruptions were reported (14 in March), while in 2007 and 2008 only 4 eruptions occurred.

Fig. 2 Style of phreatic eruptions during the 2006–2016 period of eruptive unrest. **a** Type-A eruption with column heights <50 m, **b** Type-B eruption with column heights >51 and <250 m, **c** Type-C eruption with column heights >251 m (Mora-Amador et al. 2010; modified from Rouwet et al. 2016)

On 8 January 2009, the Cinchona tectonic earthquake (M_w 6.2, 3.6 km depth with the epicenter 6.5 km E–NE of Poás, NNW–SSE oriented right lateral strike-slip, 30 cm displacement along a 15 km long fault, 29 casualties, Red Sismológica Nacional 2009) triggered landslides at the eastern and northern inner crater walls (Smithsonian 01/2009 BGVN 34:01) and on the outer flanks of Poás Volcano (Alvarado 2010). This earthquake was followed by a phreatic eruption on 12 January 2009 (Table 1), a day of abundant seismic replicas (18, Red Sismológica Nacional 2009).

In 2009, the eruption frequency did not significantly increase (5 eruptions), but the first C-type phreatic eruptions occurred (17 September and 25 December). The 25 December 2009 C-type eruption was one of the most explosive events during the ongoing eruption cycle (pre-2017, Smithsonian Bull GVN 2011) (Fig. 2c). Since September 2009, the fumarole temperature at the southern dome steadily increased from boiling point temperature to >650 °C, and manifested sulfur combustion (blue flames; Mora Amador et al. Chapter "The Extraordinary Sulfur Volcanism of Poás from 1828 to 2018"). Accordingly, throughout 2009–2010 the eruptive centre migrated from the centre of Laguna Caliente towards the dome and back. For the 110 reported eruptions (March 2006–June 2010), 85 were of A-type (77%), 19 of B-type (17%) and 6 of C-type (6%). A model on the phreatic explosion mechanism for this period (December 2005–June 2010), based on crater lake water chemistry, was proposed by Rouwet et al. (2016).

In 2010, the eruptive activity escalated and peaked in July with at least 102 phreatic eruptions, making fieldwork unsafe. The last phreatic eruption during the 2006–2014 eruption cycle occurred in October 2014. Afterwards, the volcano entered in a quiescent stage for approximately a year and a half. Minor phreatic activity resumed in May 2016.

2.5 Ongoing Magmatic Activity: 2017

On 1 April 2017 a phreatic eruption occurred on the 1953–1955 dome, west of the most active fumaroles, instead of below Laguna Caliente. A few days later (7 April 2017), a geyser-like boiling spring was born after a phreatic explosion north of the dome, showing water discharge towards Laguna Caliente from the southeast. On April 14 a major phreatomagmatic eruption blew out a large section of the dome, activity that intensified to the day of writing. The 1953–1955 dome, unaffected during the 1978–1980 and 1986–1996 unrest phases, is now mainly destroyed, 62+ years after its formation. The April 2017 activity is not dealt with in this study, also because the direct sampling method of Laguna Caliente has reached its limits for obvious safety reasons.

3 Pattern Recognition

3.1 The Method

Pattern Recognition (PR hereafter) procedures are objective statistical methods that have been used to approach problems in several fields, including geosciences (Mulargia et al. 1991; Sandri et al. 2004), searching for common patterns in precursory activity within a set of "objects" (e.g., time intervals preceding earthquakes or eruptions), and neglecting the physical peculiarity of single objects on purpose. It is a term encompassing several techniques aiming at objectively recognizing common schemes within a dataset. In PR, data are "objects" represented by vectors whose elements are observations of different "features". "Objects" are usually grouped in a number of "classes", and searching patterns means searching common schemes, or rules, distinguishing significantly the "objects" belonging to a "class" from those belonging to other classes. When more than two "features" are used, it might be very difficult to identify recurrent patterns by eye. That is why statistical algorithms for PR have been developed. Here, we analyze our dataset by means of the Binary Decision Tree algorithm (BDT hereafter) (Rounds 1980; Mulargia et al. 1991). The algorithm is based on the 2-sample Kolmogorov-Smirnoff test, iteratively searching for significant differences between the empirical cumulative distributions of two samples of data, in order to reject the null hypothesis that they are two random samples from the same distribution. The BDT algorithm is particularly effective in searching hierarchically ordered patterns, although it has proven to perform satisfactorily on synthetic datasets without such feature (Sandri et al. 2004).

Overfitting is also controlled by dividing the dataset in two: the "learning" and the "voting" sets, using the "learning" data to search and define the possible rules that divide the data in the classes (i.e., the patterns), applying such rules blindly to the "voting" data, to evaluate the performance on new data. We chose 80% of the data set to be "learning" samples (L), whereas the remaining 20% of the data set to be "voting" samples (V). An additional test serves to observe the impact of the confidence level of the procedure, where 0.01 significance (i.e. the complement of 0.99 confidence level) imposes "more severe" conditions, and 0.05 significance (i.e. the complement of a 0.95 confidence level) imposes "more flexible" conditions to the PR procedure.

3.2 Pattern Recognition for Erupting Crater Lakes: Ruapehu, Yugama and Poás

In this chapter, we apply PR methods to objectively discover possible precursory schemes for phreatic eruptions that are common among very active, and apparently different, crater lakes, on long (years) and short (months to days) term. To this purpose, we selected three crater lakes (Poás, Ruapehu and Yugama) and compiled mainly published data of physical (temperature) and chemical properties, as for these lakes (1) the available data cover several decade-long time series, (2) meanwhile, they underwent various eruptive phases, but (3) the eruption behavior and frequency are somewhat different, i.e., Poás passes prolonged periods with very frequent (weekly/daily) phreatic eruptions; at Ruapehu major phreatic and phreatomagmatic eruptions are often single events with repose periods of years; at Yugama, minor phreatic activity is of very low-frequency of occurrence. The database of chemical parameters of Laguna Caliente (Poás) is available in literature (1978–2008), (Rowe et al. 1992a, b; Martínez et al. 2000; Martínez 2008 and references therein), and from personal research for the period 2005–2010 (Rouwet et al. 2016) and for 2011–2016 (unpublished). Data for Yugama and Ruapehu crater lakes were only compiled from published data (Takano and Watanuki 1990; Christenson and Wood 1993; Christenson 1994, 2000; Ossaka et al. 1997; Ohba et al. 2000, 2008; Werner et al. 2006; Takano et al. 2008; Christenson et al. 2010). Hence, we characterized common patterns among the pre-eruptive signals in three different crater lakes.

"Objects" are vectors of values of the monitored variables at the crater lakes at different times. "Classes": Class 0 is represented by objects "not erupting" (given value 0), Class 1 "erupting" (given value 1). A value of 1 was assigned if the volcano erupted within the next month from the sampling date (date of the "object"). This defines the monitoring time window to one month, or shorter. "Features" are the monitored variables. The complete data set contains a total of 940 "objects" (i.e. temporal series of monitored parameters for the three lakes). In order to make the observed values comparable among the three crater lakes, we *standardized* the data, that is, for each *i*-th feature and *j*-th volcano separately, we transformed an observation xij into zij as:

$$zij = (xij - \theta ij)/\sigma ij \quad (1)$$

where θij is the mean of all xi at the j-th volcano, and σij is their standard deviation.

For Poás-only PR (430 "objects"), we selected the period 1978–September 2014 to establish the strongest parameters and their thresholds. The data gathered from November 2014 to February 2016 (unpublished) for Poás crater lake are retrospectively and blindly subjected to the numerical outcomes of the PR procedure (both Poás-only and the three-lakes PR), as an example of how the here proposed method could be included in future monitoring setups.

Due to the incomplete nature of the data set, for computational reasons, we were forced to combine features in subsets. Besides the recognition of common patterns in a data set, the use of subsets enables to investigate which parameter is the most indicative of all parameters in the subset. The subsets of features were chosen based on the following criteria:

(1) T versus pH versus Cl versus SO_4 versus SO_4/Cl versus Mg/Cl: these are parameters classically used in crater lake monitoring, assumed to be the most effective to recognize unrest, and possibly anticipate eruptions. Moreover, they are indicative of gas and heat input into the crater lake systems. The classical parameter Mg/Cl, is generally interpreted as being indicative for fluid migration (water-rock interaction) through a recent magma intrusion;
(2) Ca versus Mg versus Al versus Cl versus Al/Ca versus Al/Mg versus Mg/Cl: these are the most common cations, with Al normalized to Ca and Mg (a semi-random experiment), in combination with Mg/Cl;

The representativeness or confidence level of the PR procedure is expressed as #/total number of features, where # indicates the number of matching features within the two subsets. The higher #, the more indicative the PR procedure will be. For selected subset 1 the Poás-only data set contains 158 objects, whereas they are 417 if combining the three crater lakes. For the selected subset 2 the Poás-only data set contains 143 objects, whereas 406 if combining the three crater lakes. Numerical outcomes are compiled in Table 1.

Table 1 Numerical outcomes of the pattern recognition approach for Subsets 1 and 2, leading to the definition of thresholds of precursory signals for phreatic eruptions. Probabilities of the occurrence of phreatic eruptions within the next month are presented between brackets

Subsets	Thresholds
Subset 1	SO_4 > 60,000 mg/L (40.8 %) SO_4 > 60,000 mg/L & SO_4/Cl > 2.57 (22.2 %) pH ≤ -0.08 (31.7 %) pH $\leq$ 0.058 & Mg/Cl > 0.031-0.038 (33.8 − 35.1 %)
Subset 2	Mg/Cl $\leq$ 0.03 (38.1 %) Mg/Cl $\leq$ 0.03 & Ca $\leq$ 1100 mg/L & Mg $\leq$ 622 mg/L (81.8 %) Al/Mg > 3.77 & Al $\leq$ 2712 mg/L & Al/Mg $\leq$ 4.27 (69.6 %) Al/Mg > 3.75 & Mg/Cl > 0.037 (34.7 %)

4 Results and Discussion

4.1 Defining Thresholds for "The Best" Monitoring Parameters

(1) Subset 1: T versus pH versus Cl versus SO_4 versus SO_4/Cl versus Mg/Cl

After the application of the PR procedure on the Poás-only data set within a non-learning/voting approach, SO_4 happens to be the most indicative forecasting parameter. Out of 49/158 objects with a SO_4 concentration >60,000 mg/L, 20 resulted in a phreatic eruption within a month (Fig. 3). In terms of probabilities this means that when the SO_4 concentration of Poás' Laguna Caliente is >60,000 mg/L, 40.8% of the times (with a confidence level of 31%, and a significance of S = 0.01) a phreatic eruption occurred. For a significance of S = 0.05, SO_4 still happens to be the most indicative parameter, but a second pattern is recognized with the SO_4/Cl as the second most indicative parameter. For a SO_4 concentration $\leq$60,000 mg/L, and hence "non-eruptive" within the previous S = 0.01 significance PR test, an eruption can occur when the SO_4/Cl ratio is >2.57 (weight ratio) with an occurrence of 22.2% (Fig. 3, Table 1), however, with a low confidence level of 11.4%.

When the learning/voting approach is applied to the same Poás-only data set (S = 0.01), the pH of Laguna Caliente water happens to be the strongest parameter related to phreatic activity (Fig. 4). When the pH drops below −0.08 in 35% of the cases a phreatic eruption occurred within a month, for a confidence level of 38% (Table 1). Within the strict rules imposed by the S = 0.01, no further parameters were recognized as being effective.

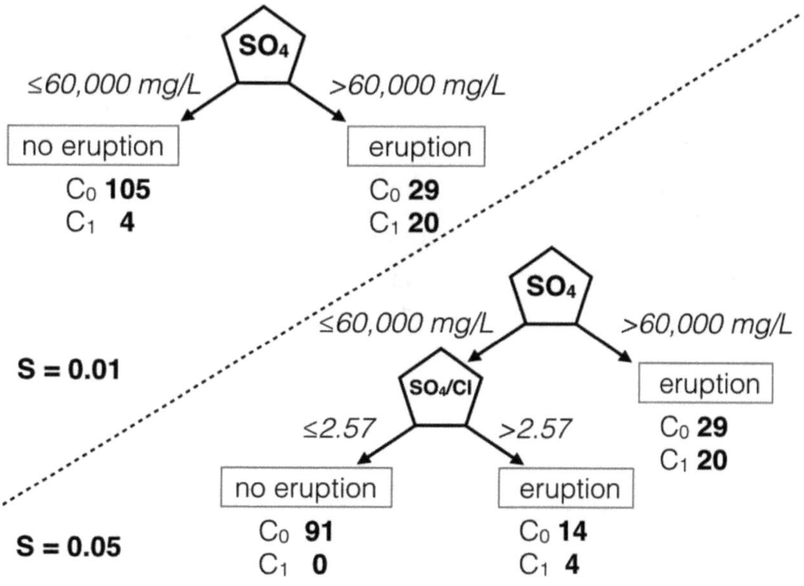

Fig. 3 Numerical outcomes of the PR approach for Subset 1 (Poás-only data set), represented as binary tracks for a significance of 0.0 1 (top), or 0.05 (bottom). Counts of success for C0 = No eruption, C1 = eruption, S = significance

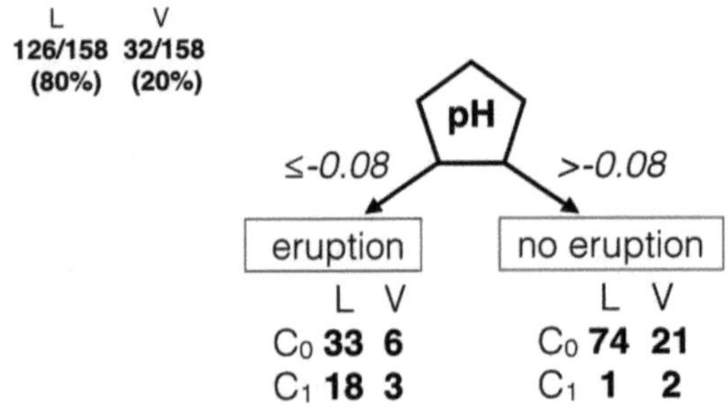

Fig. 4 Numerical outcomes of the PR approach for Subset 1 (Poás-only, learning (L)/voting (V) approach), represented as binary track. Counts of success for C0 = No eruption, C1 = eruption

Combining the data sets of Poás with those of Yugama and Ruapehu, again pH happens to be the strongest parameter followed by the Mg/Cl ratio. With or without the learning/voting approach (for a significance of S = 0.01 in both cases) very similar trends, and even similar numerical thresholds, were obtained (Fig. 5). The absolute value of the pH threshold is significantly higher than the one observed for the Poás-only PR procedure: −0.08 versus 0.058. The value of 0.058 was obtained back-calculating from the standardized value (−0.512) using Eq. (1). This higher pH is not surprising as Poás is the most acidic of the three lakes, which means that by adding the Yugama and Ruapehu data sets the average, but yet indicative, the pH value is tuned up. Important to notice is that once more pH is the most indicative parameter, despite this difference in absolute pH values for the three lakes.

Unsurprisingly, Mg/Cl is generally an indicative pre-eruptive parameter during deterministic-based monitoring. The fact that in the Poás-only data set this was not obvious is because for both Yugama (Ohba et al. 2008) and Ruapehu (Giggenbach 1974) Mg/Cl was efficiently used as an eruption precursor since the 1970's. Rowe et al. (1992a) stated that Mg/Cl was not efficient for Poás, as this parameter increases due to Cl loss from the lake as HCl, rather than due to a Mg increase in the lake. Nevertheless, Mg/Cl is a secondary parameter that seems to be efficient in phreatic eruption forecasting when >0.031 in 35.7% of the cases of the three lakes, when pH $\leq$ 0.058, though with a low confidence level of 19.1%. Further statistical evidence of the fate and role of the Mg/Cl ratio as an eruption precursor will be provided when dealing with Subset 2.

(2) Subset 2: Ca versus Mg versus Al versus Cl versus Al/Ca versus Al/Mg versus Mg/Cl

Fig. 5 Numerical outcomes of the PR approach for Subset 1 (three lakes data set), represented as binary tracks for the non-learning/voting approach (top), or learning/voting approach (bottom). Counts of success for C0 = No eruption, C1 = eruption

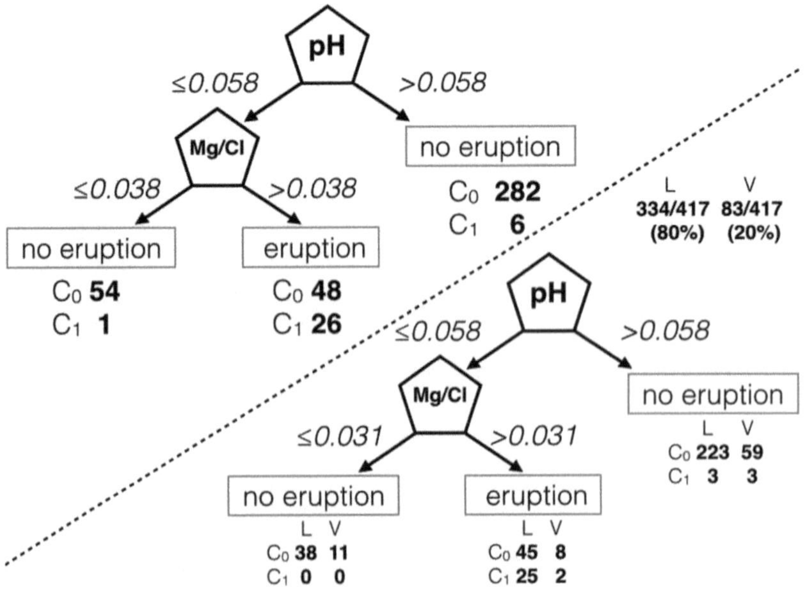

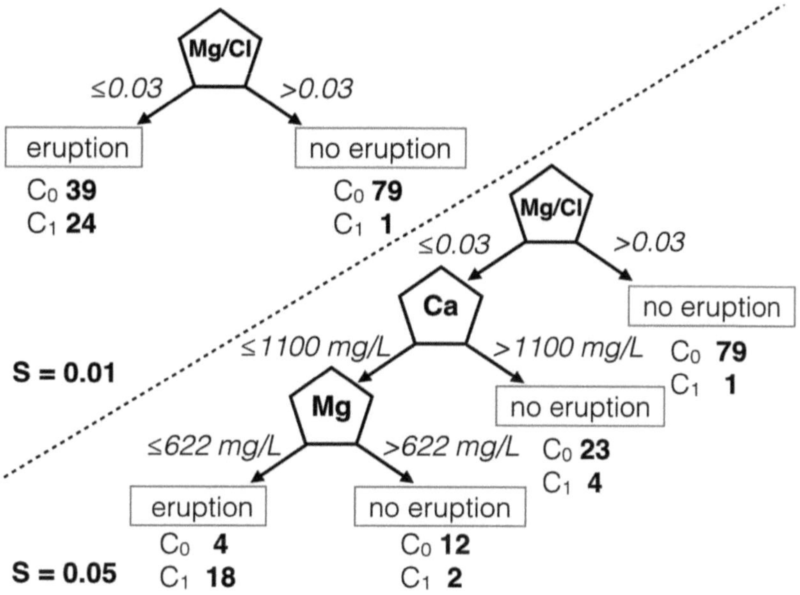

Fig. 6 Numerical outcomes of the PR approach for Subset 2 (Poás-only data set), represented as binary tracks for a significance of 0.0 1 (top), or 0.05 (bottom). Counts of success for C0 = No eruption, C1 = eruption, S = significance. Imposing less rigid rules (S = 0.05) results in secondary indicative parameters

The results of the PR procedure for the Poás-only data set for Subset 2 put Mg/Cl forward as the most indicative parameter considering strict rules with S = 0.01 (Fig. 6). In 38.1% of the cases when Mg/Cl $\leq$ 0.03 a phreatic eruption occurred within a month, although with a low confidence level of 16.8% (Table 1). Contrary to the general assumption that an increase in the Mg/Cl ratio suggests a recent intrusion of a magma batch, and contrary to the outcome using Subset 1, a Mg/Cl drop here results in a precursory signal. This could be due to the fact that an increase in Cl in the lake water, indicative of enhanced HCl degassing from the magma, could be a more dominant and arguably more sudden process, discrediting the Mg input as indicative. Within the dynamics and kinetics of phreatic eruptions, it is plausible to interpret degassing as a phreatic eruption trigger (Fischer et al. 2015; Rouwet et al. 2016; de Moor et al. 2016). Considering the same Poás-only data set, for a less rigid significance S = 0.05, secondary parameters happen to be indicative (Fig. 6): 12.6% of the phreatic eruptions have occurred within the next month, when Mg/Cl $\leq$ 0.03, Ca $\leq$ 1100 mg/L and Mg $\leq$ 622 mg/L (confidence level 12.6%). Loss of cations from the aqueous solution can occur through mineral precipitation, generally upon cooling water temperature, or through lake seepage. The former hypothesis is in favor of a sealing mechanism that could lead to phreatic eruptions.

Introducing the learning/voting approach for the Poás-only data set, regardless of the significance (S = 0.01 or 0.05), Mg/Cl ratio again results to be most indicative; when Mg/Cl $\leq$ 0.03 (note: the same numerical value as in the previous approach), 38.1% of the cases lead to a phreatic eruption within the next month (Table 1). No secondary parameters happened to be indicative (Fig. 7).

A stable pattern for the subset of features, for 406 objects of the combined Poás, Ruapehu and Yugama data set, indicates that phreatic eruptions are anticipated when the standardized Al/Mg ratio in crater lake waters is larger than $\sim$0.1. Calculating back to Al/Mg ratios for Poás, following Eq. (1) it turns out that the probability of a phreatic eruption at Poás increases when the Al/Mg > 3.77 (Fig. 8). Moreover, secondary parameters turn out indicative: for an Al concentration $\leq$ 2712 mg/L and, again, an Al/Mg ratio >4.27, 4.0% of the cases lead into eruption considering the three lakes (for S = 0.01 and non-learning/voting approach) (Fig. 8). Arguably, this is a too low percentage to be called

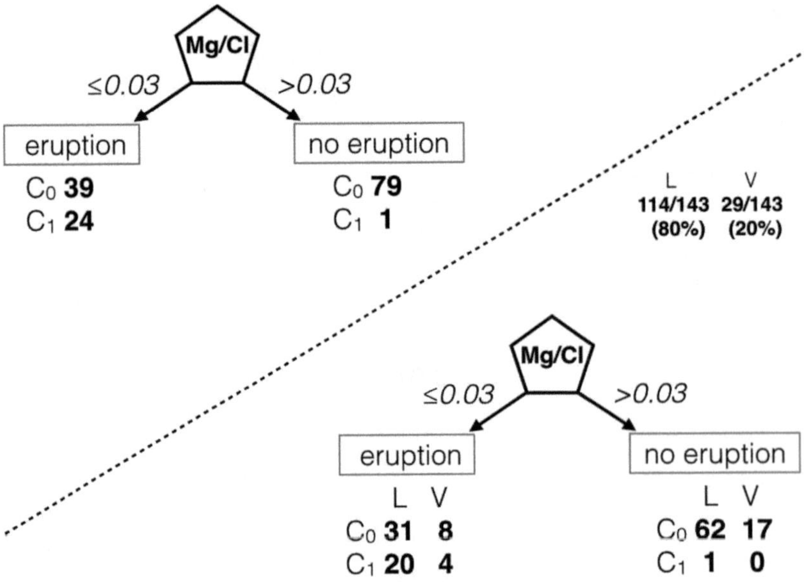

Fig. 7 Numerical outcomes of the PR approach for Subset 2 (Poás-only, learning (L)/ voting (V) approach), represented as binary track. Counts of success for C0 = No eruption, C1 = eruption

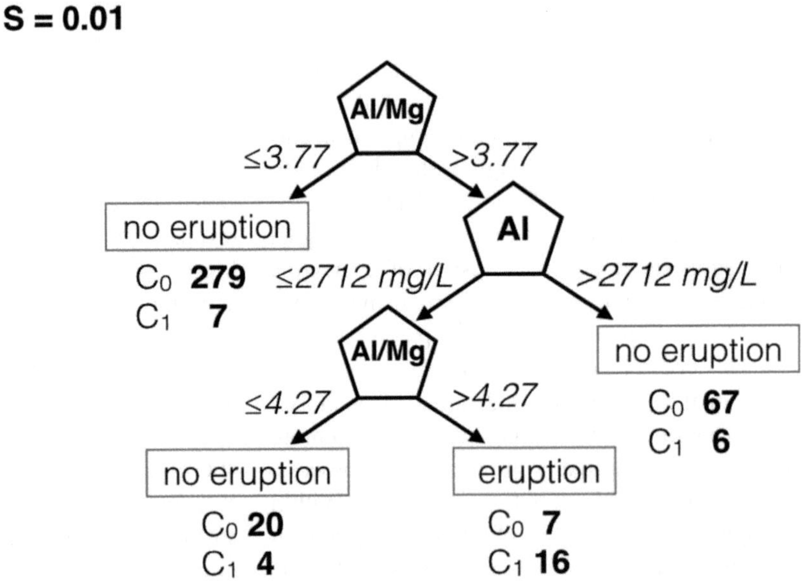

Fig. 8 Numerical outcomes of the PR approach for Subset 2 (three lakes data set, non-learning (L)/voting (V) approach), represented as binary track for a significance of S = 0.01. Counts of success for C0 = No eruption, C1 = eruption

efficient. The loss of Al, despite the increasing Al/Mg (3.77–4.27) though fits the possible scenario of alunite formation [$KAl_3(SO_4)_2(OH)_6$], and hence sealing.

When the same Poás-Ruapehu-Yugama data set is subjected to the learning/voting approach (with S = 0.01 or 0.05), Al/Mg is again the most indicative parameter, with the threshold very similar than the one in the previous application (Al/Mg > 3.75), whereas the Mg/Cl ratio pops up again as the second most indicative parameter (Fig. 9, Table 1): when Al/Mg > 3.75 and Mg/Cl > 0.037 (a value similar as the one obtained for the Poás-only data set, but with the

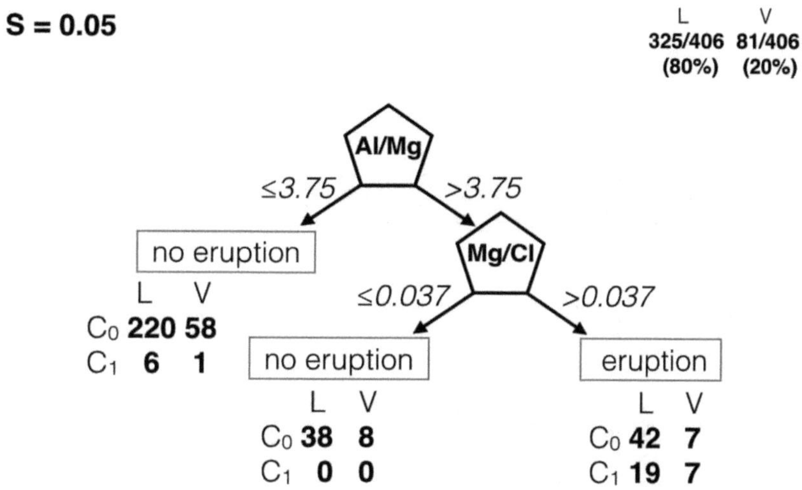

Fig. 9 Numerical outcomes of the PR approach for Subset 2 (three lakes data set, learning (L)/voting (V) approach), represented as binary track for a significance of S = 0.05. Counts of success for C0 = No eruption, C1 = eruption

opposite behavior), 34.7% of the cases lead to eruption for the three lakes (for a low confidence level of 6.5%).

Against all expectations, the Mg/Cl ratio in this subset is considered less indicative than the "randomly chosen" Al/Mg ratio. As mentioned earlier, an increase in the Mg/Cl ratio in crater lake waters generally indicates the input of a Mg-rich fluid after the flushing of hydrothermal fluids through a recently crystallized magmatic intrusion. An increase in Mg/Cl ratio for extremely acidic crater lakes (i.e. especially Ruapehu and Poás for our dataset) can also result from HCl degassing at the lake surface, making the Mg/Cl ratio at least ambiguous (Rowe et al. 1992a, b; Rouwet et al. 2016). At the moment, an explanation for the strength of the Al/Mg ratio as a potential eruption precursor based on fluid geochemistry is not provided. We suspect that Al, often the most concentrated cation in hyper-acidic lake brines, is somehow representative for the acidity of the lake water, while Mg is used, despite its high mobility, as a relatively more stable reference ion. This surprising outcome shows how a purely numerical method, such as the PR procedure, offers new tracks to discover in fluid geochemical research.

4.2 Temporal Variations Revised After Definition of Thresholds

The chemical composition of Laguna Caliente crater lake water (1) for the period 1978–2010 is based on literature (Rowe et al. 1992a, b; Martínez et al. 2000; Martínez 2008; Rouwet et al. 2016), and (2) for the period January 2011–February 2016 is presented in Table 2.

4.2.1 Methodology

Temperature of Laguna Caliente surface water was directly measured from the lake surface for the period January 2011–February 2016 (Table 2). Laguna Caliente waters were passed through 0.45 micron filters before storage in HD polyethylene bottles. Samples were not acidified for being already extremely acidic. The anion concentrations in the water samples were measured by ion chromatography (Metrohm 881 compact IC pro, accuracy ±2%), after potentiometric dilution (5000x) of the original sample. Major cation concentrations (Na, K, Ca, Mg) were determined by Atomic Absorption Spectroscopy (Thermo), whereas Al and Fe concentrations were determined by ICP-AES (Thermo

Table 2 Chemical composition of Laguna Caliente crater lake water. Concentrations in mg/L

Date	Eruption	T (°C)	pH	Na	K	Ca	Mg	Al	Fe	Cl	SO_4	SO_4/Cl	Mg/Cl	Al/Mg
19-01-11	1	–	−0.39	479	252	834	298	–	–	18700	68000	3.64	0.016	–
22-02-11	1	–	−0.40	487	286	687	269	–	–	19000	70200	3.69	0.014	–
23-03-11	1	–	−0.42	522	234	690	299	–	–	20100	74300	3.70	0.015	–
26-04-11	1	53	−0.38	593	263	739	293	–	–	19700	85700	4.35	0.015	–
23-06-11	1	–	–	–	–	–	–	–	–	24600	91800	3.73	–	–
30-06-11	1	–	−0.52	617	304	650	445	–	–	24300	92800	3.82	0.018	–
06-07-11	1	–	−0.52	746	322	569	458	–	–	24500	92900	3.79	0.019	–
30-07-11	1	–	−0.27	923	192	722	262	–	–	13400	56100	4.19	0.020	–
10-08-11	1	–	−0.55	642	379	543	435	–	–	29900	122300	4.09	0.015	–
22-09-11	0	61	−0.55	697	355	595	391	–	–	28600	118000	4.13	0.014	–
13-10-11	0	53	−0.49	566	307	687	347	–	–	26400	106900	4.05	0.013	–
27-10-11	0	60	−0.51	597	349	746	419	–	–	28800	113500	3.94	0.015	–
04-11-11	0	57	−0.49	470	398	702	411	2550	1150	29000	111700	3.85	0.014	6.20
27-12-11	1	53	−0.34	426	277	920	276	–	–	18300	79300	4.22	0.015	–
15-02-12	0	56	−0.36	460	283	817	305	–	–	18700	84400	4.51	0.016	–
21-03-12	1	54	−0.35	524	277	706	241	–	–	18700	81500	4.36	0.013	–
05-04-12	0	49	−0.40	423	330	748	280	–	–	19800	82300	4.16	0.014	–
26-04-12	1	–	−0.36	340	345	884	304	1920	840	18800	80300	4.27	0.016	6.32
30-04-12	1	–	−0.37	317	229	781	295	1900	850	19300	82500	4.27	0.015	6.44
30-05-12	1	–	−0.40	481	290	810	298	–	–	20500	89300	4.36	0.015	–
29-08-12	0	–	−0.37	323	331	765	281	1840	840	19000	84500	4.45	0.015	6.55
12-09-12	0	–	−0.37	294	211	779	269	1710	780	18400	83600	4.54	0.015	6.36
26-09-12	0	–	−0.37	316	287	726	266	1760	800	17900	81700	4.56	0.015	6.62
11-10-12	0	–	−0.36	337	304	760	250	1700	790	18400	83600	4.54	0.014	6.80
10-01-13	0	–	−0.24	265	181	785	236	1450	690	14300	64400	4.50	0.017	6.14

(continued)

Table 2 (continued)

Date	Eruption	T (°C)	pH	Na	K	Ca	Mg	Al	Fe	Cl	SO_4	SO_4/Cl	Mg/Cl	Al/Mg
23-01-13	1	–	−0.24	275	206	762	239	1460	710	14900	64700	4.34	0.016	6.11
28-01-13	1	–	−0.24	261	178	733	231	1450	680	15400	65400	4.25	0.015	6.28
27-02-13	1	–	−0.27	304	206	888	285	1500	730	15000	66300	4.42	0.019	5.26
20-03-13	1	–	−0.27	291	198	854	277	1520	740	16600	67800	4.08	0.017	5.49
03-04-13	1	–	−0.27	306	203	825	291	1510	740	15900	69600	4.38	0.018	5.19
03-05-13	1	46.1	−0.23	348	184	843	289	1500	790	15000	63800	4.25	0.019	5.19
08-05-13	1	44.6	−0.29	318	232	810	286	1560	780	16400	70600	4.30	0.017	5.45
09-05-13	1	45	−0.29	369	248	921	328	1550	770	16300	72000	4.42	0.020	4.73
14-06-13	1	46.6	−0.28	304	200	886	275	1410	750	16000	69800	4.36	0.017	5.13
27-06-13	1	47.4	−0.28	296	200	818	273	1370	740	16900	69000	4.08	0.016	5.02
04-09-14	0	45.1	−0.26	234	149	857	217	1240	710	25500	70500	2.76	0.009	5.71
17-09-14	0	47.7	−0.23	240	151	864	217	1150	640	14600	86100	5.90	0.015	5.30
24-09-14	0	48.3	−0.22	228	147	966	220	1130	630	13700	64900	4.74	0.016	5.14
27-11-14	0	34.6	−0.16	236	114	903	198	1040	560	12000	62600	5.22	0.017	5.25
13-04-15	0	–	−0.03	216	89	788	178	890	500	9200	44300	4.82	0.019	5.00
21-05-15	0	34.6	−0.02	215	87	939	179	900	490	9400	43400	4.62	0.019	5.03
14-07-15	0	39.5	0.15	226	61	908	174	720	390	6300	30000	4.76	0.028	4.14
26-08-15	0	–	0.2	214	96	895	133	690	370	5700	26800	4.70	0.023	5.19
07-10-15	0	–	0.23	140	51	912	120	590	310	5300	24300	4.58	0.023	4.92
18-11-15	0	–	0.22	143	59	866	117	600	320	5800	24900	4.29	0.020	5.13
03-12-15	0	–	0.18	163	57	907	125	650	490	9900	28200	2.85	0.013	5.20
14-01-16	0	–	0.16	172	54	979	132	630	350	6400	28600	4.47	0.021	4.77
23-02-16	0	–	0.26	118	35	495	102	530	280	5100	22400	4.39	0.020	5.20

– no data available (i.e. not analyzed or not measured)

iCAP 6000), with dilution factors of 100–1000. The pH was obtained after titration of a diluted aliquot (100x) with a 0.1 N NaOH solution (following Capaccioni et al. 2016). All analyses were elaborated at the University of Bologna, Italy (BiGEA Department). Charge balance of analyses generally closes within a 10% error. Such high error can result from the high metal content in the brines, not considered here. Laguna Caliente waters were sampled from the northern, eastern or southern (near the dome) shore. Rouwet et al. (2016) described a 20% difference in ion contents of samples from different sites during the same day.

We here plot the time series of the strongest parameters, i.e. the outcome from the PR procedure above, for the most recent eruptive period 2005–2016, with a blind and retrospective analyses for the period November 2014–February 2016 (dotted area in Figs. 10 and 11). "Blind" implies that we pretend not to know how Poás behaved during this period, in order to test the usefulness of the new method. In Figs. 10 and 11, phreatic eruptions are indicated by crosses: black for type-A, red for type-B and blue for type-C eruptions. Figure 10a shows the Poás-only results, whilst Fig. 10b shows the three-lakes results, for Subset 1. By Mid-2007, SO_4 concentrations peaked above threshold values (>60,000 mg/L) leading into the late-2008 early-2009 phreatic phase (Fig. 10a). Afterwards, SO_4 concentrations dropped below 60,000 mg/L and steadily ramped up again since early 2009, when phreatic eruptions continued. Since Mid-2009, the SO_4 threshold value was exceeded again, while the volcano escalated into the most intense phreatic phase of the 2005–2016 unrest, by mid-2010. SO_4 concentration peaked after this vigorous phase and remained high (>60,000 mg/L) until the end of 2014. During the test period (November 2014–February 2016), Poás appeared to have entered a quiescent stage, also reflected by a fast decline in SO_4 content. Phreatic activity resumed in May 2016, after a year and a half of below threshold SO_4 concentrations in the lake. As such, the May 2016 phreatic eruptions can be called "blue sky Ruapehu-type" (Christenson et al. 2010; Rouwet et al. 2014a), at the day of writing retrospectively interpreted as a different style of phreatic activity, and probably the build-up towards the most recent major phreato(magmatic) activity of April–May 2017.

Instead, SO_4/Cl ratios exceeded threshold values early on into the 2005–2016 unrest (early 2007) and remained high throughout the entire period, shown as a steady SO_4/Cl increase, peaking in October 2014. The May 2016 phreatic eruptions were still characterized by a SO_4/Cl ratio >2.57. The SO_4/Cl ratio happened to be a less efficient precursor, as the trend rather reflects long-term trends with inertia to sudden input that might trigger phreatic activity. An increase in the SO_4/Cl ratio probably states steady loss of HCl from the lake through evaporative degassing (Rouwet et al. 2016), and is not sensitive enough to be disturbed on the short-term scale. An increase in the SO_4/Cl ratio matched a decrease in Mg/Cl and pH on the long-term (Fig. 10 a vs. b). This observation suggests that the initially high Mg/Cl ratio resulted from input of a Mg-rich "magmatic" fluid, posteriorly steadily seeped out with time meanwhile HCl degassing continued (Rouwet et al. 2016). Nevertheless, the SO_4/Cl ratios >2.57 did coincide with phreatic activity, regardless of the observed trends, and are hence an indication of the continued state of unrest throughout the period 2007–2016. As variations in the SO_4/Cl ratios reflect changes in degassing regime, we support the hypothesis that phreatic eruptions are gas driven (de Moor et al. 2016).

Figure 10b (three-lakes approach) shows how a pH drop occurred before the onset of the phreatic eruptions in March 2006, although without reaching the threshold value of 0.058. Throughout 2007–2008 the pH value "oscillated" around the threshold, reflecting the unstable conditions that Laguna Caliente passed through, with sporadic phreatic eruptions (Rouwet et al. 2016). Since mid-2009, the pH dropped below the 0.058 threshold and more or less plateaued near 0.015 throughout the most intense phreatic phase of the unrest (2010–2014). After the October 2014 phreatic eruption the pH increased continuously until reaching above threshold values near mid-2015, coincident with the apparent volcanic quiescence. With such high pH

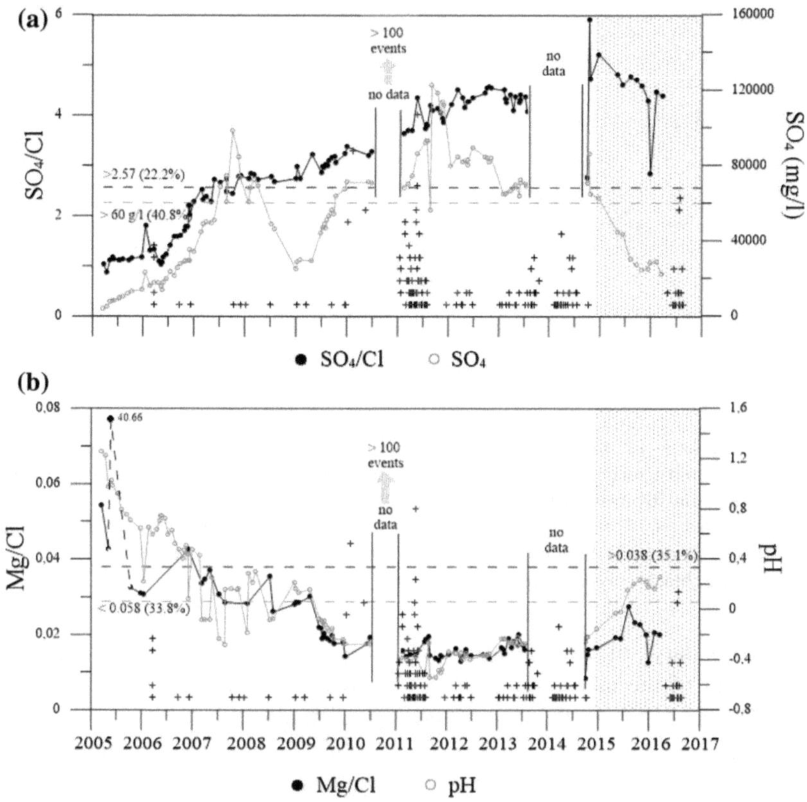

Fig. 10 Temporal variations of the most indicative precursory signals for the most recent period of eruptive unrest (2006–2016), resulting from the PR approach for the period 1978–September 2014, with their respective threshold values (dotted horizontal lines). Colored +'s indicate the eruption style: black = type-A, red = type B, blue = type-C eruptions (see Fig. 2). A "blind test" of the method is applied for the period November 2014–February 2016 (shaded background). **a** SO_4 concentration and SO_4/Cl ratios as indicative signals (Poás-only data set, Subset 1); **b** pH and Mg/Cl as indicative signals (from three lake data set, Subset 1)

values, the May 2016 phreatic eruptions can be considered "blue sky Ruapehu-type" eruptions.

With regard of the Mg/Cl ratio and the deduced threshold from the three-lakes PR approach, we disprove Mg/Cl > 0.038 as a useful forecasting threshold in the case of Poás. Rather the contrary seems true. The high Mg/Cl ratio as eruption precursor is clearly an overprinting effect originating from the Ruapehu and Yugama cases, where increasing Mg/Cl ratios were efficient precursory signals (Christenson 2000; Ohba et al. 2008). This confirms earlier findings by Rowe et al. (1992a, b) and Rouwet et al. (2016), and state that gas fluxes and dynamics in the case of Poás are a lot more significant and swifter than Mg-rich "magmatic" liquid input below the lake. This statement is confirmed during the test period (November 2014–February 2016) when the Mg/Cl ratio increased while volcanic quiescence was apparent (Fig. 10b).

In Fig. 11 (Subset 2 for the Poás-only PR approach), it is shown that after an unstable period with varying Mg/Cl ratios around the 0.03 threshold (2006–2008), the Mg/Cl ratio remained consistently below 0.03 for the most vigorous eruptive period 2010–2011, and beyond. Initially high Mg/Cl ratios (2005–2006) were caused by a high-Mg "magmatic" input (Rymer et al. 2009; Rouwet et al. 2016). This observation stresses that more Mg is lost from the aqueous solution, despite HCl degassing at the lake surface. The Ca (<1,100 mg/L) and Mg (<622 mg/L) secondary parameters confirm that cations are lost from the

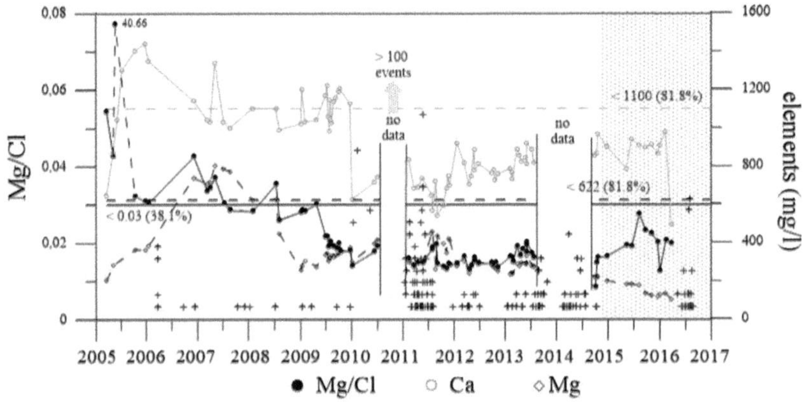

Fig. 11 Temporal variations of the most indicative precursory signals for the most recent period of eruptive unrest (2006–2016), resulting from the PR approach for the period 1978–September 2014, with their respective threshold values (dotted horizontal lines). Colored +'s indicate the eruption style: black = type-A, red = type B, blue = type-C eruptions (see Fig. 2). A "blind test" of the method is applied for the period November 2014–February 2016 (shaded background). Mg/Cl ratios and Ca and Mg concentrations are the most indicative signals (Poás-only data set, Subset 2)

solution, as (1) secondary mineral precipitates, and/or (2) flushed out of the lake through seepage. This loss of Ca and Mg from the lake water, presumably as solids, is so far the only indication that eruptive activity might not have been over after October 2014 (Fig. 11).

The too low success rate (4%, non-learning/voting approach) for the combined Al/Mg and Al set of parameters, and the earlier declared useless Mg/Cl > 0.037 threshold in the case of Poás, learns us that the three-lakes PR approach for Subset 2 results inefficient for Poás.

5 Implications for Future Monitoring of Poás (Laguna Caliente) and Other Crater Lake Bearing Volcanoes

During the past four decades, Poás has been the most active hyper-acidic peak-activity crater lake on Earth (Mora-Amador 2010), manifesting frequent phreatic eruptions during pluri-annual unrest stages. Since 1978, volcanic activity of Poás is monitored through Laguna Caliente, the reservoir collecting and preserving to more or less extent the markers from magmatic input. So far, a monitoring approach based on deterministic research has been the most obvious and straightforward means. Nevertheless, single phreatic eruptions at Poás are and remain hardly predictable.

In this revision chapter on the lake water chemistry of Laguna Caliente, we collected the available data on physical-chemical parameters of the crater lake, and subjected it for the first time to a Pattern Recognition procedure. On purpose, this approach did not consider any pre-existing conceptual model or geochemical explanation. This independent and novel method for crater lake monitoring provided remarkable outcomes, sometimes confirming earlier findings stemming from decades of deterministic research and monitoring, sometimes offering new ideas to guide future research.

Surprisingly, temperature is no key parameter in relation to phreatic activity. On the long-term, it is clearly visible that steady T increases from 20–25 to 40–50 °C mark the onset of volcanic unrest; once a high lake water temperature is reached, the PR method could not distinguish a significant relation between temperature variations and phreatic eruptive activity. Earlier findings for Poás (Rouwet et al. 2016) and Ruapehu (Werner et al. 2006; Christenson et al. 2010) described that a T-drop often anticipated phreatic eruptions, resulting from an inefficient heat and mass transfer from the underlying hydrothermal system into the lake (i.e. sealing). Nevertheless, T-logging remains a basic and simple method to timely detect volcanic unrest during transition stages from quiescence, and should hence not be discarded as a useful early-warning monitoring tool.

Despite the new findings on acid gas release from lakes (SO_2, HCl), and the lack in knowledge on the kinetics of this gas flushing, the SO_4 concentration and, to a lesser extend, the SO_4/Cl ratios still remain effective precursors. Lake water pH, an overall measure of the degassing state of the entire system caused by the entrance of acidic gas species (SO_2, HCl, HF) and enhanced by evaporation, results a strong parameter after the PR method. Surprisingly, a decreasing instead of an increasing Mg/Cl ratio results an efficient pre-eruptive signal at Poás. This outcome suggests that a Mg-loss from solution, presumably through mineral precipitation or lake seepage, is significant despite the HCl loss by evaporative degassing.

The monitoring time window in the PR approach is defined by the sampling frequency (approximately 1 month). A short-term (daily-weekly) monitoring frequency makes sense, despite the longer residence time dependent monitoring time window (Rouwet et al. 2014a), as for the PR approach the time window is independent of any physical-chemical kinetics. Fact is that the SO_4 and Cl concentrations in crater lakes are only snapshot views of a transient SO_2–HCl-charged gas flushing through the lake. Hence, for the PR approach and the gas-water geochemical monitoring it is convenient to increase the sampling frequency and tighten the time window in such a way as to match the speed of the process we aim to predict: a phreatic eruption. For now, even a monthly sampling frequency as for the three lakes studied here is utopia for many other crater lakes worldwide. The general and lake-specific thresholds defined here can be useful for deterministic, as well as probabilistic eruption forecasting (e.g. Bayesian Event Tree approaches, Tonini et al. 2016). The outcomes are highly practical for monitoring setups at Poás and at other crater lakes. The three-lakes PR approach did not always result efficient for Poás. Nevertheless, for the current sampling frequency, they certainly incorporate a "safety margin", a measure of the epistimic uncertainty (i.e. degree of ignorance of the system) of the current state of the forecasting of phreatic eruptions at peak-activity crater lakes.

Acknowledgements Diego Castioni and Piero Trentini are thanked for help in the analyses. DR and LS are supported by the FREAPROB project, INGV-Struttura Vulcani. We are grateful to Corentin Caudron and Francesco Frondini for their constructive reviews. Orlando Vaselli is thanked for swift editing. We warmly dedicate this study to our friend and co-author Bruno who suddenly passed away during the writing of this chapter.

References

Alvarado GE (2010) Aspectos geohidrológicos y sedimentológicos de los flujos de lodo asociados al terremoto de Cinchona (Mw 6.2) del 8 de enero del 2009. Costa Rica. Rev Geol Am Centr 43:67–96 (In Spanish with English abstract)

Brantley SL, Brogia A, Rowe G, Fernández JF, Reynolds JR (1987) Poás volcano crater lake acts as a condenser for acid metal-rich brine. Nature 330:470–472

Brown G, Rymer H, Dowden J, Kapadia P, Stevenson D, Barquero J, Morales LD (1989) Energy budget analysis for Poás crater lake: implications for predicting volcanic activity. Nature 339:370–373

Browne PRL, Lawless JV (2001) Characteristics of hydrothermal eruptions, with examples from New Zealand and elsewhere. Earth Sci Rev 52:299–331

Capaccioni B, Rouwet D, Tassi F (2016) HCl degassing from extremely acidic crater lakes: preliminary results from experimental determinations and implications for geochemical monitoring. In: Caudron C, Capaccioni B, Ohba T (eds) Geochemistry and geophysics of volcanic lakes. Geol Soc London, Special Publications 437. https://doi.org/10.1144/sp437.12

Casertano L, Borgia A, Cigolini C (1983) El Volcán Poás, Costa Rica: Cronologia y características de la actividad. Geofis Int 3:215–236

Caudron C, Lecoq T, Syahbana DK, McCausland W, Watlet A, Camelbeeck T, Bernard A, Surono (2015) Stress and mass changes at a "wet" volcano: example during the 2011–2012 volcanic unrest at Kawah Ijen volcano (Indonesia). J Geophys Res, Solid Earth 120. https://doi.org/10.1002/2014jb11590

Christenson BW (1994) Convection and stratification in Ruapehu Crater Lake, New Zealand: implications for Lake Nyos-Type gas release eruptions. Geochem J 28:185–197

Christenson BW (2000) Geochemistry of fluids associated with the 1995–1996 eruption of Mt. Ruapehu, New Zealand: signatures and processes in the magmatic-hydrothermal system. J Volcanol Geothermal Res 97:1–30

Christenson BW, Wood CP (1993) The evolution of a volcano-hosted hydrothermal system beneath Ruapehu Crater Lake, New Zealand. Bull Volcanol 55:547–565

Christenson BW, Reyes AG, Young R, Moebis A, Sherburn S, Cole-Baker J, Britten K (2010) Cyclic processes and factors leading to phreatic eruption events: insights from the 25 September 2007 eruption through Ruapehu Crater Lake, New Zealand. J Volcanol Geotherm Res 191:15–32. https://doi.org/10.1016/j.volgeores.2010.01.008

de Moor JM, Aiuppa A, Pacheco J, Avard G, Kern C, Liuzzo M, Martínez M, Giudice G, Fischer TP (2016) Short-period volcanic gas precursors to phreatic eruptions: insights from Poás Volcano, Costa Rica. Earth Planet Sci Lett 442:218–227. https://doi.org/10.1016/J.epsl.2016-02-056

Fischer TP, Ramírez C, Mora-Amador RA, Hilton DR, Barnes JD, Sharp ZD, de Moor JM, Barry PH, Füri E, Shaw AM (2015) Temporal variations in fumarole gas chemistry at Poás volcano, Costa Rica. J Volcanol Geothermal Res 294:56–70. https://doi.org/10.1016/j.volgeores.2015.02.002

Giggenbach WF (1974) The chemistry of Crater Lake, Mt Ruapehu (New Zealand) during and after the 1971 active period. NZ J Sci 17:33–45

Gunawan H, Caudron C, Pallister J, Primulyana S, Christenson B, McCausland W, Van Hinsberg V, Lewicki J, Rouwet D, Kelly P, Kern C, Werner C, Johnson JB, Budi Utami S, Syahbana DK, Saing U, Suparjan, Heri Purwanto B, Sealing C, Martínez-Cruz M, Maryanto S, Bani P, Laurin A, Schmid A, Bradley K, Made Agung Nandaka IG, Hendrasto M (2016) New insights into Kawah Ijen's volcanic system from the wet volcano workshop experiment. In: Caudron C, Capaccioni B, Ohba T (eds) Geochemistry and geophysics of volcanic lakes. Geol Soc London, Special Publications 437. https://doi.org/10.1144/sp437.12

Hurst T, Hashimoto T, Terada A (2015) Crater lake energy and mass balance. In: Rouwet D, Christenson BW, Tassi, F, Vandemeulebrouck J (eds) Volcanic lakes. Springer-Heidelberg, pp 307–322. https://doi.org/10.1007/978-3-642-36833-2_13

Martínez M (2008) Geochemical evolution of the acidic crater lake of Poás volcano (Costa Rica): insights into volcanic-hydrothermal processes. Ph.D. Thesis, Universiteit Utrecht, 162 p

Martínez M, Fernández E, Valdés J, Barboza V, Van der Laat R, Duarte E, Malavassi E, Sandoval L, Barquero J, Marino T (2000) Chemical evolution and volcanic activity of the active crater lake of Poás volcano, Costa Rica, 1993–1997. J Volcanol Geochem Res 97:127–141

Mora-Amador RA (2010) Peligrosidad volcánica del Poás (Costa Rica), basado en las principales erupciones históricas de 1834, 1910 y 1953–1955. M.Sc. Thesis, Universidad de Costa Rica, 115 p

Mora-Amador R, Rouwet D, Vargas P (Chapter 3) The extraordinary sulfur eruptions of Poás volcano from 1828 to 2016. In: Tassi F, Mora-Amador R, Vaselli O (eds) Poás volcano (Costa Rica): the pulsing heart of Central America Volcanic Zone. Springer, Heidelberg (Germany)

Mora-Amador RA, Ramírez-Umaña CJ (2008) Sulfur flows at Poás volcano, Costa Rica. IAVCEI Reykjavík, Iceland General Assembly

Mora-Amador RA, Ramírez C, Fernández M (2004) La actividad de los volcanes de la Cordillera Central, Costa Rica, entre 1998–2002. Rev Geol Am Centr 30:189–197

Mora-Amador RA, Ramírez CJ, González G, Rouwet D, Rojas A (2010) Laguna Caliente, Poás Volcano, Costa Rica: the most active crater lake of the world (2006–2010). Makavol Abstract, Fogo Workshop

Mulargia F, Gasperini P, Marzocchi W (1991) Pattern recognition applied to volcanic activity: identification of the precursory patterns to Etna recent flank eruptions and periods of rest. J Volcanol Geotherm Res 45:187–196

Ohba T, Hirabayashi J, Nogami K (2000) D/H and 18O/16O ratios of water in the crater lake at Kusatsu-Shirane volcano, Japan. J Volcanol Geotherm Res 97:329–346

Ohba T, Hirabayashi J, Nogami K (2008) Temporal changes in the chemistry of lake water within Yugama Crater, Kusatsu-Shirane Volcano, Japan: implications for the evolution of the magmatic-hydrothermal system. J Volcanol Geotherm Res 178:131–144

Oppenheimer C (1992) Sulphur eruptions at Volcán Poás, Costa Rica. J Volcanol Geotherm Res 49:1–21

Oppenheimer C, Stevenson D (1989) Liquid sulphur lakes at Poás volcano. Nature 342:790–793

Ossaka J, Ossaka T, Hirabayashi J, Oi T, Ohba T, Nogami K, Kikawada Y, Yamano M, Yui M, Fukuhara H (1997) Volcanic activity of Kusatsu-Shirane volcano, Gunma, and secular change in water quality of crater lake Yugama. Chikyukagaku 31:119–128 (In Japanese with English abstract)

Prosser JT, Carr MJ (1987) Poás Volcano, Costa Rica. Geology of the summit region and spatial and temporal variations among the most recent lavas. J Volcanol Geothermal Res 33:131–146

Raccichini S, Bennett FD (1978) Nuevos aspectos de las erupciones del Volcán Poás. Rev Geogr Am Central 5–6:37–53 (In Spanish with English abstract)

Raul Mora-Amador R, Rouwet D, González G, Vargas P, Ramírez C (Chapter 11) Volcanic hazard assessment of Poás (Costa Rica) based on the major historical eruptions of 1834, 1910 and 1953–1955. In: Tassi F, Mora-Amador R, Vaselli O (eds) Poás volcano (Costa Rica): the pulsing heart of Central America Volcanic Zone. Springer, Heidelberg (Germany)

Rounds EM (1980) A combined nonparametric approach to feature selection and binary decision tree design. Pattern Recon 12:313–317

Rouwet D, Morrissey M (2015) Mechanisms of crater lake breaching eruptions. In: Rouwet D,

Christenson BW, Tassi F, Vandemeulebrouck J (eds) Volcanic lakes. Springer-Heidelberg, pp 73–92

Rouwet D, Ohba T (2015) Isotope fractionation and HCl partitioning during evaporative degassing from active crater lakes. In: Rouwet D, Christenson BW, Tassi F, Vandemeulebrouck J (eds) Volcanic lakes. Springer-Heidelberg, pp 179–200

Rouwet D, Sandri L, Marzocchi W, Gottsmann J, Selva J, Tonini R, Papale P (2014a) Recognizing and tracking volcanic hazards related to non-magmatic unrest: a review. J Appl Volcanol 3:17. https://doi.org/10.1186/s13617-014-0017-3

Rouwet D, Tassi F, Mora-Amador R, Sandri L, Chiarini V (2014b) Past, present and future of volcanic lake monitoring. J Volcanol Geotherm Res 272:78–97. https://doi.org/10.1016/j.jvolgeores.2013.12.009

Rouwet D, Mora-Amador R, Ramírez CJ, González G, Inguaggiato S (2016) Dynamic fluid recycling at Laguna Caliente (Poás, Costa Rica) before and during the 2006-ongoing phreatic eruption cycle. In: Caudron C, Capaccioni B, Ohba T (eds) Geochemistry and geophysics of volcanic lakes. Geol Soc London, Special Publications 437. https://doi.org/10.1144/sp437.12

Rouwet D, Hidalgo S, Joseph EP, González-Ilama G (2017) Fluid geochemistry and volcanic unrest: dissolving the haze in time and space. In: Gottsmann J, Neuberg, J, Sheu, B (eds) Advances in volcanology. Springer-Heidelberg. https://doi.org/10.1007/11157_2017_12

Rowe GL, Brantley SL, Fernández M, Fernández JF, Borgia A, Barquero J (1992a) Fluid-volcano interaction in an active stratovolcano: the crater lake system of Poás volcano, Costa Rica. J Volcanol Geotherm Res 49:23–51

Rowe GL, Ohsawa S, Takano B, Brantley SL, Fernández JF, Barquero J (1992b) Using crater lake chemistry to predict volcanic activity at Poás Volcano, Costa Rica. Bull Volcanol 54:494–503

Rudín J, Alfaro A, Michaud G, Rudín A (1910) Gran erupcion de Ceniza del Volcán Poás. In: Vargas C (Comp) 1979, Volcán Poás Antologia (3a ed), UNED, San José, pp 75–82

Rymer H, Locke CA, Borgia A, Martínez M, Brenes J, Van der Laat R, Williams-Jones G (2009) Long-term fluctuations in volcanic activity: implications for future environmental impact. Terra Nova 21:304–309

Sandri L, Marzocchi W, Zaccarelli L (2004) A new perspective in identifying the precursory patterns of eruptions. Bull Volcanol 66:263–275. https://doi.org/10.1007/s00445-003-0309-7

Shinohara H, Yoshikawa S, Miyabuchi Y (2015) Degassing activity of a volcanic crater lake: volcanic plume measurements at the Yudamari crater lake, Aso volcano, Japan. In: Rouwet D, Christenson BW, Tassi F, Vandemeulebrouck J (eds) Volcanic lakes. Springer-Heidelberg, pp 201–218

Symonds RB, Gerlach TM, Reed MH (2001) Magmatic gas scrubbing: implications for volcano monitoring. J Volcanol Geotherm Res 108:303–341

Takano B, Watanuki K (1990) Monitoring of volcanic eruptions at Yugama crater lake by aqueous sulfur oxyanions. J Volcanol Geotherm Res 40:71–87

Takano B, Saitoh H, Takano E (1994) Geochemical implications of subaqueous molten sulfur at Yugama crater lake, Kusatsu-Shirane volcano, Japan. Geochem J 28:199–216

Takano B, Kuno A, Oshawa S, Kawakami H (2008) Aqueous sulfur speciation possibly linked to subliming volcanic gas-water interaction during a quiescent period at Yugama crater lake, Kusatsu-Shirane volcano, Central Japan. J Volcanol Geotherm Res 178:145–168

Tamburello G, Agusto M, Caselli A, Tassi F, Vaselli O, Calabrese S, Rouwet D, Capaccioni B, Di Napoli R, Cardellini C, Chiodini G, Bitetto M, Brusca L, Bellomo S, Aiuppa A (2015) Intense magmatic degassing through the lake of Copahue volcano, 2013–2014. J Geophys Res 120. https://doi.org/10.1002/2015jb012160

Tonini R, Sandri L, Rouwet D, Caudron C, Marzocchi W, Superman (2016) A new Bayesian Even Tree tool to track and quantify volcanic unrest and its application to Kiawah Ijen volcano. Geochem Geophys Geosyst. https://doi.org/10.1002/2016gc006327

Truesdell AH, Hairlip JR, Armannsson H, D'Amore F (1989) Origin and transport of chloride in superheated geothermal steam. Geothermics 18:295–304

Varekamp JC, Pasternack GB, Rowe GL Jr (2000) Volcanic lake systematics II. Chemical constraints. J Volcanol Geotherm Res 97:161–179

Vaselli O, Tassi F, Fischer TP, Tardani D, Fernandez Soto E, Martinez M, De Moor MJ, Bini G (Chapter 10) The last eighteen years (1998–2014) of fumarolic degassing at the Poás volcano (Costa Rica) and renewal activity. In: Tassi F, Mora-Amador R, Vaselli O (eds) Poás volcano (Costa Rica): the pulsing heart of Central America Volcanic Zone. Springer, Heidelberg (Germany)

Vaselli O, Tassi F, Minissale G, Montegrossi G, Duarte E, Fernandez E, Bergamaschi F (2003) Fumarole migration and fluid geochemistry at Poás volcano (Costa Rica) from 1998 to 2001. In: Oppenheimer C, Pyle DM, Barkley J (eds) Volcanic degassing. Geol Soc London 213:247–262. Special Publications

Werner C, Christenson BW, Hagerty M, Britten K (2006) Variability of volcanic gas emissions during a crater lake heating cycle at Ruapehu Volcano, New Zealand. J Volcanol Geotherm Res 154:291–302

The Last Eighteen Years (1998–2014) of Fumarolic Degassing at the Poás Volcano (Costa Rica) and Renewal Activity

Orlando Vaselli, Franco Tassi, Tobias P. Fischer, Daniele Tardani, Erick Fernández, María del Mar Martínez, Marteen J. de Moor and Giulio Bini

Abstract

This chapter reviews the geochemical and isotopic data from the fumarolic gas discharges collected in a discontinuous mode from 1998 to 2014 at Poás volcano. During this period, the "Tico" volcano experienced a renewed phreatic activity that started in 2006 after a couple of decades of relative quiescence. In January 2009, a 6.2 Mw earthquake hit the village of Cinchona, which is located a 4 km to the east of Poás. As the phreatic activity kept evolving, the hyperacidic lake ("Laguna Caliente") dried out and the high-temperature fumaroles that previously were likely entering the lake were revealed, though not accessible. The pyroclastic dome that formed in the early fifties was destroyed at the beginning of 2017 by several relatively small phreatomagmatic (strombolian and vulcanian type) small-size eruptions. The risk of sudden phreatic and phreato-magmatic events prevented the direct sampling of the fumaroles and as a consequence, no geochemical data were sampled in the last three years. Nevertheless, interesting hints were recorded by the gas geochemistry before the 2006 phreatic activity and the 2009 Cinchona seismic events, mainly based on the temporal variations of the H_2S/SO_2, H_2/H_2O, H_2/Ar, CO/CO_2, CH_4/CO_2 and HCl/HF ratios. However, in most cases the geochemical record is not complete since the gas discharging vents migrated or stopped their activity and new fumaroles formed up to the recent visual observations.

Keywords

Poás volcano · Fluid geochemistry · Volcanic hazard · High-temperature fumaroles · Volcanic gases · Phreatic eruptions

O. Vaselli (✉) · F. Tassi · G. Bini
Department of Earth Sciences, Via G. La Pira 4, 50121 Florence, Italy
e-mail: orlando.vaselli@unifi.it

O. Vaselli · F. Tassi
CNR Institute of Geosciences and Earth Resources, Via G. La Pira 4, 50121 Florence, Italy

T. P. Fischer
Department of Earth and Planetary Sciences, University of New Mexico, Albuquerque, NM 87131-1116, USA

D. Tardani
Departamento de Geología, Plaza Ercilla 803, Santiago, Chile

E. Fernández · M. del Mar Martínez · M. J. de Moor
Volcanological and Seismological Observatory, OVSICORI-UNA, Heredia 2386-3000, Costa Rica

1 Introduction

The chemical substances emitted during volcanic eruptions and harsh fumarolic activity play an important role on atmospheric and climatic

F. Tassi et al. (eds.), *Poás Volcano*, Active Volcanoes of the World,
https://doi.org/10.1007/978-3-319-02156-0_10

processes. The released fluids carry useful information about sub-surface processes and it is well established that direct gas monitoring is an important tool for the surveillance of active and quiescent volcanoes (e.g. Chiodini et al. 1995; Giggenbach 1987, 1996; Shinohara et al. 2002, 2015; Vaselli et al. 2010; Fischer et al. 2015). According to Menyalov (1975), variations in terms of "*endogenetic and exogenetic factors*" are expected to be recorded in the chemical and isotopic compositions of the discharging gases, although secondary effects related to reactions with atmospheric gases, underground waters, hosting rocks, redox conditions and bacterial metabolism may complicate the picture recorded at the fumarolic vents (e.g. Symonds et al. 2001).

Volcanic gas monitoring is based on repeated periodic (weekly, monthly, seasonally, annually) sampling of selected fumaroles since the chemical composition of a single fumarole from a volcanic edifice is not representative of the system as a whole. As a routine procedure, fumaroles characterized by different outlet temperatures and located in the surrounding areas, are to be sampled (e.g. Vaselli et al. 2006 and references therein). However, it frequently occurs that although more teams may be working in the same volcanic system, it often happens that the sampling activity is not carried out from the same fumarolic vent(s). This may pose some problems when comparing the geochemical and isotopic data gathered by several teams of volcanologists, since compositional variations of fluids emitted from different gas discharges of a volcanic system are relatively common. To make the general picture more complicated, time migration of fumarolic vents is quite common. This implies that when periodic gas sampling is planned, the selected fumaroles may have disappeared or migrated; thus the newly-born fumaroles may present chemical compositions that can be different with respect to the previously collected sampling sites without any significant variations of the magmatic plumbing system. This is the case at Poás volcano, which has shown a strong compositional variability of the fumarolic gas discharges accompanied by the appearance and disappearance or shifting of fumarolic vents in the last two decades (Vaselli et al. 2003; Fischer et al. 2015).

The present chapter is aimed to summarize and discuss the major chemical and isotopic variations that have characterized the Poás crater fumaroles from 1998 to 2014. In 1998, a bilateral project between the OVSICORI-UNA (Observatorio Vulcanológico y Sismológico de Costa Rica, Universidad Nacional) and the CNR-IGG (National Research Council—Institute of Geosciences and Earth Resources) and the Department of Earth Science (DES) of the University of Florence (Italy) started during which a periodic, though discontinuous, sampling of the Poás crater fumaroles to evaluate the status of the volcano was carried out. This collaboration continued up to 2011 when the last gas sampling was carried out. Additional and important data derived by an NSF project (Resp. TPF) are included in the chapter. Thus, the geochemical and isotopic data related to direct sampling and published by Vaselli et al. (2003), Zimmer et al. (2004), Hilton et al. (2010) and Fischer et al. (2015) were used to depict a comprehensive study on one of the most active volcanoes in Central America by also adding unpublished data produced by the Italian-Costa Rican team. In the last few years, the intense phreatic and phreato-magmatic events (see for further details Martínez et al. Chapter "Behaviour of Polythionates in the Acid Lake of Poás Volcano: Insights Into Changes in the Magmatic-Hydrothermal Regime and Subaqueous Input of Volatiles"; Mora Amador et al. Chapters "The Extraordinary Sulfur Volcanism of Poás from 1828 to 2018" and "Volcanic Hazard Assessment of Poás (Costa Rica) Based on the 1834, 1910, 1953–1955 and 2017 Historical Eruptions"; Rouwet et al. Chapter "39 Years of Geochemical Monitoring of Laguna Caliente Crater Lake, Poás: Patterns from the Past as Keys for the Future") prevented any direct gas sampling during this time by the Italian-Costa Rican team.

2 A Chronological Summary of the Fumarolic Gas Migration at Poás Volcano in the Last Eighteen Years and the Renewal Activity

The largest historical eruption at Poás volcano occurred in 1910 when a volcanic plume, up to 8 km high, was formed and about 800,000 m^2 of ash covered the volcano summit (namely, the Active Crater) and surrounding areas (e.g. Krushensky and Escalante 1967; Raccichini and Bennett 1977; Vargas 1979; Boza and Mendoza 1981; Lopes 2005; Martínez 2008; Rymer et al. 2010; Mora Amador et al. Chapter "Volcanic Hazard Assessment of Poás (Costa Rica) Based on the 1834, 1910, 1953–1955 and 2017 Historical Eruptions"). In 1953–1955 a short-lived and moderate size eruption formed a $\approx$ 25 m high intra-crater dome (commonly called, pyroclastic dome or pyroclastic cone or simply the dome) and a pit crater into which part of the dome collapsed (e.g. Krushensky and Escalante 1967; Prosser and Carr 1987; Martínez 2008; Mora Amador et al. Chapter "Volcanic Hazard Assessment of Poás (Costa Rica) Based on the 1834, 1910, 1953–1955 and 2017 Historical Eruptions"). Subsequently, a hyperacidic (pH < 1) lake (Laguna Caliente nested in the so-called Active Crater) intermittently appeared and it was relatively stable from 1965 to 1987, though alternated with sporadic geyser activity (e.g. Rymer et al. 2009; Rouwet et al. Chapter "39 Years of Geochemical Monitoring of Laguna Caliente Crater Lake, Poás: Patterns from the Past as Keys for the Future"). Presently, the latest phreatic activity (started in 2006 and still ongoing) dried the lake out and destroyed the 1953–1955 pyroclastic dome (Martínez et al. Chapter "Behaviour of Polythionates in the Acid Lake of Poás Volcano: Insights Into Changes in the Magmatic-Hydrothermal Regime and Subaqueous Input of Volatiles"; Mora Amador et al. Chapter "Volcanic Hazard Assessment of Poás (Costa Rica) Based on the 1834, 1910, 1953–1955 and 2017 Historical Eruptions; Rouwet et al. Chapter "39 Years of Geochemical Monitoring of Laguna Caliente Crater Lake, Poás: Patterns from the Past as Keys for the Future").

Explosions characterized by the emissions of sulfur-encrusted blocks (Bennett and Raccichini 1978; Bennett 1979; Francis et al. 1980) in 1978 and the presence of volcanic liquid sulfur (e.g. Oppenheimer and Stevenson 1989; Oppenheimer 1992; Mora Amador et al. Chapter "The Extraordinary Sulfur Volcanism of Poás from 1828 to 2018") was revealed at Laguna Caliente when in 1981 the outlet temperatures of the fumaroles were up to 1020 °C, suggesting that a new cycle of activity had commenced (Casertano et al. 1987; Rymer et al. 2000, 2009; Martínez et al. 2000; Martínez 2008). In 1989 the lake dried out and the crater bottom was visible (e.g. Rymer et al. 2009; Martínez et al. Chapter "Behaviour of Polythionates in the Acid Lake of Poás Volcano: Insights Into Changes in the Magmatic-Hydrothermal Regime and Subaqueous Input of Volatiles") (Fig. 1). After a sequence of phreatic eruptions (Fig. 2), the Laguna Caliente once again disappeared in 1990, 1991, 1992, 1993 and 1994 (Martínez et al. 2000, Chapter "Behaviour of Polythionates in the Acid Lake of Poás Volcano: Insights Into Changes in the Magmatic-Hydrothermal Regime and Subaqueous Input of Volatiles"; Martínez 2008). The acidic crater lake restored in mid-1995, although significant variations in terms of lake level and chemistry were observed (e.g. Rowe et al. 1992, 1995; Rowe 1994; Martínez et al. 2000; Rouwet et al. Chapter "39 Years of Geochemical Monitoring of Laguna Caliente Crater Lake, Poás: Patterns from the Past as Keys for the Future" and references therein).

The below reported observations provide details on the presence and migration of the fumarolic gas discharges and are part of the observations that the OVSICORI personnel provided in their bulletins (http://www.ovsicori.una.

Fig. 1 The crater bottom after the Laguna Caliente dried out in 1989

Fig. 2 A small phreatic eruption at Poás volcano in 1989

ac.cr/) and forwarded to the Global Volcanism Program of the Smithsonian Institute (http://volcano.si.edu/) as well as from Martínez (2008). During the period considered in this chapter, the fumarolic activity was mainly located close to the pyroclastic dome while other areas from the inner flanks of the Active Crater were affected by the presence of new fumaroles, whose activity waned with time until they disappeared and occasionally new fumaroles formed. It is noteworthy to mention that the outlet temperatures recorded after November 2008 at the interface between the Laguna Caliente and the pyroclastic dome have to be considered with caution since the fumarolic gas discharges were quite difficult to be reached and measured.

In 1998, when the periodic geochemical survey was carried out in the framework of the collaboration between OVSICORI and CNR-IGG and DES, gas bubbling was noticed around the pyroclastic dome and along the lower part of the eastern inner terrace of the Active Crater. It is to mention that vigorous discharge around the pyroclastic dome started in September 1995 accompanied by small flows of bright molten sulfur (temperature was around 113–119 °C) and strong sulfur smell, after a 7-years period of only fairly weak fumarolic activity around the pyroclastic dome. The Laguna Caliente volume increased in a steady fashion throughout 1995–1997, reaching a record level between September and December 1997, partly due to above-average rainfall. A small hydrothermal explosion occurred in the northern side of the dome on April 1996, turning the lake color from milky turquoise to grey for several weeks (Martínez et al. Chapter "Behaviour of Polythionates in the Acid Lake of Poás Volcano: Insights Into Changes in the Magmatic-Hydrothermal Regime and Subaqueous Input of Volatiles"). Interestingly, polythionates reappeared abruptly in the acidic lake in February 1996 with a 3-fold to 6-fold increase in the $S_4O_6^{2-}/S_5O_6^{2-}$ and $S_4O_6^{2-}/S_6O_6^{2-}$ ratios, respectively, after being practically absent during most the 1987–1995 period. This suggested and input of fresh S-rich magmatic gases into the Laguna Caliente. In addition, in ca. 40 year of geochemical observations at Laguna Caliente, between late 1996 and mid-2004, the lake brines unusually enriched in chloride with respect to sulfate (Martínez 2008; Martínez et al. Chapter "Behaviour of Polythionates in the Acid Lake of Poás Volcano: Insights Into Changes in the Magmatic-Hydrothermal Regime and Subaqueous Input of Volatiles"). From 1999 to 2000, fumarolic gas vents were observed around the pyroclastic dome and along the lower part of the eastern terrace at the active crater (e.g. Fumarole Norte). In March 1999, several acidic springs discharged waters along the lower part of the inner eastern terrace. On September 28, 1999, a fumarolic plume (about 2 km high) formed above the pyroclastic dome. Between December 1999 and February 2000, seismic signals corresponding to tremor increased. In February 2000, 24 h of tremor were recorded on a single day, 11 h of which corresponded to monochromatic tremor. From March 2001 to September 2001, new fumarolic vents opened in the eastern area of the inner crater and a relatively strong degassing was observed through the pyroclastic dome. Enhanced fumarolic degassing was accompanied by increasing seismic activity. For instance, in March 2001, 35 h of polychromatic tremor were recorded in a matter of 3 consecutive days. In 2000 and 2001, fumaroles and springs along the eastern terrace registered temperature of 88–95 °C and 66-89 °C, respectively. From October 2001 to February 2005, the fumarolic outgassing through the pyroclastic dome remained relatively weak and practically unnoticeable between late 2003 and throughout 2005. Similar to the fumarolic activity occurring close to the pyroclastic dome, Laguna Caliente remained relatively quiet during this time period except for some upwelling, floating molten sulfur and persistent emanation of fumes, particularly from August 2002 to December 2003.

The fumarolic activity in the eastern part of the crater maintained relatively vigorous. At the Fumarole Norte, outlet temperatures >110 °C were recorded in March 2003. Between late 2003 and early 2005, the lake volume dramatically increased, reaching its highest level since the late seventies, provoking a water overflow in the eastern edge of the lake and submerging some sub-aerial fumaroles for several weeks in early 2005.

In March 2005, a new period of strong convective activity and evaporation at Laguna Caliente commenced after 11 years of relative quiescence, changing drastically the color and chemistry of the lake, suggesting an increase in the input of heat and volatiles through the subaqueous fumarolic discharges. Simultaneously, Fumarole Norte showed a significant increase in temperature (from 105 °C in March 2005 to 203 °C in May-June 2005). Unusually high seismic activity paralleled these drastic changes, especially in 2005. Between March 2005 and early 2006 the lake level dropped by 10 m. The fumarolic activity around the pyroclastic dome showed a gradual increase in the degassing rate from April 2005 onwards although the temperature remained around 92 °C. From March 2006, the phreatic activity resumed in the Laguna Caliente after 12 years of relative quiescence. The renewal of phreatic activity was preceded and accompanied by an abrupt increase of both the seismic activity and most outlet temperatures of the fumarolic discharges. Phreatic explosions were observed in March, April, September and December 2006. No phreatic explosions were recorded in 2007, although strong degassing and convection were observed in the Laguna Caliente, producing a significant lowering of the water level and an increase of the water temperature (up to 60 °C). Springs and fumaroles that appeared in the eastern terrace in early 1999–2000 dried out or disappeared in 2007. In October 2007, fumarole temperatures at the pyroclastic cone reached 101 °C. In 2008, new phreatic eruptions occurred, generating muddy water columns reaching to ca. 200 m above the lake surface and leaving fragments of altered rocks and molten sulfur dispersed on the crater floor. In November 2008, the temperatures of the fumaroles close to the pyroclastic dome were 322 °C, although in early December 2008 they apparently went down to 95 °C. A decrease of the water temperature (45 °C in November 2008 to January 2009) of the Laguna Caliente was also recorded. On the 7th and 8th of January 2009, two shallow earthquakes (6.5–7.0 km deep) hit the surroundings of Poás volcano. The 7th of January 2009 4.5 M earthquake occurred at 10:00 a.m. local time 4 km N-NE of Fraijanes Sabanilla of Alajuela, while on the 8th of January 2009 a 6.2 M earthquake (estimated depth: 6.0 km) occurred at 13:21 local time, just 1 km south of Cinchona of Poás Alajuela (e.g. Barquero 2009; Alvarado 2010; Fernández-Arce and Mora Amadora, Chapter "Seismicity of Poás Volcano, Costa Rica"; Ruiz et al. Chapter "Geochemical and Geochronological Characterisation of the Poas Stratovolcano Stratigraphy"). Between January and July 2009 soon after these quakes hit the region, the lake showed a strong release of highly corrosive gases at its surface and large convective cells formed. A rapid uprising in acidity and concentration of dissolved volatiles and rock-forming elements occurred. The temperature fluctuated around 45–51 °C. Between January and March 2009 small phreatic eruptions were recorded. In May 2009, the pH of the lake was very low (pH = −0.72). The pH, temperature, and electrical conductivity of Laguna Caliente approached the values measured in the period 1986–1989 when the lake level was reduced and dried out completely in April 1989. Although between 2005 and 2015 the lake was affected by strong fluctuations in volume, the water levels did not reach those recorded between 1986 and 1994. The pyroclastic dome fumaroles remained around 93 °C, although in June 2009 the outlet temperature at some accessible sites around the pyroclastic dome was of 123 °C. In September 2009 the lake level dropped by about 4 m relative to that recorded in August 2009 and the temperature increased up to 55 °C. Between September and December 2009 the outlet temperatures of the pyroclastic dome fumaroles increased (to 150 and 665 °C). On the 24th and 25th of December two phreatic eruptions occurred, the later producing a phreatic plume of ca. 600 m above the surface of the lake. In January to February 2010, small to medium phreatic eruptions were reported by the park rangers. On the 23rd of February, a phreatic eruption produced a water and steam column of ca. 1 km above the lake surface. The outlet temperatures of the fumaroles around the pyroclastic dome was approaching 640 °C. Between March and May 2010, two small phreatic eruptions were observed by the park rangers and the fumaroles at the pyroclastic dome were up to 838 °C. Between June and December 2010, small to medium size phreatic eruptions were recorded (Fig. 3) and water and fumarole temperature reached to 64 and 865 °C, respectively. From January to April 2011, the fumarolic discharges close to the pyroclastic dome reached 600 °C and gas bursts and small phreatic explosions were noticed in the central part of the lake. In June 2011 the dome fumaroles reached a temperature of 663 °C and several small to moderate size phreatic eruptions (more than 100 events of a few dozens of meters of height) were registered. From July to September 2011, the dome fumaroles showed temperatures up to 890 °C (Fig. 4) and incandescence was observed even under day light conditions. Small phreatic eruptions continued to occur from July to August 2011. From October to December 2011 and in February 2012, the outlet temperature of the dome fumaroles dropped to 730 °C and no phreatic activity was observed. In March-June 2012, the dome fumaroles further dropped in temperature to 610 °C and a 500 m high phreatic eruption occurred on the 15th of May. In July 2012, the dome fumaroles showed a sharp decrease (down to 302 °C), accompanied by several small phreatic explosions occurred, the most important being the one on the 30th of July. In August to September 2012, the dome fumaroles showed temperature <200 °C and no phreatic eruptions were observed, whereas in October 2012 at least ten small to moderate size phreatic explosions were recorded. On the 20th of October 2012, about 95% of the lake surface was covered by slicks of molten sulfur. In November 2012, the temperatures at the dome fumaroles were clustering around 170 °C and small phreatic explosions occurred in the lake. The phreatic activity continued up to August 2013. In April 2013, the dome fumaroles reached a temperature of 380 °C and increased to 600 °C in August 2013. Then, at the end of 2013, the temperature was of about 400 °C whilst from January to April 2014, it was fluctuating between 600 and 720 °C. A number of phreatic eruptions occurred between February and June 2014. The lake water temperature was relatively stable, ranging between 45 and 55 °C. Between July and September 2014, the dome fumaroles showed no significant variations in terms of outlet temperatures and numerous phreatic explosions were recorded, most of them being of small size (30–50 m high), with the exception of those occurring on the 31st of July and the 27th of August, when up to 200 m high plumes were recorded. In April 2013 and July and September 2014, the flux rate of SO_2 was measured by mobile DOAS with values of 120, 530 and 240 ton/day, respectively (e.g. Martínez et al. Chapter "Behaviour of Polythionates in the Acid Lake of Poás Volcano: Insights Into Changes in the Magmatic-Hydrothermal Regime and Subaqueous Input of Volatiles" and references therein).

From October 2014 to August 2015, the phreatic activity at Poás volcano ceased. In May 2015 the fumaroles on the pyroclastic cone temperature had a temperature of 625 °C while in August 2015 it was 360 °C. In October 2014, two important phreatic episodes occurred (250 m high) a few days after the strong emission of flashes of hot gases through the pyroclastic dome. In August 2015 the lake had a temperature of only 31 °C with its water level rapidly rising up since late 2014.

In summary, the cycle of phreatic activity that started in 2006 and continued almost uninterruptedly until October 2014 coincided with a significant increase of the temperatures at the dome fumaroles. Furthermore, it has to be mentioned that over the last 40 years, Poás has experienced 3 stages of phreatic activity that were alternated with periods of quiescence in cycles of 6 to 10 years. In April 2017 a new phase opened, when phreatic and phreatomagmatic events occurred, the latter of which destroyed the 1953–1955 dome (Martínez et al. Chapter "Behaviour of Polythionates in the Acid Lake of Poás Volcano: Insights Into Changes in the Magmatic-Hydrothermal Regime and Subaqueous Input of Volatiles"; Mora Amador et al. Chapter "Volcanic Hazard Assessment of Poás (Costa Rica) Based on the 1834, 1910, 1953–1955 and 2017 Historical Eruptions"; Rouwet al. Chapter "39 Years of Geochemical Monitoring of Laguna Caliente Crater Lake, Poás: Patterns from the Past as Keys for the Future"). From the second

Fig. 3 A small phreatic eruption inside the Laguna Caliente taken on the 14th of August 2010

Fig. 4 Aerial photo of the fumarolic vents at the interface between the Laguna Caliente and the pyroclastic dome. The right-hand side fumarole shows a typical bluish color, indicative of the presence of SO_2 and fine sulfate-rich aerosols

half of 2017 to early 2018, the volcanic activity ceased and low temperature fumaroles and sporadic mud-pools at the dried bottom of the Active Crater remained (Martínez et al. Chapter "Behaviour of Polythionates in the Acid Lake of Poás Volcano: Insights Into Changes in the Magmatic-Hydrothermal Regime and Subaqueous Input of Volatiles").

3 Sampling Sites and Analytical Methods

Owing to the different status of the volcanic system from 1998 to 2014 as well as the appearance and disappearance of new fumarolic vents, there is no continuous record for most of the studied fumaroles with the exception of the pyroclastic dome fumaroles ("Dome" hereafter), although there is a gap of sampling between 2000 and 2007. Relatively long-term chemical compositions are available for the **Fumarole Norte** (namely the "Naranja" fumarole in Hilton et al. 2010 and Fischer et al. 2015), **La Niña** and **Fumarole Este** (namely, the "Official" in Hilton et al. 2010 and Fischer et al. 2015). Sporadic samplings were thus performed at the fumaroles **Pared Sur**, **Fumarole SE**, **Fumarole NE**, **Fumarole BP** (boiling pool) and **Fumarole ENE**. One gas sample was collected inside the Laguna Caliente (**Lake gas**) when a vertical profile of the crater lake was carried out (Vaselli et al. 2003). It is worth to mention that Zimmer et al. (2004), Hilton et al. (2010) and Fischer et al. (2015) have only reported the chemical and isotopic analyses of the Dome, Fumarole Norte and Fumarole Este. Moreover, with the exception of the fumaroles reported in the previously mentioned papers and those described and discussed in Vaselli et al. (2003), all the other data are unpublished. To the best of our knowledge, no gas data are available after 2015.

For the sake of clarity, the location of the fumaroles investigated is reported in Fig. 5a, b along with a detail of the Fumarole Norte (Fig. 5c) and glowing of the Dome fumarole (July 2011; Fig. 5d) with a temperature of 890 °C, the latter being sampled when the outlet

Fig. 5 Sampling location of the Poás fumaroles reported in the present chapter: **a** the Pared Sur and the Dome fumaroles (photo taken in September 1999); **b** the Fumarole Norte, Fumarole NE, Fumarole ENE, Fumarole Este, Fumarole SE, La Niña and the Lake gas (photo taken in February 2004); **c** the Fumarole Norte in March 2007 and **d** the Dome fumarole in July 2011 when the outlet temperature was of 890 °C

temperature was of 375 °C (April 2014, Fischer et al. 2015).

Gas samples for chemical and isotopic composition reported in Zimmer et al. (2004), Hilton et al. (2010) and Fischer et al. (2015) were collected by following the sampling techniques and measured by using the analytical procedures described in Giggenbach and Goguel (1989), Hilton et al. (2002), Zimmer (2002), Mitchell et al. (2010) whereas the methodologies used in Vaselli et al. (2003) and for the unpublished data were reported in Montegrossi et al. (2001) and Vaselli et al. (2006).

4 Results

The chemical (in mmol/mol) and isotopic (helium and argon, carbon in CO_2) composition of the Poás fumarolic discharges produced by the Italian-Costa Rica team (hereafter ICT), also including those from Vaselli et al. (2003), are reported while those produced by the American team (hereafter AT) can be found in Zimmer et al. (2004), Hilton et al. (2010) and Fischer et al. (2015). Most the outlet temperatures of the fumaroles were below 120 °C up to early 2005. Then, 250 °C were recorded in March 2014 (Fumarole Norte, AT). An abrupt increase was then observed at the Dome fumarole (hereafter Dome), whose outlet temperature steadily incremented since September 2009 (480 °C, ICT) and reached 763 °C in June 2010 (ICT). As previously mentioned, after the phreatic events occurring after March 2006, there were some problems in approaching the Dome due to the intense emissions and high temperatures.

All the fumaroles were dominated by steam, whose concentrations were between 725 and 998 mmol/mol with the exception of the gas collected at the surface of the acidic lake (Lake gas in Table 1), which contained the highest H_2 concentrations (February 2001, 31.58 mmol/mol; ICT) recorded at Poás from 1998 to 2014, along with the Dome gases collected in January 2009 by ICT, and that of the Fumarole NE (February 2001; ICT) when it was collected for the first time. CO_2 contents were very variable as they were ranging from 1 to 668 mmol/mol, the highest value recorded in the Lake gas. Lower concentrations were measured for SO_2 (from 0.002 to 247 mmol/mol) and H_2S (from 0.02 and 9 mmol/mol) and, when available, the H_2S/SO_2 was between 0.003 (Fumarole Este, March 2005; AT) and 822 (Pared Sur, February 1999; ICT). H_2S and SO_2 were below the detection limit in the Lake gas. ICT also measured the concentrations of elemental gaseous sulfur (Table 1), with concentrations between 9×10^{-4} and 5.8×10^{-2} mmol/mol.

HCl and HF, with subordinate contents with respect to those of other acidic gases, were varying from 0.0031 and 26 mmol/mol and 0.00001 and 4 mmol/mol, respectively. It is noteworthy to point out that, on average, the highest concentrations of the acidic gases were recorded at the Dome, although CO_2 (245 mmol/mol; ICT), SO_2 (247 mmol/mol; ICT) and H_2S (6.44 mmol/mol; ICT) showed the highest concentrations at the Fumarole ESTE when it was sampled for the first time in February 2001. The N_2/Ar ratios were spanning quite widely, being varying from 44 (suggesting the presence of an atmospheric component) to 13,715 (likely related to a contribution of thermometamorphic component), with the highest values registered at the Dome, with a clear increase with time. Setting aside the previously mentioned gas samples, the H_2 concentrations were comprised between 0.0004 and 11.8 mmol/mol. Consistently with the high concentrations of the acidic gases, the Dome also showed the highest values of H_2, which tended to increase with time. Carbon monoxide showed a behavior similar to that of H_2 and the acidic gases as the highest contents were recorded at the Dome (up to 0.999 mmol/mol). Neon and helium had contents between 4.4×10^{-8} and 5.0×10^{-5} mmol/mol and 4.0×10^{-5} and 7.96×10^{-1} mmol/mol respectively, while the He/Ne ratios were up to 1,900 and constantly higher than 10, with the exception of the Dome (February 2000; ICT).

The helium isotopic ratios (expressed as R/Ra, where R is the measured $^3He/^4He$ in the sample and Ra is the $^3He/^4He$ in the air. The values, corrected for air contamination using the He/Ne

Table 1 Temperature (in °C) and chemical (in mmol/mol) and isotopic ($\delta^{13}C$–CO_2, $^3He/^4He$ expressed as R/Ra corrected for the He/Ne ratio and $^{40}Ar/^{36}Ar$) composition of the fumarolic discharges collected by the Italian-Costa Rican Team

Poas (mmol/mol)	T (°C)	H_2O	CO_2	SO_2	H_2S	S	HCl	HF	N_2	CH_4	Ar	O_2	Ne	H_2	He	CO	$\delta^{13}C$	R/34 Ra	$^{40}Ar/^{36}Ar$
Pared Sur Feb. 1998	93	987	9.90		0.84		0.432		1.198	0.002274	0.01141	0.00001426	0.0000225	0.2793	0.00171	0.00002	n.d.		
Parete Sur Feb. 1999	89	978	18.41	0.002	1.88	0.0001			1.345	0.001163	0.00736	0.00000199	0.0000060	0.0941	0.00226	0.00044	−6.56		
Dome Feb. 1999	95	936	30.07	29.35		0.0034	1.733	0.0040	1.956	0.000226	0.00792		0.0000070	0.4542	0.00149	0.03891	n.d.		
Dome Nov. 1999	95	946	24.11	26.35	0.57	0.0372	1.563	0.0055	1.583	0.000106	0.03267			0.0518	0.00446	0.00953	n.d.	6.65	
Domo Feb. 2000	94	992	0.74	2.95	0.86	0.0165	1.430	0.0068	0.469	0.000075	0.00680		0.0000028	1.9129	0.00002	0.00495	−6.13		
Dome Oct. 2007	90	930	40.02	29.16	0.31	0.0005	0.692	0.0673	0.013	0.000021	0.00010	0.00000185	0.0000001	0.1731	0.00001	0.00050			
Dome Jan. 2008	96	926	46.55	26.45	0.41	0.0004	0.641	0.0515	0.054	0.000026	0.00021	0.00000055	0.0000001	0.1556	0.00002	0.00115			
Dome March 2008	97	912	50.32	36.34	0.35	0.0004	0.501	0.0575	0.066	0.000025	0.00032	0.00000018	0.0000002	0.2084	0.00002	0.00091			
Dome May 2008	109	891	63.37	43.57	0.53	0.0005	1.073	0.1807	0.071	0.000060	0.00008	0.00000123	0.0000000	0.1485	0.00008	0.00061			
Dome June 2008	112	891	61.46	44.85	0.51	0.0003	0.988	0.1544	0.327	0.000048	0.00025	0.00000099	0.0000001	0.2568	0.00005	0.00562			
Dome July 2008	180	872	73.46	51.42	0.56	0.0014	1.215	0.1612	0.715	0.000031	0.00069	0.00000112	0.0000003	0.4551	0.00009	0.02145			
Dome Oct. 2008	276	840	94.58	60.45	0.49	0.0014	1.744	0.1698	0.710	0.000026	0.00060	0.00000068	0.0000003	1.5410	0.00006	0.05845			
Dome Jan. 2009	322	897	61.26	31.65	3.54	0.0041	2.874	0.2045	0.898	0.000015	0.00062	0.00000024	0.0000001	2.9840	0.00009	0.10890			
Dome March 2009	93	952	26.58	15.49	3.08	0.0022	1.105	0.0215	0.665	0.000125	0.00122	0.00000046	0.0000003	0.9842	0.00022	0.02145			
Dome May 2009	93	944	33.27	16.87	3.27	0.0016	0.987	0.0256	0.649	0.000085	0.00099	0.00000039	0.0000002	1.1250	0.00016	0.02645			
Dome June 2009	123	906	61.59	25.65	2.78	0.0018	1.254	0.0985	0.771	0.000036	0.00072	0.00000026	0.0000002	1.8450	0.00011	0.07745			
Dome Aug. 11, 2009	97	931	36.13	27.49	2.56	0.0012	0.845	0.0561	0.649	0.000025	0.00051	0.00000045	0.0000003	1.3540	0.00004	0.04856			
Dome Aug. 31, 2009	116	879	61.55	54.95	0.85	0.0026	1.145	0.0815	0.595	0.000015	0.00047	0.00000070	0.0000003	1.9540	0.00006	0.07941			
Dome Sept. 2009	480	814	104.52	71.46	0.29	0.0051	2.516	0.2154	0.815	0.000005	0.00061	0.00000035	0.0000003	5.2150	0.00008	0.58790			
Dome Oct. 2009	415	822	98.75	70.46	0.42	0.0032	2.611	0.2616	0.755	0.000006	0.00056	0.00000016	0.0000002	4.7150	0.00008	0.48750			
Dome Nov. 2009	665	738	169.47	76.95	0.32	0.0042	4.155	1.6480	0.789	0.000005	0.00063	0.00000038	0.0000003	7.9540	0.00009	0.71520			
Dome Jan. 2010	650	754	156.95	72.15	0.35	0.0085	4.518	1.9450	0.812	0.000005	0.00051	0.00000030	0.0000003	8.1450	0.00007	0.89410			
Dome Mar. 2010	160	796	115.65	68.46	0.69	0.0070	5.071	2.6540	0.718	0.000006	0.00046	0.00000014	0.0000002	9.5610	0.00006	0.91560			

(continued)

Table 1 (continued)

Poas (mmol/mol)	T (° C)	H_2O	CO_2	SO_2	H_2S	S	HCl	HF	N_2	CH_4	Ar	O_2	Ne	H_2	He	CO	$\delta^{13}C$	R/34 Ra	$^{40}Ar/^{36}Ar$
Dome Mar. 2010*	480	765	138.47	75.84	0.38	0.0068	5.584	3.7410	0.670	0.000003	0.00038	0.00000013	0.0000002	9.7450	0.00004	0.88870			
Dome Apr. 2010	566	732	168.41	75.85	0.57	0.0015	6.415	3.1540	0.846	0.000002	0.00019	0.00000006	0.0000001	11.8740	0.00004	0.91480			
Dome May 2010	148	837	94.59	54.87	1.15	0.0008	2.654	0.8542	0.715	0.000002	0.00018	0.00000005	0.0000001	7.4580	0.00005	0.47850			
Dome June 2010	763	725	158.50	91.75	0.85	0.0007	7.841	3.5450	0.618	0.000001	0.00007	0.00000004	0.0000001	10.8450	0.00004	0.99860			
Dome Aug. 2010	650	741	148.46	87.65	1.41	0.0006	7.154	2.9840	0.598	0.000001	0.00007	0.00000002	0.0000001	9.4210	0.00003	0.87550			
Dome Mar. 2011	250	746	158.48	75.99	1.18	0.0004	6.485	2.1140	0.716	0.000002	0.00006	0.00000002	0.0000001	8.4570	0.00003	0.74580			
Dome Apr. 2011	250	759	146.99	76.45	1.25	0.0005	6.187	1.9850	0.698	0.000003	0.00006	0.00000003	0.0000001	6.8570	0.00004	0.69860			
Dome May 2011	280	753	149.85	77.15	1.07	0.0006	6.845	2.2180	0.658	0.000002	0.00005	0.00000002	0.0000001	8.1150	0.00003	0.81140			
Fumarole Este Feb. 2001	97	923	74.22	1.14	0.26	0.0141	0.473		0.573	0.000001	0.00627	0.00013586	0.0000036	0.0005	0.00005		−4.29		
Fumarole Este Feb. 2004	99	967	22.81	8.54	0.79	0.0006	0.751	0.0001	0.054	0.000001	0.00043	0.00000188	0.0000003	0.0012	0.00002	0.00002			
Lake gas Feb. 2001	78	282	668.00						15.899	0.000312	0.05547	0.00141086	0.0000320	31.5855	0.00001	0.02199	n.d.		
Fumarole NE Feb. 2001	101	498	245.57	247.04	6.44	0.0576	2.294	0.0002	0.758	0.000026	0.01058		0.0000063	0.0252	0.00089	0.00041	n.d.		
Fumarole NE Aug. 2001	101	794	82.12	118.22	4.30	0.0031	0.665	0.0000	0.446	0.000002	0.00767		0.0000036	0.0265	0.00030	0.00105	n.d.		
Fumarole NE Nov. 2001	104	852	117.61	29.05	0.68	0.0015	0.172	0.0001	0.959	0.000003	0.00458	0.00000921	0.0000019	0.0049	0.00025	0.00250	n.d.		
Fumarole NE Feb. 2002	103	931	34.52	26.38	0.16	0.0017	7.937	0.0366	0.076	0.000000	0.00023	0.00000010	0.0000001	0.0061	0.00006	0.00007	n.d.		
Fumarole NE Apr. 2002	98	799	140.64	57.77	1.08	0.0018	0.554	0.0022	1.083	0.000121	0.00597	0.00013264	0.0000042	0.0059	0.00020	0.00044	n.d.		
Fumarole NE Apr. 2003	118	921	40.95	35.95	1.18	0.0006	0.911	0.0006	0.085	0.000184	0.00189	0.00000368	0.0000016	0.1032	0.00003	0.00003	n.d.		
Fumarole NE Jul. 1, 2003	116	948	26.35	23.12	0.18	0.0007	1.394	0.0001	0.776	0.000004	0.00268	0.00000761	0.0000016	0.0318	0.00015	0.00094			
Fumarole NE Jul. 29, 2003	116	941	29.21	27.65	0.34	0.0003	1.899	0.0001	0.241	0.000002	0.00089	0.00000215	0.0000005	0.0067	0.00004	0.00025			
Fumarole NE Apr. 2004	115	944	26.40	21.47	0.18	0.0007	1.561	0.0001	5.976	0.000302	0.02117	0.00001473	0.0000118	0.5631	0.00016	0.00740	−3.06	7.47	325

(continued)

Table 1 (continued)

Poas (mmol/mol)	T (°C)	H_2O	CO_2	SO_2	H_2S	S	HCl	HF	N_2	CH_4	Ar	O_2	Ne	H_2	He	CO	$\delta^{13}C$	R/34 Ra	$^{40}Ar/^{36}Ar$
Fumarole NE Jun. 2007	116	907	44.99	43.82	0.47	0.0003	0.444	0.0011	0.142	0.001110	0.00079	0.00002527	0.0000005	3.0024	0.00031	0.00027			
Fumarole NE Sep. 2007	113	901	48.75	46.28	0.48	0.0004	0.550	0.0045	0.251	0.001218	0.00099	0.00002097	0.0000005	2.9326	0.00023	0.00030			
La Niña Feb. 2001	95	929	64.70		5.39	0.0029			0.409	0.000004	0.00419	0.00000663	0.0000028	0.0005	0.00015	0.00005	−2.57		
La Niña Nov. 2001	81	970	28.88		1.25	0.0006	0.003	0.0000	0.129	0.000003	0.00062	0.00000669	0.0000003	0.0084	0.00029	0.00003	−3.54		
La Niña Feb. 2002	86	963	36.22		0.60	0.0005	0.018	0.0001	0.124	0.000003	0.00042	0.00000024	0.0000003	0.0103	0.00024	0.00008	n.d.		
La Niña Apr. 2002	93	962	36.25	0.25	1.08	0.0010	0.082	0.0240	0.050	0.000001	0.00016	0.00000002	0.0000001	0.0041	0.00014	0.00001	−3.12		
La Niña Apr. 2, 2003	97	931	42.94	24.32	0.36	0.0009	1.576	0.0001	0.149	0.000024	0.00335	0.00001049	0.0000030	0.0004	0.00005	0.00009	n.d.		
La Niña 28 Apr., 2003	97	908	56.91	33.87	0.49	0.0007	0.300	0.0000	0.173	0.000123	0.00306	0.00000385	0.0000028	0.0008	0.00010	0.00017	n.d.		
La Niña Apr. 2004	94	963	31.01	0.26	0.35	0.0005	0.240	0.0000	4.622	0.000020	0.00994	0.00007624	0.0000065	0.3113	0.00014	0.00307	n.d.		
La Niña Feb. 2005	94	959	30.61	10.39	0.12	0.0006	0.038	0.0000	0.103	0.000001	0.00104	0.00000003	0.0000006	0.0132	0.00015	0.00006	n.d.		
La Niña Apr. 2005	94	984	13.29	1.97	0.38	0.0004	0.242	0.0000	0.024	0.000004	0.00053	0.00000003	0.0000003	0.0096	0.00006	0.00002	n.d.		
Fumarole Norte Feb. 2001	95	976	15.86	6.82	0.22	0.0011	0.274	0.0001	1.122	0.000016	0.00459			0.1691	0.00153	0.00097	n.d.		
Fumarole Norte Nov. 2001	101	895	63.32	40.63	0.74	0.0030	0.420	0.0002	0.197	0.000003	0.00100	0.00000019	0.0000004	0.0209	0.00066	0.00021	−2.90		
Fumarole Norte Nov. 2001	101	929	32.01	34.22	0.76	0.0016	3.490	0.0280	0.103	0.000001	0.00059	0.00000303	0.0000002	0.0085	0.00008	0.00005	−3.70		
Fumarole Norte Jan. 2002	101	929	37.33	28.98	1.03	0.0007	3.357	0.0270	0.089	0.000001	0.00042	0.00000006	0.0000002	0.0068	0.00015	0.00003	n.d.		
Fumarole Norte Feb. 2002	101	952	28.70	17.05	0.65	0.0010	1.926	0.0155	0.119	0.000001	0.00060	0.00000938	0.0000002	0.0054	0.00010	0.00032	n.d.		
Fumarole Norte Apr. 2002	94	901	62.75	31.60	0.56	0.0030	0.488	0.0124	3.001	0.000005	0.01685	0.00015727	0.0000117	0.0107	0.00030	0.00173	n.d.		
Fumarole Norte May 2002	99	943	49.12	4.25	1.33	0.0006	1.406	0.4508	0.256	0.000001	0.00132	0.00000104	0.0000011	0.0043	0.00010	0.00006	−2.90		
Fumarole Norte Jul. 2002	88	908	53.14	36.23	0.66	0.0017	1.307	0.4204	0.207	0.000001	0.00105	0.00000330	0.0000009	0.0077	0.00005	0.00025	n.d.		

(continued)

Table 1 (continued)

Poas (mmol/mol)	T (°C)	H_2O	CO_2	SO_2	H_2S	S	HCl	HF	N_2	CH_4	Ar	O_2	Ne	H_2	He	CO	$\delta^{13}C$	R/34 Ra	$^{40}Ar/^{36}Ar$
Fumarole Norte Apr. 2003	93	947	34.73	17.09	0.75	0.0008	0.120	0.0462	0.088	0.000005	0.00046	0.00000149	0.0000004	0.0035	0.00002	0.00030	−2.71		
Fumarole Norte Apr. 2004	97	928	36.00	12.37	0.87	0.0025	2.369	0.0003	19.333	0.000101	0.09864	0.00010087	0.0000547	0.7227	0.00071	0.01914	n.d.		
Fumarole Norte Aug. 2004	95	936	23.17	39.76	0.74	0.0004	0.245	0.0008	0.092	0.000001	0.00060	0.00000172	0.0000004	0.0040	0.00008	0.00009	n.d.		

ratios, were in a relatively narrow range, from 6.64 (Fumarole ESTE, November 2006; AT) to 7.47 (Fumarole NE, April 2004; ICT). The carbon isotopes in CO_2 (expressed as $\delta^{13}C–CO_2$ ‰ vs. V-PDB) showed a higher variability as they were ranging from −6.8 (Fumarole Este, March 2001 AT) to −1.3 (Eastern Flank June 2005 AT) ‰ versus V-PDB. Few samples were analyzed by AT for nitrogen isotopes in N_2 (expressed as $\delta^{15}N–N_2$ ‰ vs. NBS-AIR) with values from −3.0 (Fumarole ESTE July 2001; AT) to 1.9 (Fumarole ESTE March 2003; AT) ‰ versus NBS-AIR. The chlorine isotopes (expressed as $\delta^{37}Cl$‰ vs. SMOC), measured in the Este and Norte fumaroles (AT) varied between 10.4 and 11.1‰ versus SMOC, with the only exception of the Fumarole Norte collected in November 2006 (−0.9‰ vs. SMOC). One gas sample (Fumarole NE, April 2004; ICT) was measured for the $^{40}Ar/^{36}Ar$ isotopic ratio, showing a value of 325 (Table 1), i.e. slightly higher than that of the air (295.5).

5 Discussion

5.1 Fumarole Migration and Discontinuous Monitoring Activity

This chapter was designed to assemble the chemical variations observed in a relatively large span of time (1998 to 2014), the relationships with the relatively few isotopic data available and the volcanic activity of Poás, particularly after the new phreatic phase that started in 2006 and still on-going and the Chinchona earthquake that occurred in January 2009. Owing to safety reasons, to the best of our knowledge, no chemical and isotopic data are available after 2014.

As reported by Vaselli et al. (2003), the fumarolic activity at Poás was characterized by a large spatial variability in terms of fumarolic vents, as they migrated and/or disappeared while in other cases new fumaroles formed. Consequently, maintaining one or more fumaroles as monitoring reference sites was practically impossible. The longest period of observation was reported for the Dome from where 29 and 5 gas samples were collected by ICT and AT, respectively, between 1999 and 2014, although, even in this case, the same sampling site was not always the same. In addition, from middle 2000 to 2007 no geochemical data are available for this fumarole.

In 1998, the Dome fumarole was too weak to be sampled and only a typical hydrothermal (CO_2–H_2S dominated) gas was collected from Pared Sur that was not found in the 2000 sampling survey. Other fumaroles had a relatively short life, Fumarole Norte: from 2001 to 2008, and sampled by ITC from 2001 to 2004 and by AT up to 2009; La Niña: from 2001 to 2005; Fumarole NE: from 2001 to 2004; Fumarole Este: from 2001 to 2009. This implies that the opening of the 2006 phreatic episodes was better recorded by the NE and Este fumarole while the Dome can be considered a good proxy for evaluating the possible effects derived by the Cinchona earthquake. Another problem to be considered is when the data collected by AT and ICT are compared. Different periods of sampling, different sites from where a certain fumarole was collected and different techniques of sampling and analyses may be challenging to overlap the data despite the fact that they were likely sampled "more or less" from the same site. In addition, when new fumarolic vents open it was difficult to evaluate whether their chemical composition was to be considered chemically representative of the system since they were likely affected by hydrofracturing processes of chilled fragile margins at the surface due to volatile overpressure or uprising magma (e.g. Casertano et al. 1987; Rymer and Brown 1989; Rowe et al. 1992; Fischer et al. 2015). This means that the new fumaroles may likely need some time before reaching a steady-state status as shown by Fumarole NE (high concentrations of CO_2, SO_2 and H_2S, which from February to November 2001 dramatically decreased). A different behavior, though always indicating that few days-to-months (?) are necessary to stabilize a "young" fumarolic system, was shown by La Niña fumarole where SO_2 was not detected from February 2001 to February 2002 but appeared in

April 2002 (Table 1). During the very first sampling, this fumarole was indeed characterized by a small boiling pool that eventually dried out, thus resulting in the presence of acidic gases in April 2002. In any case, it is quite difficult in an active volcanic system such as that of Poás to attribute these initial variations and chemical modifications to the system itself or to the response of changing magmatic and hydrothermal conditions occurring at depth.

Fischer et al. (2015) examined the reactions that controlled the major components (H_2O, CO_2 and SO_2) of the Poás fumaroles, demonstrating that they were basically depending on chemical reactions of the gas phase, which involved the sulfur species at temperatures from 400 to 800 °C, whereas the rock buffer appeared to have no control on the Poás gas chemistry. This important result allowed to consider the H_2S/SO_2 ratio as an indicative parameter of possible fluctuations related to modifications of the plumbing system. Similarly, volcanic variations can be assessed by the C/S_{tot} ratio as CO_2 and S are characterized by different solubility in silicate melts, the former being less soluble (e.g. Carroll and Webster 1994, Holloway and Blank 1994, Aiuppa et al. 2005, 2009; Fischer et al. 2015).

5.2 Time Variations in the Gas Chemistry at Poás Volcano

In this section, the temporal variations of the H_2/H_2O, H_2/Ar, CO/CO_2, CH_4/CO_2 and, HCl/HF log-ratios for selected fumaroles (i.e. Dome, Norte, Niña, Este, NE and Norte fumaroles) of Poás to evaluate the modifications of the plumbing magmatic/hydrothermal system that occurred at depth from 1998 to 2014 are taken into account. As previously mentioned, the fumarole collection was discontinuously carried out due to the fact that fumarole migration in time strongly affected the sampling activity. The temporal variation of the isotopic composition of helium was also considered whereas those of carbon, nitrogen, chlorine and argon were neglected due to the paucity of the data.

Fumaroles characterized by high temperatures in active volcanic systems are mainly dominated by H_2O, CO_2, SO_2 and H_2S with minor concentrations of H_2, HCl, HF and CO and traces of noble gases, e.g. helium and argon (Giggenbach 1996). Fluids emitted by volcanoes along convergent plates also show significant N_2 contents with respect to those from systems located in divergent and hot-spot geodynamic settings, due to the involvement of oceanic organic matter-rich sediments during subduction processes (i.e. Sano et al. 2001; Fischer et al. 2002). On the other hand, in addition to the rare gases, hydrothermal discharges are mostly consisting of H_2O, CO_2, H_2S and CH_4 and other reduced species, e.g. H_2 and CO, also indicative of high-temperature conditions, whereas HCl and HF are rarely detected (Giggenbach 1996). This implies that the magmatic systems may be considered to have a more oxidant character with respect to the hydrothermal systems where reduced conditions dominate. Poás fumaroles were mainly characterized by relatively low temperatures (<100 °C) where the presence of SO_2 was however recorded along with that of HCl and HF in most gas discharges, suggesting that the hydrothermal system was not able to completely scrub the most acidic gas species, likely due to the strong flux of the uprising magmatic gases and the proximity of the magma to the surface (about 500 m; Rymer and Brow 1989). Notwithstanding, the presence and the content of CH_4 and other gases species such as H_2, H_2S and CO suggest that the hydrothermal system modified the pristine composition of the magmatic gases. As a consequence, the selected ratios (e.g. H_2S/SO_2, CO/CO_2, CH_4/CO_2) can be considered representative the hydrothermal/magmatic ratio. On the other hand, the H_2/H_2O and H_2/Ar ratios are sensitive to heat pulse events from the magmatic source, since H_2O and Ar are related to the presence of a surficial component, i.e. the air saturated groundwater feeding the system (Giggenbach and Goguel 1989), whereas H_2 readily responds to changes of the temperature (similarly to CO) and redox conditions. Considering that F is more soluble in silicate melts than

Cl, the HCl/HF ratios can provide information on the degassing state of the magmatic body at the source of the fumarolic gases.

The temporal variations of the H_2S/SO_2 ratios (expressed as log) and that of the helium isotopic values are reported in Fig. 6. From 1999 to 2005, the $\log(H_2S/SO_2)$ ratios of most fumaroles were in a relatively narrow range (−2 to −1), excepting those of La Niña, which showed strong variations likely due to the presence/absence of condensed water at the surface or just below it, which governed the solubility of SO_2. In this period, a significant variability of the $\log(H_2S/SO_2)$ ratio was also shown by the Fumarole Este that can likely be attributed to the different sites from where the gas discharge was sampled even during a single sampling session (Table 1). From the beginning of 2005, the Fumarole Norte was characterized by a sharp increase in the $\log(H_2S/SO_2)$ ratio that reached +2 in middle 2005 whereas it decreased after the beginning of the phreatic events, suggesting a more efficient discharge of the magmatic component. The Fumarole Este also showed a similar behavior. At the end of 2008, the $\log(H_2S/SO_2)$ ratio of the Dome increased quite dramatically and after the Cinchona earthquake (January 2009), an abrupt decrease with respect to those measured in 1999–2000 was recorded, reaching its lowest value in middle 2009 (ca. –2.5). This is apparently suggesting that after the partial expulsion of the hydrothermal component, the magmatic/hydrothermal ratio increased. After 2009, the H_2S/SO_2 log-ratio started to progressively increase until 2014, likely due to the restoration of the hydrothermal system. Interestingly, the helium isotopic ratio did not show any significant variation and clustered around the value of 7 with the highest value recorded in 2004 (Fig. 6).

The $\log H_2/H_2O$ ratio with time is reported in Fig. 7. The values recorded at the Dome were higher than those recorded in the other fumaroles (>−3 and down to −6.4, respectively), although they were sampled in different periods. The trends depicted by the Norte, La Niña, Este and NE fumaroles do not appear to have a clear behavior. However, the Fumarole Este showed a decrease of the $\log(H_2/H_2O)$ ratio after the end of

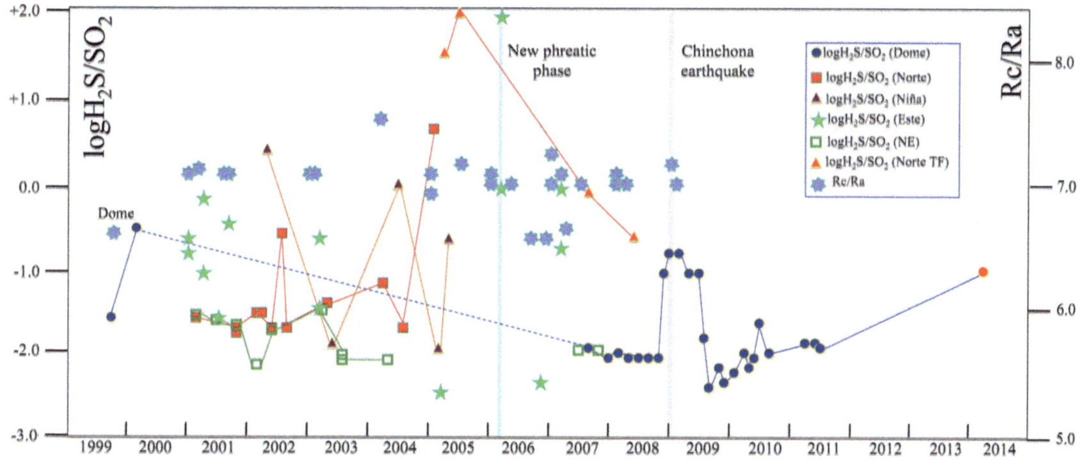

Fig. 6 Temporal variations of the $\log(H_2S/SO_2)$ ratio and the helium isotopic ratio (expressed as Rc/Ra, where Rc is the measured value corrected for the He/Ne ratio and Ra is that of the air) in selected fumaroles from Poás. The Dome is represented by blue (Vaselli et al. 2003 and unpublished data) and red (Fischer et al. 2015) circles; red squares and red triangles refer to the Fumarole Norte (Vaselli et al. 2003) and unpublished data and Fischer et al. (2015; Norte TF), respectively; the white green-bordered squares and the brown triangles represent the NE and Niña fumaroles, respectively, and the green stars indicate the Fumarole Este of Fischer et al. (2015; Este). A line connecting the Fumarole Este data was not drawn since they refer to different analyses carried out in the surroundings of such a fumarole; the blue seven-pointed stars depict the temporal variations of the helium isotopes. The renewed phreatic activity (March 2006) and the Cinchona earthquake are also indicated with light blue and light grey bars, respectively

2001, whereas from 2003 to 2007, no significant variations were recorded. The Fumarole Norte, after a progressive decrease of the log(H_2/H_2O) ratio from early 2001 to 2003, started to increase and before the phreatic activity in 2006 a sharp increase was recorded. This may suggest an efficient heat pulse by the magma chamber, although a similar behavior was not recorded by the other fumaroles. In 2007 the log(H_2/H_2O) ratio of the Dome was around −3 and then, it reached about −1.5 in correspondence of the Cinchona seismic event. Afterward, a significant decrease occurred at middle 2009. Subsequently, the log(H_2/H_2O) ratio increased sharply and in less than one year the log-ratios were higher than −1 and remained relatively stable up to the end of 2011, suggesting that well before the Cinchona earthquake the activity of Poás was affected by either an increasing heat pulse induced by magmatic source or an increasing hydrofracturation. It is to remark that the log(H_2/H_2O) ratios were in most cases significantly higher than that suggested by Giggenbach (1991) for hydrothermal volcanic systems [log(H_2/H_2O) = −2.8], indicating disequilibrium conditions related to water dissociation (e.g. Martini 1993). In 2014, the log(H_2/H_2O) ratio at the Dome was similar to that recorded in 1999 (ca. −3).

The trends of the log(H_2/Ar) ratios are reported in Fig. 8. The Dome was characterized by higher log(H_2/H_2O) values than those recorded at the other fumaroles, although no temporal overlapping measurements were carried out. Fumarole NE only approached log(H_2/Ar) ratios comparable to those of the Dome in 2007. From 2001 to 2005 the Fumarole Norte showed H_2/Ar log-ratios lower than 1.7; in correspondence of the 2006 phreatic activity a value approaching 2 was recorded. Then, it decreased to −1.6 and resumed up slightly lower than 0 in 2008 before the earthquake that stroke Cinchona. The Dome in 1999 and 2000 showed values of the log(H_2/Ar) ratios similar to those of the other fumaroles. In 2007, the log-ratio was almost 3 and increased up to 3.5 at the beginning of 2009 and after a sharp decrease, the log(H_2/Ar) ratio started to increase similarly to what recorded by the log (H_2/H_2O) ratio and values up to 5 were reached whereas in 2014, the log-ratio was slightly higher than 3. Basically, the behavior of the log(H_2/Ar) ratio intimately mimics that registered by the

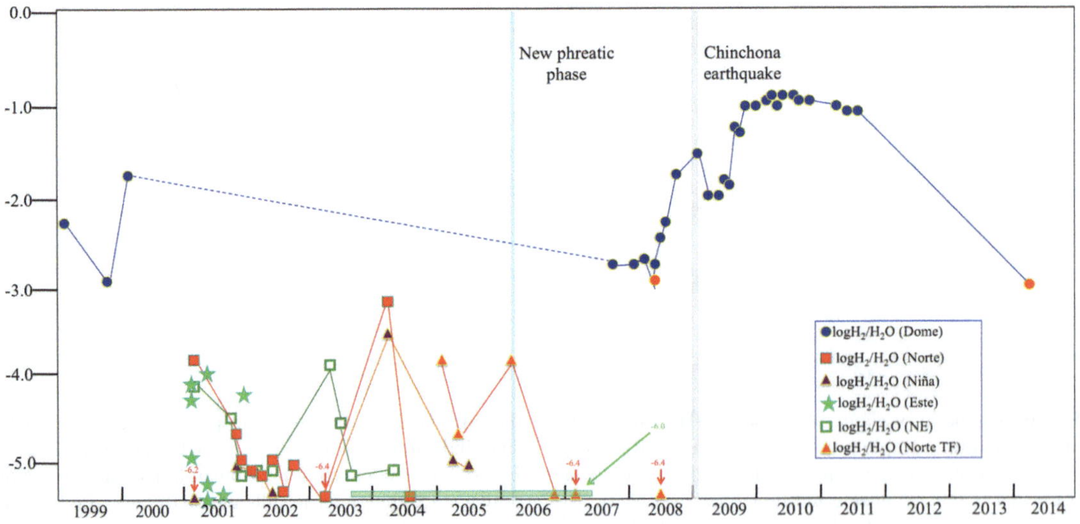

Fig. 7 Temporal variations of the log(H_2/H_2O) ratio in selected fumaroles from Poás. Symbols as in Fig. 6

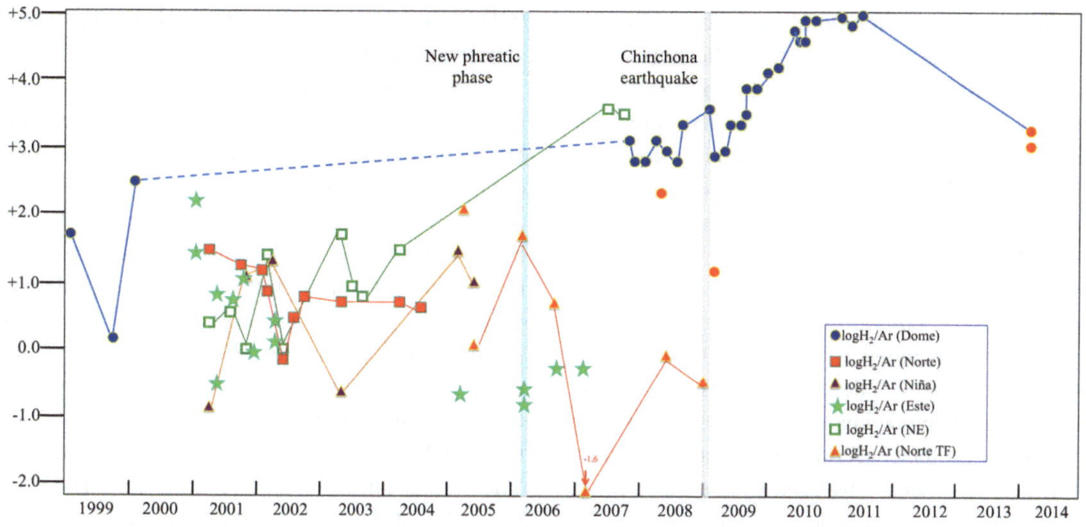

Fig. 8 Temporal variations of the log(H_2/Ar) ratio in selected fumaroles from Poás. Symbols as in Fig. 6

H_2/H_2O log-ratio, indicating an over-production of H_2 produced by H_2O thermal dissociation.

A similar picture can be observed when the CO/CO_2 log-ratios are considered (Fig. 9), carbon monoxide being more representative of a reduced (hydrothermal) environment. Most of the fumarolic gas discharges sampled from 1999 to 2006 have log(CO/CO_2) values between −6.5 and −4.0, although in 2003 a slight increase of the log(CO/CO_2) values was recorded for most fumaroles with the exception of the Dome where no data were available. As shown by the log(H_2/H_2O) and log(H_2/Ar) ratios, at the Dome those of log(CO/CO_2) are characterized by a significant increase at the beginning of 2007. From early 2008 to early 2009 the log(CO/CO_2) values steadily increased (ca. −3), preceding the further increase in this log-ratio a few months after the Cinchona earthquake. Subsequently, at the end of 2009 the log(CO/CO_2) ratio resumed by approaching −2. Eventually, in 2014 a clear drop was registered. Basically, carbon monoxide is indicative of high temperatures and tends to re-equilibrate relatively quickly.

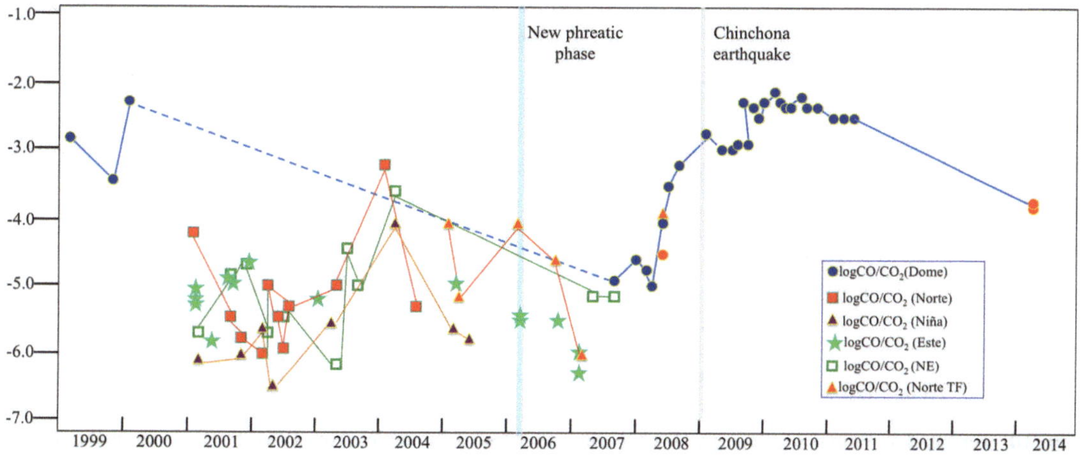

Fig. 9 Temporal variations of the logCO/CO_2 ratio in selected fumaroles from Poás. Symbols as in Fig. 6

In Fig. 10, the temporal variations of the log (CH_4/CO_2) ratios are reported. The behavior of this log-ratio refers to a hydrothermal component (CH_4) compared with CO_2 that better represents a magmatic signature, as also testified by the carbon isotopic values (Table 1). Consequently, the trends observed for the selected fumaroles of Poás strictly mimic those observed for the log (H_2S/SO_2) ratio (Fig. 6). This is particularly evident when the temporal variations recorded for the Dome are considered. In 1999–2000, the log(CH_4/CO_2) values were indeed the highest ones among all the considered dataset (up to −4), indicating that the hydrothermal component was dominating. Lower values were observed by the other fumaroles, though not sampled in the same period, which however showed an increasing trend from 2001 to 2005, few months before the renewed phreatic activity (ca. −3.5, Fumarole Norte). In 2007, the Dome was characterized by a log(CH_4/CO_2) ratio slightly lower than −6 and increased in early 2008. Then, a sharp decrease was recorded up to middle 2008. After the Cinchona event a dramatic increase of the log(CH_4/CO_2) ratio was observed that subsequently decreased down to −8.3 in 2010, supporting the presence of a magmatic component that was dominating over the hydrothermal one. In 2011 and in 2014, the log (CH_4/CO_2) recovered and reached the value showed in early 2009.

Eventually, the log(HCl/HF) ratio versus time is plotted in Fig. 11. As already mentioned, the HCl/HF log-ratio is a useful parameter to evaluate either increasing magmatic/hydrothermal ratio or increasing magma degassing. The Norte, La Niña, Este and NE fumaroles from 2001 to 2007 (when the last sample from the NE fumarole was collected), though not collected in the same periods, showed higher HCl/HF log-ratios that those recorded at the Dome, with the exception of the Fumarole Norte from the second half of 2002 to early 2003, when the log-ratios were about 0.5. The trends showed by CH_4/CO_2 and H_2S/SO_2 log-ratios at the Dome are similar to those of the HCl/HF log-ratio. The chlorine isotopes measured in several fumarolic gas discharges are not helpful in defining whether Cl is pristine, being affected by removal (distillation) of $H^{35}Cl$ from steam as Cl^-_{aq} in water (e.g. Sharp et al. 2010; Fischer et al. 2015), as supported by the high $\delta^{37}Cl$ values recorded at Poás (Table 1), although the Fumarole Norte, collected in 2006 approached the mantle value (−1.6‰; Bonifacie et al. 2008). From the end of 2003, all the fumaroles from Poás, showed a decreasing trend of log (HCl/HF) ratio that reached the lowest value at

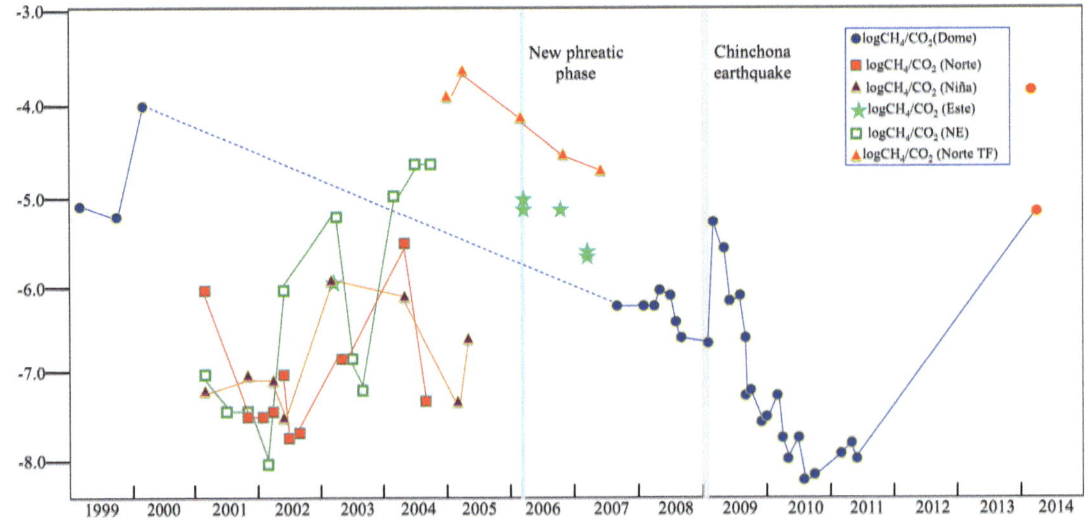

Fig. 10 Temporal variations of the log(CH_4/CO_2) ratio in selected fumaroles from Poás. Symbols as in Fig. 6

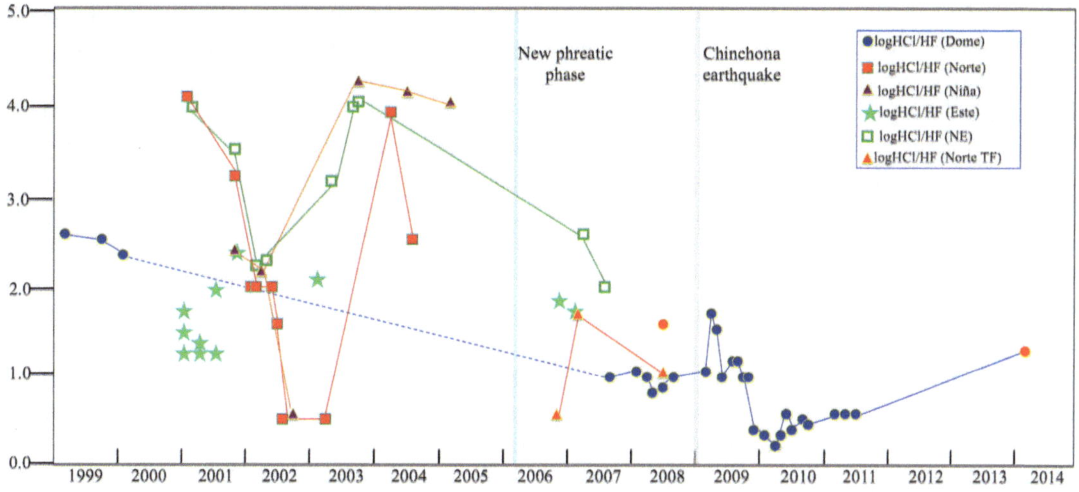

Fig. 11 Temporal variations of the log(HCl/HF) ratio in selected fumaroles from Poás. Symbols as in Fig. 6

the Fumarole Norte in November 2006 when the most primitive $\delta^{37}Cl$ value was measured. From middle 2007 to early 2009, the Dome showed a relatively constant log(HCl/HF) value that dramatically increased after the Cinchona earthquake. Then, the log-ratio started to progressively decrease up to early 2010. Subsequently, up to 2014 it constantly increased when a value slightly higher of 1 was achieved.

6 Summary and the Renewal Activity

The fumarolic activity from 1998 to 2014 at Poás was affected by strong modifications. In 1998 the fumarolic activity at Poás was relatively quiet and mainly concentrated at the interface between the pyroclastic dome and the acidic lake, whereas sporadic gas discharges were distributed throughout the inner crater with temperature never exceeding that of boiling water. From 1999 to 2000 the Dome fumarole was persisting and new fumarolic vents, not sampled, formed in the inner NE and E sectors of the crater. Despite the relatively low outlet fumarolic temperatures, concentrations of magmatic gases, such as SO_2, HCl and HF, were abundant. Significant modifications of the fumarolic fields were recorded up to 2006 (when a new phase of phreatic activity took place), although in 2005 the lake level started to decline and sub-lacustrine fumaroles turned vigorous. In 2001 the lake depth was 41 m deep (Vaselli et al. 2003), and a record level in early 2005 was achieved when the eastern bottom of the crater was over-flooded, and dropped down to 10 m in 2014 (Fischer et al. 2015; Martínez et al. Chapter "Behaviour of Polythionates in the Acid Lake of Poás Volcano: Insights Into Changes in the Magmatic-Hydrothermal Regime and Subaqueous Input of Volatiles" and references therein; Rouwet et al. Chapter "39 Years of Geochemical Monitoring of Laguna Caliente Crater Lake, Poás: Patterns from the Past as Keys for the Future"), approaching the 1989–1994 period when the lake dried out completely.

The discontinuous measurements of the fumarolic gas discharges at Poás has highlighted that: (a) useful insights inside the dynamics of the hydrothermal/magmatic system were mostly obtained by the Dome for which the geochemical record was relatively long and consistent. Nevertheless, from 2001 to 2007 no data are available and consequently, the possible geochemical variability of the gas composition prior and after the onset of the phreatic activity in 2006 was not recorded. However, the geochemical data collected for the Fumarole Norte registered an increase of the hydrothermal-related parameters (e.g. increase of the H_2S/SO_2, CO/CO_2, and

CH_4/CO_2 ratios). No data were available for this fumarole after 2008; (b) the Poás fumaroles suffered strong modifications from 1998 and time migration of the gas vents prevented the possibility of long-term gas sampling and consequently, a detailed reconstruction of the evolution of the hydrothermal/magmatic system was not allowed; (c) opening, closing and shifting of the fumarolic discharges are interesting aspects that should be investigated in more detail. As previously mentioned, the opening of a new fracture that produces a fumarolic discharge is not apparently reflecting the real chemical composition of the plumbing system since several days-to-months may be needed before a steady state status is achieved. This was particularly evident for La Niña fumarole that showed an up-and-down variation for most of the considered geochemical parameters (see Figs. 6–11) from the end of 2001 to the beginning of 2005, when it basically disappeared.

According to Vaselli et al. (2003), Fischer et al. (2015) and this work, the modifications suffered by the hydrothermal-magmatic system of Poás can be summarized in Fig. 12. Fischer et al. (2015) and Rymer et al. Chapter "(Geophysical and Geochemical Precursors to Changes in Activity at Poás Volcano" and references therein) suggested that a magma injection possibly occurred in late 2000-early 2001 on the basis of geochemical and gravity data, respectively, although on the behavior of polythionates Martínez et al. Chapter "Behaviour of Polythionates in the Acid Lake of Poás Volcano: Insights Into Changes in the Magmatic-Hydrothermal Regime and Subaqueous Input of Volatiles" and references therein) suggested the injection of fresh magma earlier (1996–1999). This implies that the chemical composition of the fumarolic gases and the changes observed from 1998 to 2014 are likely resulting by combined effects at different degrees of interaction between the magmatic and hydrothermal system and the acidic lake. Basically, from 1998 onwards the volcanic plumbing system of Poás passed from hydrothermal(magmatic)-dominated to magmatic-dominated conditions well before the Cinchona earthquake. It may speculate that the Cinchona event favored the (partial) removal of the hydrothermal system that slowly recovered until when the new eruptive phase (more violent than that commenced in 2006) took place in 2017. Injection of new and undegassed magma, heating of the hydrothermal system and gas pressure build-up, and hydrofracturing through or the rupture of the alteration-sealing cap-rock through 2005–2006

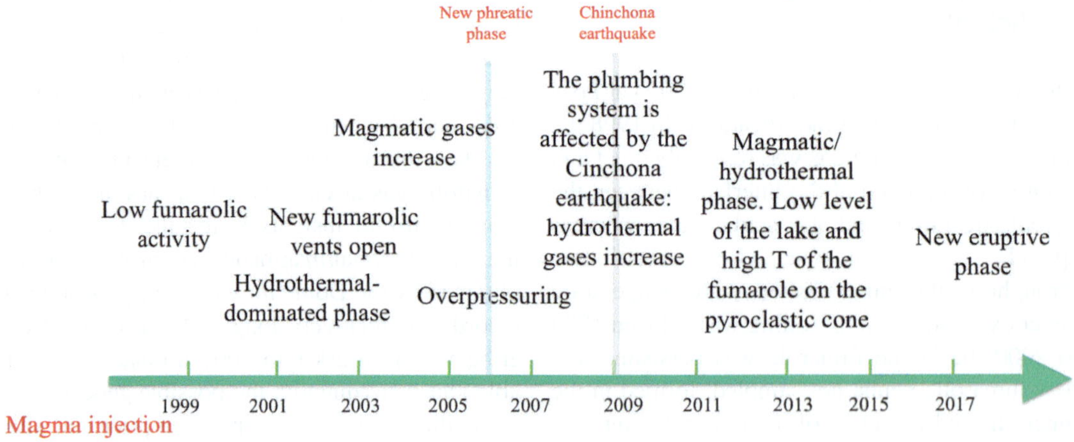

Fig. 12 Schematic diagram summarizing the most significant processes suffered by the Poás hydrothermal-magmatic system

were likely the main processes that affected the magmatic-hydrothermal system of Poás.

In late March 2017, seismic activity, ground deformation, CO_2/SO_2 ratios of the gases from the dome and the lake, lake temperature, acidity and ion composition increased significantly and sharply. In early April 2017, the lake level rose rapidly up to 2 m in two weeks, despite the scarce precipitations. Also, a 4 m wide new boiling mud-pool ("borbollón") (Fig. 13a; e.g. Martínez et al. Chapter "Behaviour of Polythionates in the Acid Lake of Poás Volcano: Insights Into Changes in the Magmatic-Hydrothermal Regime and Subaqueous Input of Volatiles" and references therein) appeared at the bottom of the crater about 100 m SE the pyroclastic dome. The fumarolic activity at the Dome fumarole increased, new fumaroles formed in the central and western sectors of the dome and some gas bubbling and steam were observed at the interface between the acidic lake and the dome, turning the lake grey and full of floating sulfur (Fig. 13b). On the 12th of April, the phreatic activity resumed and destroyed ca. 10% of the central part of the dome. The boiling mud-pool disappeared. The phreatic activity became more intense and at the end of April phreato-magmatic eruptions were recorded. In May, the repeatedly eruptive events destroyed ca. 90% of the pyroclastic dome structure. Eventually, the lake dried out completely, allowing the direct release of the fumarolic gases to the atmosphere visible (Fig. 13c), since they

Fig. 13 **a** The new mud boiling pool, a (crystobalite-rich) "borbollón" that opened in April 2017, 200 m behind the pyroclastic dome; **b** The Poás lake in April 2017 with strong upwelling; **c** the Poás lake disappeared in May 2017 showing high-temperature fumaroles around the areas occupied by the pyroclastic dome (Boca A o Boca Roja)

were previously discharging into the Laguna Caliente. Unfortunately, the hostile and dangerous environment prevented regular direct geochemical sampling. Nevertheless, once the still on-going volcanic activity will be over, gas sampling should be carried out as fast and often as possible in order to verify whether the pre-eruptive conditions are to be restored and the geochemical behavior of the remaining or new fumaroles.

Acknowledgements We are indebted with the OVSICORI personnel for the help during the many sampling campaigns that we carried out in the Costa Rica volcanoes, particularly at Poás. This work was financially supported by a bilateral project between CNR and OVSICORI and by the Laboratories of Fluid Geochemistry and Stable Isotope Geochemistry of the Department of Earth Sciences (University of Florence) and CNR-IGG of Florence. OV and FT wish to express their gratitude to Eliecer Duarte (OVSICORI) for the sampling activity and the nice time we had together inside the crater of Poás. We hope that our collaboration can keep going on.

References

Aiuppa A, Federico C, Giudice G, Gurrieri S (2005) Chemical mapping of a fumarolic field: La Fossa Crater, Vulcano Island (Aeolian Islands, Italy). Geophys Res Lett 32. http://dx.doi.org/10.1029/2005GL023207

Aiuppa A, Federico C, Giudice G, Giuffrida G, Guida R, Gurrieri S, Liuzzo M, Moretti R, Papale P (2009) The 2007 eruption of Stromboli volcano: insights from real-time measurement of the volcanic gas plume CO_2/SO_2 ratio. J Volcanol Geotherm Res 182:221–230

Alvarado G (2010) Hydrological and sedimentological aspects of the mudflows related to the Cinchona Earthquake (Mw 6.2) of January 8, 2009. Rev Geol Am Central 43:67–95 (In Spanish with English abstract)

Bennett FD (1979) Fumarolas y pozos subacuáticos de azufre en el Volcán Poás. Costa Rica. Rev Geogr Am Central 11(12):125–130 (In Spanish with English abstract)

Bennett FD, Raccichini S (1978) Subaqueous sulphur lake in Volcán Poás. Nature 271:342–344

Barquero J (ed) (2009) El terremoto de Cinchona del 8 de enero 2009. In: National seismological network (RSN-ICE:UCR), 138 pp

Bonifacie M, Jendrzejewski N, Agrinier P, Humler E, Coleman M, Javoy M (2008) The chlorine isotope composition of Earth's mantle. Science 319:1518–1520

Boza MA, Mendoza R (1981) The national parks of Costa Rica. Industrias Gráficás Alui, Madrid, p 310

Carroll MR, Webster JD (1994) Solubilities of sulfur, noble gases, nitrogen, chlorine, and fluorine in magmas. In: Carroll MR, Holloway JR (eds), Volatiles in magmas. Reviews in mineralogy. Min Soc Am pp 232–279 (Fredericksburg, VA)

Casertano L, Borgia A, Cigolini C, Morales LD, Montero W, Gómez M, Fernández JF (1987) An integrated dynamic model for the volcanic activity at Poás volcano, Costa Rica. Bull Volcanol 49:588–598

Chiodini G, Cioni R, Marini L, Panichi C (1995) Origin of the fumarolic fluids of Vulcano island, Italy and implications for volcanic surveillance. Bull Volcanol 57:99–110

Fernández-Arce M, Mora-Amador R (Chapter 5) Seismicity of Poás volcano. In: Tassi F, Mora-Amador R, Vaselli O (eds) Poás volcano (Costa Rica): the pulsing heart of Central America Volcanic Zone. Springer, Heidelberg (Germany)

Fischer TP, Hilton DR, Zimmer MM, Shaw AM, Sharp ZD, Walker JA (2002) Subduction and recycling of nitrogen along the Central American Margin. Science 297:1154–1157

Fischer TP, Ramírez C, Mora-Amador RA, Hilton DR, Barnes JD, Sharp ZD, Le Brun M, de Moor JM, Barry PH, Füri E, Shaw AM (2015) Temporal variations in fumarole gas chemistry at Poás volcano, Costa Rica. J Volcanol Geotherm Res 294:56–70

Francis PW, Thorpe RS, Brown GC, Glasscock J (1980) Pyroclastic sulphur eruption at Poás volcano Costa Rica. Nature 283:754–756

Giggenbach WF (1987) Redox processes governing the chemistry of fumarolic gas discharges from White Island. New Zealand Appl Geochem 2:143–161

Giggenbach W (1991) Chemical techniques in geothermal exploration. In: D'Amore F. (coordinator), Application of geochemistry in geothermal reservoir development. UNITAR/UNDP Publication, Rome, Italy, pp 119–142

Giggenbach WF (1996) Chemical composition of volcanic gases. In: Scarpa R, Tilling RI (eds) Monitoring and mitigation of volcanic hazards. Springer Verlag, Berlin-Heidelberg, pp 221–256

Giggenbach WF, Goguel RL (1989) Methods for the collection and analysis of geothermal and volcanic water and gas samples. CD 2387, Department of Scientific and Industrial Research, Chemistry Division

Hilton DR, Fischer TP, Marty B (2002) Noble gases and volatile recycling at subduction zones. In: Porcelli D, Ballentine CJ, Wieler R (eds) Noble gases in cosmochemistry and geochemistry. Mineralogical Society of America, Washington, D.C

Hilton DR, Rámirez CJ, Mora-Amador R, Fischer TP, Furi E, Barry PH, Shaw AM (2010) Monitoring of temporal and spatial variations in fumarole helium and carbon dioxide characteristics at Poás and Turrialba volcanoes, Costa Rica (2001–2009). Geochem J 44:431–440

Holloway JR, Blank J (1994) Experimental results applied to C–O–H in natural melts. In: Holloway JR, Blank JC (eds) Volatiles in magmas. Reviews in mineralogy. Mineralogical Society of America, Fredericksburg, VA, pp 187–230

Krushensky RD, Escalante G (1967) Activity of Irazú and Poás volcanoes, Costa Rica, November 1964–July 1965. Bull Volcanol 31:75–84

Lopes R (2005) The volcano adventure guide. Cambridge University Press, p 362

Martínez M (2008) Geochemical evolution of the acid crater lake of Poás volcano (Costa Rica): insights into volcanic-hydrothermal processes. Ph.D. Thesis, Utrecht University, The Netherlands

Martínez M, Fernández E, Valdés J, Barboza V, Van der Laat R, Duarte E, Malavassi E, Sandoval L, Barquero J, Marino T (2000) Chemical evolution and volcanic activity of the active crater lake of Poás volcano, Costa Rica, 1993–1997. J Volcanol Geotherm Res 97:127–141

Martínez M, van Bergen MJ, Takano B, Fernández Soto E, Barquero Hernández J (Chapter 7) Behaviour of polythionates in the acid lake of Poás volcano: Insights into changes in the magmatic-hydrothermal regime and subaqueous input of volatiles. In: Tassi F, Mora-Amador R, Vaselli O (eds) Poás volcano (Costa Rica): The pulsing heart of Central America Volcanic Zone. Springer, Heidelberg (Germany)

Martini M (1993) Water and fire: vulcano island from 1977 to 1991. Geochem J 27(297):303

Menyalov IA (1975) Prediction of eruptions using changes in composition of volcanic gases. Bull Volcanol 39:112–125

Mitchell EC, Fischer TP, Hilton DR, Hauri E, Shaw AM, de Moor JM, Sharp ZD, Kazahaya K (2010) Nitrogen sources and recycling at subduction zones: insights from the Izu–Bonin–Marianas arc. Geochem Geophys Geosyst 11, Q02X11. http://dx.doi.org/10.1029/2009GC002783

Montegrossi G, Tassi F, Vaselli O, Buccianti A, Garofalo K (2001) Sulfur species in volcanic gases. Anal Chem 73:3709–3715

Mora-Amador R, Rouwet D, González G, Vargas P, Ramírez C (Chapter 11) Volcanic hazard assessment of Poás (Costa Rica) based on the 1834, 1910, 1953–1955 and 2017 historical eruptions. In: Tassi F, Mora-Amador R, Vaselli O (eds) Poás volcano (Costa Rica): the pulsing heart of Central America Volcanic Zone. Springer, Heidelberg (Germany)

Mora-Amador R, Rouwet D, Vargas P, Oppenheimer C (Chapter 3) The extraordinary sulfur volcanism of Poás from 1828 to 2018. In: Tassi F, Mora-Amador R, Vaselli O (eds) Poás volcano (Costa Rica): The pulsing heart of Central America Volcanic Zone. Springer, Heidelberg (Germany)

Oppenheimer C (1992) Sulphur eruptions at Volcan Poás, Costa Rica. J Volcanol Geotherm Res 49:1–21

Oppenheimer C, Stevenson D (1989) Liquid sulphur lakes at Poás volcano. Nature 342:790–793

Prosser JT, Carr MJ (1987) Poás volcano, Costa Rica: geology of the summit region and spatial and temporal variations among the most recent lavas. J Volcanol Geotherm Res 33:131–146

Raccichini S, Bennett FD (1977) Nuevos aspectos de las erupciones del volcán Poás. Rev Geogr Am Centr 5–6:37–53 (In Spanish with English abstract)

Rouwet D, Mora-Amador R, Sandri L, Ramírez-Umaña C, González G, Pecoraino G, Capaccioni B (Chapter 9) 39 years of geochemical monitoring of Laguna Caliente crater lake, Poás: patterns from the past as keys for the future. In: Tassi F, Mora-Amador R, Vaselli O (eds) Poás volcano (Costa Rica): the pulsing heart of Central America Volcanic Zone. Springer, Heidelberg, Germany

Rowe GL (1994) Oxygen, hydrogen, and sulfur systematics of the crater lakes system of Poas Volcano, Costa Rica. Geochem J 28:263–287

Rowe GL Jr, Brantley SL, Fernández M, Fernández JF, Borgia A, Barquero J (1992) Fluid-volcano interaction in an active stratovolcano: the crater lake system of Poás volcano, Costa Rica. J Volcanol Geotherm Res 49:23–51

Rowe GL Jr, Brantley SL, Fernández JF, Borgia A (1995) The chemical and hydrologic structure of Poás Volcano, Costa Rica. J Volcanol Geotherm Res 123:223–267

Ruiz P, Carr MJ, Alvarado GE, Soto GJ, Mana S, Feigenson MD, Sáenz LF (Chapter 5) Coseismic landslide susceptibility analysis using LiDAR data PGA attenuation and GIS: The case of Poás volcano, Costa Rica, Central America. In: Tassi F, Mora-Amador R, Vaselli O (eds) Poás volcano (Costa Rica): the pulsing heart of Central America Volcanic Zone. Springer, Heidelberg (Germany)

Ruiz P, Mana S, Gazel E, Soto GJ, Carr M, Alvarado GE (Chapter 2) Geochemical and geochronological characterisation of the Poas stratovolcano stratigraphy. In: Tassi F, Mora-Amador R, Vaselli O (eds) Poás volcano (Costa Rica): the pulsing heart of Central America Volcanic Zone. Springer, Heidelberg (Germany)

Rymer H, Brown GC (1989) Gravity changes as a precursor to volcanic eruption at Poás volcano, Costa Rica. Nature 342:902–905

Rymer H, Cassidy J, Locke CA, Barboza MV, Barquero J, Brenes J (2000) Geophysical studies of the recent 15-year eruptive cycle at Poás Volcano, Costa Rica. J Volcanol Geotherm Res 97:425–442

Rymer H, Locke CA, Borgia A, Martínez M, Brenes J, van der Laat R, Williams-Jones G (2009) Long-term fluctuations in volcanic activity: implications for future environmental impact. Terra Nova 21:304–309

Rymer H, Locke CA, Borgia A, Martínez M, Brenes J, van der Laat R (2010) Geophysical and geochemical precursors to the current activity at Poás volcano, Costa Rica. Book of abstracts of the IAVCEI Commission of Volcanic Lakes 7th Workshop on Volcanic Lakes Costa Rica 2010

Rymer H, Martínez M, Brenes J, Williams-Jones G, Borgia A (Chapter 8) Geophysical and geochemical precursors to changes in activity at Poás volcano. In: Tassi F, Mora-Amador R, Vaselli O (eds) Poás volcano (Costa Rica): the pulsing heart of Central America Volcanic Zone. Springer, Heidelberg (Germany)

Sano Y, Naoto T, Nishio Y, Fischer TP, Williams SN (2001) Volcanic flux of nitrogen from the Earth. Chem Geol 171:263–271

Sharp ZD, Barnes JD, Fischer TP, Halick M (2010) An experimental determination of chlorine isotope fractionation in acid systems and applications to volcanic fumaroles. Geochim Cosmochim Acta 74:264–273

Shinohara H, Kazahaya K, Saito G, Matsushima N, Kawanabe Y (2002) Degassing activity from Iwodake rhyolitic cone, Satsuma-Iwojima volcano, Japan: formation of a new degassing vent, 1990–1999. Earth Planets Space 154:175–185

Shinohara H, Ohminato T, Takeo M, Tsuji H, Kazahaya R (2015) Monitoring of volcanic gas composition at Asama volcano, Japan, during 2004–2014. J Volcanol Geotherm Res. https://doi.org/10.1016/j.jvolgeores.2015.07.022

Symonds RB, Gerlach TM, Reed MH (2001) Magmatic gas scrubbing: implications for volcano monitoring. J Volcanol Geotherm Res 108:303–341

Vargas CA (1979) Antología: el volcán Poás. Universidad Estatal a Distancia, San Jose, Costa Rica

Vaselli O, Tassi F, Minissale A, Montegrossi G, Duarte E, Fernández E, Bergamaschi F (2003) Fumarole migration and fluid geochemistry at Poás Volcano (Costa Rica) from 1998 to February 2001. In: Oppeneheimer C, Pyle DM, Barclay J (eds) Memoirs of Geological Society London, Special Issue on: "Volcanic degassing", vol 213, pp 247–262

Vaselli O, Tassi F, Montegrossi G, Capaccioni B, Giannini L (2006) Sampling and analysis of fumarolic gases. Acta Vulcanol 18:65–76

Vaselli O, Tassi F, Duarte E, Fernández E, Poreda TJ, Delgado A (2010) Evolution of fluid geochemistry at the Turrialba volcano (Costa Rica) from 1998 to 2008. Bull Volcanol 72:397–410

Zimmer MM (2002) Volatile chemistry of the Costa Rican segment of the Central American volcanic arc. M.Sc. Thesis, University of New Mexico, Albuquerque, 87 pp

Zimmer MM, Fischer TP, Hilton DR, Alvarado GE, Sharp ZD, Walker JA (2004) Nitrogen systematics and gas fluxes of subduction zones: insights from Costa Rica arc volatiles. Geochem Geophys Geosyst Q05J11. https://doi.org/10.1029/2003gc000651

Volcanic Hazard Assessment of Poás (Costa Rica) Based on the 1834, 1910, 1953–1955 and 2017 Historical Eruptions

Raúl Alberto Mora Amador, Dmitri Rouwet, Gino González, Priscilla Vargas and Carlos Ramírez

Abstract

Poás is a complex stratovolcano with an altitude of 2,708 m a.s.l., located in the Cordillera Volcánica Central of Costa Rica. Prior to 2017, the last three historical eruptions occurred on 7th of February 1834, between January and May 1910 and during the period 1953–1955. Very few information exists on the 1834 eruption. The only references state that: it was an important event; ash reached >53 km W–SW of Poás, and it harmed the grasslands around the volcano. Related deposits of this eruption suggest phreatic activity, which launched bombs and blocks. Moreover, there is evidence of pyroclastic flow deposits near the crater. The 1910 eruption is better described. Despite the fact that ash fall is only reported near the volcano, a volume of the deposit of 1.6×10^7 m^3 was estimated. Deposits of the eruption are white in color with many hydrothermally altered, and minor presence of juvenile fragments (vesicular lapilli). The eruption is classified as vulcanian, with deposits of ash fall and pyroclastic flows close to the crater. A Volcano Explosivity Index 3 (VEI 3) is estimated. The eruption affected agriculture. The 1953–1955 eruptions had a longer duration. Various ash fall deposits at several sites were reported. Deposits of this eruption, easily distinguished in the field, are black scoria lapilli, bombs with, sometimes fusiform, bread crust textures. In the eastern sector of the crater bombs can reach meters in size; such large bombs near the eruption centre at one side suggest the inclination of the eruptive conduct, or an asymmetrical vent-crater system. Inside the crater a 40 m-high dome and a lava flow were extruded during the eruption. Towards the eastern side of the current Laguna Caliente crater lake, relicts of a 8.5 m thick lava pool are found. During the entire eruptive episode, the acid lake presumably lacked. The eruption is described to be of a mixed type: strombolian, phreatomagmatic, vulcanian and dome extrusion eruptions. Considering the characteristics of this eruption, the height of the eruption column, ejected volume (2.1×10^7 m^3), and its presumed duration, a VEI 3 is estimated. The eruptions damaged agricultural activity (including cattle), and forced the spontaneous evacuation of some people. In April 2017 magmatic eruptions followed a

R. A. Mora Amador (✉)
Escuela Centroamericana de Geología, University of Costa Rica (UCR), San José, Costa Rica
e-mail: raulvolcanes@yahoo.com.mx

R. A. Mora Amador · P. Vargas
Laboratorio de Ecología Urbana, Universidad Estatal a Distancia (UNED), San José, Costa Rica

D. Rouwet
Istituto Nazionale di Geofisca e Vulcanologia (INGV), Sezione di Bologna, Italy

G. González · C. Ramírez
Volcanes Sin Fronteras (VSF), San José, Costa Rica

F. Tassi et al. (eds.), *Poás Volcano*, Active Volcanoes of the World,
https://doi.org/10.1007/978-3-319-02156-0_11

decade-long period of intense phreatic activity. These eruptions destroyed the 1953–1955 Dome and led to the complete dry out of Laguna Caliente. Pyroclastic cones and sulfur volcanism manifested at the bottom of the former crater lake bottom. The 2017 eruption severely affected touristic activities at and near Poás, with an estimated economic loss of 20 million dollars. By May–August 2018 Laguna Caliente reappeared. The volcanic hazards related to the three studied historical eruptions are: pyroclastic flows (at least 1 km from the eruptive centre, including reaching the current mirador sector), ballistics (bomb ejections up to 2 km from the emission centre), dispersion and fall of pyroclasts (tens of kms), gas emission and acid rain, dispersed by WSW dominant winds, and lahars in most of the river canyons SW of the volcano.

Keywords
Poás volcano · Laguna caliente · Sulfur · Phreatic eruption · Volcanic hazard

1 Introduction

The society often forgets how vulnerable we are as humans, and simply procrastinates lessons from memories into the future. Beck (2000) describes how the concept of risk inverts the relationship between past, present and future. The cause of an experience in the present is often framed in the future, and is therefore inexistent, invented and fictitious. In volcanology, the reconstruction of the eruptive history of a volcano based on writings from the recent history intrinsically implies the concept of risk, as impressions of the observers strongly bias the scientific concept of hazard. Reinterpretation of past writings with scientific rigor is necessary to translate the risk concept into useful information for risk assessment and reduction at active volcanoes (Bretón et al. 2002; Cashman and Cronin 2008; Del Gaudio et al. 2010; Todesco et al. 2015; Caudron et al. 2015). Using the information from the geological record, compiling details from writings helps constructing the eruptive history and unrest signals of volcanoes (Phillipson et al. 2013). Such catalogues are the basis for probabilistic hazard assessment (e.g. BET, pattern recognition, stochastical analyses, Marzocchi et al. 2004; Marzocchi and Bebbington 2012), besides being highly insightful on how the volcano will behave in the future on the long-term (e.g. largest expected eruption).

Poás volcano, Costa Rica, is a complex stratovolcano with an altitude of 2,708 m a.s.l., with three main structures: the Main Crater, the Von Frantzius cone and Laguna Botos (Fig. 1). The Main Crater hosts the hyper-acidic and hot crater lake Laguna Caliente, rimmed to the south by a dome extruded in 1953. Laguna Caliente has a diameter of 280–300 m and a depth between 23 and 60 m, depending on the state of activity (Brown et al. 1989; Rowe et al. 1992a, b; Martínez et al. 2000; Rouwet et al. Chapter "39 Years of Geochemical Monitoring of Laguna Caliente Crater Lake, Poás: Patterns from the Past as Keys for the Future"). Poás is the most visited volcano in Central America with over 400,000 visitors a year.

This chapter compiles bibliographic data from ancient writings (newspaper articles, scientific reports) for the three major historical eruptions of Poás volcano (1834, 1910 and 1953–1955), covering a period before modern volcanology came up, and prior to the most recent phreatic eruption cycles of 1985–1994 and 2006–2016. The most recent "historical" activity reappraisal occurred at the moment of writing (April 2017). A geological survey of the deposits and stratigraphy related to these three eruptions aims to reconstruct the eruptive events. The information from ancient writings is extremely helpful to reconstruct the eruptive history of Poás generally characterized by low-volume eruptions whose deposits are largely weathered or even eroded from the geological record, introducing large uncertainties in the interpretation of the eruptions and the eruptive models of the 1834, 1910 and 1953–1955 eruptions. A major question is answered in this contribution in order to assess the volcanic hazard, deduced from the four eruptions in the recent, or very recent past: How

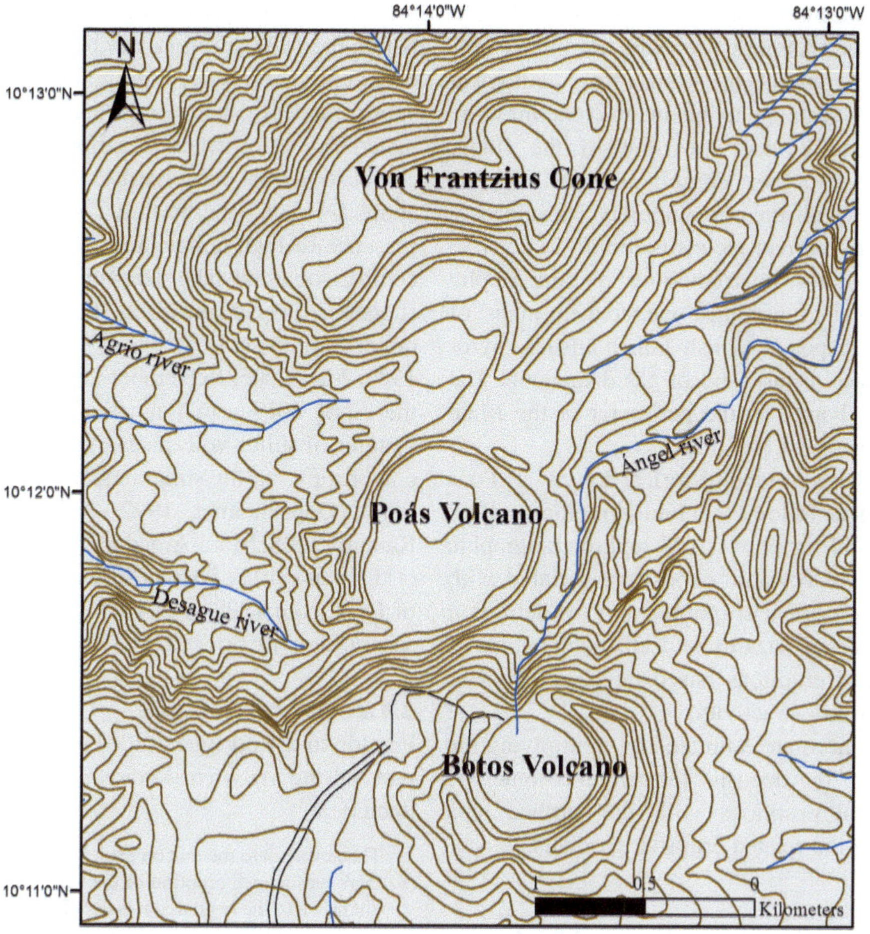

Fig. 1 Location map of the summit area of Poás Volcano, with topographic curves every 10 m

will the next eruption be like? Models of the four historical eruptions are presented here. This chapter can be helpful in creating a hazard map, as being based on the interpretation of the ancient writings and the geological record, it can be a useful document for authorities to take risk-reducing measures in the short-term, and orient human activities near Poás on the long-term.

2 Methodology

The first step is a thorough search of bibliographic information on the major historical eruptions. Ancient writings (newspaper articles, scientific reports) are gathered at the libraries *Carlos Monge*, *Federico Tinoco* and the library of the *Escuela Centroamericana de Geología* (all at the *Universidad de Costa Rica*). Other information was collected at the *Biblioteca Nacional de Costa Rica*, the *Biblioteca del Museo Nacional de Costa Rica* and the offices of the *Archivo Nacional*. Moreover, an unpublished letter by a researcher that visited the area during the activity of the 1950s was used. Parts of the narrations from ancient writings are reported as such in this study (in italic), and literally translated from Spanish into English. The literal translation will often demonstrate that interpretations in a volcanological framework are not always straightforward.

Based on information derived from the data compilation of ancient writings, a geological survey of recent volcanic deposits around Poás was executed. As many deposits are highly eroded (yearly rainfall of 4,000 mm), geological descriptions resulted to be insufficient in some cases to totally meet the aims of the study. Therefore, a proper combination of all this information with the contemporary chronicles is the key for a correct reconstruction.

Based on the interpretations of stratigraphy and the ancient writings, we created maps of maximum dispersal of ash fall, ballistics, pyroclastic surges, lahars, and of gas dispersion. The location of bombs in the perimeter of the Main Crater is recorded by GPS.

A map based on the historical activity of Poás was drawn, as follows: (1) bibliographic study of the historical eruptions; (2) tephrostratigraphic survey of the volcanic deposits associated with the major historical eruptions, and (3) definition of the expected hazards and characterization of the expected eruptions based on hazard assessment of historical activity.

Ash volumes are estimated by multiplying the thickness of the deposits for reference outcrops of historical eruptions with the surface area covered by the related eruptive products.

3 Historical Volcanology of Poás

3.1 The 7 February 1834 Eruption

3.1.1 Pre-eruptive Stage

The only document available for the pre-eruptive stage was written by Von Frantzius (1861), who reported some observations made by Miguel Alfaro, who descended the crater in 1828:

> … during the previous years Miguel Alfaro himself had seen a vapor column, which also threw stones upwards, and around those sulfur accumulations and in the volcanic ash a blue flame burned. In the surroundings ash also found conical masses of pure sulfur, with an altitude of 3 to 4 feet. The lake then was a lot smaller; nevertheless, the water roared with more power and appeared more acidic and hotter than now. It seemed that at that time the volcano would have been a lot more active in throwing ash, and that it went gradually slowing down

The description of Miguel Alfaro, six years before the 1834 eruption, can be interpreted as the occurrence of phreatic eruptions (explosion of vapor, mud or other non-incandescent or melted material), the formation of sulfur cones and sulfur combustion, that are related to exhalation temperatures >444 °C (the boiling T of sulfur), and the presence of high-temperature fumaroles. Alfaro's comparison of the lake size with the 1861 situation suggests a drop in lake level. This type of activity is rather common at Poás (Fig. 2). Recently, in April 1989, May 2005, September 2009, and following the April 2017 eruptions (see Sect. 3.4) similar features of sulfur volcanism were observed: sulfur combustion, sulfur structures, and phreatic eruptions (Oppenheimer 1992; Mora-Amador and Ramírez 2008; Mora-Amador 2009; Mora Amador et al. Chapter "The Extraordinary Sulfur Volcanism of Poás from 1828 to 2018"). No information on seismic activity prior to the 1834 event exists.

3.1.2 Eruptive Stage

Peraldo and Rodríguez (2001) textually reported part of the *Archivo Nacional Municipal, 778*, document:

> The news got to me that on Friday seven in the night there was an ash eruption of the los Votos volcano, unknown to the level that it is certain that in the area near Púas, like the site of Agualate, has been buried the graze lands leaving it without resources for the animals to provide for themselves and how in this site there was the cattle that belongs to the parish that was auctioned the day 2 of the current (year), I believe it is essential on the moment that I've known, to manifest to You this occurrence that in similar (…unreadable…) and to have lost at the end of the last auction with the possible promptness to tell me to who it has been auctioned to proceed with the delivery to find oneself in the case that soon resign to pay for the animals, and that it doesn't result in loss of the parish or to who will be (… unreadable…) and that opportunely will dispose of them. Barva, 15 February 1834.

The report above is the only document mentioning the exact moment of the event: February 7, 1834 in the night. The key point is understanding the meaning of the word "aterrar" in Spanish. In the *Nuevo Diccionario de Costarriqueñismos* (Quesada 2007), the word "aterrar" means

(a) (b)

Fig. 2 Sulfur volcanism at Poás volcano. **a** 5 m-long blue flame stating sulfur combustion (9 September 2009, inside the active crater). **b** Photograph by Clive Oppenheimer (April 1989) showing a 3 m-high "small sulfur volcano". Both features should have been similar as the ones witnessed by Miguel Alfaro in 1828

"obstruct" or "filled with ground", although not necessarily with ground ("tierra"), hence, "aterro" in the context of the writing means obstruction caused by a ground avalanche. We conclude that the report refers to ash (and possibly lapilli) fall out, in some ranches near Poás. The corresponding areas at present should be Poásito, Vara Blanca and surroundings. From the report, the graze lands were "buried" to such an extent that animals cannot graze. Following Blong (1984), a thickness of 10–15 mm of ash caused the loss of graze land. Regardless of the terminology of "aterrado", we infer a significant impact by ash and lapilli fall out with a thickness in the order of centimeters between 5 and 10 km around the emission centre. Von Frantzius (1861) is the first to write a scientific report on the 1834 eruption:

> In 1834 a strong ash rain destroyed graze lands from the city of Alajuela on the south flanks of the volcano.

He continued:

> …when in the year 1834 the ash rain occurred, the sky suddenly darkened, strong detonations were heard and the ash fell in southwestern direction until close to Esparza, at a distance of ten miles (diez leguas).

The Spanish word "legua" is an ancient unit to express the distance a person or a horse can cover within one hour. As such, a "legua" is a distance varying from 4 to 7 km. Hence, the distance described by Von Frantzius (1861), varies from 40 to 70 km from the emission centre (i.e. 55 km on average).

In 1963 Oersted mentioned:

> … in the volcanic ashes dispersed in the surroundings, small pieces of native sulfur are found, and this substance should have been more abundant, as the crater was frequently visited by sulfur explorers. This volcano does not seem to be completely extinct; fact is that, in 1834, a strong eruption occurred accompanied by underground detonations and ashes vomited by the volcano launched up to 30 miles (millas) distance.

Oersted (1863) spoke about a distance of 30 miles ("millas"), but did not mention the direction the ash clouds travelled. The roman mile is 1,480 m. Based on this conversion, the author estimated an emission distance of 44.4 km from the emission centre. The distance between Poás and Esparza is 53.1 km, as such, the estimated distances by Von Frantzius (1861) is similar as the one reported by Oersted (1863). The underground detonations mentioned by Oersted (1863)

Fig. 3 Part of the sketch of the Cordillera Volcánica Central of Costa Rica, drawn by Oersted (1863). The sketch shows the active crater of Poás volcano liberating gas, and, at a higher elevation Botos volcano

can be interpreted as thunder or possibly earthquakes during the eruption (Fig. 3).

3.1.3 Effects of the Eruption

The report of the *Archivo Nacional, Municipal, 778*, mentioned how graze lands were "buried", causing the mobilization of cattle to other places, due to the lack of food. During this period, there were very few villages near Poás. Basically, the area near the crater area was occupied by cattle ranches and damage was only reported for those ranches. No loss of human lives is reported.

3.1.4 Eruption Model

We can infer an initial phreatic phase, with the eruption of lake sediments and hydrothermalized material saturated in molten sulfur, after which a major phreatic event occurred that launched bombs and blocks to a distance still close to the crater (700 m from the possible emission site). Deposits show parallel lamination that represent ash fall out, as well as inclined lamination with wedging structures indicating pyroclastic surges in the sector close to the crater. A minimum column elevation of 5 km is estimated.

Afterwards, the wind took care of transporting the fine material, following the information obtained until the town of Esparza, up to 55 km distance from Poás volcano, in southwestern direction.

3.2 The January–May 1910 Eruption

3.2.1 Pre-eruptive Stage

Leiva (1904) reported a lake water temperature of 42 °C, besides sporadic phreatic eruptions:

> …this feature is generally pre-announced with agitation of the waters very near the centre of the lake. After a noise bursts while a dense column of whitish gases violently rises up to many hundreds of meters to open up into crests of brilliant white that dissolve by the wind or that otherwise fall as condensates of fine mists.

Rudín (1905) witnessed several phreatic eruptions (Fig. 4):

> …these appeared without preliminary signs of any kind and consisted in a small bubble of mud, a bit off the centre of the lake, followed by a small column of water vapor.

Moreover, he described other mayor ones:

> …when suddenly an immense black column rose from the lake that seemed boiling mud that, while falling back down, broke in thousands of fragments, giving the eruption a magnificent and impressive aspect. In a few seconds the black column was surrounded by water vapor, emanating from the column itself, followed by the release of the vapor mass that was expulsed from the chimney of the volcano, leading to the disappearance of everything enveloped in a cloud that completely filled the crater.

Rudín (1905) estimated a diameter of Laguna Caliente of 500, and 600 m height and 100 m in diameter of the column during the largest eruption. During the 2006–2016 phreatic eruption cycle, phreatic eruptions reaching 500 m in

Fig. 4 Photograph showing excursionists escaping during an eruption (Rudín 1905)

height (200 m above the crater rim) can be affected by the wind direction. In some of these recent cases, erupted ash reached the tourist "mirador", Laguna Botos, and the visiting centre of the *Parque Nacional Volcán Poás* (PNVP hereafter).

The last report prior to the 1910 eruption was written by Leiva (1906), who observed during his visit that the level of Laguna Caliente had decreased. He reported continuous phreatic eruptions; moreover, some people living near reported, prior to his visit, flames produced by sulfur combustion (Fig. 2a).

3.2.2 Eruptive Stage

On 25 January 1910 in the afternoon, one of the most important historical eruptions of Poás started. Hereafter, a compilation of public news from newspapers is presented in chronological order, together with reports from a group of researchers sent by the president of the Republic of Costa Rica:

On Wednesday 26 January 1910, *La República* informs:

> Eruption of Poás Volcano. Strong ash rain-very beautiful spectacle". This information arrived at the Telegraph from Alajuela the previous night at 6:15 p.m., stating: "Poás Volcano in eruption, very beautiful spectacle in the cordillera, ash rain in abundance, people alarmed. To north side of the city, the sky seems reddened.

Brenes (1932) reported that the eruption was perfectly visible from the faraway province of Guanacaste. On Thursday 27 January 1910, *La Prensa Libre* entitled: "*The eruption of Poás*", informed by Alberto Rudín:

> We arrived at the volcano at eight a.m. In the "Lechería" (i.e. dairy farm) lots of sand, rain, small stones, up to the size of corn, in Potrero Pequeño mud and stones fell, up to a hundred weight. The crater and surroundings entirely covered by a layer of volcanic mud. The quantity of mud thrown out was tremendous as it entirely covered the crater, the northern hill and the surroundings of the west and south. Signs are obvious that it occurred in real streams. Lots of stones of five to forty cm that opened, while falling, holes of a yard deep.
> The stones should have come from the interior, as they show molten sulfur. Today there was not even one eruption until 2 p.m. it seems to me there was still no reason for serious fear. The waters of all the rivers had a remarkable sour taste and the roads were found entirely covered by a layer of ash that seemed aluminum.

"Potrero Pequeño", mentioned by Rudín is the current site of the entrance gate to the PNVP, which is also known locally as "Lechería". This site is currently inhabited and known for commercial activities for tourists. Interpreting the news that this event showered more than 2 km distance from the emission centre, lapilli and decimeter-sized blocks generated impact craters of nearly a meter deep, while at a distance of 4–5 km, ash and lapilli fell. The description of the eruptive products suggests that much of the material originated from Laguna Caliente (mud, molten sulfur). Sulfur in the rock pores can be explained: at the bottom of Laguna Caliente subaqueous pools of molten sulfur form (Oppenheimer 1992; Takano et al. 1994; Rowe et al. 1992a, b; Christenson et al. 2010; Rouwet and Morrissey 2015; Mora Amador et al. Chapter "The Extraordinary Sulfur Volcanism of Poás from 1828 to 2018"). The presence of sulfur in pore spaces of many rocks suggest an emission temperature of at least 116 °C, the melting point of sulfur (Takano et al. 1994). More recently, Bennett and Raccichini (1978) and Francis et al. (1980) have reported the same feature for Poás, also observed since March 2006 following phreatic eruptions (Fig. 5).

The next day, on 28 January *La Prensa Libre* informed: "*More ash rain*", detailing that:

> This morning, between 9:30 and 10:30, a fine dust fell, not as the white ash such as on the 25th in the afternoon, but rather a rose ground, in small quantities; they were very small grains that reduced into fine dust when crushed. This fact was known because in a patio a billiard cloth was left to dry, that, when removed, it was noted that the small grains fell on it. It was shaken off and relocated with the other side up and some grains continued to fall.

On Saturday 29 January, *Prensa Libre* informed: "From Grecia, ash rain", and detailed:

> It was six in the afternoon the day before yesterday when, above the summit of Poás, a huge plume of smoke, gases and ashes that rose to prodigious heights was soon disintegrated into an ash rain that

Fig. 5 Various hydrothermalised blocks with sulphur at the "playón" to the south inside the crater, found weeks after the phreatic eruptions of March 2006. **a** Millimeter-sized "volcanoes" emerge from the pores of the rock. **b** Detail of the sample showing solidified sulfur forming micro-stalactites. **c** One of the tens of blocks found at the southern "playón" inside the crater. Rudín should have observed something similar in 1910, with the difference that he found these features outside the crater indicating that the event should have been more violent

> caused alarm among the inhabitants of the town. The streets, roofs and fields were whitened by this phenomenon that was never contemplated before. The course from the day before yesterday till today, without anything regrettable to occur, faded off fears that these natural features gave rise to, whereas the common people blamed it to the appearance of the comet Halley.

We can deduce that at least two eruptions were reported after the one of 25 January: (1) on the 27th of January early-morning minor ash fall in the centre of Alajuela, and (2) in the afternoon ash fall in Grecia during the same day, both of minor size and intensity than the first reported eruption.

An important detail was revealed on 2 February: "*The ashes on the La Paz river, a tributary of the Sarapiquí, poisoned the water, causing the death of fish in this river.*" Possibly the type of oblique eruptions, described by Casertano et al. (1983), or strongly wind-directed eruptions are responsible for the large quantities of ash and sulfur in these rivers, besides the ash fall transported by directed winds.

On the 4th of February 1910, Juan Rudín, Anastasio Alfaro, Gustavo Michaud and Alberto Rudín, presented a report entitled: "*Great ash eruption of Volcán Poás, 25 January, 4 h. 45 m. p.m*". A transcription of their observations on the summit of the volcano, after their expedition on request by the President of the Republic, Cleto González Víquez, followed (Rudín et al. 1910):

> On 25 January, shortly before five in the afternoon, an immense column could be observed from San José, apparently as a smoke plume, which actually was a mixture of water and ashes that rose above the summit of Volcán Poás, up to a height that we could estimate as 4000 meters, and that consequently, through evaporation, extended towards the flanks and upwards to the prodigious heights of

approximately 8000 meters… The column initially appeared to be of an intense grey color, whereas afterwards, due to the evaporation, an immense cloud of a light grey dye extended and slowly changed its nuances. The apparent shape was the one of a giant fungus, or maybe better, of an unleafed cauliflower, of colossal dimension, wide at the top and reposing on a relatively tight base.
When moved by higher winds, the cloud extended for the entire Meseta Central, and produced the reported ash rains between 6 and 8 in the evening of 25 January. After posterior observation made in San José, in San Pedro de Poás and on the summit of the volcano, it is not exaggerated to assure that the quantity of ash thrown out by Poás in the afternoon of the 25th can be estimated to be 800,000 cubic meters, with a mass of 640,000 tons, that is a sufficient amount to cover la Sabana (cfr. an area in San José, now a park) with a 1-kilometer thick layer. Commissioned by the Ministry, we left this city the 28th in the morning and could observe that the quantity of ash increased progressively until covering the cultivated lands in the region of San Joaquín de Heredia, with a fine grey-colored layer as if the coffee plantations were intentionally covered with the famous fertilizer Albert. On the side of the roads of Alajuela and its surroundings there were rests of ashes that gave a precious aspect in the shadows, as if there were aluminum rocks. The appearance between Alajuela and the river of Poás seems uniform, full of small tributaries that accidentally search their way between highs and lows until the river banks.
Along the entire trajectory the fall of small stones in sufficient abundance can be noted, and following on what the people living at La Lechería told, these falling small stones gave the impression of a heavy hailstorm; some of those stones reached a diameter of one to two centimeters. Higher up on the mountain small stones of higher magnitude are disseminated, that often remained, together with the ashes, on leafs of trees and bushes, and that fell on the soil when shaken by the wind or intentionally by hands.
Close to the second ranch, some pieces of stones of three to four centimeters size, generally light and porous, were found. In the proximity of the crater some stones fell that increased in size when nearing the crater rim. In the morning of the 26th, it was entirely covered by a uniform layer of ash that gave the weird appearance for those people used to admire its nuances. Afterwards, water and wind cleaned some sites, showing spots of other colors that break the monotony of the whole. The first impression that one had when reaching the crater is that the lake at the bottom became bigger, without being possible to say how much and from which side. Moreover, some considerable collapses occurred on the north side, having enlarged the Laguna Caliente towards this side in estimable way.
The entire surroundings of the crater received ash in the form of mud, as signs are shown that this occurred on the branches covering them also on the lower parts.
To a distance that varies from 150 to 200 meters from the upper rim of the crater, numerous stones had fallen, of which its origin will be explained hereafter: they are present in all sizes, from 5 to 48 cm in thickness, some light but the major part with a solid and heavy consistency. Almost all of them had fallen with their direction inclined towards the exterior, as such describing a parabola in their trajectory, as can be proven by the direction of the holes in which they are buried; they also should have to be fallen from high elevation, as the big ones staved into the soil for more than a meter depth, and as they kept their power to cut roots and easily break branches thicker than a human arm. The holes are most abundant on the SW side, while at the SE, towards the cold lake, they become less numerous, until complete disappearance in this direction.

In another report entitled "some new data on Volcán Poás", Alberto Rudín (1910) details:

Already going down the playones, we could notice that the amount of volcanic mud thrown out in this area was a lot more than other areas received: soil, branches, and three leafs are completely covered, forming a layer of nearly a decimeter. I have to advise that because of the thick fog that day we got lost and went down a lot more than we generally do, until close to river Ángel, after having crossed the myrtle bushes. These are literally laying on the soil, squeezed by the weight of the ash and the mud that covers them; apparently they died as a consequence of the large amounts of acids they received. In this entire trajectory only very few stones had fallen. In the eastern playones, in addition to mud, a true rain of stones of all sizes has fallen; but contrary to what is observed on the south side, they shouldn't have fallen from very high, the majority settled on the soil, whereas only some opened a rather large gap.
As on the south side, the stones must have fallen while hot, as sulfur that impregnated them poured out in melted state falling on the soil and sometimes forming precious stalactites and stalagmites and even perfect columns, while other times, probably when the rocks fell being hotter, or with more sulfur in them, it occurred at the surface that large plates formed that were similar to palm trees. The amount of this deposit that could be collected in the available short moments was enough to instigate several residents of San Pedro to create a commercial business. For this reason, at this

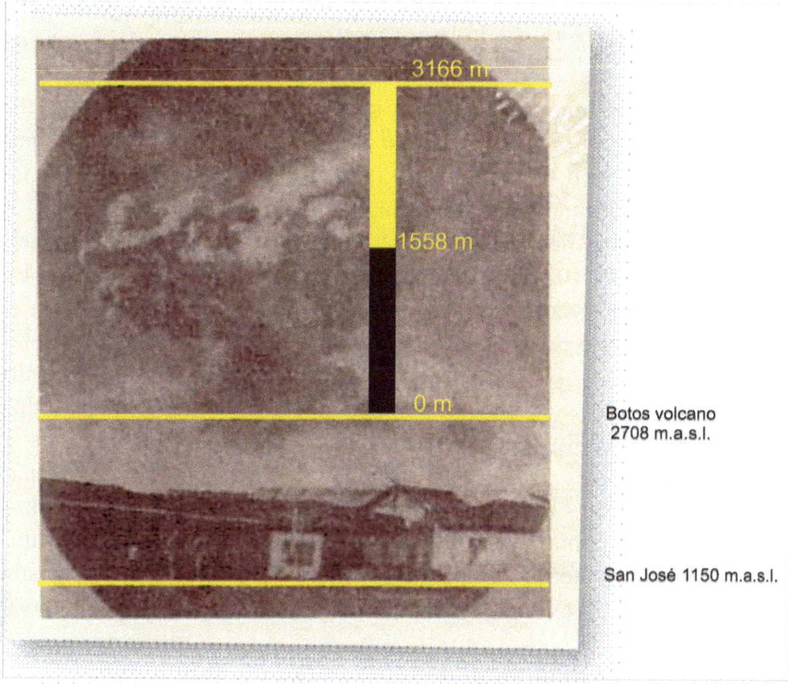

Fig. 6 Historical photograph of 25 January 1919 taken by Rudín. The height of the ash and gas plume is estimated at a little more than 3 km

> moment it is very difficult to still find those stalactites of regular shape that remained intact.
> In the northern playones the amount of mud that fell was that voluminous that it is absolutely impossible to verify if stones also fell. The thickness of the layer, measured in the sites where it could not have been accumulated by the action of water, reaches in many sites up to half a meter.
> We also descended the crater from the west side following a rope that started at the arrival. This descent is maybe even worse than from the front side, and surely a lot longer, as it has many difficult and dangerous passages. In this area very few stones fell, and in spite of the ash being present, there is no sign of mud. As noted, mud is distributed very irregularly; it fell abundantly on the north side, a little less on the east side, little in the south and nothing in the west.

Analyzing the historical photograph taken during the eruption from San José (Fig. 6), the height of the ash and gas cloud above Poás volcano is estimated at a little more than 3,000 m. There is a possibility that afterwards the ash column became larger because of later events and because of the wind forces being able to transport the ash to the heights as reported by Rudín et al. (1910). Nevertheless, in the photograph it is noted that the fungus starts to collapse due to the decoupling of the ash from the gas motion, and ash starts to fall.

3.2.3 Effects of the Eruption

On Wednesday 9 February, República reported dead fish in Sarapiquí River:

> On the 26th of last January, all day long immense amounts of fish descended from Sarapiquí river, some were already dead others were dying, of all types and colours, small and large ones. The neighbors of the river collected large quantities of dying Joturos. The river banks were invaded by many death animals. Until Sunday this gave off an unpleasant smell. Some people say the water of Sarapiquí is sour: I neither wanted to taste it, nor did I go near as the stink was unbearable. It is thought that the sis due to the fact that maybe the lake of Poás flooded into Los Ángeles driver and from there into Sarapiquí.

The first of February 1910, La Prensa Libre informed: "The ash rain", reporting on damages due to the eruption:

> The eruption of Poás volcano, that brought ash rains, resulted in the fact that some ranches became useless as the cattle stopped grazing, because the

grassland remained covered with a layer of fine ash, leaving an acre and sour taste. In the gardens of this city, the flowers that received this ash were burned.

La República informed on the second of February: "*The ash of Poás turns out to be poisonous*" and detailed on how the ash falling into La Paz river, a tributary of Sarapiquí, poisoned the waters causing the death of fish.

Rudín et al. (1910) reported a total destruction of the vegetation near the active crater. Neither deaths, nor damage to infrastructure were reported.

3.2.4 Deposits Related to the 1910 Eruption

Field work indicates that the deposits related to the 1910 eruption are characterized by whitish-grayish material composed of ash with lithic fragments, occasional juvenile scoria with a diameter up to 2 cm (especially on the rim of the active crater), as well as andesitic lapilli with a diameter up to 5 cm, with both punctual and floating contacts; the contacts of the upper and lower layers are irregular (Fig. 7). Only inside the crater and in the proximal area an inverse grading could be observed, with less ash and more lithics at the punctual contact. A larger thickness is noted to the north, east and south of the active crater, whereas thickness decreases towards the west. The maximum thickness reaches 85 cm in the southeastern sector. Occasionally, at less than two kilometers from the active crater, it is difficult to distinguish these deposits. We observed light grey ashes with parallel lamination and diffuse inclination, and lateral thinning with desiccation cracks (Fig. 8).

In general, a small layer of red-brown soil overlies these deposits. Despite the short distance

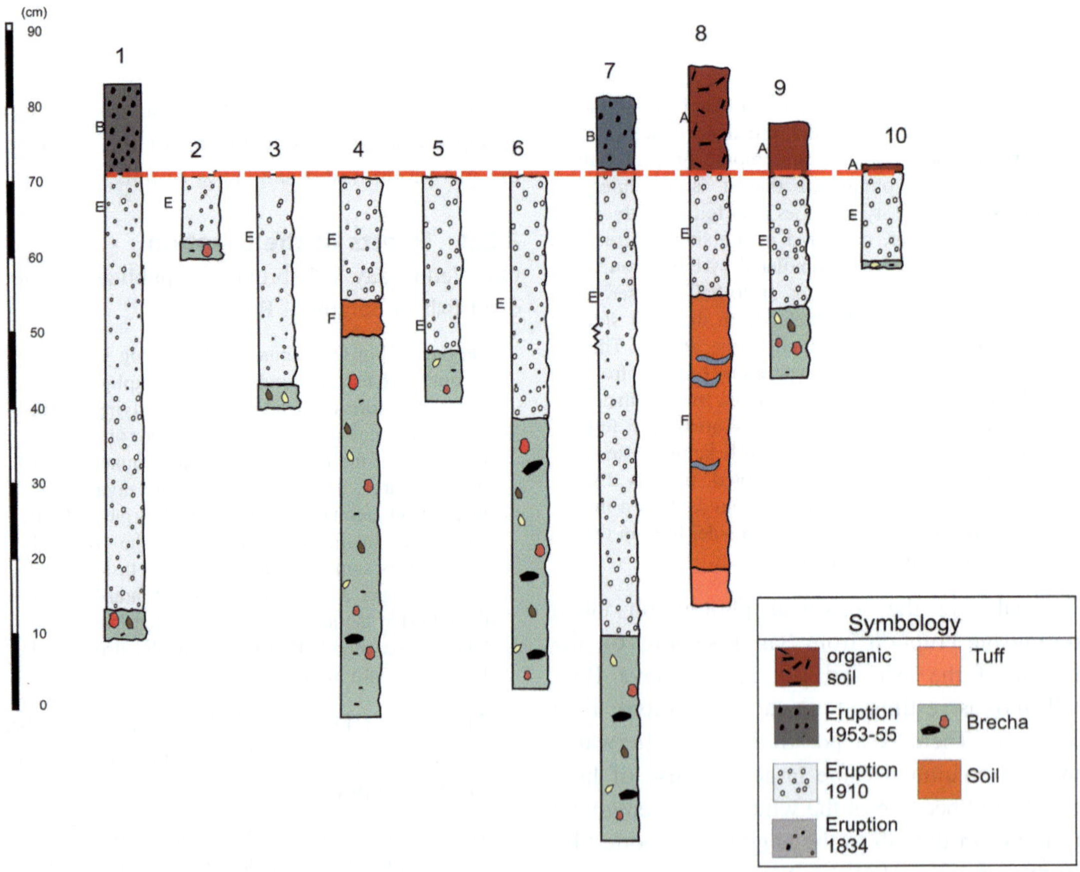

Fig. 7 Tephrastratigraphic columns performed during this work that display the deposits of the 1834, 1910, 1953–1955 eruptions

Fig. 8 **a** Deposits associated to the 1910 eruption. **b** Grey ash with parallel and cross laminations; dark grey pumiceous scoria correspond to juvenile, andesitic material. **c** Stratigraphic column 15, located inside the active crater. The black lapilli correspond to the 1953–1955 deposits, while the white colored deposits, representing ash with hydrothermalised lithic with point-point and floating contacts with inverse grading, correspond to the 1910 eruption. **d** Sector "Potrero", 2 km south of the active crater, showing the white layer in stratigraphic column 21, of only 5–7 cm thickness

from the emission centre, there is no regularity on the layer thicknesses. That is why it is thought that the events in this period were oblique eruptions, as described by Rudín et al. (1910) and Casertano et al. (1983).

3.2.5 Eruption Model

A magma rise generated the heating of the magmatic-hydrothermal system and triggered phreatic eruptions accompanied by high-temperature fumaroles with sulfur combustion. It should be stated that the 1910 deposits show fragments of scoracious rocks, hence probably indicating a juvenile origin, and a transition from phreatic to phreatomagmatic activity.

The deposits evidence different eruptive events, with fall out and pyroclastic surge material close to the crater area. In one outcrop (column 186) up to five different events can be described, with parallel and inclined lamination, and lateral wedging).

Based on the column height, on the field data gathered and on the estimated volume of the erupted ash column into the atmosphere, the

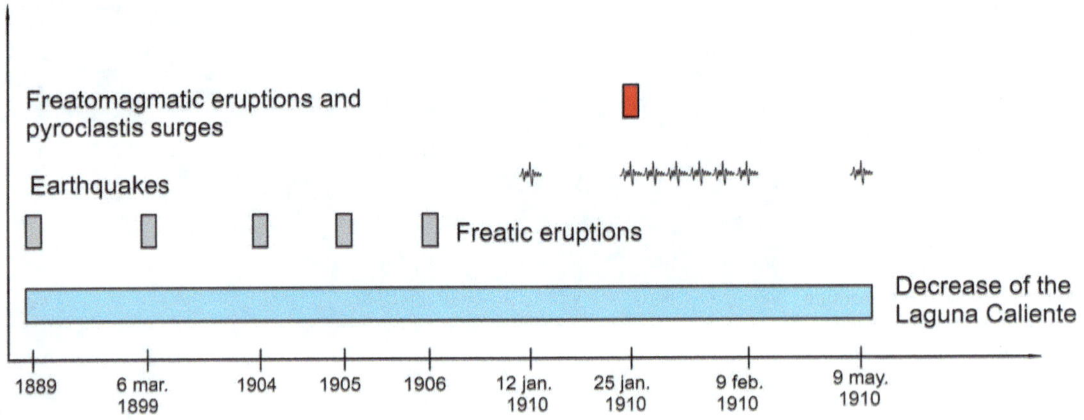

Fig. 9 Summary of the pre- and syn-eruptive phases of the 1910 eruption

1910 eruption represents a Volcanic Explosivity Index of 2 (VEI, Newhall and Self 1982).

Figure 9 compiles the pre- and syn-eruptive phases of the 1910 eruption, based on information from literature and field data from the period 1889-May 1910. It is worth noting that prior and during the eruption the Laguna Caliente crater lake did not disappear.

3.3 The 1953–1955 Eruptions

3.3.1 Pre-eruptive Stage

Between 1934 and 1950, Hantke (1951, 1953) reported weak fumarolic activity in the main crater. Specifically for 4 February 1947 he observed a rather low lake level for Laguna Caliente (Fig. 10). Fernández (1961) is the one to describe the first drastic changes at Poás, starting in 1952.

Monestel (1953) mentioned that Dr. Frank Turnes of Abott Laboratorios witnessed a fall out event when he walked near Laguna Botos on 18 June 1952:

> …Being close to the cold water lake, they were surprised by tremors, of which the "biggest" one he estimated to be IV on the modified scale [unreadable] and later they were wetted by an eruption of hot water and grey mud.

He added the following:

> On the 3th of March 1953 he witnessed an event… at 3:30 in the afternoon, on a clear day, calm wind and ambient temperature of 19 °C. The eruption was of short duration and the intercrateric lake - calm in the morning when he arrived with Dr. Turner-, dropped its water level. He said it was strange that the blue-green color of the lake, that in other occasions (1952) he'd seen grey, had changed into reddish as of oxidation. He informed Don Fausto that the first day of March, when he descended to study the plants, from Bajos del Toro, wading through Desagüe river, when starting the rough job, and passing the waterfall, he noted the prevalence of waters contaminated by volcanic materials, grayish-yellows, with a strong smell of sulfur. The river dragged masses of grey clays with volcanic ash, plastic sands as well as some tree trunks [unreadable] and chunks of black and grey very porous and light rocks.

In this note (Monestel 1953), refers to possibly mud flows in Desagüe river that have been confirmed by inhabitants of Bajos del Toro.

Before the 16th of May 1953 eruptions, at least three possibly phreatic eruptions were reported: 18 June and 9 September 1952 and another one on 2 March 1953. Some of them were preceded by seismic movements. These eruptions should have had an altitude of at least 300 m, in order to make the ashes reach Laguna Botos, 300 m higher in elevation; moreover, the felt earthquakes could be related to magma ascent.

3.3.2 Eruptive Stage

The third important eruptive activity at Poás in historical times started between 15 and 16 May 1953. By 20 May 1953, the newspaper La

Fig. 10 Photograph taken by the family Acosta Rodríguez from the current "mirador" in 1947. Note the low lake level and degassing at the site of the current dome that was emplaced in 1953

Nación mentioned ash fall in Vara Blanca and surroundings. They reported that since Monday 18 at mid-day ash was falling, with a thickness up to a "quarter", affecting the people, crops and cattle. It is mentioned that the cattle cannot use the pasture, get fed and drink water, as the ash covered the terrain with an "*immense grayish sheet that releases a strong sulfur smell*".

The same day they published that the river waters came down contaminated with ash smelling of volcanic gases that cause dizziness and vomit to whom inhales them. As this event was immediately recognized as harmful, it can be interpreted as directed, oblique eruptions that Poás launched towards the headwaters of Ángel river, apart from the ash fall.

On Friday 22 May, La Nación reported a "*New and big eruption of Poás*", and that the entire city of Grecia woke up covered by a thin layer of ash, and that the eruption was easily visible. They continued that the Jefe Político, Sr. Ricardo Barquero, informed that the big eruption was accompanied by ash rain, and mentioned that the day before the city woke up covered by a thin ash layer that covered roofs and houses, trees, streets, etc., which caused unrest in the community. They explained:

> The eruption was seen during almost every hour of the mourning. Yesterday in the capital, at about seven thirty in the mourning a tremor of very low intensity and short duration was felt.

On 24 May 1953, La Nación published that volcanic activity increased:

> In the early morning, the neighbors in the surroundings of the volcano evacuated their homes in a hurry, helped by workers of the Ministerio de Obras Públicas that are building the road to Poás. The neighbors only grabbed the essentials to pass these days while quiet returns. At three in the mourning three strong tremors could be felt at regular intervals that created a nervous tension to all the people in the area. Meanwhile, the rain of ash and volcanic material continued, thrown out by the crater, being dispersed by the wind towards all the cardinal points.

On 25 May 1953, Yeudi Monestel, journalist of La Prensa Libre, sent a letter to Dr. César Dóndoli (founder of the Central American School of Geology, *Universidad de Costa Rica*), explaining what he observed during his visit to Poás. In a modified sketch (Fig. 11) he shows how far the volcanic material reached in the entire area. To continue some parts of this document:

> …in the row before "Potrero del Hotel", the guide Nazario Arias Piedra showed us on the right hand side of the gauge two impact craters:
> One of three meters in diameter and another one of only 1.3 meters, in which at their bottom two enormous rocks had been embedded, one of them angular with a weight of at least a ton. One of the crater - the one with the largest rock- was 60 cm deep.

Monestel (1953) mentioned that from "Potrero del Hotel" on, a coarser and more sandy material was observed, with many chunks of porous grayish lavas. His sketch (Fig. 11) clearly shows where the ash, and material with a larger diameter-lapilli-type, bombs and blocks- fell. Again, like in the 1910 eruption, they mention that the Ranch is a site where coarser material fell. It is worth noting that they mentioned the falling of large-size bombs and blocks in the row before "Potrero del Hotel".

On the other hand, following the report, the ash fall continued during days and weeks, affecting graze fields and, hence, cattle, causing illness and ulcerous to the animals.

On Thursday 28 May 1953, La Nación published:

> Rumbles and smoke from Poás volcano yesterday. They report on the areas where ash fell that weighs the critical situation presented in the first days, luckily no animal had died, moreover they informed that the region of Sarapiquí was notified on the rumbles of Poás volcano with regular intervals and that a thick smoke column rose from its crater that can be observed from many kilometres distance.

On 6 June 1953, La Nación published: "*Ash rain in Río Cuarto*". In the article it is mentioned that an eruption of smoke and ash on 5 June covered the region of San Carlos with a layer of volcanic detritus. The article continued:

> The police officer of Grecia, Sir Ricardo Barquero, sent yesterday late a telegram to the Ministerio de Gobernación, reckoning on the alarming information that he received from Río Cuarto, in relation to the eruptions of Poás volcano since several weeks. The mentioned official said that the local authority reported him that yesterday at nine o'clock a heavy ash fall took place that covered practically every pasture in the Río Cuarto region, and that this mainly affected the cattle.

The 5 June eruption was particularly important, because it could be observed from considerable distance. In Fig. 12 the photograph published by La Nación on 6 June 1953, shows an eruption with a plume of ash and gas from a distance, but, unfortunately, they did not specify if the photograph was taken from Río Cuarto, from Grecia or from the Central Valley, San José.

La Nación published the following on 6 June 1953: "*On Thursday of Corpus Cristi (cfr. 40 days after the resurrection), the eruption of Poás volcano was gigantic, rising up to twenty-three thousand feet height.*" This is equivalent to 7,000 m height. Nevertheless, following the analysis of Fig. 12, assuming it was taken from (1) San José, the eruption height was

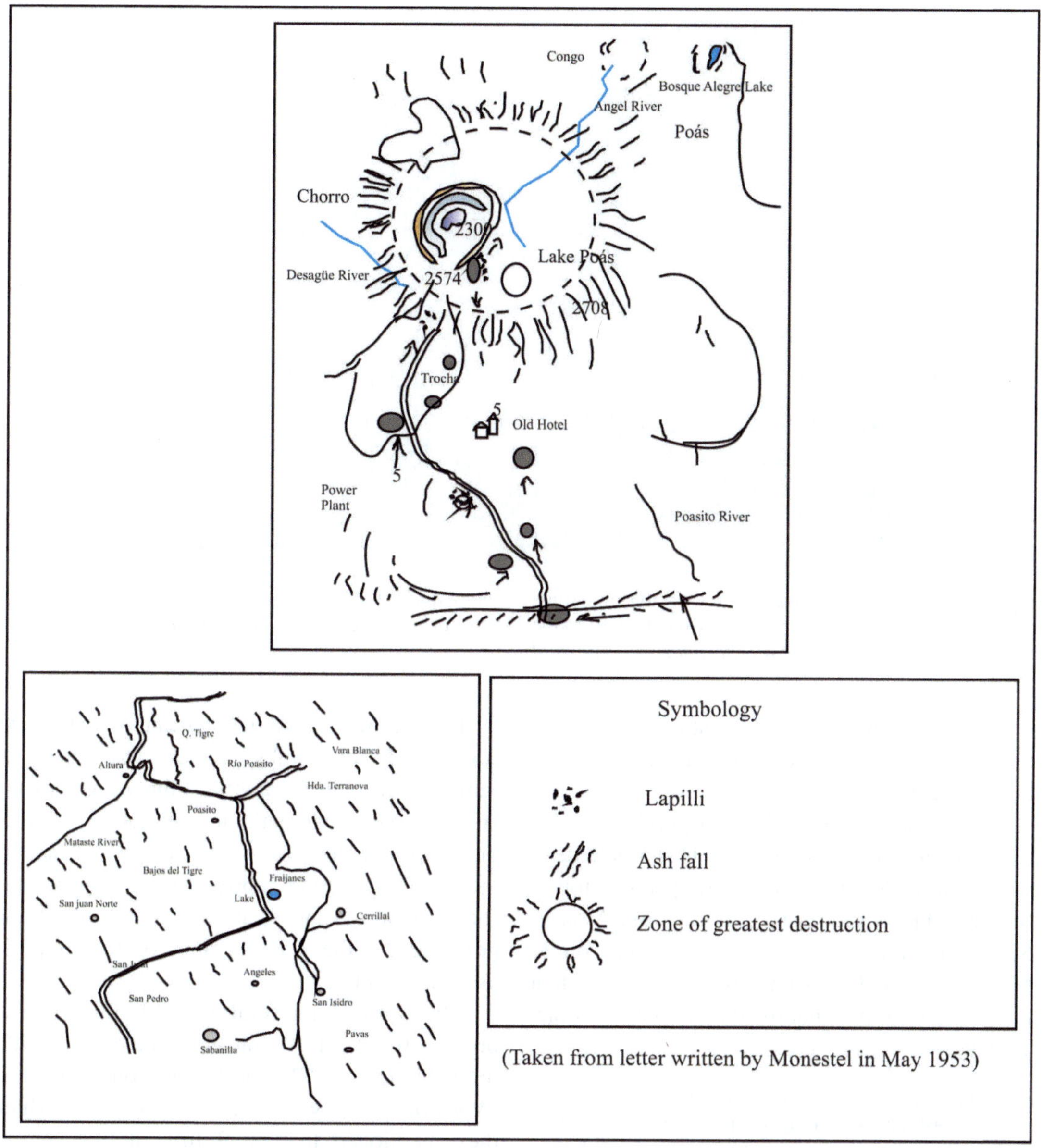

Fig. 11 Sketch modified from Monestel (1953), indicating the spot of the pyroclastic fall out as observed during the day of field-work on 24 May 1953

approximately of 3,400 m, and (2) Río Cuarto, it is estimated at 4,800 m height.

On Wednesday 10 June 1953, the same newspaper informed: "*Poás swallowed the immense lake in its crater*". This particularly long article commented how "*the geyser of Poás, the largest in America*" had disappeared. Moreover it added how active the lake was a few weeks before, and how the recent activity managed to make it disappear. It described how the "columns of white smoke, above the summit" were visible and how they transformed into "immense black cauliflowers. Terribly black like a wolf's mouth."

Fig. 12 Photograph published by La Nación on 6 June 1953, corresponding to the eruption the day before. At the footnote of the picture it reads: "*The amazing eruption of smoke and ash yesterday of Poás with which the region of San Carlos was covered with a layer of volcanic detritus.*"

The black and dark colorations might be related to juvenile material of magma liberated during the event.

On 12 June 1953, La Nación published: "*We could see the fury of the giant*". This article dealt with a chronicle sent to the newspaper's editorial, by Sir Rafael Sanabria B. and don José Manuel Salazar, narrating their visit to the "giant" on 24 May 1953. In this article they mentioned that they were informed that the day before (Saturday 23 May)…

> …the imposing eruptions, how they threw out sandy ash and smoke up to high elevations, and how they launched incandescent stones of big dimension that fell towards the rim of the crater, from its bottom, during the moments when the base of the enormous fungus formed by the eruptions remembered atomic explosions, red lightning, impossible to photograph, very quickly left the crater. Preceded by thunderous rumbles, that asked for huge respect, thick clouds of sandy ash and sulfurous vapors, that to our great luck, if not, we wouldn't be able to witness this marvellous show, were pushed to the north by the wind. It was about 11 o'clock in the morning when we had the opportunity to see the largest eruption and we felt how the soil, without reaching tremor.

The earlier mentioned incandescent rocks correspond to bombs, which are made of semi-molten material that was thrown out of the crater at near-magmatic temperatures, whereas the shapes at the base of the fungus formed by the eruptions, could correspond to pyroclastic surges that advanced at the higher parts and perimeter of the active crater, or to a rapidly ascending plume.

Afterwards, for several weeks no further news on the activity of Poás volcano could be found; it is until 10 July 1953 that La Nación, published "*Volcanic activity resumes*", an article in which they mentioned that on 9 July of this year, a spectacular eruption of Poás volcano could be observed from the capital. The articles mentioned:

> This eruption drew the attention as it had the peculiarity that the smoke was elevated above the colossus having a black opaque color, contrary to the regularly clear color. The eruption was produced by large puffs, assuming they were the result of internal explosions of the volcano. The show was imposing.

Bullard (1956) in his article "*Volcanic activity in Costa Rica and Nicaragua in 1954*" mentioned that on 19 November 1953 the eruption was particularly violent, and launched a large quantity of pyroclastic material. This activity resulted in: "*on the crater rim scorias and bomb-shaped fragments, with the size of a human head, are abundant.*"

The phenomenon described between 14 and 19 November 1953 is related to the launching of juvenile pyroclastic material, lapilli and bombs. It is possible that during this phase the major parts of the dome south of Laguna Caliente were emplaced.

On 2 June 1957, Burgos mentioned in La República that in April 1955:

> At a few meters in front of our amazed pupils the colossus vomited fire, large whirlwinds of smoke, ash, launching up scorias and incandescent material, in the middle of subterraneous detonations and frightening roars. The majority of it fell noisily in the crater itself, and the rest on the playones. Its flanks gave way to an inflamed material that flowed, illuminating everything, while the earth trembled.

Burgos also referred to Strombolian activity. It is possible that he was also talking about a lava pool or lively red cracks of the recently emplaced dome. Describing that "*the flanks opened*", it could be related to the behavior of lava accumulated in a depression.

Dóndoli (1965) mentioned that during 1956 the volcano resumed its normal activity of intense fumarolic emissions, with intermittent throw outs of ashes and sands. He noted that in the surroundings of the cone at the bottom, cake-shaped material of scoriaceous lava was launched by the volcano. Moreover, he indicated that Laguna Caliente was reformed in the crater during periods of intense rainfall, but that it easily evaporated while generating the expulsion of mud. This feature was repeated in various occasions between 1956 and 1965. We can deduce that for the duration of an entire decade Laguna Caliente was formed and disappeared, suggesting a strong relation with the rainy periods and with variations in volcanic activity. Generally speaking, at least phreatic eruptions must have occurred during this entire period.

Vargas (1967) mentioned that in the Sarchí, Toro Amarillo and Tres Amigos rivers, during the period 1953–1955, repeated mud fronts were observed.

3.3.3 Effects of the Eruption

On 18 July 1953 Prensa Libre informed: "*Hundreds of cattle in danger.*" The article continued with a report by R. Calderón Monge, political chief:

> Dense layers of, firstly, ash, and later on, black dust, launched by Poás volcano left in trouble the ranchers of the higher parts of San Pedro de la Unión and La Luisa, and especially those of the district of Toro Amarillo. It will become worse if the summertime of the last couple of days will continue, as the volcanic material will not be washed away and the cattle will perish of hunger. Yesterday I arrived at Trojas, in the way to Toro Amarillo and I'll tell you that the roads where it didn't rain became black colored. I think that the Ministerio de Agricultura should take measures to save hundreds of cattle.

On Friday 15 January 1954, the newspaper La República published: "*Large losses for coffee plantations due to the eruption of Poás volcano.*" In this article, the Chief of the Coffee Section of the *Ministerio de Agricultura e Industria* informed on the state in which the coffee plantations remained in the areas of San Miguel and San Roque de Grecia, as a consequence of the strong eruptions of Poás. It also informed that the ashes burnt the leaves of the coffee trees, impairing its future flourishing, destroying its vivariums and leaving the trees that serve as cover completely leafless.

3.3.4 Deposits Related to the 1953–1955 Eruption

The 1953–1955 eruption can be recognized in the field by the presence of a layer of black scoria of basaltic-andesitic composition, made of ash and cauliform or fusiform bombs, with a maximum size of 32 cm. This layer is exposed inside the crater, and on the north and east of the crater rim (Fig. 13).

In 1953 the crater lake completely disappeared and a dome started to grow reaching 40 m in height. This dome is the one that still degasses/erupts at present although it was destroyed in April 2017, now the locus of major magmatic eruptions (see Sect. 3.4). The sequence of layers of the eruptive products of the 1953–1955 eruptions is composed of lapilli of 0.5–4 cm, with a clast-supported contact. Moreover, there are fusiform bombs (Figs. 13 and 14) and andesitic blocks of variable dimensions (between 10 and 160 cm), with afanitic and afanitic-porfiritic textures and with some mm-size crystals of dark grey plagioclase.

The thickest deposits (~50 cm) are found towards the south inside the crater. These appear as dark smudges or patches of several decimeters to meters on the inner crater walls. Towards the north, this layer is almost completely absent, except for some centimetre-sized bombs. Towards the east side of the crater occurs the highest amount of bombs, with centimeter to meter sizes. Porous black-grey colored blocks and bombs with a diameter of decimeters to meters, of mainly andesitic-basaltic composition, and with afanitic-porfiritic textures and some fusiform textures, are recognized.

3.3.5 Dome Emplacement, Lava Flow and Lava Pool

Bullard (1956) mentioned that as a result of the 1953 eruption a cone of ash and pyroclastic products started to form in the place previously occupied by the lake.

During their field exploration, Racicchini and Bennett (1978) determined that the pyroclastic material covered a merely compact, though fractured, igneous mass. They suggested this was related to an extrusion of a viscous melt, that ended with the formation of an endogenous

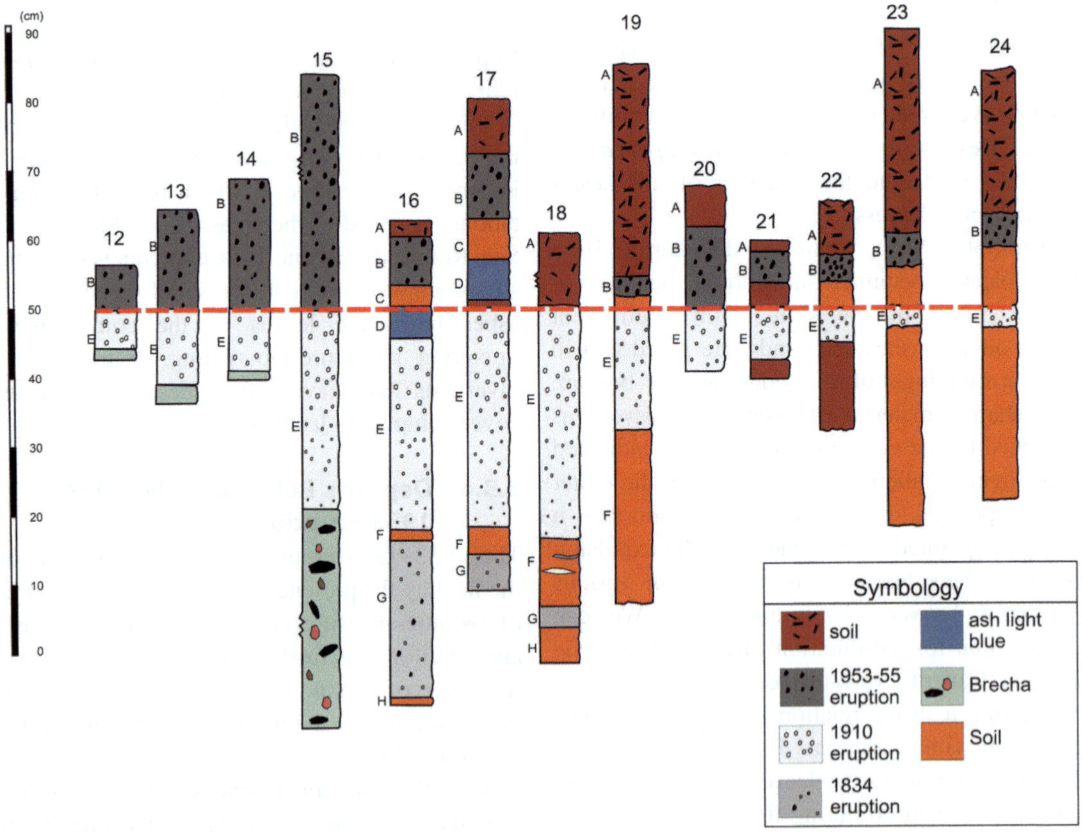

Fig. 13 Tephrostratigraphic columns of deposits related to the 1953–1955 eruption sequence

Fig. 14 Volcanic deposits of the 1953–1955 eruption. **a** Meter-sized bomb inside the active crater. **b** Fusiform bomb in the "terrazas" sector, i.e eastern rim of the active crater. **c** Dark grey lapilli, upper part of stratigraphic column 185. **d** Meter-sized fusiform bomb inside the active crater. **e** Bread crust texture bomb in the "terrazas" sector

dome, and that during the 1953 eruption part of this dome was buried and part was destroyed. Francis et al. (1980) and Casertano et al. (1983) described the 1953 activity as eruptive episodes of incandescent pyroclastic products (Strombolian eruptions) with lava effusion inside the crater. They estimated that the dome was emplaced on 9 November 1953.

A massive lava is observed in the field, manifesting fracture cooling during the emplacement of the dome. Out of these fractures fumaroles exhale high-temperature gases that in various occasions have reached nearly 1,000 °C. In the upper part and south end of the dome a decimeter-thick layer of hydrothermalized bombs, blocks, lapilli and ash cover the dome structure.

The north face of the dome is uncovered, and when gases are transported vertically by the wind, a massive lava-like structure can be clearly distinguished. At the base of this structure a colluvium of eruptive products is present, derived from the erosion and landslides of these rocks.

At the east and south parts at the shores of Laguna Caliente, a massive, fractured, andesitic lava flow is exposed. When its interior crops out it has a dark grey color and a cap of whitish weathering. The maximum thickness of the flow is 8.5 m (Figs. 15 and 16). It is possible that in this site a lava pool was formed, filling a pre-existing depression prior to the emission of the flow.

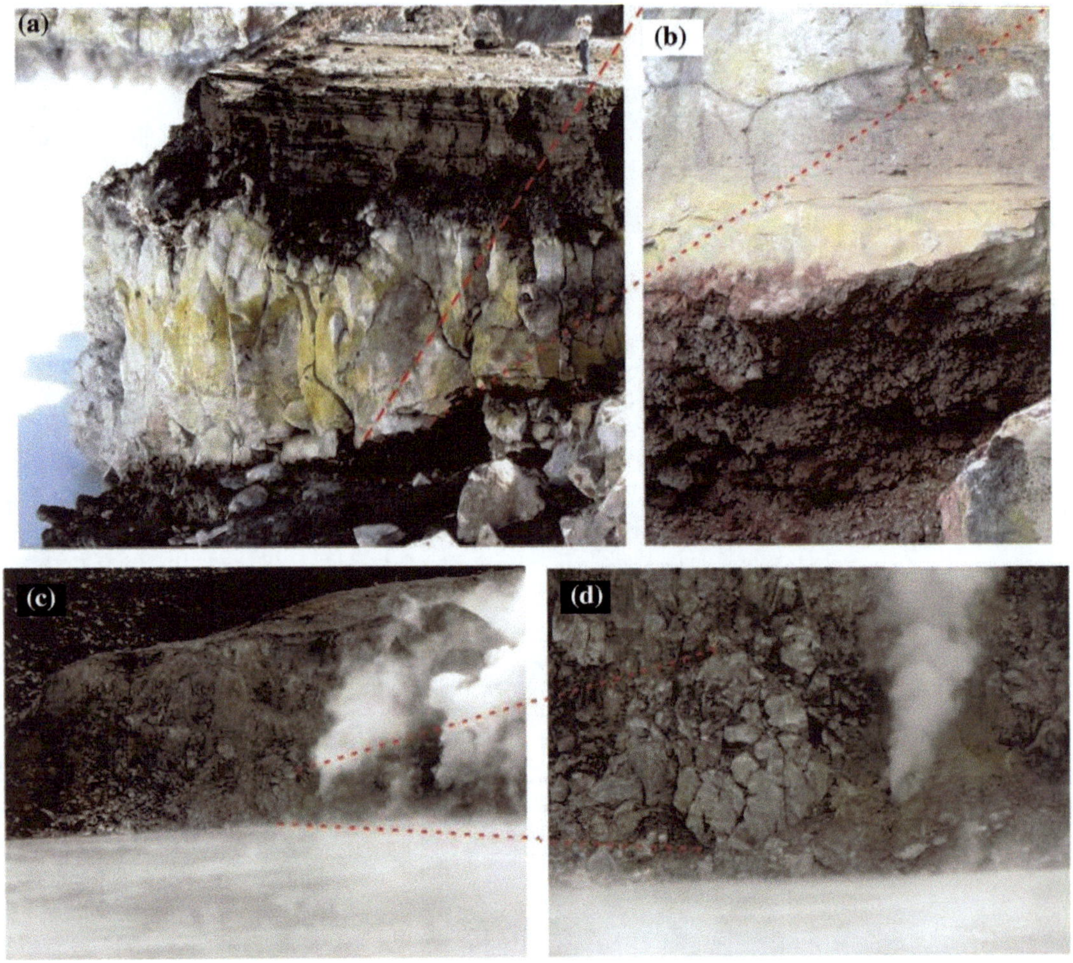

Fig. 15 Details of the intracrateric lava flows and dome. **a** The unconsolidated deposits, on top of the lava flow, with a maximum thickness of 4 m correspond to products of various more recent eruptions. Here the thickness of the lava flow is 8.5 m, suggesting that a lava pool was formed here. **b** At the bottom of the flow, a dark-red breccia is found. **c** The dome with strong fumarolic activity. **d** Massive lava, fractured upon cooling

Analyzing the historical information of the dome emplacement together with the lava flow and lava pool, the information that we gathered suggest that this activity occurred in three episodes:

(1) At the end of May 1953, intense eruptions produced incandescent pyroclasts;
(2) At the beginning of November 1953, eruptions of pyroclasts are reported. Eyewitnesses described reddish radiance seen from San Miguel de Grecia, Poasito and Vara Blanca;
(3) In April 1955, when the journalist Vargas (1967) witnessed a natural spectacle: "*Its flanks gave way to an inflamed material that flew, illuminating everything, while the soil rumbled.*" It is possible that Vargas referred to the lava flow emplacement at the crater bottom.

3.3.6 Fall Out of Bombs

Bombs with a diameter larger than half a meter were found in the upper part of the crater ("terrazas" sector) and were localized with GPS. A higher concentration of bombs was found in

Fig. 16 **a** Lava dome seen from the north, and opposite side of the lava flow. This flow thickens towards the north. **b** Detail of Laguna Caliente and the degassing dome. **c** Increase in thickness of the lava flow towards the north, where a small lava pool was formed

the eastern sector. This bomb distribution may indicate the conduit through which the 1953–1955 eruptions were generated was tilted towards the east (Fig. 17).

3.3.7 Eruptive Model

In the 1940s, when fumarolic activity was constant, an increase in heating of Laguna Caliente was observed. In 1952, the first phreatic eruptions were reported, with falling ash and mud in the immediate surrounding of the crater (close to Laguna Botos), accompanied by earthquakes perceived by people living in the surroundings. It is possible that at the beginning of 1952 a magma started to rise causing the complete dry-out of Laguna Caliente, followed by eruptions of juvenile material, block to ash particles.

Afterwards the dome and lava emplacement took place, reaching a thickness of circa 40 m. The most important eruptions of juvenile material were reported at the end of March, June, November and December 1953. In 1954 no Strombolian eruptions were reported. The last Strombolian-type eruption was reported in April 1955. However, pyroclastic surges that affected the crater area during this period cannot be excluded, although field evidence is absent.

This eruption was complex because it involved four types of activities: Strombolian, phreatomagmatic, dome-extrusion, Vulcanian. Considering the characteristics of the eruptions, the eruption column height and ejected volume a VEI of 3 is estimated (Newhall and Self 1982).

Figure 18 provides a summary of the phases described prior to and during the 1953–1955 eruptions. The figure shows that Laguna Caliente was subjected to dry-out, disappearance and ephemeral existence during a period of two decades. This period of time was the most unstable phase of Laguna Caliente during historical times.

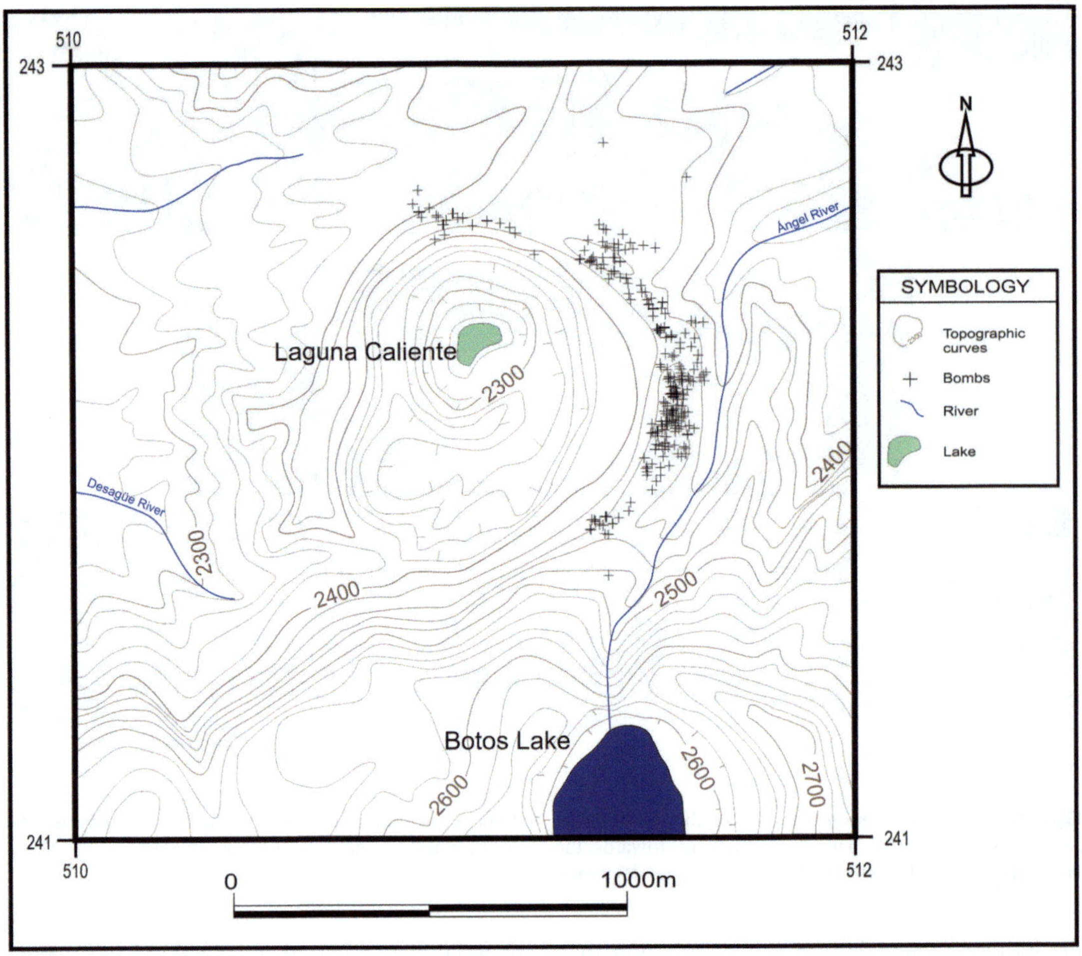

Fig. 17 Location of bombs in the upper part of the active crater

Eruptive processes that transient from phreatomagmatic explosions to lava flows had a duration of two years (May 1953–April 1955). These events were sometimes accompanied by pyroclastic surges. It is quite possible that phreatic eruptions occurred during the entire eruptive period.

3.4 The 2017–2018 Eruptions

3.4.1 Pre-eruptive Stage

Between 1986 and 1994 Poás manifested phreatic activity and the dry-out of Laguna Caliente (Mora Amador et al. Chapter "The Extraordinary Sulfur Volcanism of Poás from 1828 to 2018"; Martínez et al. Chapter "Behaviour of Polythionates in the Acid Lake of Poás Volcano: Insights into Changes in the Magmatic-Hydrothermal Regime and Subaqueous Input of Volatiles"; Rouwet et al. Chapter "39 Years of Geochemical Monitoring of Laguna Caliente Crater Lake, Poás: Patterns from the Past as Keys for the Future"; Vaselli et al. Chapter "The Last Eighteen Years (1998–2014) of Fumarolic Degassing at the Poás Volcano (Costa Rica) and Renewal Activity"), followed by more or less a decade of quiescence. In 1998, five new fumarolic fields were observed in the southeast inside the crater with temperatures between 70 and 140 °C, and the liberation of gases on the north face of the 1953 dome.

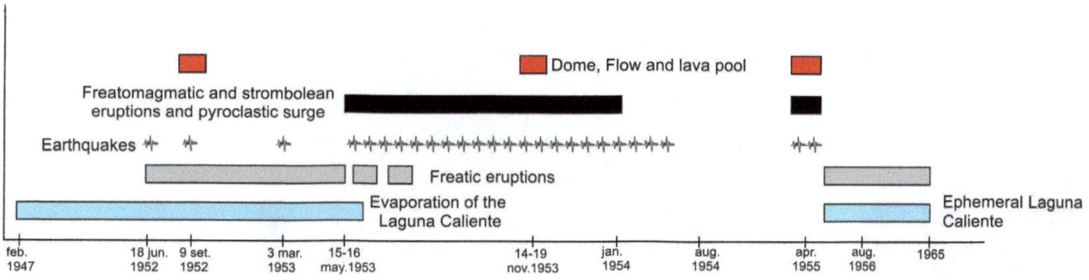

Fig. 18 Summary of the phase prior to and during the 1953–1955 eruptions (modified from Soto and Alvarado 1989)

Since 2000, fumarolic activity has migrated towards the east of the crater, where gases escaped from at least four sites. One of them was very vigorous and named Fumarola Naranja (see Mora Amador et al. Chapter "The Extraordinary Sulfur Volcanism of Poás from 1828 to 2018"). In January 2005 Laguna Caliente reached its maximum level since 1950. Since February-March 2005 Laguna Caliente manifested changes in color and temperature (Rouwet et al. 2016). In May 2005 an 82 m-long sulfur flow was emplaced from Fumarola Naranja; contemporaneously, sulfur slick floated on the surface of Laguna Caliente, originating from the sub-lacustrine molten sulfur pool (Mora-Amador and Ramírez 2008; Mora Amador et al. Chapter "The Extraordinary Sulfur Volcanism of Poás from 1828 to 2018"). On 25 and 26 December 2005 tailed floating sulfur spherules were observed, often a precursory signal for phreatic activity on crater lake bearing volcanoes (Takano et al. 1994; Mora et al. this issue). During the same period, low amplitude seismicity increased (Fernández and Mora Amador Chapter "Seismicity of Poás Volcano, Costa Rica"). On 24 March 2006 the first phreatic eruptions occurred after almost 12 years of quiescence.

This resumed phreatic activity culminated during the next 10 years, converting Poás' Laguna Caliente into the most active crater lake on Earth (Fischer et al. 2015; de Moor et al. 2016; Rouwet et al. 2016) (Fig. 19a). Incandescence of the dome fumaroles (Fig. 19b) and floating sulfur on Laguna Caliente were common features (Fig. 19c). The gas geochemistry of the Fumarola Naranja, Dome fumaroles and gas plume liberated by the 1953 dome and Laguna Caliente are reported by Hilton et al. (2010), Fischer et al. (2015), de Moor et al. (2016), and Vaselli et al. (Chapter "The Last Eighteen Years (1998–2014) of Fumarolic Degassing at the Poás Volcano (Costa Rica) and Renewal Activity"). A detailed analysis of the changes in water chemistry of Laguna Caliente are documented by Rouwet et al. (2016) (for the period 2005–2010) and Rouwet et al. (Chapter "39 Years of Geochemical Monitoring of Laguna Caliente Crater Lake, Poás: Patterns from the Past as Keys for the Future") (for the period 2010–2016). Despite being a decade-long period of phreatic eruptive activity, similar as the 1986–1994 period, within the context of this study the 2006–2016 period is considered a pre-eruptive stage. By October 2016, phreatic activity finished and Poás was apparently set for a new period of quiescence. This was also suggested by the observations made in January–February 2017: the absence of sulfur spherules (i.e. absence of sub-lacustrine molten sulfur), low degassing pressures from the dome fumaroles.

3.4.2 Eruptive Stage

After six months of apparent calmness, dynamic changes have occurred since early April 2017. We here report:

(1) On 1 April 2017 a new fumarole appeared SW of the active fumarolic field on the 1953–1955 dome in a tephra deposit (possibly an old pyroclastic cone). The gas escape during one day.

Fig. 19 **a** Phreatic eruptions were frequent at Laguna Caliente between 2006 and 2016. This eruption was documented on 25 May 2011. **b** Between July and October 2011 incandescence was observed at the dome (picture on 11 September 2011, a dome temperature of 800 °C was measured). **c** Sulfur slick originating from the sub-lacustrine molten sulfur pools at Laguna Caliente are often precursory signals of enhanced phreatic activity. **d** On 7 April 2017 the breaching of a new thermal spring located in the south of the intracrater was observed. Water was expulsed from this geyser-like spring at high pressure up to 3 m height. A 400 m-long stream was formed consequently draining into Laguna Caliente. During the first week of April 2017 a lake level rise of a few centimetres was observed, arguably caused by a rushing vapor front topping a rising magma beneath the lake area. **e** High pressure water discharges from the new thermal spring. **f** On 10 April 2017 black sulfur spherules were sampled. Moreover, the surface of Laguna Caliente was covered by black floating sulfur

(2) On 7 April 2017 renewed activity is reported: a geyser-like spring opened in the southern sector of the crater at an outcrop of Holocene tephra deposits (Fig. 19d, e). From this boiling spring (approximately 90 °C, at crater height), thousands of cubic meters of hot acid (pH = 0.5) water was liberated, forming a 400 m-long creek eventually draining into Laguna Caliente. The following days intense exhalative degassing activity occurred from Laguna Caliente, covered with sulfur slick. Black spherules with a diameter of 1.5 mm were observed (Fig. 19f).

(3) On 12 April (5 pm) the first major phreatomagmatic eruption occurred (Fig. 20a). The tourist mirador was covered in ash and mud (Fig. 20b), and the dome was broken and fractured (15 m long fractures, Fig. 20c). By 12 April (mid-day), only fine mud and gas exhaled from the spring, without a clear water discharge (Fig. 20d). Moreover, part of the northern wall of the dome, at the locus of the 1st of April new fumarole exhalation, was destroyed by an explosion. An eruptive plume of 700 m was observed, while Laguna Caliente exhibited constant vigorous degassing.

(4) On 14 and 15 April Poás generated eruptions with an eruptive column of more than 5 km (Fig. 21a), launching bombs and blocks of more than 50 kg, and ash until the tourist mirador (Fig. 21b–d). The dome was almost completely destroyed.

(5) After the 14 April eruption, at the site previously occupied by the dome, a scoria cone was formed, up to a pyroclastic cone of at least 40 m height, from where small Strombolian eruptions occurred with columns heights of 200–1,000 m. A N–S trending fracture, starting from the previous playón until the intracrater is evident.

(6) During 21 and 22 April 2017 multiple eruptions occurred. Shock waves could be perceived up to a distance of 2 km. The gas and ash plume rose up to >1 km height.

(7) On 22 April the major eruption of this cycle occurred at 7 pm (Fig. 22a), generating a pyroclastic flow emplaced inside the active crater, with a ash cloud reaching the tourist mirador (southern sector) and bombs reaching the eastern "terrazas" (Fig. 22b). The eruption launched bombs and blocks up to 1.5 km from the vent (tourist centre parking lot). Inside the active crater and on the "terrazas" sector 1–8 m sized bombs were deposited (Fig. 22c, d). Moreover, the pyroclastic surge destroyed vegetation of the Botos forest (southeast sector) (Fig. 22e).

(8) From May till September 2017 sporadic eruptions of gas, ash and pyroclasts occurred (Fig. 23a). These eruptions were rather small and the majority of the eruptive material remained inside the crater. Different centers of activity can be recognized inside the crater (Fig. 23): (a) the main pyroclastic cones, (b) the small sulfur volcano, (c) a small gas exit structurally similar to a small "castle", and (d) a diffuse fumarolic field towards the west. In August 2017 eruptions of ash, aerosols and brown colored gases from the pyroclastic cone were frequent, together with yellow colored gases from the small sulfur volcano.

(9) On 8 January 2018 Carlos Cordero, Park Ranger, reports on the formation of a small lake in the NE sector if the intracrater (i.e. previously dried out Laguna Caliente lake basin). This lake had a celestial color and a diameter of approximately 50 m.

(10) On 20 January 2018, Cordero observed the appearance of a larger lake, arguably formed earlier but unobserved due to bad weather conditions (Fig. 23b). The newly formed lake locally reached depths of 15–20 m. Laguna Caliente officially reappeared, with a temperature of 40 °C and a pH near 1. The formation of the lake occurred in 12 days. Intense degassing occurred from the entire lake surface. Moreover, it is possible to observe mud deposits and floating sulfur on the lake surface forming strings. In the eastern sector

Fig. 20 **a** On 12 April 2017, Poás resumed magmatic activity with its first phreatomagmatic eruption in 64 years. In the picture (13 April 2017) the north face of the dome manifests intense exhalative activity and fracturing. **b** The eruption launched ash up to 1 km distance. Plants known as "poor man's umbrellas" were covered by acid mud. **c** The south face of the dome shows new fractures of 15 m long, caused by the 12 April 2017 eruption. **d** After this eruption, the thermal spring formed on 7 April 2017 stopped its water discharge

of the intracrater a fumarolic field led to the emplacement of small molten sulfur deposits and liberated gas at high pressure.

(11) In the eastern sector of the new Laguna Caliente a 15 m long fumarolic field is observed, in which small amounts of molten sulfur was emplaced, two boiling pools were present and gas exhaled at high pressures.

(12) Laguna Caliente dried out again by the third week of February 2018. In March drone

Fig. 21 **a** Picture taken from Naranjo de Alajuela, showing the 14 April 2017. This eruption reached a column height of 5 km that was observed from the central valley of Costa Rica. **b** Blocks of 50 kg were launched against the tourist mirador, 800 m from the emission vent, destroying armed concrete. **c** and **d** Ash and mud reached the visiting centre, 1.4 km from the emission vent

footage documented minor eruptive activity from three main vents at the former lake bottom (Fig. 23c).

(13) During April 2018, exhalative activity has continued. Monitoring authorities reported very low seismic activity. In May 2018, minor phreatic eruptions occurred from the new pyroclastic cones. Due to the start of the rainy season, small lagoons were formed in the northern sector of the former crater lake basin. The activity of the sulfur cones increased, and temperatures above 100 °C were measured.

(14) Between June and August 2018 (i.e. rainy season), the crater lake basin was re-filled almost entirely, manifesting evaporative degassing from the lake surface. The lake level rose several meters, to almost completely cover the relicts of the pyroclastic cones.

Fig. 22 **a** Vigorous degassing from the new vent at the 1953 dome 6 h prior to the large eruption of 22 April 2017. **b** Juvenile material impacted preferentially east of the crater, similar as during the 1953–1955 eruptions. These bombs show "bread crust" and fusiform textures. **c** Meter-sized blocks reached the eastern "terrazas", leaving impact craters of up to 8 m. **d** Inside the crater, 15 m large bombs show bread crust textures, and red coating. **e** The forest southeast of the crater was hit by pyroclastic flows and surges

3.4.3 Effects of the Eruption

The PNVP is one of the most visited parks in Costa Rica, and Poás is the most visited volcano in Central America (Mora Amador et al. Chapter "History, legends, customs and traditions of Poás volcano, Costa Rica"). Due to the closing of the park since the 12 April 2017 eruption until August 2018 (the moment of writing) the economic loss is estimated at, at least, 20 million dollars. This estimate considers (1) the drop in economic activities of the surrounding people that generally thrive from national and international tourism, (2) the damage to the infrastructure of PNVP (Fig. 24), and (3) the direct loss due to the absence of the entrance fee for this prolonged period. Ash fall and acid gases mainly affected the areas of the PNVP. Moreover, the infrastructures of the PNVP were recently renovated and were planned to be inaugurated on 28 April 2017.

At least three lahars were registered in the surrounding rivers Ríos Agrio and Desagüe (Western flank of Poás), with the most important one following the 12 April 2017 eruption. Touristic activity in this area was suspended for several weeks. The impact of volcanic bombs and blocks, ash fall, the pyroclastic flow of 22 April 2017 and acid gases seriously damaged the mirador, and the tourist service areas (Fig. 24a–e), as well as the trails to Laguna Botos (southeast sector) (Fig. 24f).

3.4.4 Eruptive Model

The magmatic eruptive stage was preceded by almost 11 years (March 2006–October 2016) of phreatic eruptions of variable sizes and explosivities (up to eruption columns of approx. 1 km). During this pre-eruptive stage sulfur volcanism manifested in its various forms: molten sulfur pools and spherules, sulfur flows and pyroclasts (Mora Amador et al. this issue).

A relatively sudden increase (early April 2017) in the temperature of the magmatic-hydrothermal system, due to a new magma emplacement, fractured the crater floor and destroyed the 1953–1955 Dome. Afterwards, during the various eruptive pulses small ash cones were formed and posteriorly destroyed, and eventually built up to

Fig. 23 **a** In August 2017 small eruptions were frequent, with ash eruptions (brown plume) from the pyroclastic cone (left), and sulfur eruptions (yellow plume) originating from the previous sub-lacustrine sulfur pools (right). **b** In January 2018 Laguna Caliente reformed, without any reported eruptive activity. On the left are the remnants of the rim of the pyroclastic cone, partially submerged by the newly formed Laguna Caliente. **c** Drone footage above Laguna Caliente in early March 2018, showing the pyroclastic cone (left) formed during the 2017 eruptions, and active sulfur pools (right)

larger pyroclastic cones. The eruptive activity logically led to the dry out of Laguna Caliente by July 2017, mainly caused by the temperatures above boiling point for the entire lake basin. Since October 2017, exhalation temperatures and seismicity have declined. In January 2018, Laguna Caliente reappeared, disappeared four weeks later, and returned again between May and August 2018 (rainy season) with a maximum temperature of 62 °C.

Fig. 24 **a** The recently restructured sector of the tourist mirador was damaged most by the April 2017 eruptions. The ramps and stairs were destroyed by the impact of bombs. **b**, **c** and **d** Examples of the damage to metallic structures in the mirador area. **e** Vegetation ("poor man's umbrellas") at the tourist mirador were "burned" by pyroclastic surges. **f** The concrete trail to Laguna Botos was punched by blocks and bombs. The 22 April 2017 eruption launched juvenile material up to at least 1.5 km distance. Rocks possibly reached the forest areas beyond 1.5 km

4 Volcanic Hazards of Poás

Prosser and Carr (1987), Paniagua and Soto (1988), Soto and Paniagua (1992), Alvarado et al. (2000, 2007) made a review of the volcanic hazards related to Poás volcano. A new map of Poás, based on the historical record and field observations (from 1995 to 2010) of the 1834, 1910, 1953–1955 and 2017 eruptions, is proposed in this study.

Major hazards related to the historical eruptions of Poás volcano are pyroclastic surges, fall out of ballistics and tephra, lahars, and degassing of acidic gases, generating acid rain. In the map (Fig. 25) the largest polygon corresponds to the sites where ash fall (with fragments <2 mm) was reported during the first three studied eruptions.

A yellow circle with a radius of 6.6 km around the active crater indicates the area where the fall out of lapilli and volcanic material with a grain-size between 2 mm and 6.4 cm is reported. The pink circle, with a radius of 2.5 km around the active crater corresponds to the sites where fall out of material with fragments larger than 6.4 cm occurred. Moreover, a circle with a radius of 1 km around the emission center, indicates the area affected by pyroclastic surges.

This new map is of highly practical use as during weekends or holidays, up to 100 people can contemporaneously visit the "mirador" at the crater rim. Details of the areas that could be affected within the PNVP are presented in Fig. 26.

4.1 Pyroclastic Surges

Paniagua and Soto (1988) mentioned that pyroclastic surges produced by explosive blasts during phreatomagmatic cycles could occur, but that these will possibly only affect the inner crater area. Nevertheless, during field-work, deposits related to pyroclastic surges were observed not only in the perimeter of the active crater (in the south, i.e. tourist mirador sector of the active crater and, in the southeast, i.e. mirador of Laguna Botos), but also along the trail towards Laguna Botos, and the eastern sector (i.e. "terrazas" sector). These deposits were emplaced during the first three studied eruptions (1834, 1910 and 1953–1955), with a maximum runout of 1 km from the emission centre (Figs. 25 and 26).

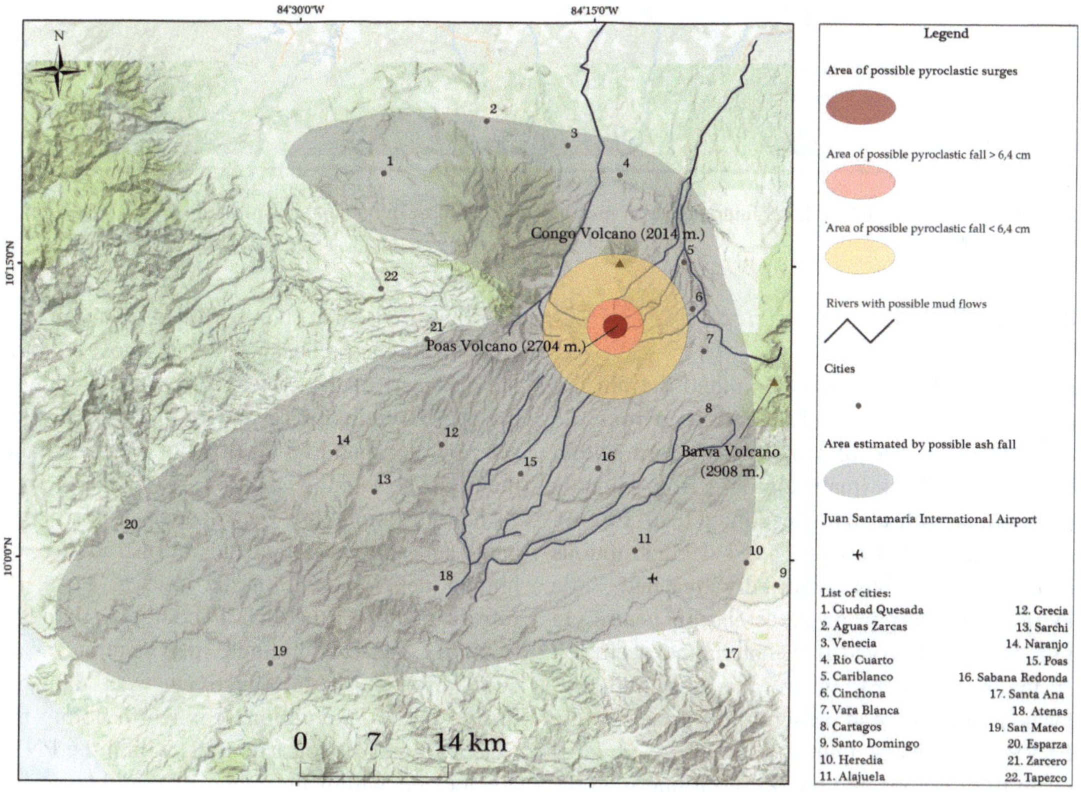

Fig. 25 Volcanic hazard map based on the three major historical eruptions of Poás volcano

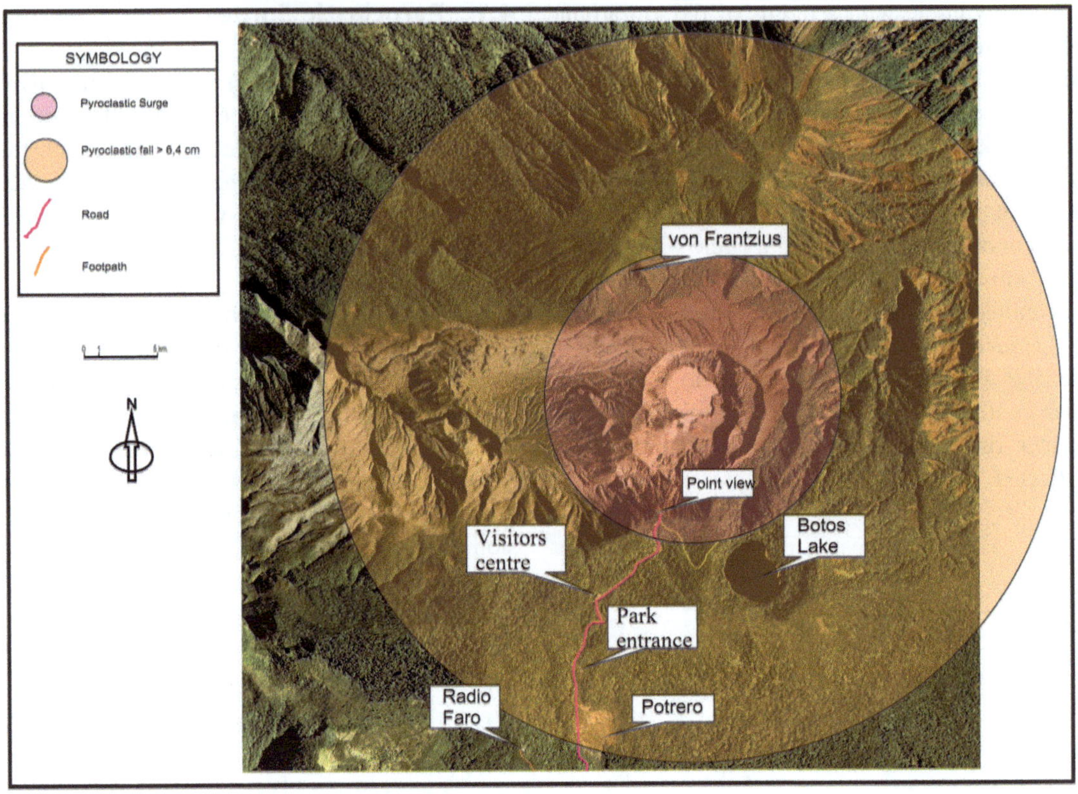

Fig. 26 Detail of the effect that a future eruption may have on the most important sites of the PNVP (LIDAR image, October 2009)

4.2 Ballistics

Paniagua and Soto (1988) delimited the fall out of the prehistorical and historical pyroclastic deposits to a zone of maximum hazard within 8 km around Poás (mainly blocks from phreatic explosions, with a diameter larger than 30 cm). During the 1834, 1910 and 1953–1955 eruptions ballistic projections reportedly reached little more than 2 km from the emission centre. In Fig. 11 it is observed that material >6.4 cm fell at the site Potrero, where the old hold was located, and where at present the entrance of the PNVP is located. Other sites where a large quantity of bombs and blocks are found is in the eastern and northern sector of the active crater (Fig. 17).

Figure 17 shows the perimetrical area of the active crater that can be affected by bombs with dimensions of at least half a meter. More than two hundred such bombs were distinguished with the highest concentration of those towards the east part of the crater. A person standing at the tourist "mirador" at the southern rim of the active crater could be hurt by hot bombs smaller than half a meter.

Falling bombs are very dangerous for people and can seriously damage infrastructure and human development. In the case of Poás, the infrastructure of the PNVP could be damaged (visitors centre, buildings, park entrance and access roads), as it effectively was damaged during the 2017 eruption. It is worth noting that practically all the touristic trails within the PNVP are located within the range of possible projections of bombs and blocks.

Lapilli fall out can also damage windows and other vulnerable structures. Its accumulation on roofs can cause their collapse. Lapilli fall out could affect villages towards the south of the volcano, such as Poasito, Altura and some dairy farms located in this sector, as well as some restaurants and tourist centers. The communities of Isla Bonita, Cinchona and Vara Blanca could be affected by this phenomenon, if a change in wind direction occurs at the moment of the eruption. Towards the north and west, within this range, there are no settlements (Fig. 25).

With respect to the direct impact on human lives, the probability of an impact is low, especially compared to the impact of pyroclastic density currents. Evidently, the direct impact of a bomb will cause injury or even death. Nevertheless, the distribution density of the bombs at a certain distance from the volcano is low enough; hence, the probability of direct impact will be low. Wearing a helmet offers the sufficient protection for lapilli fall out. Bomb impacts in the vicinity of the crater are a common phenomenon at volcanoes with moderate activity. This process causes deadly victims among tourists worldwide almost every year, due to the fact that they approach the volcano too much seeking to be near an eruption. This situation is particularly worrisome in the case an eruption occurs during the visiting hours of PNVP.

4.3 Dispersion and Fall Out of Pyroclasts

Ortiz and Araña (1996) mentioned that, of all the volcanic hazards in the world, dispersion and fall out of pyroclastics are those that affect the largest area. The fine material is dragged into the convective column up to high elevations, where it is dispersed due to the combined effect of diffusion and wind transportation. Ash particles are tremendously abrasive, for which they can damage all types of machinery, from airplane engines to printers or hard discs, etc. Motor vehicles suffer significant damage when driving on ash-covered roads.

The fine material rapidly decreases the filtering capacity if soils, tap pipes and blinds water flow paths, hence considerably increasing the probability of flooding, and possibly triggering mud flows. As the ash is extremely fine, it easily penetrates into clean rooms, such as surgery rooms, pharmaceutical, precision, mechanical, optical, and industrial alimentation laboratories.

4.4 Volcanic Gases and Acid Rain

Volcanic gases generally cause problems near eruptive vents or fumarolic fields, as they rapidly dilute into the atmosphere till concentrations below the toxicity threshold (Ortiz and Araña 1996). Nevertheless, in some circumstances they result toxic or even lethal. Some of the gases, like CO_2, are denser than the air. Hence, they tend to accumulate in depressions, such as valleys and canyons. Moreover, CO_2 is one of the major species in volcanic gases, as such, it can reach high concentrations and move downwards like a dense flow, asphyxiating every living being along its track. In the case of Poás, this problem can manifest when the wind direction is north-south, creating curtains of gases, especially SO_2, besides the odorless CO_2, at the tourist mirador and at the trails of the PNVP. The smell of gases can be strongly perceived within a radius of 2.5 km from the emission centre, and even weakly until 6.5 km.

Another hazard factor is constituted by acid rain, produced when water drops condense on volcanic aerosols that act as nucleus. Paniagua and Soto (1988) mentioned that the dominant wind from the WSW and the Cordillera Central, push these gases forward, for which these flanks of Poás are characterized by "burned" vegetation and acidic water in streams (e.g. Río Desagüe). This situation causes the dying of vegetation or inhibits its growth at those sites continuously exposed to acid rain. Roofs with metal covers suffer important corrosion. The loss of electric or electronic components, engines, etc, caused by volcanic gases and derivatives can be considerable.

At Poás, the grade of harm by acid rain largely depends on the wind direction. As such, the preferential sector of harm is NE–NW. Communities such as Bajos del Toro, San Miguel de Grecia, San Juan del Norte, Trojas and San Luís are directly affected by this phenomenon.

The last severe crisis caused by acid rain occurred in 1994, when Laguna Caliente disappeared for the last time. Mora (1997) carried out a survey in the areas affected by the gases, with the results summarized.

4.5 Lahars

The presence of a crater lake in the summit of a volcano, is a primary factor for the production of lahars, providing the necessary amount of water, although they can also be triggered by intense rain fall, frequent in the tropics (Carrivick et al. 2009; Procter et al. 2010; Rouwet et al. 2014b; Manville 2015 and references therein). Lahars are reported to have occurred in the riverbeds originating at Poás' summit area, such as (1) La Paz, Ángel, that on their turn are drained by Sarapiquí river, and (2) Río Desagüe, Anonos, and Agrío that are drained by Río Toro, and (3) the rivers located on the south flank of Poás: Sarchí, Achiote, Prendas and Poásito rivers. Figure 25 indicates the rivers that have presented similar events, or that could present them in the future.

5 Summary

Due to the heavy rain regime at the higher elevations of Poás volcano (approximately 4,000 mm/year) a lot of the fall out material, ash and lapilli has been eroded. This is why the in situ geological record is not sufficient to create adequate hazard maps based on historical activity of Poás. Hence, the historical record (news papers, journals, letters, technical reports, etc) becomes a valuable tool to line out and characterize recent eruptive activity.

Major volcanic hazards related to the historical eruptions of 1834, 1910, 1953–1955 and most recently in 2017 are: pyroclastic surges, ballistics projectiles, tephra fall out, lahars, earthquakes, acid gases and rain.

Very few information exists on the 1834 eruption. Nevertheless, a phase of clear precursory signals occurred: phreatic eruptions and an increase in fumarolic activity. The main event of this eruption was phreatomagmatic that launched bombs within a diameter close to the active crater. Ash fall was reported until Esparza (>50 km W of Poás) while pyroclastic surges occurred near the crater rim. The eruption is catalogued as Vulcanian.

The 1910 eruption was detailed by Rudín et al. (1910). This eruption has premonitory signals through phreatic eruptions and an increase in the crater lake temperature. Besides the 25 January 1910 main eruption, at least two eruptions occurred afterwards. Analyzing the bibliographic information and volcanic deposits, we conclude that this eruption was phreatomagmatic. In the eruptive products some juvenile fragments were detected, although hydrothermalized lithics were most abundant, immersed in a matrix of fine and coarse ash. There is evidence of pyroclastic surges near the active crater. It is possible that the differences in thickness of the deposits of this eruption are due to the explanation by Casertano et al. (1983): this eruption occurred with an oblique eruptive centre. These evidences, together with the estimated column height (3–4 km), suggest the 1910 eruption was a Vulcanian with a VEI 2.

The eruptive period of 1953–1955 had the longest duration of the three studied historical eruptions. This eruption was anticipated by phreatic eruptions, easily perceived earthquakes and the evaporation of Laguna Caliente crater lake due to increased activity. Since May 1953, several Strombolian eruptions took place. The crater lake dried out totally and a dome was emplaced in the centre of the crater. During this phase, Vulcanian, Strombolian and dome extrusion eruptions occurred. The VEI for this eruption is estimated at 3.

The largest area affected by ash fall resulted from the 1834 eruption, with 1469 km^2; in second place, the 1953–1955 eruption that affected 1,212 km^2 and, at last, the 1910 eruption that affected only an area of 432.6 km^2.

Currently, the cantons of Alajuela, Grecia and San Pedro de Poás are potentially the ones to be most affected by future volcanic activity similar to the one described in this study. Medical hospitals and clinics, fire department buildings, educational centers in these cantons deserve major attention and should be prepared to cope with cases of asthma, dermatitis, ocular problems and burnings following volcanic activity.

At the moment of writing this chapter, the fourth historical eruption of Poás started, after a phreatic phase that started in March 2006. Major and ongoing magmatic activity was preceded by dynamical changes inside the active crater with similar dynamics as during the three previous historical eruptions. As of August 2018, the 2017 eruption is not the major "historical" eruption of the four described here.

The here produced map for the medium to short term activity of Poás volcanic, based on the 1834, 1910, 1953–1955 and 2017 eruptions, is a tool that will orientate authorities in their decision making process and control human development, following restricted and regulated land use, create adequate facilities for human activity, promoting development while respecting the peculiarity of the natural site.

References

Alvarado GE, Pérez W, Sigarán C (2000) Vigilancia y peligro volcánico. In: Denyer P, Kussamul S (eds) Geología de Costa Rica. Editorial Tecnológica de Costa Rica, Cartago, pp 251–272 (In Spanish)

Alvarado GE, Soto G, Pullinger C, Escobar R, Bonis S, Escobar D, Navarro M (2007) Volcanic activity, hazards and monitoring. In: Bundschuh J, Alvarado GE (eds) Central America: geology, resources and hazards. Taylor and Francis, Londres, pp 1155–1186

Beck U (2000) Retorno a la teoría de la sociedad del riesgo. Bol A.G.E. 30:9–20

Bennett FD, Raccichini SM (1978) Subaqueous Sulphur lake in Volcan Poás. Nature 271:342–344

Blong RJ (1984) Volcanic hazards: a sourcebook on the effects of eruptions, 1st ed. Ed Academic Press, Sydney, 424 pp

Brenes R (1932) Macizo del Poás. In: Obregón M (ed) Geografía General de Costa Rica. Imp. Lines A. Reyes, San José, pp 127–128. (In Spanish)

Bretón M, Ramírez J, Navarro C (2002) Summary of the historical eruptive activity of Volcán de Colima, Mexico 1519–2000. J Volcanol Geotherm Res 117:21–46

Brown G, Rymer H, Dowden J, Kapadia P, Stevenson D, Barquero J, Morales LD (1989) Energy Budget analysis for Poás Crater lake: implications for predicting volcanic activity. Nature 339:370–373

Bullard F (1956) La actividad en Costa Rica y Nicaragua en 1954. In Vargas C, (1979) Volcán Poás Ed UNED, San José, pp 109–118

Carrivick J, Manville V, Cronin SJ (2009) A fluid dynamics approach to modelling the 18th March 2007 lahar et Mt. Ruapehu, New Zealand. Bull Volcanol 71:153–179

Casertano L, Borgia A, Cigolini C (1983) El Volcán Poás, Costa Rica: Cronologia y características de la actividad. Geofis Inter 3:215–236

Cashman KV, Cronin SH (2008) Welcoming a monster to the world: Myths, oral tradition, and modern societal response to volcanic disasters. J Volcanol Geotherm Res 176:407–418

Caudron C, Syahbana DK, Lecocq T, Van Hinsberg V, McCausland W, Triantafyllou A, Camelbeeck T, Bernard A, Surono (2015) Kawah Ijen volcanic activity: a review. Bull Volcanol 77:16. https://doi.org/10.1007/s00445-014-0885-8

Christenson BW, Reyes AG, Young R, Moebis A, Sherburn S, Cole-Baker J, Britten K (2010) Cyclic processes and factors leading to phreatic eruption events: insights from the 25 September 2007 eruption through Ruapehu Crater Lake, New Zealand. J Volcanol Geotherm Res 191:15–32

Del Gaudio C, Aquino I, Ricciardi GP, Ricco C, Scandone R (2010) Unrest episodes at Campi Flegrei: A reconstruction of vertical ground movements during 1905–2009. J Volcanol Geotherm Res 195:48–56

de Moor JM, Aiuppa A, Pacheco J, Avard G, Kern C, Liuzzo M, Martínez M, Giudice G, Fischer TP (2016) Short-period volcanic gas precursors to phreatic eruptions: insights from Poás Volcano, Costa Rica. Earth Planet Sci Lett 442:218–227. https://doi.org/10.1016/j.epsl.2016.02.056

Dóndoli C (1965) Volcanismo reciente de Costa Rica. In: Vargas C (1979) Volcán Poás Ed UNED, San José, pp 127–132

Fernández R (1961) La actividad del volcán Poás en el año 1953 y su transformación de seudogéiser en volcán humeante. In: Vargas C (1979) Volcán Poás Ed UNED, San José 119–126

Francis PW, Thorpe RS, Brown GC (1980) Pyroclastic sulphur eruption at Poás Volcano, Costa Rica. Nature 283:754–756

Fernández-Arce M, Mora-Amador R Seismicity of Poás volcano, Costa Rica (Chapter 5). In: Tassi F, Mora-Amador R, Vaselli O (eds) Poás volcano (Costa Rica): the pulsing heart of Central America Volcanic Zone. Springer, Heidelberg (Germany)

Fischer TP, Ramírez C, Mora-Amador RA, Hilton DR, Barnes JD, Sharp ZD, Le brun M, de Moor JM, Barry PH, Füri E, Shaw AM (2015) Temporal variations in fumarole gas chemistry at Poás volcano, Costa Rica. J Volcanol Geotherm Res 294:56–70. https://doi.org/10.1016/j.jvolgeores.2015.02.002

Hantke G (1951) Übersicht über die Vulkanische Tätigkeit, 1941–1947. Bull Volcanol 14:161–210

Hantke G (1953) Übersicht über die Vulkanische Tätigkeit, 1948–1950. Bull Volcanol 14:151–185
Hilton DR, Ramirez CJ, Mora-Amador R, Fischer TP, Füri E, Barry PH, Shaw AM (2010) Monitoring of temporal and spatial variations in fumarole helium and carbon dioxide characteristics at Poás and Turrialba volcanoes, Costa Rica (2001–2009). Geochem J 44:431–440
Kempter KA, Benner SG, Williams SN (1996) Rincón de la Vieja, Guanacaste province, Costa Rica: geology of the southwestern flank and hazards implications. J Volcanol Geotherm Res 71:109–127
Leiva E (1904) El volcán Poás. In: Vargas C (1979) Volcán Poás (ed) UNED, San José, pp 63–66
Leiva E (1906) Una excursión al volcán Poás. In: Vargas C, (1979) Volcán Poás (ed) UNED, San José, pp 69–70
Manville V (2015) Volcano-hydrologic hazards from volcanic lakes. In: Rouwet D, Christenson B, Tassi F, Vandemeulebrouck J (eds) Volcanic Lakes, Springer, Heidelberg, pp 21–71. https://doi.org/10.1007/978-3-642-36833-2_2
Martínez M, Fernández E, Valdés J, Barboza V, Van der Laat R, Duarte E, Malavassi E, Sandoval L, Barquero J, Marino T (2000) Chemical evolution and activity of the active crater lake of Poás volcano, Costa Rica, 1993–1997. J Volcanol Geotherm Res 97:127–141
Martínez M, van Bergen MJ, Takano B, Fernández E, Barquero J (2008) Long term behaviour of polythionate polymers in the hyperacid crater lake of Poás volcano: insights into the subaqueous input of magmatic gases (Chapter 7). In: Tassi F, Mora-Amador R, Vaselli O (eds) Poás volcano (Costa Rica): The pulsing heart of Central America Volcanic Zone. Springer, Heidelberg (Germany)
Marzocchi W, Bebbington M (2012) Probabilistic eruption forecasting at short and long time scales. Bull Volcanol 74:1777. https://doi.org/10.1007/s00445-012-0633-x
Marzocchi W, Sandri L, Gasparini P, Newhall C, Boschi E (2004) Quantifying probabilities of volcanic events: the example of volcanic hazard at Mt Vesuvius. J Geophys Res 109:B11. https://doi.org/10.1029/2004jb003155
Mora M (1997) Informe de la actividad de los volcanes Poás e Irazú, 1994–1996. RSN UCR 52 pp
Mora-Amador R (2009) Volcán Poás, 18 de setiembre del 2009, combustión de azufre y erupción freática. RSN UCR (informe interno)
Mora-Amador R, Rouwet D, González G, Vargas P, Ramírez C Volcanic hazard assessment of Poás (Costa Rica) based on the major historical eruptions of 1834, 1910 and 1953–1955 (Chapter 11). In: Tassi F, Mora-Amador R, Vaselli O (eds) Poás volcano (Costa Rica): The pulsing heart of Central America Volcanic Zone. Springer, Heidelberg (Germany)
Mora-Amador R, Ramírez C (2008) Sulfur flows at Poás volcano, Costa Rica. IAVCEI General Assembly 2008, Reykjavík Iceland:85
Mora-Amador R, Rouwet D, Vargas P (2016) The extraordinary sulfur eruptions of Poás volcano from 1828 to 2016 (Chapter 3). In: Tassi F, Mora-Amador R, Vaselli O (eds) Poás volcano (Costa Rica): the pulsing heart of Central America Volcanic Zone. Springer, Heidelberg (Germany)
Newhall CG, Self S (1982) The volcanic explosivity index (VEI): an estimate of explosive magnitude for historical volcanism. J Geophys Res 87(C2):1231–1238
Oersted AS (1863) L'Amerique Centrale: Recherches sur sa flore et sa Géographie Physique. Imprenta Blanco Luna, Copenhague (Denmark)
Oppenheimer C (1992) Sulfur eruptions at Volcán Poás, Costa Rica. J Volcanol Geotherm Res 49:1–21
Ortiz R, Araña V (1996) Daños que pueden producir las erupciones. In: Ortiz R (ed) Riesgo, Volcánico edn. Servicio de publicaciones, Lanzarote, España, pp 37–66
Paniagua S, Soto G (1988) Peligros volcánicos en el valle central de Costa Rica. Ciencia y Tecnología 1:145–156
Peraldo G, Rodríguez A (2001) Documentos alusivos a las amenazas naturales en Costa Rica y sus efectos en la sociedad del siglo XIX. Inf Sem I.G.N. 37:89–101
Phillipson G, Sobradelo R, Gottsmann J (2013) Global volcani unrest in the 21st century: an analysis of the fist decade. J Volcanol Geotherm Res 264:183–196
Procter JN, Cronin SJ, Fuller IC, Sheridan M, Neall VE, Keys H (2010) Lahar hazard assessment using Titan2D for an alluvial fan with rapidly changing geomorphology: Whangaehu River, Mt. Ruapehu. Geomorphology 116:162–174. https://doi.org/10.1016/j.geomorph.2009.10.016
Prosser JT, Carr MJ (1987) Poás Volcano, Costa Rica geology of the summit region and spatial and temporal variations among the most recent lavas. J Volcanol Geotherm Res 33:131–146
Quesada MA (2007) Diccionario de Costarriqueñismos, 4th edn. Tecnológica de Costa Rica, San José, 413 p (In Spanish)
Rouwet D, Morrissey MM (2015) Mechanisms of crater lake breaching eruptions. In: Rouwet D, Christenson BW, Tassi F, Vandemeulebrouck J (eds) Volcanic lakes. Springer, Heidelberg (Germany). https://doi.org/10.1007/978-3-642-36833-2_3
Rouwet D, Mora-Amador R, Sandri L, Ramírez CJ, González G, Pecoraino G, Capaccioni B (2014a) 39 years of geochemical monitoring of Laguna Caliente crater lake, Poás: Patterns from the past as keys for the future (Chapter 9). In: Tassi F, Mora-Amador R, Vaselli O, (eds) Poás volcano (Costa Rica): The pulsing heart of Central America Volcanic Zone. Springer, Heidelberg (Germany)
Rouwet D, Sandri L, Marzocchi W, Gottsmann J, Selva J, Tonini R, Papale P (2014b) Recognizing and tracking volcanic hazards related to non-magmatic unrest: a review. J Appl Volcanol 3:17. https://doi.org/10.1186/s13617-014-0017-3

Rouwet D, Mora-Amador R, Ramírez CJ, González G, Inguaggiato S (2016) Dynamic fluid recycling at Laguna Caliente (Poás, Costa Rica) before and during the 2006—ongoing phreatic eruption cycle (2005–10). In: Ohba, T, Capaccioni B, Caudron C (eds) Geochemistry and Geophysics of Active Volcanic Lakes. Geol Soc London, Special Publication, 437. https://doi.org/10.1144/sp437.11
Rowe GL, Brantley SL, Fernández M, Fernández JF, Borgia A, Barquero J (1992a) Fluid-volcano interaction in an active stratovolcano: the Crater Lake system of Poás Volcano, Costa Rica. J Volcanol Geotherm Res 64:233–267
Rowe GL, Ohsawa S, Takano B, Brantley SL, Fernández JF, Barquero J (1992b) Using crater lake chemistry to predict volcanic activity at Poás Volcano, Costa Rica. Bull Volcanol 54:494–503
Rudín M (1905) Algunos detalles sobre el volcán Poás. In: Vargas C (1979) Volcán Poás Ed UNED, San José, pp 67–68
Rudín J, Alfaro A, Michaud G, Rudín A (1910) Gran erupción de ceniza del volcán Poás. In: Vargas C (1979) Volcán Poás Ed UNED, San José, pp 75–82
Soto G, Alvarado GE (1989) Procesos volcánicos asociados con el agua subterránea: El caso de los volcanes Arenal y Poás, Costa Rica. Congr Nac Recur Hídr, San José, Costa Rica, pp 249–261
Soto G, Paniagua S (1992) La Cordillera Volcánica Central (Costa Rica): sus peligros potenciales y prevenciones. Rev Geogr Amér Central 25–26:291–304 (In Spanish with English abstract)
Takano B, Saitoh H, Takano E (1994) Geochemical implications of subaqueous molten sulfur at Yugama crater lake, Kusatsu-Shirane volcano, Japan. Geochem J 28:199–216
Todesco M, Rouwet D, Nespoli m, Bonafede M (2015) How steep is my seep? Seepage in Volcanic lakes. Hints from numerical simulations, In: Rouwet D, Christenson BW, Tassi F, Vandemeulebrouck J (eds) Volcanic lakes. Springer, Heidelberg. https://doi.org/10.1007/978-3-642-36833-2_3
Vargas MJ (1967) Aspectos de la actividad volcánica de Costa Rica en los últimos tiempos. Efimérides Costarricenses 3:18–25 (In Spanish)
Vaselli O, Tassi F, Fischer TP, Tardani D, Fernandez Soto E, Martinez M, De Moor MJ, Bini G The last eighteen years (1998–2014) of fumarolic degassing at the Poás volcano (Costa Rica) and renewal activity (Chapter 10). In: Tassi F, Mora-Amador R, Vaselli O (eds) Poás volcano (Costa Rica). The pulsing heart of Central America Volcanic Zone. Springer, Heidelberg (Germany)
Von Frantzius A (1861) Aporte al conocimiento de los volcanes de Costa Rica, Escalamiento al volcán Poás, marzo 1860. In Vargas C, (1979) Volcán Poás Ed UNED, San José 11–34. (In Spanish)

References Newspapers and Journals

La República, 26 enero 1910: Erupción del Volcán Poás
La Prensa Libre, 27 enero 1910: Erupción del Poás
La Prensa Libre, 28 enero 1910: Más lluvia de ceniza
La Prensa Libre, 29 enero 1910: De Grecia, Lluvia de Ceniza
La Prensa Libre, 1 febrero 1910: La lluvia de ceniza
La República, 2 febrero 1910: La ceniza del Poás resulta ser venenosa
La República, 9 febrero 1910: Peces muertos en el río Sarapiquí
La Nación, 20 mayo 1953, pág. 1: Lluvia de ceniza del Volcán Poás en Vara Blanca y alrededores
La Nación, 22 mayo 1953, pág. 5: Nueva y grande erupción del Poás
La Nación, 24 mayo 1953, pág. 43: Evacuación de viviendas en alrededores del Volcán Poás
La Nación, 28 mayo 1953, pág. 11: Retumbos y humo del Volcán Poás ayer
La Nación, 6 junio 1953, pág. 20: Lluvia de ceniza en Río Cuarto
La Nación, 10 junio 1953, pág. 1: El Poás se trago la laguna inmensa de su cráter
La Nación, 12 junio 1953, pág. 1: La furia del gigante
La Nación, 10 julio 1953, pág. 1: Vuelve la actividad volcánica
La Prensa Libre, 18 julio 1953, pág. 1: Centenares de cabeza de ganado en peligro
La República, 15 enero 1954, pág. 3: Grandes pérdidas cafetaleras por erupciones del Poás
Archivo Nacional de Costa Rica, Serie Municipal, No. 778
Monestel Y (1953) Carta a César Dóndoli, Inédito

History, Legends, Customs and Traditions of Poás Volcano, Costa Rica

Raúl Alberto Mora Amador, Mario Fernández and Dmitri Rouwet

Abstract

This chapter compiles known information on the history, legends and traditions regarding Laguna Caliente of Poás volcano. We describe the evolution since the first ascends in the 16th century, until the massive incurrence of touristic activities. The exciting "Legend of Rualdo" on a bird that sacrificed its sweet song to save its pretty maiden, who was going to be sacrificed to stop the fury of the volcano, is described. The "Legend of Rualdo" narrates supernatural facts and explains the formation of Laguna Caliente thanks to a pact between a bird and the powerful volcano. Moreover, the origin of the yearly tradition of climbing Poás volcano on March 19th, the day of the patron Saint of San José, is documented. Legends and traditions can be useful tools to decipher past activity of volcanoes, in this case Poás.

Keywords

Poás volcano · Legends · Traditions · History · Laguna Caliente

1 Introduction

Local traditions can provide a valuable community education tool as well as an important means of aiding the psychosocial recovery of individuals and communities after volcanic disasters (Cashman and Cronin 2008). According to Gregg et al. (2008), cultural aspects of a society can be used in assessments and mitigation plans. Breton et al. (2002) indicated that much of the knowledge of the earlier volcanic events of Colima volcano is from non-scientific writings. To explain the deadly event of Lake Nyos in 1986, the predicates of the indigenous view of the universe and several examples given of lake myths, including strange phenomena seen in Lake Nyos were examined (Shanklin 2007). Regarding the veracity of the legends and stories on volcanic activity, Swanson (2008) assured that recent geologic studies confirm the essence of the oral traditions.

Every year approximately 400,000 tourists visit Poás Volcano National Park (Arias 2017). This high number of annual visits creates a continuous feedback of traditions and customs that oversteps nationalities, ages and ethnic groups. Laguna Caliente of Poás volcano is not deprived

R. A. Mora Amador (✉)
Laboratorio de Ecología Urbana, Universidad Estatal a Distancia, San José, Costa Rica
e-mail: raulvolcanes@yahoo.com.mx

R. A. Mora Amador
Escuela Centroamericana de Geología, Universidad de Costa Rica, San José, Costa Rica

M. Fernández
PREVENTEC, Escuela de Geografía, Universidad de Costa Rica, San José, Costa Rica

D. Rouwet
Istituto Nazionale di Geofisica e Vulcanologia, Sezione di Bologna, Bologna, Italy

F. Tassi et al. (eds.), *Poás Volcano*, Active Volcanoes of the World,
https://doi.org/10.1007/978-3-319-02156-0_12

of such myths, legends and histories, information that is dispersed in diverse documents and hence in need to be compiled uniformly. The scope of this research is to shed light on those legends, myths and customs that elucidate on true facts, possibly inspired by past volcanic events, enabling to better understand eruptive processes.

The current study is based on a bibliographical review of the histories, legends, customs and stories on Poás volcano; hence, scientific articles, books, magazines and newspapers (e.g. from the national archives) were consulted. Moreover, people who shared key information on the topic were interviewed. Among the most significant results we emphasized that the "Legend of Rualdo" could be associated with an eruptive event of Poás.

2 History of Human Settlements Near Poás

2.1 Pre-Columbian and Colonization Period: 14th–19th Centuries

During the pre-Colombian period, the territory now occupied by the Poás Volcano National Park, was inhabited by an indigenous group called Botos, who paid taxes to the Huetares group of the Western Kingdom, led by Cacique Garabito. The Poás area was then a transit area between the Central Valley and the Northern Plains (Rodríguez Argüllo 2008).

In 1569, Francisco de Marmolejo and a group of men visited the region of the Botos. Later, Perafán de Rivera mandated 35 people of the town of the Botos to Francisco Magariño. A possible result of this fact was that many of the natives who lived in territory of Poás fled from the conquerors to the north (Rodríguez Argüllo 2008). Due to the opening of the Muías Way in 1601 the zone was affected by more frequent transit as a result of increasing commercial activities. In 1605 Alonso de Bonilla visited the community of the Botos (Rodríguez Argüllo 2008).

In the middle of the 17th century, the name Botos was replaced by Poás or Puás, which, apparently, was registered for the first time in 1663. The existence of a resting area for the mules called "Potreros de Púas" was well-known since 1662. This was a place of commercial importance where the prongs (i.e. "púas") or thorns abunded, which probably gave origin to the name of the volcano (Rodríguez Argüllo 2008).

The first settlers arrived in the region in 1806, originating from the villages of Barva and Heredia. The first church of the zone was built in 1837, in a place that was known as San Pedro de la Calabaza, the current San Pedro de Poás. In 1850, the small village of Poás appeared as District of the Lajuela (Alajuela), and by the 15th of October 1901 it was decreted as Canton Poás (Sedó 2011).

The first expedition to the crater of Poás was carried out in 1828. Miguel Alfaro descended into the crater and described the activity of Laguna Caliente. The details of this visit, and others made by some European naturalists during the second half of the 19th century, provided the first scientific observations on Poás (Mora Amador et al. Chapter "Volcanic Hazard Assessment of Poás (Costa Rica) Based on the 1834, 1910, 1953–1955 and 2017 Historical Eruptions").

2.2 Beginnings of the 20th Century

In the beginning of the 20th century, the visits to the volcano increased significantly, not only by inhabitants of the Central Valley, but also by foreigners. Moreover, many people interested in the mineral wealth of the volcano also approached the zone in search of sulfur (Rodríguez Argüllo 2008). There were several other reasons to visit the volcano. One of them is the great facility to travel to the zone since there was a train service from San José to Alajuela, where people could rent a horse to travel to San Pedro de Poás. In that town, another horse could be rented and a guide could be hired to accompany the visitors to the volcano summit. Between 1904 and 1920, the volcano was very active, and hence attractive for the visitors.

The economic progress of the country and increased ability for people to spend in entertainment also had an important impact in the stream of tourists to the area of San Pedro de Poás. The amount of visitors peaked during the weekends, holidays and summer. This is probably the reason why a trip to Poás became popular on the specific date of 19 March, holiday day for the "josefinos" (people from San José city), being the day of patron saint of the capital, San José Obrero.

In 1913, Magdaleno Ugalde, neighbor of San Pedro de Poás and Trino Araya of San José, joined their capital and built the "Hotel of Poás Volcano" in a place called "Potrero del Volcán" or "Potrero Grande". This hotel opened on Christmas Eve of 1913 and inaugurated on the 19th of March 1914. It had ten dormitories of several sizes and included dining room and kitchen. It was also surrounded by a large terrace. The hotel had all the services and comforts for national and foreign tourists. Some outstanding visitors were Otilio Ulale and Federico Tinoco, presidents of Costa Rica years later. The meal service included several liquors, tortillas, custard, hot coffee, tamales and other more delights.

The natural landscapes were the main attraction for visitors to the volcano. The exploitation of the volcano wealth for commercial purposes bloomed, like the case of Tila Ugalde, sister of the hotel's owner, who was going to the crater to collect up to eight acid water bottles, which she sold in the pharmacies as "teeth remover". This sulfured liquid was used to treat ached teeth; a few drops of "teeth remover" and a while later the tooth fell out.

The adventure of visiting the volcano was very attractive. The travelers hired the guide and went into the muddy trails surrounded by vegetation. During the heavy journey, the travelers listened to the stories of the guide on the wonders of Poás. When arriving at the "Cuesta de los arrepentidos" (the ridge of the regretted), the story of the rancher Marcos Ugalde, one of the first settlers in the area, was always told. It tells that a tiger was killing almost all of Marcos's cattle. Then, the rancher decided to kill the animal. He constructed a cage and hung a chicken inside. Night after night he was waiting until the tiger appeared in search of food, to easily kill it. Many other histories and anecdotes were told to the travelers by the local guides. After four or five hours of long walk, they arrived at the "Lechería", a small shed where the owners sold milk. After another hour, they reached the Potrero of the volcano and the hotel. At dawn the following day, the traveler initiated the long walk to arrive at the crater, which was a kilometer uphill. The cost of one night at the hotel was five colones.

The "Hotel of Poás Volcano", located towards the north of Potrero Grande, deteriorated after 1918, when Magdaleno Ugalde, the only owner, died. Later, the business passed to other investors and the quality of the service diminished. Around 1930, Don Adam Saborío bought the rights of the hotel and turned it into his house, although he continued providing accommodations to the travelers climbing the volcano.

3 The 20th and 21st Centuries

Details on the historical eruptions of Poás in the 20th century can be read in Mora Amador et al. (Chapter "Volcanic Hazard Assessment of Poás (Costa Rica) Based on the 1834, 1910, 1953–1955 and 2017 Historical Eruptions"). It is to mention that from late 20th century to early 21st century the visits to the Poás Volcano National Park increased significantly (Fig. 1). Since 1995

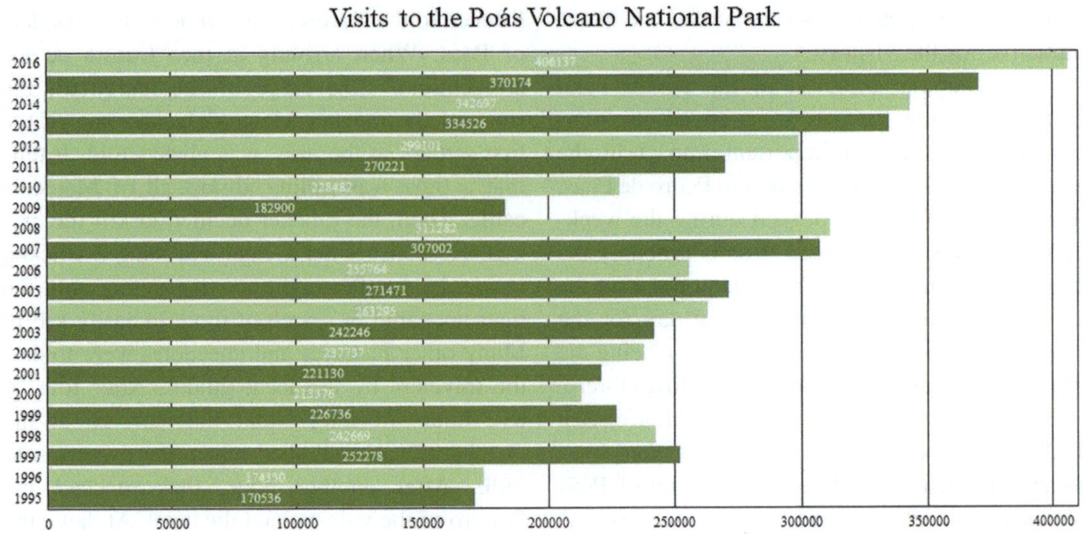

Fig. 1 The Poás Volcano National Park is the most visited active volcano of Central America and probably one of the most visited in Latin American. After the April 2017 eruptions the park has remained closed (as of February 2018)

the number of tourists/year increased constantly, until 2009 when the Cinchona earthquake caused road damages. Three years later, the increasing trend was re-established. According to data provided by SINAC (National System of Areas of Conservation), the proportion between national and foreign tourists is approximately 1:1. A tourist trip to Poás volcano has become a true tradition for the families of the Central Valley of Costa Rica. They often visit it to have a picnic afterwards in areas outside the park, in the bordering properties that have spectacular views, which include Barva volcano and the top of Irazú volcano on clear-skied days.

Witnessing the degassing Laguna Caliente is an experience that foreign tourists exported to the World. Now a feedback of histories and experiences exists, that probably will continue enriching the traditions around Poás volcano. A curiosity, Laguna Caliente was pictured on the 1940 20 colones, and on the 1990 10,000 colones bills. Moreover, the crater of Poás is imaged numerous times in both telephone cards and commercial adds (Fig. 2).

4 The Sacrifice "Legend of Rualdo"

Zeledón (2007) transcribed the legend entitled: The Sacrifice of Rualdo, which we here translated literally and present in the next paragraphs.

> …both main protagonists of this strange story are nothing less than a bird and Poás volcano.

"In the forests that extend underneath infinity of singer birds live, some with very peculiar names: picudos, caciquitas, viudas, cardenales, monjitas. Only one, the most beautiful for its colors, is

Fig. 2 **a** The 1940 20 colones bill shows an image taken from the present tourist viewpoint, which includes a typical phreatic eruption of Laguna Caliente. **b** The 1997 10,000 colones bill shows 4 active volcanoes of Costa Rica, from right to left: Rincón de la Vieja, Irazú, Arenal and Poás (down to the right). The pre-2017 central dome did not exist by that time. **c** 140 colones postal stamp with an image of Laguna Caliente and Laguna Botos. **d** Telephone card with a photograph of the active crater from the tourist viewpoint. **e** Packaging of a famous coffee brand with the images and name of Poás volcano

completely dumb; its name is Rualdo and it is the other protagonist of this history.

Many centuries ago, before the arrival of the Spaniards, Rualdo was a bird of vulgar plumage, but its song was the most beautiful of the entire jungle. In the limits of the jungle, near the flank of the volcano, there was an Indian town. A pretty and orphaned girl lived in that place and Rualdo was her sole company. The parents of the bird nested near the girl's hut; they went and they came, glad to take the food for their little bird, but a day they did not return. The little girl grabbed the young bird, she took care of it, and when she grew older she did not want to separate from it. The Indian told it her secrets and the bird brightened the girl's loneliness with its songs.

At sunset, Poás blushed still more and covered the sky with its flames. The Earth shook and down its slopes descended doughy torrents that devoured the forest. The entire town jumped to adore him. The wizards pronounced unintelligible prayers and they offered animals and fruits to him. Only the girl remained curled up and trembling deep inside her hut. The volcano roared every time with more ferocity; his spirit was not satisfied to the offerings; it was hungry of human meat. The oldest of the wizards came near the lava, and the monster trusted its desire to him.

The Indians arrived at the door of the hut; they were quiet and seriously watching inside. The girl curled up herself even more and her terrorized eyes shone in the darkness. They found her on surprise, and the priests raised her by the flanks of the volcano mumbling chanting. Finally, they left her alone: she had to advance and to be given to the flames for the welfare of her town. She did not have salvation, as death was also waiting behind her in the flint knives.

She moved with the vacillating steps of a sleepwalker. She suddenly stopped: Rualdo flew on the reddish smoky clouds. Evading the fiery flames, sit, sang to the volcano, Rualdo spoke to him in the

Fig. 3 Picture of a Rey Rualdo or Rualdo bird (*Chlorophonia callophrys*). In English: Golden-browed Chlorophonia, photographed in the neighborhood of the volcano. Through general agreement of the local government ("municipalidad") the Rualdo was declared "Ave Símbolo del Cantón de Poás", on the 26th of April 2016. Picture by Mario E. Campos Sandoval

mysterious language of Nature. It requested pity for its soul and offered the volcano the most beautiful it possessed: the wonder of its voice.

Poás softened, the sweetness of Rualdo's songs made sprinkle its tears that filled the crater. The activity extinguished and the lake was born. In front, the town sunk in religious silence as the bird returned on the shoulder of the Indian girl. The torrid emanations of the fire had dried his voice forever; but the heat had turned its plumes golden. Rualdo left the crater as an enamel bird, out of the mouth of the furnace of a jeweler.

The bird does not sing anymore, but it is green, azure and yellow (Fig. 3). The mountain still releases its hot spurts: they are delayed weepings of the volcano."

The "Legend of Rualdo" has transpired into the daily life of Costaricans. It was first published in diverse books of Costarican legend compilers, such as Elías Zeledón (Zeledón 2007).

During the 80s, the newspaper La Nación published a comic on the "Legend of Rualdo" every Wednesday for several weeks. In 2014, Franco Céspedes published a graphic novel entitled the Sacrifice of Rualdo.

5 Religious Customs

5.1 Walking from San Pedro de la Calabaza to Poás Volcano

Sedó (2011) described a traditional religious custom that relates the catholic to Poás volcano, following the details.

At the late 19th and early 20th century, the "josefinos" celebrated the holiday in commemoration of the patron Saint of San José city. The tradition was to go on a stroll to the volcano and nearby zones. People thought that March was the moment when the sky was clearest and the crater was most imposing. For this reason, a great amount of people peregrinated until the town San Pedro de la Calabaza (presently, San Pedro de Poás), walking from there by trails until the volcano crater (Fig. 4).

Fig. 4 Each March 19, catholics peregrinate to Poás volcano on the day of San José

The traditional walk is carried out in March and since 2002 it is considered a true tradition of San Pedro de Poás (Chaves 2010). More than 250 people meet in center of the town and undertake the long walk towards the volcano, doing the route by the old trails that the travelers used in the past. Such routes have popular names like La Avispa or La Legua. Whatever trail they choose, firm steps along steep trails within a landscape full of greenness and beauty are necessary. Well-organized neighbors install canopy stands along the track to welcome the travelers with music and provide them with water, fruit and other typical meals.

6 Summary

As evidenced in this study, the first visit to Poás occurred in 1828. Already at the beginning of 20th century the visits increased significantly: both national and foreign tourists were attracted by the natural beauty and the mineral wealth of the volcano. As a reaction to this tourist activity associated with Poás, the first hotel of the area, called "Hotel del Volcán Poás", was built in 1913.

The sacrifice "Legend of Rualdo" suggests that it was associated to a significant and important eruptive event. The legend mentions the fury of the volcano, fire and incandescence, which is suggestive to magmatic activity.

The tradition of climbing Poás on the 19th of March started at the end of 19th century, due to the celebration of the day of the San José patron saint. This is why many people peregrinate until San Pedro de Poás from where they climb until the crater.

It is a mistake to ignore the lessons of the ancestral towns. The historical memory of the different societies is a fundamental tool to possibly reconstruct volcanic activity. We must study this information carefully because behind the legends, myths and traditions it is possible to find useful tracks to understand the behavior of active volcanoes.

References

Arias G (2017) Sistema Nacional de Áreas de Conservación. Parque Nacional Volcán Poás (in Spanish)

Breton M, Ramírez J, Navarro C (2002) Summary of the historical eruptive activity of Volcán De Colima, Mexico 1519-2000. J Volcanol Geotherm Res 17:21–46

Cashman K, Cronin S (2008) Welcoming a monster to the world: myths, oral tradition, and modern societal response to volcanic disasters. J Volcanol Geotherm Res 176:407–418

Chaves L (2010) Caminata desde la Calabaza hasta el Poás. En www.cantonpoas.com. 23 de marzo 2010 (in Spanish)

Gregg C, Houghton B, Paton D, Swanson D, Lachman L, Bonk W (2008) Hawaiian cultural influences on support for lava flow hazard mitigation measures during the January 1960 eruption of Kīlauea volcano, Kapoho, Hawai'i. J Volcanol Geotherm Res 172:300–307

Raul Mora-Amador R, Rouwet D, González G, Vargas P, Ramírez C (Chapter 11) Volcanic hazard assessment of Poás (Costa Rica) based on the 1834, 1910, 1953–1955 and 2017 historical eruptions. In: Tassi F, Mora-Amador R, Vaselli O (eds) Poás volcano (Costa Rica): the pulsing heart of Central America Volcanic Zone. Springer, Heidelberg (Germany)

Rodríguez Argüllo P (2008) Historia del cantón de Poás. Comisión del Centenario del cantón de Poás y de la creación de la Municipalidad 1901-2001, 109 p (in Spanish)

Sedó P (2011) Festividades con encanto tico. Universidad de Costa Rica, pp 94–98 (in Spanish)

Shanklin E (2007) Exploding lakes in myth and reality: an African case study. Geol Soc London, Special Publications 273:165–176

Swanson D (2008) Hawaiian oral tradition describes 400 years of volcanic activity at Kīlauea. J Volcanol Geotherm Res 176:427–431

Zeledón E (2007) Leyendas Costarricenses. Universidad Nacional Costa Rica, EUNA, 286 p (in Spanish)

Poás Volcano Biodiversity

Frank González, Carolina Seas, Zaidett Barrientos, Sergio Gabriel Quesada-Acuña and Raúl Alberto Mora Amador

Abstract

Stablished in 1971, Poás Volcano was the first Costa Rica National Park. Rich in biodiversity and landscapes due to the physiognomic characteristics and its outstanding attraction it has as an active volcano, Poás is the main protected area of the country and the Central American region in terms of touristic visitation. As protected area it is also important its contribution to the preservation of the Cordillera Volcánica Central native forest and the role that forest coverage plays, in the generation of water for human consumption.

Keywords

Poás volcano · Costa rica national parks · Biodiversity

1 Introduction

Poás Volcano, one of the Costa Rican active volcanoes, is located in the western part of the Cordillera Volcánica Central. This volcano was the first Costa Rican National park, it was established in 1971. It has an extension of about 16,000 acres. Jointly with other 175 protected areas, it conforms the National System of Conservation Areas of Costa Rica (hereafter SINAC, Spanish acronym for Sistema Nacional de Área de Conservación). This protected area has become the main attraction of the country and the whole Central American region in terms of touristic visitation, also due to the fact that the park hosts an active volcano. In addition, this National Park is important for the conservation of biodiversity since it protects wildlife species in an altitudinal interval that ranges from 1,200 to 2,708 m. It includes five life zones and protects important areas of water recharge (SINAC 2008).

Heavy rainfall (ca. 3,700 mm per year) characterizes the local climate. Rainfall occurs throughout the year, although the rainiest period is between May and October, with a relatively dry period from December to April. Mean annual temperature is relatively stable (from 9 to 13 °C), although during the day temperature may fluctuate considerably (from 0.5 to 18 °C) (Macey 1975; SINAC 2008). The warmer period is from May to July, and the coldest one is from December to January, although higher temperatures occur on clear days in February and March.

F. González (✉) · C. Seas · Z. Barrientos · S. G. Quesada-Acuña · R. A. Mora Amador
Laboratorio de Ecología Urbana, Vicerrectoría de Investigación, Universidad Estatal a Distancia (UNED), 2050 San José, Costa Rica
e-mail: fgonzalezb@uned.ac.cr

F. Tassi et al. (eds.), *Poás Volcano*, Active Volcanoes of the World,
https://doi.org/10.1007/978-3-319-02156-0_13

Fig. 1 The canopy, showing an emergent tree with epiphytic plants of the family *Bromeliaceae*. By courtesy of Julián Monge

2 Flora

Poás Volcano has many physiognomic features similar to other massifs from the Cordillera Central and the Cordillera de Talamanca. Most peaks are covered with forest vegetation except where they are destroyed by volcanic eruptions. Owing to the soil characteristics, vegetation plays a very important role in soil retention (Macey 1975).

Canopy usually does not exceed 20 m, but some emergent trees can reach 30 m (Fig. 1). Epiphytic plants are abundant, and most trees are covered with mosses, bromeliads, orchids, and ferns (Fig. 1). The undergrowth is dense, though composed by relatively few species. The mulch is thick, porous, and mainly covered by a discontinuous layer of mosses and ferns (SINAC 2008).

According to the Biodiversity National Institute of Costa Rica (Spanish acronym INBio 2017), 69 families, 111 genera, and 150 species of flora are represented in the protected area. According to Macey (1975), there are three different associations of flora in Poás, largely delineated by altitude. Between 2,400 and 2,500 m there are 79 species reported: Oaks (*Quercus costaricensis*, *Quercus seemannii*), cypress (*Podocarpus oleifolius*) and Magnolias (*Magnolia poasana*) dominate this association. It is also common to find species from the genus *Ocotea* and the species *Weinmannia pinnata*. The trees belonging to this group have relatively narrow terminal cups and a straight and well-differentiated shaft. The extension of this association was strongly reduced in all the Cordillera Central since most areas were transformed into pastures, crops and, more recently, into dispersed urban settlements.

Fifty-eight species of plants are reported between 2,500 and 2,600 m and there is a distinct association dominated by *Quercus spp. Schefflera spp.* and *Clusia spp.* According to Macey (1975), Costa Rican *Podocarpus* and

Fig. 2 Pernettya prostrata (*Ericaceae*). By courtesy of Tomás Alfaro

Fig. 3 Escallonia myrtilloides (*Escalloniaceae*). By courtesy of Tomás Alfaro

Brunellia are also found. Unlike the previous altitudinal band, here no emergent trees are observed.

Above 2,600 m, the dwarf cloud forest, with about 13 species reported, is dominated by the same species that characterize the previous band, although the individuals are smaller. Macey (1975) reports for these elevations the species *Weinmannia trianae* (although a misidentification with *Weinmannia pinnata* is possible). Another family of shrubs present is *Ericaceae* (especially *Vaccinium consanguineum* and *Pernettya prostrata*) (Fig. 2). These species are growing in the areas surrounding the main crater. The undergrowth is thin and dominated by juveniles of canopy species. Other very common shrubs are *Monochaetum vulcanicum* and *Escallonia myrtilloides* (Fig. 3).

In the latter two associations, trees have proportionally wider crowns, lower branching, and many of them have very irregular ridges (SINAC 2008). In the crater and its surroundings, there is very little vegetation; *Pernettya prostrata* and ferns of the genus *Elaphoglossum* are the most tolerant species to the local extreme conditions of temperature changes and strong winds.

Ferns are one of the better represented groups with approximately 22 families, also orchids are very abundant. Within the reported groups, some species of ferns and orchids such as *Acineta chrysantha*, *Epidendrum polyclamys*, *Masdevallia chontalensis* and *Podocarpus oleifolius* are included in the CITES list of endangered species (INBio 2017).

A very common and striking species in the park is *Gunnera insignis*, locally known as “the poor’s umbrella”, because of its large and perforated leaves. *Archibaccharis jacksonii* is a Costa Rican endemic herb that can be found in the protected area as well. Another species to be mentioned is the exotic (introduced) *Ulex europaeus*. This yellow-flowering shrub originated in Europe has invaded parts of the park (SINAC 2008).

Two important aspects to consider in relation to this protected area are: (1) its contribution to the preservation of the Cordillera Volcánica Central native forest; and (2) the role that forest coverage plays in the generation of water for human consumption. The water generated from Poás supplies part of the Central Valley population, the most populated region of the country.

3 Mammals

Highlands biodiversity is low in comparison to lowlands biodiversity, but at the same time there is a lot of endemism and Poás Volcano is not the exception (Chaverri-Polini 1998; SINAC 2008). The national park management plan indicates that research-monitoring studies are not a priority for the park administration, and this is the reason why there is a lack of knowledge about the fauna (SINAC 2008). Nevertheless, there are at least 40 mammal's species recorded in the area, but sampling of small mammals and bats has to be improved.

Table 1 shows a list of the most representative species of mammals found in the park (Rojas and Barboza 2007; SINAC 2008; CRBio 2017). It is important to mention that Syntheosciurus brochus poasensis, also known as Bangś squirrel (Fig. 4) is a subespecies of squirrel endemic to Costa Rica and only found at Poás (Goodwin and Underwood 1943; Giacalone et al. 1987).

Table 1 Representative species of mammals found in Poas Volcano National Park

Species	Common name
Bassaricyon gabbii	Northern olingo
Potos flavus	Kinkajou
Bassariscus sumichrasti	Cacomistle
Conepatus semistriatus	Striped hog-nosed skunk
Bradypus variegatus	Three-toed sloth
Choloepus hoffmanni	Two-toed sloth
Canis latrans	Coyote
Cebus capucinus	White-faced capuchin
Dasypus novemcintus	Long-nosed armadillo
Gallictis vittata	Greater grison
Urocyon cinereoargentus	Grey fox
Herpailurus yagouaroundi	Jaguarundi
Leopardus pardalis	Ocelot
Leopardus wiedii	Margay
Peromyscus mexicanus	Mexican deer mouse
Nasua narica	White-nosed coati
Sciurus deppei	Deppés squirrel
Sciurus granatensis	Red tailed squirrel
Syntheosciurus brochus poasensis	Bangś (Poás) squirrel
Tamandua mexicana	Northern tamandua

Fig. 4 Squirrel at Poas Volcano (*Sciurus* sp.). By courtesy of Julián Monge

4 Birds

The major group of vertebrate fauna within the park is represented by birds. Around 220 species have been sighted and at least 30 of those are endemic to Costa Rica and western Panamá (Barrantes and Loiselle 2010; Chavarría-Pizarro et al. 2010).

Some species are very common all around the year, such as *Chlorospingus pileatus* (Sooty-capped Bush Tanager), *Selasphorus flamula* (Volcano Hummingbird), *Turdus nigrescens* (Sooty Thrush; Fig. 5), *Pseilliophurus tibialis* (Yellow-Thighed Finch; Fig. 6), *Pezopetes capitalis* (large-footed Finch) and *Myioborus torquatus* (Collared redstart). Other birds have become less common specially because of poaching, e.g. *Myadestes melanops* (Black-faced solitarire) (Arévalo 2010; Chavarría-Pizarro et al. 2010; eBird 2017).

5 Herpetofauna

The herpetofauna of the Poás Volcano National Park is still poorly studied, possibly due to the difficulty of access to some areas or the risks associated with an active volcano (SINAC 2008). However, the altitudinal range of the park allows the presence of five high humidity life zones (SINAC 2008). According to their natural history and the historical record of their geographical distribution, Poás Volcano is suitable for the presence of at least: 5 salamander species, 30 anuran species, 10 lizard species and 30 species of snakes (Leenders 2001; Savage 2002; Solórzano 2004; Muñoz and Johnston 2013).

Some species that are observed in the area are, as follows: *Bolitoglossa robusta* (the ring-tailed salamander), *Nototriton abscondens* (concealing moss salamander), *Diasporus diastema* (common dink frog), *Craugastor podiciferus* (piglet litter

Fig. 5 Sooty Thrush (*Turdus nigrescens*). By courtesy of Cristhian Ureña

Fig. 6 Yellow-Thighed Finch (*Pseilliophurus tibialis*). By courtesy of Cristhian Ureña

Fig. 7 Costa Rican wormsnake (*Geophis brachycephalus*). By courtesy of Laura Artavia

frog), *Smilisca sordida* (drab treefrog), *Hyalinobatrachium fleischmanni* (Fleischmann's glass frog), *Sceloporus malachiticus* (green spiny lizard), *Norops tropidolepis* (cloud forest anole), *Clelia scytalina* (highland mussurana), *Geophis brachycephalus* (Costa Rican wormsnake; Fig. 7), *Lampropeltis triangulum* (tropical milksnake), *Ninia maculata* (banded coffee snake), *Atropoides picadoi* (Picado's jumping pitviper), *Bothriechis nigroviridis* (black-speckled palm pitviper) (Savage 2002; Solórzano 2004; SINAC 2008; Muñoz and Johnston 2013).

6 Other Groups

At least two dinoflagellates, six diatoms and four algae inhabit the Botos lagoon (Hargraves and Víquez 1981). Forty-seven diatoms species were also reported for Río Agrio (Wydrzycka and Lange-Bertalot 2001). To the best of our knowledge, there are published studies on fungi, land or freshwater mollusks, insects, crustacean and arachnids, among others. However, many specimens were collected in the national inventory performed by the Biodiversity National Institute of Costa Rica (INBio). Unfortunately, many of them were not yet identified and several new taxa are waiting for description. Hynek et al. (2018) studied the microbial diversity, limited to a single species of the bacterial genus *Acidiphilium*. This organism likely gets its energy from oxidation of reduced sulfur in the lake (Laguna Caliente), including elemental sulfur. Given Mars' propensity for sulfur and acid-sulfate environments, this type of organism is of significant interest to the search for past or present life on the Red Planet.

7 Summary

The Poás National Park is a major attraction and is the most visited protected area in Costa Rica and Central America (Mora Amador et al. Chapter "History, Legends, Customs and Traditions of Poás Volcano, Costa Rica"). The fabolous and astonishing feeling to be on the breathtaking view of Laguna Caliente couples with the biodiversity of this site where a high number of animals and plants live in a unique environment. Most areas of the park are covered by verdant foliage dominated by mountain forest. Coffee and strawberry crops and plant farms decorate the lower flanks of the Poás volcano. Contrarily, the effect of acidic gases released by the Laguna Caliente can be seen on both the bonsai sized trees and resilient ferns and schrubs, the latter being widely distributed throughout the inner flanks of Poás. In addition, as reported by Hynek et al (2018): "*Laguna Caliente is one of the most extreme habitats on our planet and may well represent the edge of the habitable range, likely due to the extremely low pH, coupled to moderate but fluctuating temperatures and volcanic dynamics, as well as very high concentrations of dissolved ions*", the hyperacidic lake of Poás was used as analog of Mars. Nonetheless, it was a surprise when the expected (microbial) biodiversity at Laguna Caliente turned to be characterized by a single species. This contrasts with the large-scale biodiversity recorded in the national park, making this area more intriguing and interesting.

References

Arévalo J (2010) Evaluación de las aves silvestres mantenidas en cautiverio en comunidades cercanas al Volcán Póas, Costa Rica. Zeledonia 14:1–11 (In Spanish)

Barrantes G, Loiselle B (2010) Reproduction, habitat use and natural history of the black and yellow silky-flycatcher (Phainoptila melanoxantha), and endemic bird of the western Panama-Costa Rica highlands. Ornitol Neotrop 13:121–136

Chavarría-Pizarro T, Gutiérrez-Espeleta G, Fuchs EJ, Barrantes G (2010) Genetic and morphological variation of the sooty-capped bush tanager (Chlorospingus pileatus), a highland endemic species from Costa Rica and western Panama. Wilson J Ornithol 122:279–287

Chaverri-Polini A (1998) Las montañas, la diversidad biológica y su conservación. Unasylva 195:47–54 (In Spanish)

CRBio (2017) Atlas of Living Costa Rica. http://www.crbio.cr/crbio/?page_id=61&lang=en

eBird (2017) An online database of bird distribution and abundance. http://www.ebird.org

Giacalone J, Wells N, Willis G (1987) Observations on Syntheosciurus brochus (Sciuridae) in Volcán Poás National Park, Costa Rica. J Mammal 68:145–147

Goodwin GG, Underwood CF (1943) Two new squirrels from Costa Rica. Am Mus Nat Hist 1218:1–2

Hargraves PE, Víquez R (1981) Dinoflagellate abundance in the Laguna Botos, Poás Volcano, Costa Rica. Rev Biol Trop 29:257–264

Hynek B, Rogers K, Antunovich M, Avard G, Alvarado G (2018) Lack of microbial diversity in an extreme mars analog setting: Poas Volcano, Costa Rica. Astrobiol 18:923–933

INBio (Biodiversity National Institute of Costa Rica) (2017) Atta: biodiversity information system. http://atta.inbio.ac.cr

Leenders T (2001) A guide to amphibians and reptiles of Costa Rica. Zona Tropical Publications, 305 p

Macey A (1975) The vegetation of Volcán Poás National Park, Costa Rica. Rev Biol Trop 23:239–255

Mora-Amador R, Fernández M, Rouwet D History, legends, customs and traditions of Poás volcano, Costa Rica (Chapter 12). In: Tassi F, Mora-Amador R, Vaselli O (eds) Poás volcano (Costa Rica): the pulsing heart of Central America Volcanic Zone. Springer, Heidelberg (Germany)

Muñoz F, Johnston RD (2013) Amphibians and reptiles of Costa Rica: a pocket guide. Cornell University Press, Ithaca, 215 p

Rojas L, Barboza M (2007) Ecología poblacional del ratón Peromyscus mexicanus (Rodentia: Muridae) en el Parque Nacional Volcán Poás, Costa Rica. Rev Biol Trop 55:1037–1050

Savage JM (2002) Amphibians and reptiles of Costa Rica: a herpetofauna between two continents, between two seas. University of Chicago Press, 934 p

SINAC (2008) Plan de Manejo del Parque Nacional Volcán Poás. ACCVC-SINAC, 101 p

Solórzano A (2004) Serpientes de Costa Rica: Distribución, taxonomía e historia natural. INBio, 792 p

Wydrzycka U, Lange-Bertalot H (2001) Las diatomeas (Bacillariophyceae) acidófilas del río Agrio y sitios vinculados con su cuenca, volcán Poás, Costa Rica. Brenesia 55–56:1–67

MIX
Papier aus verantwortungsvollen Quellen
Paper from responsible sources
FSC® C105338

If you have any concerns about our products,
you can contact us on
ProductSafety@springernature.com

In case Publisher is established outside the EU,
the EU authorized representative is:
Springer Nature Customer Service Center GmbH
Europaplatz 3, 69115 Heidelberg, Germany

Printed by Libri Plureos GmbH
in Hamburg, Germany